THE HENRY HOLT
Handbook
of Current
Science
&
Technology

D1212790

THE HENRY HOLT

Handbook of Current Science & Technology

A SOURCEBOOK OF FACTS AND ANALYSIS
COVERING THE MOST IMPORTANT EVENTS
IN SCIENCE AND TECHNOLOGY

BRYAN BUNCH

A Henry Holt Reference Book

Henry Holt and Company

New York

REF
Q
158.5
.B86
1992

LIBRARY
FLORIDA KEYS COMMUNITY COLLEGE
5901 West Junior College Road
Key West, Florida 33040

Copyright © 1992 by Bryan Bunch. All rights reserved, including the right to reproduce this book or portions thereof in any form. A Hudson Group Book, produced by Scientific Publishing, Inc., for Henry Holt Reference Books. Published by Henry Holt and Company, Inc., 115 West 18th Street, New York, New York 10011. Published in Canada by Fitzhenry & Whiteside Limited, 91 Granton Drive, Richmond Hill, Ontario L4B 2N5.

Library of Congress Cataloging-in-Publication Data

Bunch, Bryan H.
    The Henry Holt handbook of current science and technology : a sourcebook of facts and analysis covering the most important events in science and technology / by Bryan Bunch.—1st ed.
        p.   cm. — (A Henry Holt reference book)
    Includes bibliographical references and index.
    1. Science.   2. Technology   I. Title.   II. Series.
    Q158.5.B86   1992
    500—dc20

                                                            92–6119
                                                            CIP

ISBN 0-8050-1829-8

Henry Holt Reference Books are available at special discounts for bulk purchases for sales promotions, premiums, fund-raising, or educational use. Special editions or book excerpts can also be created to specification.

For details contact: Special Sales Director, Henry Holt and Company, Inc., 115 West 18th Street, New York, New York 10011.

First Edition—1992

Copyediting and Index: AEIOU, Inc.
Design and Production for Scientific Publishing: G&H SOHO, Ltd.

Printed in the United States of America

Recognizing the importance of preserving the written word, Henry Holt and Company, Inc., by policy, prints all of its first editions on acid-free paper. ∞
1 3 5 7 9 10 8 6 4 2

Contents

CHEMISTRY

EARTH SCIENCE

ENVIRONMENT AND ISSUES

LIFE SCIENCE

MATHEMATICS

PHYSICS

TECHNOLOGY

APPENDICES

INDEX

THE HENRY HOLT

Handbook of Current Science & Technology

HOW TO USE THIS BOOK

The Henry Holt Handbook of Current Science and Technology is designed to provide easy access to major developments in science and technology that occurred during 1990 and 1991. It also presents the background needed to interpret them. In addition, it serves as an up-to-date handbook of basic scientific information. Data included have been chosen to cover facts that are frequently needed while reading or writing. The same data also: form an essential aid to understanding new science devices, experiments, and theories wherever encountered.

As the following plan for the book indicates, *The Henry Holt Handbook of Current Science and Technology* (or *Current Science,* for short) is designed so that it can be read from start to finish—perhaps skimming or skipping some of the tabular material—to obtain an overview of science and technology in the early 1990s. Many will prefer to treat *Current Science* like a magazine, however, browsing among the articles that are of particular interest. In either case, the structure and content of the book also make it a reference work to which one can return again and again.

Plan of the Book

The book begins with an introduction describing the current state of the scientific enterprise. Following that, eight broad subject areas are covered in alphabetical order:

Astronomy and Space; Chemistry; Earth Science; Environment and Issues; Life Sciences (including physical anthropology); Mathematics; Physics; and Technology.

The book concludes with several appendices devoted to data that are likely to be referred to frequently and that are common to all the main subject areas.

Within each broad subject area, the same basic plan is followed.

State of the Subject Each subject-area section begins with an article on the state of that subject in the early 1990s. The author of *Current Science* provides a personal evaluation that is based on years of work in analyzing current science and on the extensive research that was used in developing this volume.

Timetable of the Subject This is followed by an historical timetable through 1989 consisting of short entries describing major developments in the subject area in chronological order. The timetables are designed both for easy reference and so that the reader can develop a feeling for how a subject has evolved.

Section Topics The bulk of this section consists of articles on the scientific developments of 1990 and 1991. Some events of the late 1980s also are included, especially if their main impact has been in the early 1990s.

The individual articles are grouped into several sections on a given theme. For example, "What's New in the Solar System" deals with developments in astronomy concerning the Sun, planets other than Earth, and such smaller bodies as meteoroids, asteroids, and comets—arranged roughly from the center of the solar system outward.

Articles and Subtopics Within each section, individual topics are grouped into one or more articles. "What's New in the Solar System," for example, includes such articles as "Solar Neutrinos" (offering new results on the problem of why neutrino telescopes do not find as many neutrinos produced by the Sun as calculated), "Venus Laid Bare (giving an account of the findings of the Magellan space probe and associated Earth-based studies of the second planet), and "Weird Volcanoes of the Solar System" (mainly telling about the unusual volcanoes of Jupiter's moon Io and Neptune's moon Triton, with new ideas about what they are). Some of these articles are further divided into sections on specific subtopics. For example, an account of problems with various scientific space missions includes a separate heading for each mission considered as well as a general introduction to the whole issue. A few articles are labeled "UPDATE" to indicate that they summarize recent gains in a specific area that is currently very active.

References Articles conclude with a listing of periodical references and suggestions for additional reading. A number of basic reference books that cover the subject areas of *Current Science* are listed in an appendix, but none of these books handles the very recent developments of 1990 and 1991. That material can be found only in magazines and newspapers. The magazines listed are commonly available in general libraries in the United States. Foreign periodicals are excluded.

<u>Set-Off Background Material</u> An important feature within many articles is background material that is set off from the body of the article.

The main focus of this book is on current science and technology, but some readers may not have the prerequisite knowledge to understand new technical developments. Sometimes a simple glossary is all that is needed to clarify an article or a table. Other times it may be a paragraph or two that describes a brief history of such a key development as plate tectonics. For some topics in mathematics or physical sciences, such a background explanation can run longer than the actual article describing new developments.

Science periodicals or yearbooks, which constitute the most accessible source of current scientific and technological information outside of this book, either omit such background material or incorporate it into the body of articles. *The Henry Holt Handbook of Current Science and Technology* is unique in allowing the reader a choice.

Here is an example of how this feature may be used: In the article discussing findings of new scientific satellites such as the ROSAT and Gamma Ray Observer satellites, it is assumed that the reader knows what gamma rays are—very short electromagnetic waves or, equivalently, very high-energy photons. Set off from the article, but embedded in the discussion of the two satellites, are two paragraphs explaining exactly what gamma rays are, including specific numerical definitions. A reader who already understands gamma rays can skip this material. A reader who is unsure will want to read it, although the article by itself contains as much coverage as and more details than would be in most magazine articles on either of these space vehicles. Finally, a reader who knows in a general way what gamma rays are, but who wants to learn more about them, would find details in the two set-off paragraphs that are seldom spelled out in general articles.

<u>Tables, Charts, and Lists of Useful Data</u> Most tables that have to do with a particular subject matter are grouped at the end of the broad subject sections. Sometimes, when it is particularly useful with regard to a specific article, a table is included in or immediately following the article. Every effort has been made to include in considerable detail the data that might be most useful.

Who Needs This Book

Everyone who regularly uses current scientific data that span several fields or that cover a particular subject needs *The Henry Holt Handbook of Current Science and Technology*. Among those people are science writers and reporters, science teachers, and investors in new technology. This group also includes any others who have to know details of recent developments in science.

Science as an International Enterprise

Every effort has been made to include not only the discoveries and developments of science around the world, but also the viewpoints of scientists in various nations. A work published in English in the United States tends to have a North American point of view. This is compounded in science and technology because, despite many problems with science education in the United States, a great many of the advances in current science take place in the United States. Scientists and institutions are identified by nationality throughout the book to help the reader from outside the United States keep track of the progress of international science.

THE STATE OF SCIENCE TODAY

The scientific enterprise in the early 1990s continues to increase its power to provide a clearer understanding of the universe and to affect the lives of people around the world. Only those living in rain forests or other fringe environments remain outside the scope of its influence, and that number is receding rapidly.

Growth in science has proceeded along a steeper than exponential curve since its inception as a way of trying to understand the universe, traditionally 600 BC. A pure exponential curve has a constant doubling time; the quantity being measured doubles every so many years. For scientific knowledge, growth has been so fast that the doubling time itself may have a halving time. That is, it seems to take half the time for knowledge to double as the previous amount of time required.

Scientists themselves have suggested since late in the nineteenth century that this kind of expansion cannot continue. In many instances of various types of growth, scientists observe what biologists call the reproduction curve for a growing population; it begins as an exponential curve and then flattens out as the resources the population depends on are depleted. The reproduction curve eventually becomes an elongated S. Many scientists expect that growth of knowledge will follow such a reproduction curve in the long run. The unanswerable question is when the flattening off will start.

Despite nineteenth-century and more recent pessimism over the possibility of further scientific progress, there seems to be no serious depletion of the underlying resource for the growth of science: ignorance. If we have reached the flattening out of gains in knowledge, it is not apparent yet. Perhaps the day will come when the object of scientific enterprise will be only that of adding an additional decimal place to what we already know. However, in the 1990s, that day does not seem to have arrived.

Common Concerns

Running out of ignorance may not be imminent, but at the start of the 1990s, scientists have many other problems.

There is considerable concern in the United States, the world's leader in science for the past fifty years, over whether its young people are attracted to science in sufficient numbers or are obtaining the necessary foundation in science. This is especially true with regard to women and minorities. Despite these legitimate worries, the international scene in science and, to a lesser extent, technology remains dominated by the United States, especially since the collapse of the USSR. The ongoing rise of European unity and of the nations of the Pacific rim may result, however, in a change in the location of the active center of international science away from the United States and whatever the successors to the Soviet Union turn out to be.

On the other hand, the vagaries of international politics have suggested in the early 1990s that scientists may be able to worry less that their endeavors would lead to the destruction of life on Earth. On Nov 26 1991, for example, atomic scientists turned back their clock on the cover of the *Bulletin of the Atomic Scientists,* which is intended to show the imminence of nuclear doom, to seventeen minutes to twelve. In the year or so prior to that it had been at fifteen minutes to twelve. In its earliest incarnation in the 1940s, the clock was set at eight minutes to twelve. After the US announcement on Sep 23 1949 that the Soviet Union had exploded an atom bomb, the clock was set forward to three minutes to twelve. Subsequently, it was reset as events seemed to dictate. The current setting of seventeen minutes to twelve reflects the most hope in almost 50 years.

Scientific misconduct is also a current issue. As competition increased in science, not only among nations but also among individual scientists, the attraction of fraud has also grown. In an earlier era, when the most that could be gained from successful scientific fraud was the knowledge of putting something over on fellow scientists, fakery was probably rare. When careers came to depend on the production of new science, the possibility that some might use less-than-honest means to advance their fortunes became greater. At least, that's how a lot of people saw it at the start of the 1990s, resulting in investigations and the creation of new watchdogs for the scientific community.

In a somewhat related problem, as scientific funding increasingly came from governments rather than other institutions, the question of what kind of science should be funded has become more critical. The issue of "big science" or "small science," problematic since the success of the Manhattan Project during World War II, has become increasingly important to scientists in search of funds.

Science Education and Policy

A rash of books about scientific literacy in the United States provoked discussion in the early 1990s, but few practical suggestions for improving either knowledge of science by nonscientists or better education for potential scientists. In Sep 1991 the report of a task force from the Carnegie Commission on Science, Technology, and Government headed by Lewis Branscomb of Harvard's Kennedy School of Government, called *In the National Interest,* proposed that every science agency in the US federal government devote about 10 percent of its budget to education in science and mathematics, about the share that the National Science Foundation now allocates.

Not everyone thinks that more science education is a good thing. In Apr 1991 economist Paula Stephan of Georgia State University spoke at the annual meeting of the American Association for the Advancement of Science in favor of fewer Ph.D.'s in science. Like many scientists, Stephan feels that the intellectual level of the current crop of scientists is too low; the brightest people are entering other fields. She cites as evidence statistics showing 320 doctoral students and engineers per million adults in the United States in 1940 as compared with 2000 per million in the early 1990s. That high a percentage of scientists and engineers suggests to Stephan that not all of them are as bright as scientists were in the 1940s. Her remedy would be to wait and do nothing. She believes that in a decade or so the scarcity of good scientists will encourage people to change in various ways that will result in higher quality. A Gallup poll taken in 1991 showed that scientists in general also believe that the intellectual quality of their fellow scientists is in decline or at best holding steady; fewer than one in five polled think that the quality level is rising.

Revisions Suggested for the Operation of Science

Sociologist and historians who specialize in the scientific enterprise have long known that the fellowship of science has evolved its own structure and folkways in the post-Galileo period. Some have gone so far as to say that these ways and mores determine what science is and what it is not, rather than the objective criteria about knowledge of the universe that we expect from science. Among the important defining characteristics of science is communication of results (general principle) by way of journals (specific application) in articles that are refereed by other scientists (a corollary known as peer review). Since this apparatus is considered basic to modern science, any suggestion that there is something wrong with it produces strong reactions in scientists.

Psychologist Domenic V. Cicchetti of the Veterans Administration Medical Center in West Haven CT reported in the Mar 1991 *Behavioral and Brain Sciences* on the problems that he found with the peer review system after a study of 20 years of such reviews. Many of the scientific journals he studied were in fields related to behavioral sciences, but Cicchetti also looked at *Physical Review* and at grant reviews conducted by the US National Science Foundation and National Academy of Sciences in chemistry, economics, and physics. Cicchetti found that reviewers disagreed with each other more often than expected, possibly resulting in lack of funding or absence of publication for important research considered to be outside the mainstream of science. Such significant misses seem even more likely since Cicchetti found more agreement about research that was funded or published and more disagreement about that which was rejected. Funding of research is more important than publication, because articles rejected by one journal are usually published elsewhere, although the second-choice journal may be less prestigious.

Cicchetti was not the only one concerned by the peer review process. More than a year before his report, in early 1990, the National Science Foundation completely overhauled its peer review system for grants.

International Science in the Early 1990s

The breakup of the USSR, preceded by the freeing of its satellite states, affected a great many scientists. Many East German scientists lost their jobs when East and West Germany merged; one estimate in Mar 1991 held that nine out of ten Eastern scientists would be unemployed by 1992.

In the United Kingdom, funding by successive Conservative budgets for science has been so low that by the late 1980s as many as one in four Fellows elected to the Royal Society actually lived in or were employed in countries other than the United Kingdom, the majority in the United States.

The British government, however, insists that there is no "brain drain," even as its Science and Engineering Research Council, the main source for funding in the United Kingdom, cuts its 1992 budget for new grants by 50 percent.

In the United States, the executive office proposed budget increases of 8 percent for basic research for 1992, with less than 2 percent of that going to direct military spending. The National Science Foundation, the US leader in basic research, did even better, with an 18 percent increase scheduled, mostly for basic research. Not all of that money was approved by Congress by the end of 1991, but science was doing fairly well despite loud complaints over funding by some, notably president of the American Association for the Advancement of Science (AAAS) Leon Lederman [US: 1922-],. In a Jan 1991 report mailed to all 140,000 members of AAAS, Lederman called for an immediate doubling of federal funding for science. By way of comparison, from 1985 through 1990 the budget for the National Science Foundation grew an average of 7.1 percent a year unadjusted for inflation; that corresponds to a doubling time of 10 years.

Funding for scientific research in capitalist societies, which is beginning to be all societies, comes not only from governmental bodies, but also from corporations, especially very large and successful ones in technological fields. Large enough corporations can afford to fund basic research in the hopes that some of it may be applied, and some of them actually do fund such efforts. Smaller companies usually have only enough money to fund research directed toward specific money-making products. In the past, notable research, such as that leading to radio telescopes, transistors, information theory, and confirmation of the big bang theory, originated in AT&T's Bell Labs; more recently, some of the main theoretical and practical advances have come from various research laboratories funded by IBM, notably the Thomas J. Watson Laboratory in Yorktown Heights NY and IBM's Zurich laboratories that developed the scanning tunneling microscope in 1982 (see "Scanning Tunneling Microscope Finds New Uses," p 596). Like most of the large computer companies, including Fujitsu, Digital Equipment, Apple, and Sun, IBM has been spending about 10 percent of its revenues on research and development.

Since the recent breakup of AT&T, Bell Labs has split in two, with noticeable loss of effectiveness. A restructuring of IBM announced in Dec 1991 may lead to similar problems with their research efforts, although not necessarily. In 1991 IBM was spending about $6 billion each year in research and development funds. As other companies grow, they may produce their own new research facilities. The growth of Microsoft Corporation in the early 1990s, for example, led in May 1991 to the beginnings of a new laboratory devoted to basic research.

Citation analysis is a technique for examining the state of science that was suggested by Yale historian Derek J. de Solla Price and Eugene Garfield in 1955. Garfield later started the Institute of Scientific Information in Philadelphia PA. The technique analyzes with computers the references that new research papers in science make to the papers that preceded them; such references are called citations. The original idea was that the more citations to a specific paper by others, the more important the cited paper is to the development of science. While this simple concept has had to be refined in a number of ways—articles reviewing a field or articles describing some laboratory technique can be cited disproportionately to their importance, for instance—it is still considered important by many scientists, although a few say that citation analysis is useless.

According to Garfield and his institute, citation analysis shows that the United States continued to be the world leader in science in 1991, advancing by 6.9 percent in the number of citations during the 1980s, while

France also advanced (by 1 percent) and the United Kingdom dropped back (3.4 percent). The actual average of the number of citations per paper by the ten leading nations, measured between Sep 1987 and Aug 1990 inclusive, was as follows:

Nation	Number of citations per paper	Nation	Number of citations per paper
1. Switzerland	7.33	6. United Kingdom	5.62
2. Sweden	6.72	7. W Germany	5.47
3. United States	6.65	8. Belgium	5.38
4. Denmark	6.22	9. Australia	5.36
5. Netherlands	6.01	10. Canada	5.31

These figures require some interpretation. Switzerland is credited with many papers in particle physics arising from work at CERN (*Centre Européen de Recherce Nucléaire*), which is staffed with scientists from many nations. Sweden's number is inflated by many short papers in biology, a field in which heavy citing is common. The noticeable number of English-speaking nations in the list may reflect the number of scientists who use English as their scientific language of choice; while the noticeable absence of the USSR and Japan may reflect an international lack of knowledge of Russian and Japanese.

Another measure of US strength is that only 15 percent of US papers are not cited at all (disregarding abstracts), while the world average outside of the United States is 33 percent.

Fraud and Other Unsavory Allegations

Various allegations of fraud in the 1980s continued to make news throughout the early 1990s. One measure of scientists' concern about the issue was about a half-dozen seminars on the subject in 1991, including three run by the US National Institutes of Health.

One notorious fraud allegation from 1986 continued to make headlines in the early 1990s and may have been resolved by 1991, although the alleged perpetrator was still claiming innocence even after her principal sponsor retracted the original paper. In the midst of the apparent resolution of this problem, controversy suddenly surrounded the US National Institutes of Health's Office of Scientific Integrity, which had become the main arbitrator of the problem. Along the way a Nobel Laureate was sufficiently discredited by the affair that he had to resign a major post.

The Timetable on the next two and one-half pages is one way to make sense of the long-lasting drama, often known as the "Baltimore Affair."

Timetable of a Fraud Investigation

1970	David Baltimore is one of two scientists who independently discover reverse transcriptase, the enzyme that is the key to genetic engineering
1975	Baltimore receives a share of the Nobel Prize in physiology or medicine for his work on reverse transcriptase
1983	About this time Baltimore begins collaborating with various other biologists on a study of transgenic mice

1985

Jun 20-22	According to biologist Thereza Imanishi-Kari, she performs critical subcloning of genes from transgenic mice at this time while at MIT's Center for Cancer Research, leading to results reported in 1986
	Postdoctoral student Margot O'Toole begins working with and disagreeing with Imanishi-Kari at MIT

1986

Apr 25	Paper by David Weaver, Imanishi-Kari, Baltimore, and three others ("Altered repertoire of endogenous immunoglobulin gene expression in transgenic mice containing a rearranged Mu heavy chain gene") is published in *Cell*; it claims foreign genes transplanted into a line of mice can indirectly change the antibodies produced by the mice's own genes
May	O'Toole discovers 17 pages of supporting data from Imanishi-Kari's work that contradict claims in Apr paper and reports this to Henry Wortis, her senior advisor at Tufts University, but he finds no evidence of misconduct after a brief investigation with two other researchers
Jun 6	O'Toole prepares a formal list of her objections to the *Cell* paper for an investigation into her charges by Herman Eisen of MIT; this investigation is inconclusive
Jun 16	According to O'Toole, she and Imanishi-Kari tell Baltimore, Weaver, and Eisen that some of the data in the paper are false; Baltimore says that does not call for retraction of the paper or even an erratum
	O'Toole abandons her career in science, resigning from MIT

1988

Jan	US National Institutes of Health (NIH) appoints a panel of three immunologists to review the dispute between O'Toole and Imanishi-Kari; later two of the panelists are accused of having ties to Baltimore
Apr 11	Congressman Ted Weiss (D-NY) of the House Government Operations Subcommittee holds the first hearing on "scientific fraud and misconduct" to mention the Imanishi-Kari affair
Apr 12	Congressman David Dingell (D-MI) and his subcommittee on oversight and investigations of the Energy and Commerce Committee conducts their first hearing on the Imanishi-Kari matter
Aug	Dingell asks US Secret Service to examine Imanishi-Kari's scientific notebooks for evidence of fraud

1989

Jan 31   The NIH panel investigating the O'Toole charges finds "no evidence of fraud"

Mar   NIH establishes an Office of Scientific Integrity (OSI) as a branch of the Public Health Service (PHS)

May 4   At a hearing of the Dingell subcommittee, Baltimore attacks the subcommittee for the way it is conducting the investigation and defends Imanishi-Kari

May   OSI begins operation; Suzanne Hadley is deputy director and will supervise the Imanishi-Kari matter

1990

Apr 30   Baltimore is interviewed by OSI teams and says that incorrect data in Imanishi-Kari's notebooks are not "fraud" because the data were not published

         O'Toole is hired by Genetics Institute after nearly 4 years of being unable to get a job in science

May 14   US Secret Service agents testify to Dingell hearing that notes purporting to be Imanishi-Kari's work of Jun 20-22 1985 are falsified

Jul 1   Baltimore takes office as president of Rockefeller University

Dec   Federal judge Barbara Crabbe rules that OSI guidelines for misconduct were not drawn up in accordance with US law since they were not open to public comment

1991

Jan   US Health and Human Services Secretary Louis Sullivan appoints an NIH committee to look into the operations of OSI

Mar 21   Report from OSI concludes that Imanishi-Kari fabricated data and calls O'Toole a hero; Baltimore asks *Cell* to withdraw 1986 paper, which it agrees to do on Mar 22

         National Academy of Sciences' twenty-five-member panel, chaired by Edward E. David, issues draft report calling for an independent body to prepare standards for and monitor scientific misconduct

Apr   Bernadine Healy becomes director of NIH

May 3   Baltimore's fourteen-page apology to O'Toole is revealed to the press

May 15   Imanishi-Kari calls the whole affair a witch-hunt that arose from O'Toole's misunderstanding of her research notes

Jun 13   The NIH Office of Scientific Integrity Review publishes the rules under which OSI operates for public comment

Jun   A group of 143 scientists, many of them eminent immunologists, complain that OSI has denied basic rights to Imanishi-Kari

Jul   Hadley who has supervised the Imanishi-Kari investigation and also the Gallo probe (see below) is removed from both cases by NIH director Healy

Aug	Anthony Cerami, who opposed Baltimore's appointment at Rockefeller University, leaves the University, taking his thirty-person team to the Picowar Institute for Medical Research in Manhasset NY
Sep	PHS receives thousands of letters, including a joint one from various associations of universities, attacking OSI rule-making procedures
Oct	Nobel Prize winner Gerald M. Edelman, who led the forces at Rockefeller University against hiring Baltimore, announces that he and his lab are moving to the Scripps Research Institute
Oct 17	David Rockefeller gives $20 million to Rockefeller University to indicate his satisfaction with Baltimore as its president
Nov	The NIH advisory committee appointed by Secretary Sullivan in Jan urges OSI to tighten its definition of misconduct and to allow scientists under investigation to defend themselves
Dec 2	Baltimore announces his resignation as president of Rockefeller University because of the "climate of unhappiness" created by the Imanishi-Kari investigation
1992	
Mar	Scheduled meeting of NIH Office of Scientific Integrity Review is to consider new rules for OSI

At the same time as the Baltimore and Imanishi-Kari case was coming to a head, an even older controversy over scientific integrity was also in the news. Ever since the original announcements of discovery of the virus now known as HIV, knowledge of its role in AIDS, and development of a probe for its detection, Robert Gallo of the US National Institutes of Health and Luc Montagnier of France's Pasteur Institute have engaged in an on-again, off-again battle for credit. To further complicate matters, a draft report from OSI attacking the original 1984 HIV paper by Gallo and Mikulas Popovic on several technical grounds was leaked and reported on Aug 11 1991. The report concluded that Popovic was guilty of scientific misconduct, a charge that Popovic vigorously denies. At the end of 1991, this controversy was still unresolved.

Another fraud case was handled quickly by Caltech. As soon as there was a hint that some of his postdoctoral students were altering data, biologist Leroy Hood started retracting papers that might have been tainted. He also telephoned colleagues whose work might be affected. Caltech's internal investigation proceeded swiftly and without much controversy at first. The accused and their friends, however, continued to claim innocence and protest what they saw as as unfair procedures—such as moving too swiftly—before the investigation was complete.

When called to the attention of investigative bodies, most accusations of misconduct are rejected after study. Of the first 110 investigations completed by OSI, only 16 cases, or 1/7 of them, were ruled to be fraud. Furthermore, not all OSI investigations concern fraud at all. About 40 percent of the investigations so far have centered on other forms of misconduct, such as corruption of the peer review system or theft of intellectual property.

Big vs. Small Science, Round by Round

The early 1990s saw a continuation of the debate on the relative importance of big and little science to the scientific enterprise and to the world in general.

"Big science" is experimental work that requires enormously expensive, high-technology equipment tended by scores or perhaps hundred of scientists. The classic model for big science is the large particle accelerator, such as the proposed $8.4 billion Superconducting Supercollider (SSC), which meets both criteria. A $40 billion space station is certainly "big," although some say that the planned US space station is not science at all, just a display of technology. Large astronomical observatories on the ground and in space are probably "big" since they use expensive telescopes, even though most individual experiments with such equipment are done by individuals or by small groups. The Human Genome Project (see "Human Genome Project UPDATE," p 446) is big science for the opposite reason; the equipment is not especially big or expensive, but the number of experimenters collaborating is enormous and therefore the cost of the project is great. There are also borderline cases. Large-scale programs of ocean drilling are almost "big," while superdeep boreholes are probably "big."

One tries to make such categorizations because the 1980s, especially toward their end, and the early 1990s have been rife with complaints by proponents of "small science" that big science is taking most of the money without producing much in the way of return. At the same time, proponents of individual "big" projects are claiming that one has to spend vast sums on their project because the project is the next important step or next logical step in science.

Small science advocates complain that the evaluation procedures for large, expensive projects are flawed and that once the projects are initially funded, it is almost impossible to get rid of them. Evaluation of small projects is traditionally conducted by peers—people whose own projects in the same field are fighting for the same funds in about the same amounts. The peers are thought to be much harder nosed when evaluating a project that is directly competing with them for scarce funds; so projects that are approved by the peers are expected to be good ones. Furthermore, the peers are scientists who know the field in detail; it would be difficult to fool such experts with half-baked or poorly designed concepts. On the other hand, very expensive projects are evaluated by politicians, people with no vested interest in or depth of knowledge about the proposal seeking funding.

Once a large project is approved, it has its own momentum. No one wants to waste the first few billion dollars spent because the idea is not working out. Small projects run through their smaller investments completely in a few years. If the concept for a small project was no good, it no longer matters financially because all the money appropriated has already been spent and no one is going to spend any more money on such a dumb idea.

Let's look at the record. Here is a summary of what seem at this time to be the greatest accomplishments in science during the 1980s and early 1990s categorized as "big" and "small."

"BIG"	*"SMALL"*
1980	
Voyager reaches Saturn	Quantum Hall effect discovered
	K/T iridium layer found
	Inflationary model of universe developed
	Scanning tunneling microscope invented
	Genetic engineering made possible
1981	
	Catalytic action of RNA discovered
	Quadratic-sieve method of factoring developed
1982	
Soviet soft-lands space probes on Venus	Bell's theorem confirmed by experiment
Artificial heart employed	
1983	
IRAS finds evidence of planets forming around stars	Dinosaur fossils found near Arctic Circle
	Development of each cell of *Caenorhabditis elegans* mapped
CERN finds W and Z particles	Polymerase chain reaction invented
	Gene markers for diseases found
1984	
	Apes and chimps found to have nearly identical genes
	Quasicrystals discovered
	Soft-bodied fossil deposits found at Chenjiang, China

(Continued)

"BIG"	*"SMALL"*
1985	
	Buckminsterfullerene discovered
1986	
Comet Halley investigated	Gene for muscular dystrophy discovered
	High-temperature superconductors invented
1987	
Supernova 1987A detected by optical and neutrino telescopes	Mitochondrial genes used to find African "Eve"
	Gravitational lenses observed
1988	
Voyager reaches Neptune	
1989	
	Whale with legs found
1990	
COBE shows background radiation is smooth	Gene for neurofibromatosis discovered
	Gene for maleness located
	Impact crater at Chicxulub, Mexico, that was probable cause of K/T mass extinction, found
1991	
Magellan orbits Venus	Gene therapy begins
Radar studies made of Mercury	

While one could argue about some of the specific items listed or not listed, or even the categories in which they are placed, the pattern suggests that the small science advocates can point to a large number of previously unpredictable advances, but the big science people can claim that their accomplishments could be achieved no other way. One expects that this dichotomy will persist into the 1990s and beyond.

Robert Laughlin of Stanford University has estimated the entire cost associated with a particular but typical small science project, the discovery and understanding of the quantum Hall effect. Discovery, which occurred more or less by chance favoring the prepared mind of Klaus von Klitzing [German: 1943-], who with fellow researchers was making routine measurements on semiconductor devices, was paid for by the routine work. Since then, however, studying the quantum Hall effect has cost various governments around the world about $10 million in all according to Laughlin, most of the money coming from grants in the range of $50,000 to $100,000. The contrast between $10 million over 10 or 12 years and $8.4 billion for the SSC over a similar time period is striking. The SSC will cost roughly a thousand times as much as the discovery and understanding of the quantum Hall effect.

Advocates of small science also argue that the particles that are the target of the SSC will exist for hundreds of billions of years (if they are there at all), and it is hoped that the targets of the Human Genome Project will be around for a long while as well. In the meantime, it might be well to spend money on science projects related to evanescent situations, such as the apparent imminent extinction of vast numbers of species.

If you define the "big" projects on the objective grounds that they cost more than $25 million—about 80 such projects exist in the United States— they ate up 10 percent of the nondefense research and development (R&D) funds in the mid-1980s. That rose to 15 percent in fiscal 1991 and was budgeted to be 22 percent of the nondefense research and development budget in fiscal 1992. (Figures were even higher—up to 35 percent of the nondefense R&D budget back in 1983—mainly due to the cost of development of the space shuttle.)

The US Congressional Office of Technology Assessment reported in Mar 1991 after a 2-year study of science funding that unless big science projects were budgeted separately from other science projects, the other projects would suffer financially. It suggested that big science projects need to be "added on" rather than "included in" the science budget; otherwise, science spending for all but the megaprojects would be flat in the 1990s.

Special Concerns

Many major trends in science tend to fall more heavily on one discipline than on science as a whole. Special concerns that affect particular branches of science are in articles on "the state of" each discipline later in the book.

(Periodical References and Additional Reading: *Science News* 4-14-90, p 234; *Science* 5-18-90, p 809; *New York Times* 12-22-90; *Science* 1-11-91, pp 152 & 153; *Science* 1-18-91, p 266; *Science* 2-1-91, p 507; *New York Times* 2-4-91; *Science* 2-8-91, p 616; *Science* 2-15-91, p 740; *New York Times* 2-21-91, p C1; *Science* 3-1-91, pp 1014, 1015, 1016, 1017, & 1019; *Science* 3-8-91, p 1168; *Science* 3-15-91, pp 1307 & 1308; *New York Times* 3-26-91, p C1; *New York Times* 3-28-91; *Science* 3-29-91, pp 1552 & 1555; *New York Times* 3-31-91; *New York Times* 4-1-91, p A1; *The Sciences* 5/6-91, p 18; *New York Times* 5-4-91, p 1; *Science* 5-10-91, p 768; *Science News* 5-11-91, p 294; *New York Times* 5-17-91, p A19; *Scientific American* 6-91, p 34; *Science* 6-21-91, p 1607; *Science News* 6-22-91, p 394; *Science* 7-5-91, p 24; *Science* 7-19-91, p 261; *Science* 7-26-91, p 372; *New York Times* 8-1-91; *Science* 8-9-91, p 618; *Science* 9-6-91, pp 1084 & 1086; *Science* 9-13-91, p 1205; *Science* 9-20-91, p 1344; *Science* 10-11-91, p 186; *Science* 11-22-91, p 1099; *New York Times* 11-29-91, p A35; *Science* 11-29-91, p 1287; *New York Times* 12-3-91, p A1; *New York Times* 12-4-91, p B1; *New York Times* 12-8- 91, p F1)

Astronomy and Space

THE STATE OF ASTRONOMY

Astronomy is the oldest science. As far back as we can go in the records of rocks and monuments, we can discern astronomy being practiced. Recognition of the cycles of the sky and their influence on events on Earth led to better farming as well as to the psuedoscience of astrology. Early Greeks and Hellenic scholars were able to measure the distances to the Moon (fairly accurately) and the Sun (not so accurately) and to predict the positions of planets in the sky (with great success).

Progress in Observation

In 1609 Galileo [Italian: 1564-1642] turned a telescope on the heavens for the first time and opened the study to new scientific feats—the observation of new moons and planets, the recognition that the Milky Way is a vast collection of stars, and, within a couple of hundred years of that first astronomical telescope, the knowledge of a vast universe populated with other collections like the Milky Way. All advances in purely optical astronomy are just improvements on what Galileo did.

Another whole set of observations began in 1665 or 1666 when Isaac Newton [English: 1642-1727] first studied the rainbow, called the spectrum, produced when sunlight is passed through a prism. Although scientists mapped the lines that appear in the spectra of the Sun and other stars throughout the 19th century, it was halfway through the century before they realized that these lines are the key to something that they had believed would always be unknown—the composition and nature of the stars. With the invention of interferometry (See "Successes with Improved Optical Telescopes," p 32) at the end of the nineteenth century and further understanding of the causes of the spectral lines early in the next century, the spectrum revealed more and more—temperatures of stars, their proper motions, the presence of unseen stars, even star masses. Like the improvement in optics, the continuing advance in spectroscopy is a welcome expansion of what is by now an old idea.

James Clerk Maxwell [Scottish: 1831-1879] paved the way in 1873 for the third observational revolution in astronomy with the recognition that visible light and its then known companions, infrared and ultraviolet light, are only a small part of the spectrum. Although radio waves were recognized in 1888 and X rays in 1895, it was not until an accidental discovery in 1931—radio waves from space—that anyone realized that useful information about the universe might be found in parts of the electromagnetic spectrum beyond light. Even after the first radio telescopes were built in the 1940s and infrared astronomy became routine, the shorter wavelengths emitted by the universe were mostly beyond our reach because short electromagnetic waves are absorbed in the atmosphere long before they reach Earth's surface. High-altitude flights and later space-based observations finally made the whole electromagnetic spectrum available. Rightly enough, observation over the whole spectrum has come to be called the "new astronomy." It is new not only because of new methods, but also because it has uncovered quasars, pulsars, possible black holes, and some very strange and energetic objects that no one can identify.

Just as access to space has opened up the short end of the electromagnetic spectrum to us, it has also brought us close to the bodies of our own solar system. We have landed spacecraft on two planets besides Earth, taken humans to the Moon, and flown missions near a comet, an asteriod, and all the planets but Pluto. These encounters have in the past twenty years completely changed our views of the solar system. This will be discussed further in "The State of Earh's Space Programs" on p 104.

There are a few other ways that we can obtain information from space. Since 1912 scientists have known about cosmic rays, high-energy subatomic particles from space. Most primary cosmic rays, the ones from outer space, break up on entering the atmosphere, allowing us to observe secondary rays only. These have been studied to some degree, but more for interest in the subatomic particles involved than in the information the rays reveal about their origins. A notable exception is the neutrino (usually not thought of as a cosmic ray, although technically it is), which has been studied since 1965. In the early 1990s several new neutrino telescopes were built or planned. Space-based studies of other cosmic rays are in a primitive state so far, but advances will surely be made.

Another useful source of information consists of actual pieces of matter other than cosmic rays that sometimes come to Earth. Meteorites have been identified from the Moon and Mars. It appears that some meteorites also contain material ejected by nearby stars.

A final planned window on the universe will come with the detection of gravity waves, an article of faith to most physicists despite failure so far to observe them. While past detectors have failed, several scientists are working toward new instruments. As these are expensive, they may not be constructed for many years.

As in many other branches of science, computers are enabling astronomers to accomplish studies that would be unthinkable without them. These studies revolve around models of stellar behavior or gravitational interactions of many bodies, processes that are too complex to work out by hand or to unravel with clever mathematics. Such computer models can then replace observation for these kinds of study.

Progress in Theory

Despite the long history of improvement in observation, most progress in the basic theories of astronomy did not come until the twentieth century, the main exception being Newton's laws that explained most of the dynamics of the solar system and multiple star systems. Central to the key developments of the twentieth century was the work of Albert Einstein [German-Swiss-American: 1879-1955], who showed that matter and energy are one in 1905 and developed the modern theory of gravity in 1915. Einstein's 1905 work eventually led to the understanding of how stars get their energy, while the gravity theory, known as general relativity, has helped in understanding the development and the construction of the universe.

Einstein's insights have been enriched and often explained in detail by a theory that Einstein himself found pernicious in many of its aspects, although he was one of the originators: quantum theory. There are several branches of quantum theory, but they are all more-or-less integrated into a structure of particle physics called the standard model. Many scientists would like to combine the standard model or an improved form of it with Einstein's gravitational theories to produce what is called a "theory of everything." None has succeeded.

Theory continues to attempt to deal with phenomena, whether it is neutrinos in the wrong amounts or galaxies in the wrong places. It is difficult to pin down progress in theory because typically there are many theories competing to explain the same phenomenon, and many years may pass before one of the competitors is generally accepted. Even then, the theory is subjected to ongoing criticism and review.

Despite this problem, observation seldom runs far ahead of theory. From the expanding universe theory and the cosmic background radiation, neutron star and black hole hypotheses, theory has existed before observation—and generally been ignored. Not until observation and theory coincide do we see real progress.

What Can We Expect in the 1990s?

A wish list might include

- correction of all the errors on the Hubble Space Telescope so that it can reach full potential;
- installation of adaptive optics and other new technologies to make ground-based optical telescopes the equal of Hubble in optical light and lower frequencies;
- installation of new radio telescope arrays that will make practical interferometry based on thousands of kilometers and more;
- understanding of the geology of Venus and the weather of the gas-giant planets;
- better understanding of the relationship among comets, asteroids, and meteoroids;
- proof of an explanation of why there are so few recognizable neutrinos from the Sun;
- development of neutrino telescopes for observation of bodies other than the Sun;
- detection of gravity waves;
- location of definite brown dwarfs and planets orbiting stars other than the Sun;
- discovery of a black hole everyone can agree upon;
- agreement on definite evidence for the composition and extent of the "missing mass" in the universe;
- a detailed map of a large section of the universe;
- an explanation of how galaxies, clusters of galaxies, and superclusters of clusters were formed;
- and determination of an age for the universe with hard evidence.

Current progress toward these goals, as well as related matters, is detailed in the following pages. Astronomy in the twentieth century has a track record of having at least one totally unexpected observational or theoretical discovery each decade: 1900s, interstellar matter; 1910s, cosmic rays and distance estimates from Cepheid variables; 1920s, the expanding universe; 1930s, radio astronomy; 1940s, the big bang theory and radio galaxies; 1950s, the Oort cloud and superclusters of galaxies; 1960s, quasars and pulsars; 1970s, rings around Uranus and Jupiter, the mysterious Chiron; 1980s, dust clouds around stars. No doubt the 1990s will also bring unexpected discoveries and ideas.

Timetable of Astronomy and Space to 1990

2296 BC   Chinese astronomers begin recording the appearance of "hairy stars" (comets)

585 BC   The scientist Thales [Greek: 624 BC-546 BC] predicts the solar eclipse that appears this year in the Near East

480 BC   Astronomer Oenopides of Chios [Greek: b. c 480 BC] discovers that Earth is tilted with respect to the Sun

410 BC   The first horoscopes become available in Mesopotamia (now Iraq, Syria, and Turkey)

352 BC   The Chinese report the first known "guest star," or supernova

340 BC   Astronomer Kiddinu [Mesopotamian: b. c 340 BC] discovers the precession of the equinoxes, the apparent change in the position of the stars caused by Earth's wobbling on its orbit

300 BC   Chinese astronomers compile accurate star maps

240 BC   Chinese astronomers observe Halley's comet

Astronomer Eratosthenes [Hellenic: c 276 BC-c 196 BC] correctly calculates the size of the Earth

165 BC   Chinese astronomers are the first to notice sunspots

130 BC   Astronomer Hipparchus of Nicea [Hellenium: c 190 BC-c 120 BC] correctly determines the distance to the Moon

140   The *Almagest* of Ptolemy [Hellenic: c 100-c 170] develops the astronomy of the solar system in a form based on the Sun and planets rotating about Earth

1543   *De Revolutionibus* by Nicholas Copernicus [Polish: 1473-1543] presents convincing arguments that Earth and the other planets orbit the Sun

1577   Tycho Brahe [Danish: 1546-1601] proves that comets are visitors from space, not weather phenomena, as previously believed

1592   David Fabricius [German: 1564-1617] discovers a star, later named Mira, that gradually disappears; when it is studied by Phocyclides Holawarda in 1638, he recognizes that it appears and reappears on a regular basis—the first known variable star

1609   Galileo [Italian: 1564-1642] makes the first astronomical telescope

Johannes Kepler [German: 1571-1630] discovers that the planets move in elliptical orbits

1610   Galileo observes the moons of Jupiter, the phases of Venus, and (although he does not recognize what they are) the rings of Saturn

1611   Several astronomers in the West discover sunspots about the same time for the first time

1633   Galileo is forced by the Roman Catholic Church to recant his support of Copernicus's theory that Earth revolves about the Sun

1668	Isaac Newton [English: 1642-1727] invents the reflecting telescope
1671	Giovanni Domenico Cassini [Italian-French: 1625-1712] correctly determines the distances of the planets from the Sun
1682	Edmond Halley [English: 1656-1742] describes the comet now known as Halley's comet and in 1705 correctly predicts it will return in 1758
1718	Halley discovers that stars move with respect to each other
1755	Immanuel Kant [German: 1724-1804] proposes that many nebulas are actually composed of millions of stars and that the solar system formed when a giant cloud of dust condensed
1758	Halley's comet returns as predicted by Halley
1773	William Herschel [German-English: 1738-1822] shows that the solar system is moving toward the constellation Hercules
1781	Herschel discovers the planet Uranus
1785	Herschel demonstrates that the Milky Way is a disk- or lens-shaped group of many stars, one of which is the Sun
1801	Guiseppe Piazzi [Italian; 1746-1826] discovers the first known asteroid, Ceres
1838	Freidrich Bessel [German: 1784-1846] is the first to determine the distance to a star other than the Sun
1846	Johann Galle [German: 1812-1910] discovers the planet Neptune using predictions of Urbain Leverrier [French: 1811-1877] and John Couch Adams [English: 1819-1892]
1924	Edwin Hubble [US: 1889-1953] shows that galaxies are "island universes"—giant aggregations of stars as large as the Milky Way
1929	Hubble establishes that the universe is expanding
1930	Clyde Tombaugh [US: 1906-] discovers the planet Pluto
1931	Karl Jansky [US: 1905-1950] discovers that radio waves are coming from space, leading to the founding of radio astronomy
1948	George Gamow [Russian-US: 1904-1968], Ralph Alpher [US: 1921-], and Robert Herman [US: 1914-] develop the big bang theory of the origin of the universe
1957	The Soviet Union launches Sputnik I, the first artificial satellite
1959	The Soviet Union launches a rocket that hits the moon
1961	Cosmonaut Yuri Gagarin [USSR: 1934-1968] becomes the first human to orbit Earth
1962	The United States space probe Mariner 2 is the first artificial object to reach the vicinity of another planet, Venus
1963	Maarten Schmidt [Dutch-US: 1929-] is the first astronomer to recognize a quasar

1965	Arno Penzias [German-US: 1933-] and Robert Wilson [US: 1936-] find radio waves caused by the big bang, proving to most astronomers that the big bang actually occurred
1966	A space probe launched by the Soviet Union is the first object produced by humans to land on the Moon (the USSR crashed a probe into the Moon in 1959); another becomes the first to orbit the Moon
1967	Jocelyn Bell [English: 1943-] discovers the first known pulsar while working for Anthony Hewish [English: 1924-]; Hewish later gets the Nobel Prize for the discovery
1969	The United States lands two people, Neil Armstrong [US: 1930-] and Buzz Aldrin [US: 1930-], on the Moon and returns them to Earth
1971	A US spacecraft is the first to orbit another planet, Mars
1975	A USSR space probe transmits pictures from the surface of Venus
1976	The US Viking space probes begin transmitting pictures of the surface of Mars; they do not succeed in detecting life on the planet
	The rings of Uranus are discovered
1979	The US space probe Voyager 1 discovers that Jupiter also has rings
1980	Alan Guth [US: 1947-] develops the theory of the inflationary universe, an explanation of how the big bang occurred
1981	The United States introduces a reusable spacecraft, the space shuttle
1986	The space shuttle *Challenger* blows up, killing all seven aboard
1987	Supernova 1987A, the nearest supernova visible from Earth since 1604, is observed
1988	On Aug 29 an error in a computer command causes Phobos 1, a USSR space probe to Mars, to tumble so that all communication with it is lost
1989	On Mar 27, just as USSR space probe Phobos 2 is reaching its destination, the Martian moon Phobos, where it is to deploy two surface landers, contact with the spacecraft is suddenly lost; 17 minutes of garbled, but decipherable, data are returned before the shutdown
	On Aug 24, the US space probe Voyager 2 flies by Neptune, then the planet farthest from the Sun (Pluto is farthest most of the time), its rings, and its moons; Voyager's images of the planet show Neptune's rings to be "clumpy" but complete; it is not clear how many rings there are, with some scientists counting three, while others see five; finding the magnetic poles gives scientists the first accurate measurement of Neptune's rate of rotation, about 16 hours, 3 minutes, give or take four minutes
	In Sep, Martha P. Haynes [US: 1951-] of Cornell University and Riccardo Giovanelli [Italian: 1946-] announce their observation of a large cloud of gas that they believe is a galaxy in formation

M E T H O D S   O F   O B S E R V A T I O N

HUBBLE TROUBLE AND SUCCESSES DESPITE IT

The Hubble Space Telescope (HST), the first space-based optical telescope, was designed to carry out 467 projects over 15 years that would determine the correct age of the universe, observe the absorption spectra of quasars, survey the universe for faint objects, find accurately the distance to stars, make observations in the ultraviolet range, provide good views of the planets, study black holes, and investigate the origins and composition of comets. It was launched from the space shuttle *Discovery* on Apr 25 1990, after 7 years of delay and a cost of $2.1 billion. The launch involved the highest flight ever by a space shuttle—610 km (380 mi).

Problems

From the first attempts to use the space telescope, however, it suffered from a number of what appeared to be minor problems. One of the antennas that relays information back to Earth jammed, and the telescope wobbled. The engineers decided to live with the defective antenna until a space shuttle mission makes general repairs on the telescope. A computer program will restrict the antenna's motion to angles that should prevent further jamming.

The wobble is caused by changes in temperature when the space telescope leaves or enters Earth's shadow, resulting in brief periods of unequal heating on the solar panels. Heat expands most solids and cold causes them to contract, resulting in unbalanced forces on a satellite. In Jul 1990 National Aeronautics and Space Administration (NASA) engineers tried a computer fix for the wobble. It failed, however, and the telescope continued to lose an eighth of its potential viewing time during the period when the wobble was being steadied. Furthermore, a computer simulation in 1991 suggested that the constant bending was likely to break off the solar panels, which would leave the spacecraft without power and completely useless.

The wobble is not even considered the major problem, however. Images taken with the spacecraft's main telescope are blurred, no matter how carefully the telescope is focused. The blurring is of a type called spherical aberration, which gives each image a fuzzy halo. Spherical aberration results when rays from a telescope's central part and edges do not come to the same focus point. "The outside part of the mirror is lower than the rest," as James Westphal of Caltech explains it. The aberration results from an incorrect shaping of the 2.4-m (94.5-in.) primary parabolic mirror. This was suspected as early as May 20 1990, but announced to the public on Jun 27.

From the first observation with the faulty mirror, NASA officials suspected that "the methods used to measure the figure of the mirror during manufacture ... resulted in the mirror being very precisely made, but to the wrong figure." This suspicion was verified as subsequent investigations revealed that a crucial testing device had been constructed with a built-in error of 1 mm (1/20 in.). This basic problem limits the telescope's usefulness until NASA can fit the device with a corrective lens sometime in late 1993 or early 1994. Maintenance from the space shuttle and installation of improved equipment had been previously scheduled for 1993 in any case. The Ball Corporation of Broomfield CO was given the $30.4 million contract for the corrective lenses, named the Corrective Optics Space Telescope Axial Replacement, or Costar, in Oct 1991.

Another set of problems arose as HST continued operations. This time the gyroscopes that were supposed to ensure stability began to fail one at a time. Gyroscopes are used to keep airplanes and ships on course because the spinning body that is the essence of a gyroscope resists any change in orientation. This is why children's tops do not fall over so long as they continue spinning fast enough. A body in space needs at least three gyroscopes, each at right angles to the others, to maintain its orientation, essential if the body is a telescope that must be pointed in specific directions.

HST started off its trip with six gyroscopes, twice as many as it needed. All six were scavenged from instruments made in 1975 as backups for another astronomical satellite, the International Ultraviolet Explorer (IUE), which flew with its six first-string gyroscopes in 1978. By 1983 only two of the gyroscopes aboard IUE had failed, a good record for electromechanical parts in space. As of 1991 two of the IUE gyroscopes still worked, and astronomers could use these in conjunction with a Sun sensor to keep the satellite stable.

The IUE backups were improved before being slated for HST, but their basic electronics were still more than 15 years old by the time HST flew. Nevertheless, NASA thought the gyroscopes would last another 14 years in space. In any case, the original plan was for frequent space shuttle repair missions to HST, during which defective parts would be replaced. This plan was long abandoned by the time HST was launched, for by then it was clear that there would be many fewer shuttle flights annually than originally planned.

About 9 months after launch, the first HST gyroscope stopped working on Dec 3 1990. The second dropped out for good on Jun 29 1991, after twice having temporary malfunctions. On Jul 26 a third began to have problems. If the third one malfunctions completely, NASA may have to consider scheduling an earlier repair flight. Although the telescope can be stabilized with three gyroscopes, having only three leaves no margin for error. The Sun sensor on HST is not as good as the one on IUE and cannot be substituted for a missing gyroscope. An earlier repair flight might also be able to solve the problem of the solar panel wobble, although the fix for the main mirror would not be ready until the scheduled 1993 maintenance.

One of two spectrographs aboard HST, the Goddard High Resolution Spectrograph, had to be shut down in Sep 1991 after problems with its power supply surfaced on Jul 24 and Aug 5. Engineers think that a defective solder joint caused the problem. Although the loss of power by itself would still permit some use of the spectrograph, it seemed more prudent to shut it down completely and allocate its observing time to other instruments. Bypassing the flaw from the ground would entail a risk to the other four instruments aboard HST. Correcting the problem in space during the planned 1993 shuttle mission to the space telescope would mean adding another day to the mission. In the meantime, the Faint-Object Spectrograph was found to be limited by a build-up of aluminum oxide on its mirror. Apparently, the problem developed while the mirror was in storage during the long wait for launch.

Possibilities Amid the Problems

Although the main telescope could not take clear images, it could still gather light, important for various studies. In particular, the ability of HST to observe ultraviolet radiation, screened from ground telescopes by Earth's atmosphere, was only slightly affected. The mirror could also focus light on the Wide Field/Planetary Camera, but not as efficiently as planned. While still obtaining better pictures with this camera than any ground-based telescope could, the pictures were only marginally better, not spectacularly better.

Since all the instruments see through the main mirror, all had some problems mixed with possibilities for useful science. The Faint Object Camera provides resolution in visible light no better than from ground-based telescopes, so its work will be limited to ultraviolet light only. The Faint Object Spectrograph, designed to study spectra and polarization of very weak visible and ultraviolet images, was slightly degraded, causing the faintest objects to be lost from sight. Until it was shut down in Sep 1991, the Goddard High Resolution Spectrograph also had problems with getting good resolution of very dim objects, but it had only moderate difficulty with brighter ones. The Goddard had been less impacted than other instruments, because it operated solely in ultraviolet light.

The High Speed Photometer that measures light intensities also suffers from loss of sharp images of faint objects. In one plan for correcting the aberration, the High Speed Photometer would be sacrificed to make room for corrective optical equipment. The advantage of this plan is that all four of the remaining instruments could retain the abilities they were originally planned to have.

The only part of the operation completely unaffected is the set of Fine Guidance Sensors used to determine the positions of stars precisely. Their primary purpose is to determine parallax, used to find the distance to stars. The same studies may also reveal unambiguous evidence of planets around other stars by detecting periodic wobbles in a star's position caused by the planet and star orbiting about a mutual center of gravity.

Almost all of the planned experiments can be improved to nearly their original expectations by taking a longer time to gather radiation. Thus, another consequence of the aberration of the main mirror is that schedules were thrown into disarray.

Results

Despite all its flaws, the Hubble Space Telescope was showing astronomers some new views as early as during its engineering tests. In the ultraviolet, the Wide Field/Planetary Camera was able to resolve R136, until recently believed by many astronomers to be the largest known single star, into at least sixty giant stars. This had not been possible with ground-based telescopes. Another early feat in which HST outpaced ground-based observations was the first photograph that resolves Pluto and its satellite Charon into two distinct objects. HST also was able to rule out some black hole candidates while finding other ones. Another important success was obtaining the best measurement yet to the nearby galaxy the Large Magellanic Cloud, 169,000 light-years away. This is an important step toward determining the distance scale and age for the universe (see "Cosmology UPDATE," p 83). On Mar 11 1991 HST produced the clearest and most detailed image of Jupiter seen since Voyager space probes were within five days of their planetary flybys of Jupiter in 1979.

An interesting Hubble result was the confirmation of a theory about unusual stars called blue stragglers. These hot blue stars are called stragglers because they are found in regions where all other stars are cool and red. Often blue stragglers are found in 150-odd roughly spherical swarms of several hundred thousand stars that stud our galaxy. These are called globular clusters.

Such clusters are thought to have formed early in the history of the Milky Way. According to theories of stellar evolution, red stars are old and blue stars are young. But in these clusters, red and blue stars are neighbors that should have formed at the same time. Stars in globular clusters move slowly with respect to one another. The theory is that blue stragglers form when two such stars gently bump into or graze one another. In the process, one of the stars acquires a lot of hydrogen from the other, rekindling its fires and making the cannibal star appear young. Or the two stars could fuse into one large star, which would also cause the fused star to burn with the brightness of youth.

Another idea is that tidal forces between the close stars cause both to burn hydrogen more efficiently. This last hypothesis was first proposed in 1964 independently by Fred Hoyle [English: 1915-] and William H. McCrea [English: 1904-]. Hubble has shown that this scenario is very likely. HST examined a crowded field of more than 600 stars in a very dense globular cluster with its Faint Object Camera. The camera picked out twenty blue stragglers, a number so high that the collision or cannibalism theory seems to be the only possible explanation. Francesco Paresce and coworkers at the Space Telescope Science Institute in Baltimore think that most of the blue stragglers are double stars whose greater mass has caused them to drift toward the center of the cluster.

Another surprising result was the discovery that there are a lot more clouds of intergalactic hydrogen than were anticipated between Earth and the quasar 3C 273. Instead of the one or two that were expected to show up in the quasar's ultraviolet spectrum, there were—depending on who does the counting—at least five and possibly as many as sixteen. These high numbers are closer to what one would have expected when the universe was half as old as it is today. Further observations by HST of other quasars may help to account for the large number of clouds.

(Periodical References and Additional Reading: *Astronomy* 1-90, p 38; *Science News* 1-6-90, p 8; *Science News* 5-5-90, p 276; *Science News* 5-26-90, p 325; *Astronomy* 7-90, p 30; *Discover* 7-90, p 8; *Science* 7-6-90, p 25; *Science News* 7-7-90, p 4; *Science* 7-13-90, p 112; *Science News* 7-14-90, p 21; *Science* 7-20-90, p 242; *Astronomy* 8-90, p 38; *Science* 8-31-90, p 987; *Astronomy* 9-90, p 44; *Astronomy* 10-90, p 20; *Astronomy* 11-90, pp 22 & 30; *Science News* 11-10-90, p 295; *New York Times* 11-26-90; *Astronomy* 12-90, p 22; *Science News* 12-8-90, p 359; *New York Times* 1-17-91; *Astronomy* 2-91, pp 24 & 30; *Discover* 2-91, p 46; *Science* 2-1-91, pp 507 & 522; *Science* 6-1-91, p 1074; *Science* 8-9-91, p 627; *Science News* 8-10-91, p 86; *New York Times* 8-27-91, p C2; *Astronomy* 9-91, pp 42 & 46; *Science News* 9-21-91, p 182; *Planetary Report* 9/10-91, p 3; *New York Times* 10-23-91, p A19; *Scientific American* 11-91, p 112)

START OF THE KECK ERA

For more than 40 years, since 1948, the 5-m (200-in.) Hale Telescope on Mt Palomar CA has reigned as the leading optical astronomical instrument available. In the early 1990s its preeminence was challenged by a host of new devices, ranging from the flawed Hubble Space Telescope (HST) to the New Technology Telescope (NTT). But the real challenge in light-gathering comes from the new record-holder in size, the W.M. Keck Telescope on Mauna Kea on the "big island" of Hawaii. The $94-million Keck is twice the size of the Hale in diameter, which means that it has four times Hale's light-gathering power, although it will not have the resolving power of the NTT. Light-gathering power is important because among other reasons, it can speed up operations by factors of six or more, meaning more data for more astronomers. The Keck can achieve such a large size without sagging under gravity because it is built in thirty-six segments, each 1.8 m (6 ft) across, 400 kg (880 lb), and separately supported. Each segment is supported by three movable pistons that continually adjust the segment to align it properly with the others.

Furthermore, its location on Mauna Kea ensures much better viewing than on Mt Palomar, both because it is higher—Keck is at 4145 m (13,600 ft)—and because it does not have the distraction of lights from a nearby major city.

Keck was supported by a grant of $70 million to Caltech and grew out of an ongoing large telescope project of the University of California, which funds Keck's operating costs. The University of Hawaii is also a partner in the enterprise. Keck's main instruments include a near-infrared camera that will use wavelengths ranging from 1 to 5 microns, a long-wavelength infrared camera for wavelengths between 8 and 20 microns, and both low- and high-resolution spectrographs. Its specialties will be the near infrared wavelengths, where the youngest and most distant quasars and galaxies are, and spectroscopic analysis of very faint objects.

In Nov 1990, when enough of its mirror segments were in place to produce a light-gathering power equivalent to that of the Hale Telescope, astronomers obtained the first images—known as first light—from the Keck. The actual object imaged was galaxy NGC 1232, thought to be about 65 million light-years away. Completion of the Keck was scheduled for the end of 1991.

Encouraged by the promising results of the first tests, a combine of the Keck Foundation, Caltech, the University of California, and NASA started planning in Mar 1991 for Keck II, which would be an identical twin 85 m (275 ft) away from what has already been renamed Keck I: object, optical interferometry (see following article). When used as an interferometer, the Kecks will be able to resolve the headlights of an oncoming car 25,750 m (16,000 mi) away. With a grant of $74.6 million from the Keck Foundation covering 80 percent of the cost, Keck II is scheduled for completion in 1996.

The Keck era may not last so long as Hale supremacy did, however. The European Southern Observatory, a consortium of eight nations, is building four optically linked telescopes based on the NTT design. The four will each have a mirror slightly smaller than those of a Keck, just 8.2 m (27 ft) in diameter, but the four are designed to be used together. In combination, they will produce light-gathering ability equivalent to a single telescope 16 m (52.5 ft) in diameter, with more than two-and-a-half times the light-gathering power of a single Keck or about 25 percent more than the Kecks combined. The new telescope, to be called the Very Large Telescope and to be located on the summit of Cerro Paranal in the Atacama Desert of northern Chile, is expected to begin operations in 2000, making the Keck era only 9 years long.

(Periodical References and Additional Reading: *Science* 9-14-90, p 1244; *Science News* 12-1-90, p 348; *New York Times* 12-5-90, p B13; *Discover* 1-91, p 25; *Discover* 3-91, p 16; *Science* 3-15-91, p 1301; *Astronomy* 8-91, p 18; *Discover* 11-91, p 40; *Scientific American* 11-91, p 112)

SUCCESSES WITH IMPROVED OPTICAL TELESCOPES

When the Hubble Space Telescope (HST) was planned, it seemed to be the only way that astronomers were ever going to be able to get the clear, crisp images they need to advance science. Ground-based optical telescopes have to deal with Earth's atmosphere, which not only absorbs radiation but also causes images to twinkle. By the time HST was built, however, clever ways had been found to make optical telescopes, ways that produce some images that are as good as any the HST was planned to achieve and better than its flawed actual accomplishments, although there are many tasks that still can be accomplished only from space.

Bigger Is Better

Although HST has the best viewing location of any telescope, its flawed main mirror is only 2.4 m (94.5 in.) in diameter, half the diameter of the Hale Telescope at Mt Palomar, the largest fully operating successful ground-based telescope (a larger one in the USSR has been plagued with problems; see "Major Telescopes," p 97) and a fourth the diameter of a Keck. The larger a good mirror is, the more light it can collect, the bottom line for any telescope.

One group of large telescopes that has been years in the making is a set of nine or so that will rely on advanced technology developed by Roger Angel and his staff at the Steward Observatory Mirror Laboratory in Tucson AZ. Angel's telescopes use a honeycomb backing to reduce the weight of the mirror and a novel method of spin casting to make the blank for the mirror. So far the laboratory has built three 3.5-m (138-in.) mirrors that are expected to be installed in the early 1990s, with first light at the ARC Telescope at Apache Point Observatory in Sunspot NM scheduled for 1991. Next the laboratory is planning to spin a 6.5-m (256-in.) mirror to replace the six small ones at the Multiple Mirror Telescope on Mt Hopkins in Arizona. It then plans to make a series of five new 8-m (315-in.) mirrors for projects that will probably achieve first light during the latter half of the 1990s.

While none of these will be as large as the Keck twins (see above), they individually and collectively will provide major new capabilities for the world's astronomers. Roger Angel and coworkers estimate that the 1990s will see four times the total worldwide light-collecting ability from the world's optical telescopes than in previous years.

Images Instead of Photographs

The science of astronomy was given an enormous advance when the first astronomical photographs began to be used in the mid-1850s. For example, photography pioneer George Phillips Bond [US: 1825-1865] was the first to observe in 1856 that photographs of stars revealed their magnitudes by the size of the images. As photography advanced, it was clear that photographs revealed much more than that, since film could capture hours of light from a single object, while the human eye was limited to instants.

Although photography continues to improve, it has in many ways run into technological walls. Photographs are limited by the chemical nature of the reactions used in making them. A photon of light has to have sufficient energy to change a molecule from one state to another. Furthermore, the size of the molecules is the ultimate limit on the graininess of the image.

Astronomers have therefore turned to new ways of capturing the light that is gathered by their telescopes. Instead of the coated film of photography, they utilize the silicon wafers of modern electronics. In particular, electronic light detectors called charge-coupled devices can count individual photons of light as they strike the detector surface and record the data in digital form. The resulting images often seem to make all objects appear square to the uninitiated.

A new telescope, planned for 1995, will use the largest array ever of such charge-coupled devices to map the entire universe in three dimensions. Such a map will be a major tool in cosmology. The thirty charge-coupled devices in the digital sky survey telescope, supported by the University of Chicago, Princeton, and the Institute for Advanced Study, will be able to record as many as 120 million points of light at a time from a unique wide-field 2.5-m (100-in.) telescope. It will be installed in the telescope complex at Apache Point in the Sacramento Mountains of New Mexico. The Apache Point complex also involves New Mexico State University, Washington State, and the University of Washington.

Active Optics and Flexible Mirrors

Perhaps the most successful new approach that is already in place is the one called active optics. The main idea is to make a thin mirror that can be continually adjusted by computer-directed actuators so that the best possible image is obtained. Such thin mirrors are called meniscus mirrors. Instead of the typical 6 to 1 or 8 to 1 ratio of diameter-to-thickness of traditional mirrors, a meniscus mirror may have a ratio of 40 to 1 or more.

The prototype of active optics, the New Technology Telescope (NTT), is 3.58 m (141 in.) in diameter. NTT uses seventy-eight actuators to adjust the mirror about once a second. It also has an improved mounting and better thermal control than other large mirrors. The result is that the mirror is three times as precise as it would be if conventionally made.

NTT is run by the European Southern Observatory and located at 2400 m (7875 ft) in La Silla, Chile. It began basic operations on Mar 23 1989, and was formally dedicated on Feb 6 1990. Soon after, operated by a team headed by Joseph Silk of the University of California, Berkeley, it captured images of the faintest galaxies ever recorded. This was achieved by aiming the telescope at an apparently empty patch of sky and exposing light detectors for 6 continuous hours.

Because of its immediate success, NTT will soon be followed by several telescopes using meniscus mirrors, including more for the European Southern Observatory and telescopes built by Japan and the United Kingdom.

Interferometry and Aperture Synthesis

One technique for making more of visible light at ground level is called interferometry, which can be used in either a traditional form or in a special technique called aperture synthesis that is described below.

One two-telescope interferometer is the Mark III on Mt Wilson CA. It has been used to measure the diameters of nearby stars directly. Its success has led to plans to build arrays of telescopes that will be able to make actual images of stars. Until recently, the most one could see of any star other than the Sun was a point of light. Telescopes planned for 1992 or 1993 will include arrays that can produce images by interferometry.

In the meantime, images have already been produced by a variation of interferometry called aperture synthesis. In aperture synthesis, two sides of the same star are used as two different sources of light, collected at two different locations. The two different locations for the collectors make the distance the light travels from each source different, thereby creating an interference pattern in two directions, instead of the single direction of conventional interferometry. A computer is then used to convert the interference pattern back into an image. When used in radio telescopes, two or

INTERFEROMETRY

The basic idea of interferometry is that the same wave from two slightly displaced sources (in time or in position) will interfere with or reinforce itself, making a pattern that is much larger and more observable than the wave itself. A wave is a sequence of peaks and valleys. When two waves—or two versions of the same wave, with one displaced—are superimposed, the combination of two peaks or two valleys will create higher peaks or deeper valleys, while the combination of a peak and a valley will tend to cancel each other. The reinforced sections of light appear as visible regions called fringes, which are separated by cancelled sections. If you know how far apart the wave and its displacement are, the fringes can be used to construct the original wave. The way to accomplish this for astronomical objects is to collect the wave at two places, using the distance between them as the displacement.

Interferometry was used by Albert A. Michelson [German-US: 1852-1931] and Edward W. Morley [US: 1838-1923] in 1887 to measure the speed of light precisely, and is used by radio astronomers to locate exactly where radio waves originate as well as to provide other information about the source.

more different telescopes placed apart are used and different times of observing are used to produce the additional direction; for both radio and optical aperture synthesis, the farther apart the receivers are, the sharper the reconstructed image.

So far, astronomers have not been able to obtain two different mirrors for which they know the exact distance apart to the precision needed for light waves, which are much smaller than radio waves. Instead, they have cleverly masked individual large mirrors so that the light is collected separately on two sides of the same mirror and then combined in the interferometer. The first experiment was accomplished by a group headed by John Baldwin of Cambridge University, who used the 419-cm (165-in.) William Herschel Telescope in the Canary Islands to produce detailed images of Betelgeuse. Another team, led by Shrinivas Kulkarni of Caltech, applied aperture synthesis to the Hale Telescope at Mt Palomar, improving its resolution by about thirty times. Both groups then turned to the famous variable star Mira, a red giant about 200 light-years away. Mira has been observed to change from very bright to complete disappearance according to an irregular cycle of about 330 days. The result was a surprise. Mira is extremely oblate, more like a football than a baseball. Whether the shape has anything to do with its variability is not clear, but astronomers will continue to observe it through its cycle to see if they can determine what is happening.

The Cambridge group is also proceeding to the next logical step—construction of independent mirrors about 100 m (328 ft) apart to improve the technique. As with radio astronomy interferometry, the farther apart the collectors of waves are, the more precise the images.

Adaptive Optics

Adaptive optics has been called by *Science* magazine "untwinking the stars," since its main purpose is to remove the twinkling that is caused by distortion of light as it passes through the atmosphere. If Earth's atmosphere were a perfectly homogeneous gas, untwinking would not be necessary, but air is filled with dust and different parts of the atmosphere have different densities due to variations in heat or humidity. The result is that starlight is bounced about in an essentially unpredictable way during its passage from space to Earth. If one knows how the light has changed in a given instant, it is possible to compensate for the twinkling and produce a steady image. But the change itself is varying from instant to instant.

At least one astronomer thinks that for infrared wavelengths, twinkling is not a factor. T. Stewart McKechnie of Lentec Corp in Albuquerque NM has calculated that turbulence cells in the atmosphere are of a size that allows infrared light to slip right past them. Furthermore, he attributes most blurring found in mirrors at all wavelengths to a combination of sloppy optics and shaking of the mirror. Astronomers are studying his ideas, which were first put forward in Nov 1990.

In the meantime, astronomers recently have developed several crude devices to compensate for the variations caused by the atmosphere, based on ideas first put forward in 1953 by Horace W. Babcock [US: 1912-]. All rely on mechanical actuators to change slightly the shape of the mirror in response to computer commands.

Some systems require using a nearby bright guide star whose twinkling is compensated for on the assumption that light from other fainter nearby objects has passed through nearly the same column of the atmosphere and therefore will require virtually the same compensation. The main problem with these systems is that they can be used only when a sufficiently bright guide star is in the same visual field as the object of interest. The star must be very bright because the atmospheric column changes on the order of every 10 milliseconds, requiring a very bright source to be measurable in such a small time interval. There are more suitable bright stars in the infrared region of the spectrum than in the optical. The first serious test of the new technology known to astronomers came late in 1989 with successful imaging in the infrared region at the Haute Provence Observatory in France; that imaging reached the diffraction limit of the 1.53-m (60-in.) mirror used.

The NTT in the early 1990s installed a simple adaptive optics system that uses a small deformable mirror to cancel out twinkling. A similar system already installed on a smaller telescope used in the infrared region reached the mirror's diffraction limit—that is, it produced results that are theoretically as good as possible from that particular mirror.

On May 27 1991 the astronomy community learned that the US military's Star Wars (officially SDI, but no one calls it that) program had developed a system of adaptive optics that is better than any previously employed by astronomers. Developed to help correct the paths of laser weapons, the system was declassified after the Star Wars laser weapon program was terminated. The basic method used in the Star Wars procedure—creation of an artificial guide star with a laser and adjustment of a mirror to eliminate distortion—is no different from ones previously proposed by astronomers but never built because of lack of money. With much greater sums at their disposal, however, the Star Wars team developed working equipment. The laser is reflected by sodium ions high in the atmosphere. A computer can make corrections in the image as often as thirty times a second.

A test installed on a moderate-sized 60-cm (152-in.) telescope produced images right at the mirror's diffraction limit, the fundamental physical wall against which all reflecting telescopes, including the HST, operate. Edward Kibblewhite of the University of Chicago is working at MIT's Lincoln Laboratory to convert the Star Wars technology to astronomical use. He believes that "if it lives up to its tremendous progress [adaptive optics] will revolutionize astronomy." At longer wavelengths (infrared and longer), the new technology should produce better results than the HST.

An important feature of adaptive optics is that it can be retrofitted to existing telescopes at a reasonable cost. Another virtue is that it will improve even further the results that can be obtained with optical or infrared interferometry. The only drawback of adaptive optics is that it works only for a small viewing area; it cannot correct for the image of an extended nebula, for instance.

In Feb 1991 John Torny of MIT using the Canada-Hawaii-France Telescope on Mauna Kea, photographed the most distant stars ever observed—individual stars in the NGC 4571 galaxy in the Virgo cluster, 50 million light-years away. The picture was not only remarkable as a record; it also suggested the universe is younger than previously believed. The achievement was made possible by a prototype adaptive optics system, the location of the telescope, and a very clear night.

(Periodical References and Additional Reading: *Astronomy* 2-90, p 14; *Science* 2-23-90, p 917; *Astronomy* 3-90, p 10; *Astronomy* 7-90, p 38; *Science* 7-20-90, pp 223 & 253; *Astronomy* 11-90, pp 34 & 44; *Physics Today* 3-91, p 22; *Physics Today* 4-91, p 48; *Science* 6-28-91, p 1786; *Discover* 7-91, p 17; *Science* 7-12-91, p 138-39; *New York Times* 8-6-91, p C1; *Scientific American* 11-91, p 112)

DETERIORATION OF RADIO TELESCOPES

Optical mirrors are made of brittle glass that needs to be protected from breaking, but given such protection, glass mirrors last virtually forever. Radio telescopes face maintenance problems. Each telescope is a giant structure of metal girders. A steerable radio telescope can be compared to a large bridge that has to be moved about from time to time in a precise fashion. Like bridges, radio telescopes can deteriorate unless given constant maintenance. Also, they can collapse from metal fatigue, as the 985-m-diameter (300-ft-diameter) Green Bank radio telescope did on Nov 15 1988, after 26 years of operation.

Late in 1990 the National Science Foundation awarded contracts for a replacement for the Green Bank radio telescope. The new instrument will be fully steerable, which the original was not, and have an elliptical antenna with a major axis of 120 m (394 ft) and a minor axis of 100 m (328 ft). A maintenance contract will likely follow.

The premier radio telescope operating in the United States combines the maintenance problems of a bridge with those of a railroad. The Very Large Array (VLA) near San Augustin NM consists of twenty-seven 25-m-tall (82-ft-tall) radio telescopes mounted on railway tracks so that they can be moved to appropriate positions along 65 km (40 mi) of track in the shape of a giant Y. Maintenance problems at the VLA are compounded by shortcuts taken during its construction to save money. For example, the railroad track and ties were all secondhand, with track dating back as far as 1902 and ties mostly from the 1930s. The ties in particular are falling apart at a great rate, much faster than they can be replaced. It is therefore possible that one of the telescopes could pitch over someday as it is moved along a bad piece of track.

In another cost-saving measure, inspection manholes were built from stacked burial vaults with their ends removed. These are now collapsing and need to be replaced with more conventional reinforced concrete.

Even equipment that used state-of-the-art materials when the VLA was built in 1980 has deteriorated, notably the electric cables. These were insulated with polyethylene, which breaks down after a few years. As a result, the cables short out fairly often, especially during the power surges caused by frequent thunderstorms.

In another problem, one of the twenty-seven telescopes has had its main ball bearings go out. It currently cannot be moved and is useless.

The problems at Green Bank and the VLA are, unfortunately, typical of radio telescopes. Without continual and expensive maintenance, these great instruments can fall apart.

(Periodical References and Additional Reading: *New York Times* 1-1-91, p 140; *Science* 7-19-91, p 268)

WHAT'S NEW IN THE SOLAR SYSTEM?

SOLAR NEUTRINOS

For the past 20 years there has been an unsolved mystery about the neutrinos produced by fusion reactions in the core of the Sun. All efforts to detect them find far fewer than predicted by fusion theory, the theory that describes how light atoms combine, releasing energy in several forms, including neutrinos. This process provides the energy released by the Sun.

Since 1968, the main "neutrino telescope" has been a 375,000-L (100,000-gal) tank of cleaning fluid (perchloroethylene) at the bottom of the Homestake Gold Mine in Lead SD. It has consistently recorded from one-fourth to one-third the expected number of high-energy neutrinos. Another experiment, on line since 1988, the Japanese Kamiokande II, uses a different method to detect high-energy neutrinos, but has gotten the same results as the Homestake telescope. With that in mind, several new telescopes that can detect solar neutrinos at lower energy levels have been or will be constructed.

The USSR-US gallium experiment, known as SAGE, consists of a tank of 30 tons of liquid gallium instead of cleaning fluid. In principle, the gallium should be able to record low-energy neutrinos. According to theory, since low-energy neutrinos are five thousand times as abundant as high-energy ones, SAGE should detect about a neutrino a day. In practice, the first several months of operation in 1990 revealed no neutrinos above the background level that would be expected if the Sun were nonexistent. A twin to SAGE went on line in Italy in 1991; it may help clarify the situation.

One possible explanation for the results so far is the popular theory called MSW (after Stanislav Mikheyev and Alexi Smirnov of the Soviet Academy of Sciences and Lincoln Wolfenstein of Carnegie-Mellon). It proposes that neutrinos change flavor as they pass through the Sun and travel out into space. According to MSW calculations, a substantial portion of the electron neutrinos that are produced in the Sun's core become either muon neutrinos or tau neutrinos before they reach Earth. All of the detectors now on line can only observe electron neutrinos, so if MSW is correct, the missing neutrinos can be explained. A new neutrino telescope in a nickel mine about 320 km (200 mi) north of Toronto, Canada, called the Sudbury Neutrino Observatory (SNO), will use 1000 tons of heavy water in an experiment that should be able to observe muon and tau neutrinos as well as electron neutrinos. In addition to using Cerenkov radiation from scattered electrons, it will also detect the breakup of the deuterium caused by the muon and tau neutrinos. When SNO begins operations in 1995, MSW will be put to the test.

Background

The most common fusion reaction in the Sun's core that produces a neutrino occurs when two protons form a deuterium (heavy hydrogen) nucleus, a positron, and a neutrino. This happens a lot, but neutrinos from the deuterium reaction are not very energetic. Sometimes, at a later stage in the cycle, a beryllium nucleus is formed. As the beryllium decays, it can take two routes, each of which releases a neutrino. The rarer mode that leads ultimately to two alpha particles (helium nuclei) releases a high-energy neutrino. This accounts for about one out of five thousand solar neutrinos. The neutrinos stream away at nearly or exactly the speed of light (depending on whether they have mass or not, which has not been fully determined) and take part in no further reaction.

Sometimes one of the high-energy neutrinos from the rare beryllium decay is captured by a chlorine atom, which causes it to decay into an electron plus radioactive argon-37. It is the chlorine in the cleaning fluid at the Homestake Mine that is the sensor for the neutrino telescope. Three-fourths of the time, the chlorine interacts with the high-energy neutrinos, and only a fourth of the time with the more abundant neutrinos at lower energies.

An entirely different method for detecting neutrinos is used by the Japanese Kamiokande II experiment. It is looking for flashes of light in pure water formed when a neutrino slams into an electron and jolts it into briefly moving faster than the speed of light in water. The resulting flash, which is called Cerenkov radiation, cannot be mistaken for anything else. This radiation is produced only by the high-energy solar neutrinos.

Other detectors use gallium. Even a low-energy neutrino can convert gallium-71 to radioactive germanium-71. Measurement of radioactivity can be used to determine the number of neutrinos that have caused such conversions.

In the meantime, there is a new mystery. In Apr 1990, Kenneth Lande, working on the Homestake solar neutrino telescope, reported that the number of neutrinos drops drastically at sunspot maximums and climbs to its highest levels when sunspots are inactive. This result also goes against theory, since neutrinos from the Sun's core should be able to pass through sunspots, which are a surface phenomenon, without even noticing them, just as they pass through everything else most of the time. Lande's observation was largely confirmed by two separate studies released later in 1990. Work by Lawrence M. Krauss of Yale University also showed a connection between the number of neutrinos and solar oscillations.

One possible explanation for a connection between sunspots and neutrinos has been offered by Lande. Since sunspots are magnetic storms, perhaps their intense magnetic fields somehow flip the left-handed neutrinos so that they become right handed. Right-handed neutrinos cannot be detected even in theory, although one theory endows such neutrinos with mass, while keeping left-handed neutrinos massless. The problem with Lande's idea is that neutrinos are not known to interact with the electromagnetic force in any way, the only force involved in magnetism.

Neutrinos come in three flavors: electron, muon, and tau. Like all particles, they also have antiparticles, although it is possible that the neutrino, like the photon, is its own antiparticle. Since neutrinos have little or no mass and interact very little with matter, they are extremely difficult to detect. All known ways of detecting a neutrino depend on the weak interaction, a force that acts only on particles with a left-handed spin. If neutrinos with a right-handed spin exist, instead of interacting with matter slightly, by the weak interaction, they fail to interact at all. Thus, right-handed neutrinos, if they exist, are undetectable.

(Periodical References and Additional Reading: *Astronomy* 3-90, p 40; *Science News* 4-21-90, p 245; *Science* 4-27-90, p 444; *Science* 6-29-90, p 1607; *Science News* 9-1-90, p 141; *Science News* 12-8-90, p 358; *New York Times* 1-22-91, p C5; *Physics Today* 4-91, p 55; *Discover* 5-91, p 66)

THE SUN PUTS ON A SHOW

Although the Sun is 150 million km (93 million mi) away from Earth, we bask in its light and warmth. The more we learn about the Sun, however, the more we find that it affects us on Earth in ways that go far beyond supplying photons. Our changing relationship with the Sun is thought to bring on ice ages on a regular basis (see "Dating the Ice Ages," p 229). There is a persistent belief, but little evidence that stands up, that the 22-year sunspot cycle affects weather in the short run. Once in a while the Sun's effect on Earth is a bit more noticeable than usual. A rare event of this type and a more common one both occurred in 1991.

The Great Solar Flare

Although Earth's magnetic field gradually shifts with time and even reverses every 100,000 years or so (see "Reversals of Magnetism," page 241), it is quite tame compared with the magnetic field of the Sun. The Sun's magnetic field, probably because the Sun does not rotate as a whole, almost always has regions of high magnetism, where the magnetic field interacts with the electrically charged plasma that surrounds the Sun to form the storms we observe as sunspots. The Sun's overall magnetic field resembles Earth's with a north and south pole (which reverse every 11 years instead of every 100,000 or so), but this field is weak compared with those of the sunspots. In 1991 Philip R. Goode of the New Jersey Institute of Technology and W.A. Dziembowski of the Polish Academy of Science proposed that the different rates of rotation within the Sun could be used to account for the 22-year sunspot cycle.

From time to time, a region of magnetic storms, or sunspots, builds up such magnetic tension that the field lines appear to snap. These crash into the Sun with such force that they send plasma all over, out from the Sun in a great flare and into the Sun's surface, where, of course, the plasma encounters matter that was there to begin with. The collisions between the incoming plasma and the interior of the Sun result in the production of gamma rays (photons); the more violent collisions break molecules apart, releasing neutrons. Some of the plasma flung into space, called the solar wind, travels the 150 million km (93 million mi) to Earth, where it interacts with Earth's magnetic field and atmosphere. Typically, a flare goes on for about 20 minutes; some solar flares have previously been observed to last as long as 4 hours.

Thus, when gamma rays were detected on the departing limb of the Sun on May 18 1991 by the GOES-7 satellite, whose job it is to monitor the Sun, scientists thought they knew what was happening. A solar flare was starting and it would be long gone by Jun 1, when the Sun's rotation would return that part of the sphere to a position facing Earth. But on Jun 1, even before the region was fully observable from Earth, a blast of gamma rays came from over the rim; it was so powerful that it saturated the satellite's measuring equipment—that is, the intensity was too great to measure. An example of saturation would be a thermometer that registered up to 50° C (122° F) measuring temperature of a hot syrup as 50°. The temperature might be 50° C or it might be higher. The thermometer would be saturated. That is what happened to the measuring equipment for gamma rays on GOES-7.

As the Sun turned, six flares appeared and were visible from Earth, although not all at one time; five of the six were each strong enough to saturate the GOES-7 gamma-ray detector. Furthermore, flares continued until the flaring region passed out of view on Jun 15, when the region was once again carried to the other side by the Sun's rotation.

Seeing that something new and strange was afoot, ground controllers pointed one of the Gamma Ray Observatory (GRO) detectors, the Oriented Scintillation Spectrometer Experiment, at the Sun on Jun 1. On Jun 7 they rotated the whole satellite so that all four of its gamma-ray detectors faced the Sun. Not only were they able to record the spectrum of gamma rays, they also were able to detect neutrons. Analyses of both emissions should reveal more about composition of the upper layer of the Sun.

Meantime, a dramatic amount of the plasma that was thrown into space was beginning to hit earth, even though, since the plasma was not aimed, our planet received only a little bit from the edges of the flare. The particles were the same as those that form the solar wind—mostly protons and ions—but there were a lot more of them. Like the solar wind, flare particles are channeled by Earth's magnetic field toward the poles. When they reach the atmosphere, their interaction produces the auroras. During the Great Solar Flare, as the June sequence of flares was often called, the aurora borealis was observable as far south as the middle of the United States, from Washington DC through Colorado to Eureka OR. Many people stayed up to the middle of the night to have what for most would be a once in a lifetime experience.

Auroras are nice, but some of the things that go with them are less so. The particles disrupt the ionosphere, playing havoc with radio communications. They also cause power surges of the kind that can disrupt computers. These surges can be powerful enough to cause transformers and generators at power plants to burst into flame. Such surges are caused when the particles push so hard on Earth's magnetic field that they move it. The basic principle of electromagnetism is that a moving magnetic field induces an electric current (and vice versa).

Airline passengers and flight crews get undesirably large doses of radiation during a flare. Satellites can be damaged by radiation or even pushed off course. All of these have happened at one time or another in the past. The Great Solar Flare, which was probably not as large as one in 1956 that took place before satellite measurements could be made, disrupted radio signals and probably knocked out three generators at a power plant in Virginia. Other effects no doubt occurred, but they were not associated with the flare or not widely reported. Another large solar flare late in Oct 1991 also caused problems on Earth. It knocked out the power transmission line between Quebec and New England and disturbed electric power systems all over Earth.

The Great Solar Eclipse

The longest and most observable solar eclipse of the century followed the Great Solar Flare by about a month. On Jul 11 1991, the moon's shadow covered a region from the Pacific through Hawaii, Baja California and farther into Mexico, parts of Central America, and down into South America. The eclipse was notable because of its long duration in Mexico, where skies are almost always clear, and because the centerline of totality passed almost directly over Mauna Kea on Hawaii, where the greatest collection of optical and infrared astronomical instruments anywhere on Earth—four large optical telescopes, two infrared devices, and two telescopes that observe waves in the millimeter region—reside. The shadow of totality would also pass across Mexico City, possibly Earth's most populous urban region. In terms of duration alone, the 1991 eclipse was the second longest of the twentieth century and will not be exceeded for the next 140 years, until Jun 13 2132. The longest a total solar eclipse can continue at any single site is 7 minutes 31 seconds. This one lasted 6 minutes 56 seconds at its midpoint. More people viewed it than any previous eclipse.

The actual eclipse began in the mid-Pacific just at dawn, starting at 6:38 a.m. local time. Then the shadow raced across a long section of Earth at a speed of about 8000 km (5000 mi) per hour. The shadow left Earth 3 hours 20 minutes later at local sunset in a remote part of Brazil.

There were viewing problems in Hawaii. Anticipated, but beyond correction, was high haze from a dust cloud from the eruption of Mt Pinatubo in the Philippines. Not anticipated, but not surprising, were high, thin clouds of water vapor over Mauna Kea and worse clouds over most of the rest of the island. Totally unexpected was a large vent of steam from neighboring Mauna Loa, which prompted an unnecessary volcano alert. Finally, as the eclipse approached, the cloud cover *below* the observatories on Mauna Kea started to creep upward toward the summit. Clouds managed, however, to stay below the observers for the 4 minutes 12 seconds of totality at that site.

Despite all this, most of the observations at Mauna Kea were completed. There were nine major experiments conducted by fifty astronomers. These included a search for a ring of dust believed to be left over from the formation of the solar system (preliminary report: not found); studies of the individual layers of the Sun's atmosphere at millimeter wavelengths; better measurements of the strength of the solar magnetic field; and still and motion pictures of the corona that will be used by Serge Koutchmy of the Institute of Astrophysics in Paris and coworkers in an effort to determine why the corona is so much hotter than the Sun's surface. Giovanni Fazio, Eric Tollestrup, and coworkers from the Harvard-Smithsonian Center for Astrophysics also produced the first infrared images of a solar eclipse.

Those who traveled to Baja California got a longer eclipse, but some of them also encountered clouds. The 12 million people in heavily polluted Mexico City, however, had unexpectedly clear skies.

An interestingly different viewpoint was achieved as a result of a suggestion by Bruce Bierman of La Mesa CA. He proposed that appropriately placed Earth-orbiting satellites photograph the Sun's shadow as it raced over the planet. As a result, photographs taken both by the European Space Agency's Meteosat 3 and NASA's GOES-7 showing the eclipse were published.

Total solar eclipses are not rare. On the average, there is one every 16 months or so. But they reach only a small fraction of Earth's surface during each occurrence, hitting a given location about once every 350 years. Thus, unless it is your practice to travel to where they will be, you are lucky to see one during a lifetime. Of course, many people, for personal or professional reasons, travel to where the eclipses are—Siberia, Indonesia, Borneo, and the Philippines have been a few recent destinations.

SOME FORTHCOMING ECLIPSES

There are four types of eclipses, depending on the exact relationship of the Sun, Moon, and Earth. In a total eclipse, the Moon appears from Earth to cover the Sun completely. An annular eclipse occurs when the Moon is somewhat farther from Earth, so that even if the shadow is observed in a

Date	Type	Approximate Path	Longest Duration
Jan 4 1992	A	Pacific S of Hawaii	11 m 41 s
Jun 30 1992	T	S Atlantic and below S Africa	5 m 21 s
May 10 1994	A	S Pacific, Mexico, US, Canada, Atlantic to N Africa	6 m 13 s
Nov 3 1994	T	Pacific and across middle of S America	4 m 23 s
Apr 29 1995	A	S Pacific through northern S America	6 m 37 s
Oct 24 1995	T	Iraq, Iran, Pakistan, northern India, SE Asia, Indonesia	2 m 9 s
Mar 9 1997	T	Siberia	2 m 50 s
Feb 26 1998	T	Mid-Pacific through extreme N of S America and Caribbean	4 m 9 s
Aug 11 1999	T	N Atlantic	2 m 23 s
Jun 21 2001	T	S Atlantic, Africa, Madagascar, Indian Ocean	4 m 56 s
Dec 14 2001	A	Across Pacific to Central America	3 m 53 s
Jun 10 2002	A	N Pacific to Mexico	23 s
Dec 4 2002	T	S Atlantic through Africa and Indian Ocean to Australia	2 m 4 s
Nov 23 2003	T	Antarctica	1 m 57 s
Apr 8 2005	A/T	S Pacific to northern S America	42 s
Mar 29 2006	T	Atlantic through N Africa across Mediterranean into USSR	4 m 7 s
Aug 1 2008	T	N of Hudson Bay region	2 m 27 s
Jul 22 2009	T	India into S China into S Pacific	6 m 39 s
Jul 11 2010	T	S Pacific to Chile	5 m 20 s
May 20 2012	A	N Pacific to US Northwest and Midwest	5 m 46 s
Nov 13 2012	T	Australia through S Pacific	4 m 2 s
Nov 3 2013	A/T	Atlantic through central Africa	1 m 39 s
Mar 20 2015	T		2 m 47 s
Mar 9 2016	T	Indonesia	4 m 9 s
Aug 21 2017	T	N Pacific across US into Atlantic	2 m 40 s
Jul 2 2019	T	S Pacific into southern S America	4 m 32 s

direct line with the Moon and Sun, the Moon's disk does not cover the entire Sun, leaving a ring of sunlight around the Moon. A partial eclipse occurs when only part of the Moon's shadow touches Earth, while the rest passes beyond one of the poles. The Sun's disk appears to have a bite taken out of it to observers on Earth. An uncommon type of eclipse is the annular/total, which is annular at each end of the path across Earth, but total in the middle of the path. Partial eclipses are omitted in the table, along with many whose entire path of totality is in the eastern hemisphere. Annular eclipses are labeled A and total eclipses T in the table.

(Periodical References and Additional Reading: *New York Times* 6-13-91, p A23; *New York Times* 6-16-91, p IV5; *Science News* 6-22-91, p 368; *Astronomy* 7-91, p 62; *New York Times* 7-12-91, p A1; *Science* 7-12-91, p 152; *Science* 7-26-91, p 386; *Science News* 7-27-91, p 52; *Planetary Report* 9-10-91, p 16; *Astronomy* 10-91, p 64; *Scientific American* 10-91, p 24; *New York Times* 10-29-91, p A19; *Astronomy* 11-91, pp 14, 17, & 32)

MERCURY'S HOT SPOTS AND COLD SPOTS

Mercury is the second smallest planet. Because of its closeness to the Sun, unusual orientation, and possible composition, it may be the oddest planet, a place of weird extremes. Yet, although fairly easy to reach by space probe—it is "downhill" from Earth—it remains almost as unknown as far-off Pluto. The only space probe to the planet was in 1974, a mission that sent back fuzzy pictures of about half the planet's surface as well as other data. The early 1990s, however, revealed a few new oddities that could be observed from Earth.

The Hot Spots

On Jun 12 1990, Jack O. Burns, Michael J. Ledlow, and coworkers reported that Mercury has two "hot poles"; these hot spots on the planet's equator receive more sunlight than other parts of the surface. The radio images that were used to discover the hot spots do not measure temperature at the exact top of Mercury's surface, where immediate cooling or solar heating affects temperature, but about 1 m (3 ft) below the surface, a region where the insulating property of the loose outer covering of Mercury maintains a memory of past heating and cooling.

All planets have orbits that are ellipses, but the degree of flattening, called the eccentricity, varies. Mercury's orbit is second only to that of Pluto in its deviation from a purely circular path. Combined with its proximity to the Sun, this eccentricity means that there is twice as much solar heating when Mercury is closest to the Sun (perihelion) as when it is far-

---

THE ADVANCE OF THE PERIHELION

Followers of Einstein's theories may recall that the explanation of a change in the perihelion of Mercury was one of the triumphs of Einstein's 1915 theory of general relativity. Einstein's theory explained only 43 seconds of arc of advance each century, however. The perihelion advances, combining Newtonian and Einsteinian reasons, by a total of 574 seconds of arc each century, about 16 percent of a degree. During a century, Mercury makes about 415 revolutions, so each revolution has a perihelion advance of less than 1.5 seconds. This is not enough of a change to affect the location of the hot spots, each of which covers about 7.5 percent of Mercury's surface. Over several thousand years, however, the hot spots migrate around the equator.

---

thest away (aphelion). Combined with the 3 to 2 ratio of Mercury's year to its day, this orbit causes the hot spots. If at one perihelion, hot spot A is facing the Sun, then at the next perihelion, one revolution and one and a half rotations later, the location directly opposite A is facing the Sun; two revolutions match three rotations, which brings A back into the heavy heating. As a result, A and its opposite get hotter than the surrounding regions. One estimate is that the hottest temperatures reach more than 425°C (800° F).

An Unexpectedly Cold Heart

The same measurements of heat flow that detected the hot spots produced a mystery. Since 1974, when space probe Mariner 10 discovered an unexpectedly large magnetic field around Mercury, it has been assumed that Mercury must have a core of liquid iron, since current theories attribute Earth's magnetism to movements of electrically active liquid iron in Earth's core. Because Mercury is a very small planet, it would have to consist mainly of such a liquid core to produce the magnetic field. But a large pool of molten iron inside a small planet should produce a lot of internal heat, some of which should be observed escaping from the surface. But Burns and Ledlow found no heat coming from the interior, suggesting that there is no pool of molten iron. If this is so, then it is not clear how to account for the magnetic field.

And the Cold Spots

In Aug 1991 Duane Muhleman and Bryan Butler of Caltech and Raymond Jurgens and Martin Slade of the Jet Propulsion Laboratory used the Goldstone radio facility in the Mojave Desert of California to bounce radar signals off Mercury. The echoes were received by the Very Long Array of 26 radio telescopes near Socorro NM. The main target was the region near Mercury's north pole. Because the inclination of Mercury's axis to the Sun is only 2 percent, some valleys near the poles may receive no sunlight all year, in which case temperatures may fall as low as -113° C (-235° F). The radar probe revealed that the region around the pole is much brighter than other parts of the planet, with characteristics that suggest fields of water ice. Earlier studies had also shown hints of ice at the south pole.

One interesting implication of the possibility of ice at the poles of Mercury is that Earth's Moon might also have polar ice caps. If so, this could be important to any future attempt to colonize the Moon.

(Periodical References and Additional Reading: *Astronomy* 10-90, p 24; *Discover* 1-91, p 42; *Planetary Report* 9-10-91, p 8; *New York Times* 11-7-91, p A22)

VENUS LAID BARE

Venus is often referred to as Earth's twin planet, since it is similar in size and is made largely from rock (like Earth and Mars, and unlike most other planets); it has a substantial atmosphere; and its orbit is right next to Earth's. But there are more differences than anyone can account for, and the differences are more interesting than the similarities. One of the differences is that the cloud cover on Venus is almost complete, instead of being broken like that of Earth, which makes observation of Venus difficult. A number of major studies have tried to understand these differences in the hope of better understanding both Earth and Venus. The Magellan mission to map Venus was the major effort of 1990 and 1991.

The Magellan space probe was launched in May 1989, went into orbit about Venus in Aug 1990, and began its first mapping cycle on Sep 15 of that year. Its first mapping cycle of one Venusian year of 243 days was completed late in May 1991. During that first cycle it used its special radar system to penetrate the cloud cover and map 84 percent of the planet's surface in detail, showing features as small as a football field. Then Magellan was shifted slightly to a new orbit and began mapping for the second Venusian year, bringing the amount of the surface mapped to 95 percent. This second tour was scheduled to last until Jan 15 1992.

Magellan's goal is a complete map of Venus, and it has been remarkably successful. According to Joe Boyce, NASA program scientist for Magellan, Magellan has produced a better map of Venus than any we have of Earth. Scientists still do not have as good maps of the ocean basins on Earth—more than half the planet—as they now have of four-fifths of the surface of Venus.

Perhaps the principal object of mapping Venus is to find out what kind of tectonics, if any, it has. Are there moving plates, as on Earth? What about volcanoes? Earthquakes? What causes mountains or mountain chains? Is Venus separated into different kinds of crust, as Earth's surface is divided between continental crust and oceanic crust? What can Venus's crust tell about the planet's extraordinary atmosphere?

Volcanic activity appears to be dominant on Venus. All its rocks seem to be volcanic in origin. About 90 percent of the surface resembles the recent lava fields of Hawaii or Iceland rather than the typical topography of the rest of today's Earth. On the whole, Venus is thought to be what Earth was like during the Archaean era, the period from about 4.5 to 2.5 billion years ago when Earth was dominated by volcanic processes.

Although Magellan failed to glimpse a volcano in actual eruption, it saw much evidence of the effects of volcanism. Large lava channels, longer and more regular than flows on Earth, were observed. Domes, some as much as 1.6 km (1 mi) high and 60 km (95 mi) in diameter, also larger than similar features on Earth, are also thought to be volcanic features. When the domes and lava channels combine with circular and oval features around the domes, the result is a characteristic Venusian feature scientists have named arachnoids because of their spiderlike appearance. There is some evidence that suggests that Maat Mons, the second-highest mountain on Venus, may be an active volcano.

It is not clear that anything like Earth's plates can be found on Venus, but there is a lot of evidence of crustal stretching and compression. Although the crust of Venus appears to be more like oceanic basalt in composition, it behaves as if it were much weaker, probably because the high temperature of Venus keeps the rock about halfway to its melting point. Some mountain ranges run in straight lines, a feature different from those of mountain ranges on Earth. Elevated regions of crisscrossing fractures and ridges, called tesserae, from the Greek for "tiles," seem to have been formed by stretching and compression, perhaps from several directions at once. There are various steep slopes found in several places on the planet that are difficult to explain. Also, early on its second pass at mapping the planet, Magellan found evidence of a giant landslide. On Aug 29 1991 Jeffrey Plaut of the Jet Propulsion Laboratory in Pasadena CA observed the change in images taken 8 months apart. If confirmed by the third pass, the landslide would be in the same class with the largest known on Earth (in the Hawaiian Islands). By Oct, however, Plaut and most other scientists had come to believe that the apparent evidence was an artifact of the radar imaging process.

There is little erosion on Venus, which lacks both water and strong winds, although winds of a couple of km (1 mi) per hour have effects like winds of 25 km (15 mi) per hour because of the dense atmosphere. The small amount of erosion observed in Magellan images suggests to some scientists that the surface is on average less than 400 million years old, with no part of the crust older than about a billion years. Probable explanations focus on the volcanic eruptions that have resurfaced the entire planet.

In addition to other surface features, Magellan is finding impact craters, but all are at least 6 ½ km (4 mi) in diameter. No smaller ones exist because any objects that could cause them burn up in the thick atmosphere before reaching the surface of the planet. Some objects, however, have produced shock waves that cause smudges on the planet's surface. The total number of craters located in the first mapping cycle was about a thousand, more than are known on Earth, where erosion removes signs of craters relatively quickly, but far fewer than are seen in terms of total surface area on Mercury, the Moon, or Mars. All large craters have their interiors partially flooded with lava.

Plate tectonics as it is known on Earth appears to be absent. There seems to be nothing like Earth's worldwide network of rift valleys and ridges or the deep trenches into which plates plunge. One theory in this regard is that the lack of surface water on Venus makes the difference, since there is considerable evidence that Earth's surface water percolates deep into the crust and mantle, causing some of the observed effects of plate tectonics (see "Understanding Earth's Mantle and Core," p 256).

Although the most dramatic new information about Venus comes from the Magellan probe, other studies of the planet also enrich our understanding of our sister world.

Before the first space probes of the United States and the Soviet Union to Venus, which started as early as 1962, people generally believed Venus to be covered with water. This was inferred from the clouds that cover the planet, erroneously thought to be water clouds. Space probes and better Earth-based observations soon revealed that there is little water in the clouds and that the temperature of the planet's surface, around 457° C (854° F), is far too high for liquid water. However, space probes have revealed that the ratio of heavy hydrogen, or deuterium, to ordinary hydrogen in the Venusian atmosphere is about 120 times as great as that ratio in Earth's atmosphere. The accepted explanation is that there was once more hydrogen on Venus and at that time probably more of it was combined with oxygen to form water; but with the passage of time, the light, ordinary hydrogen boiled out of the atmosphere, leaving the heavy stuff behind. Although data from the International Ultraviolet Explorer satellite indicate that the deuterium was not there after all, a study announced in Feb 1991 by a multinational team headed by French astronomer Catherine de Bergh using the Canada-France-Hawaii Telescope confirmed the deuterium ratio. Most astronomers think that Venus once had oceans, but as its temperature increased, these boiled away. The unresolved questions now relate to the source of the water. A minority view is that the water was brought to Venus over time by comet impacts and that there never was enough of it at any one time to form oceans (see "Where Does the Water Come From?," p 242).

Another study of Venus from near space was conducted by the Galileo space probe on its twisted path to Jupiter. Along the way, it passed by Venus on Feb 10 1990. As Galileo came closer, observatories all over Earth trained their instruments on Venus so that Earth-based observations could be combined with the Galileo data. One study, using the Infrared Telescope Facility on Mauna Kea in Hawaii, was made of the night side of Venus by a team primarily from the University of Hawaii. The team observed phenomena that they attributed to regions of high water vapor density in the atmosphere and corresponding regions of low water vapor density that resulted in "dry spots" and "cold spots" that could be caused by different amounts of water vapor at different levels in the atmosphere.

Galileo also was able to map the temperature of the lower regions of the atmosphere. Comparison of Galileo's atmospheric hot spots with Magellan's geologic features may produce the first clues to active volcanoes on Venus. Another coup was the detection of the strongest evidence yet for lightning on Venus.

(Periodical References and Additional Reading: *Science News* 2-17-90, pp 103 & 111; *Astronomy* 3-90, p 18; *Science* 3-9-90, pp 1163 & 1191; *Science News* 6-23-90, p 392; *Astronomy* 7-90, p 42; *Science News* 8-18-90, p 100; *Science* 8-31-90, p 977; *Discover* 9-90, p 18; *Science News* 9-22-90, p 181; *Science News* 9-29-90, p 199; *Astronomy* 12-90, p 48; *Astronomy* 1-91, pp 24 & 34; *Science* 1-11-91, p 180; *New York Times* 1-26-91; *Astronomy* 2-91, p 44; *Science* 2-1-91, pp 495 & 547; *Science* 3-1-91, p 1026; *Science* 4-12-91, pp 187, 213, & 247-312; *Astronomy* 5-91, p 26; *Science* 5-3-91, p 651; *New York Times* 5-30-91, p B11; *Science* 5-31-91, p 1293; *Planetary Report* 5-6- 91, p 8; *Science* 6-7-91, p 1372; *Science News* 6-8-91, p 359; *EOS* 6-18-91, p 265; *Science News* 7-20-91, p 39; *EOS* 7-22-91, p 313; *Astronomy* 9-91, p 32; *Science* 9-13-91, p 1208; *Science* 9-27, pp 1463, 1492, & 1516; *EOS* 10-1-91, p 426; *Science News* 10-12-91, p 239; *Science News* 10-26-91, p 269; *New York Times* 10-30-91, p A21)

WEIRD VOLCANOES OF THE SOLAR SYSTEM

Although there is no hard evidence yet, many scientists think that Venus shares volcanic activity with Earth. No other planets seem to have or seem likely to have active volcanoes. Except for Mercury, Mars, and Pluto, other planets are balls of gas, not a suitable medium for volcanic activity. Mercury is not even releasing heat from its interior (see above), so volcanoes are certainly not in the picture there. Mars presents the most likely opportunity for flowing lava, but good observations suggest that any volcanism it has had is in the past. Pluto is too far away for Earth-based science, but current thinking is that it is a frozen ball of ice. If warmed a bit, Pluto might produce some sort of eruption, but in its distant neighborhood, it's just too cold.

There is something like volcanism elsewhere, however. It is certain that Jupiter's moon Io has eruptions that can only be called volcanic. Also, something that certainly looked like a kind of volcano on Neptune's moon Triton was spotted by Voyager 2 as it flew by.

Evidence for Io's volcanoes was first gathered in 1973, but no one knew what to make of it. Io's volcanoes produce sodium atoms, which can be observed around the little moon. At the time the sodium was first detected, the most popular theory to explain it was that radiation from Jupiter somehow knocked sodium atoms off the surface of Io, which is the closest of Jupiter's sixteen moons to the giant planet. In Mar 1979 Stanton Peale of the University of California, along with Patrick Cassen and Ray Reynolds of NASA's Ames Research Center, proposed the surprising idea that Io might have a hot, molten core and volcanoes. This was a bold concept, as the Voyager space probes were due to reach the vicinity of Io at about the same time as their proposal was put forth. Their theory was that tidal forces caused by Jupiter and the larger Jovian moons heat Io's interior beyond the melting point, and that molten material then breaks through the crust to form volcanoes, just as it does on Earth.

Both Voyagers spotted the volcanoes of Io as they flew by on Mar 8 and Jul 9 1979. Voyager 1 picked up eight volcanoes visible enough to be given such names as Loki and Pele. Most were still erupting when Voyager 2 passed, although Pele, which had produced a plume more than 320 km (200 mi) high, had become dormant.

Until 1990 the Voyager images were the only ones showing Io's volcanoes, but in 1990 University of Hawaii astronomers John Spencer, Mark Shure, and Michael Ressler, blessed with good viewing conditions from Mauna Kea, were able to capture three eruptions from Earth. One of these was a continuation of one of the eruptions previously observed by the Voyagers. They were able to capture the eruptions using NASA's 3-m (120- in.) infrared telescope and a new, more sensitive camera.

Although this was the only observation of the actual eruptions since 1979, a closely related feature was also imaged from Earth during 1990. Michael Mendillo, Jeffrey L. Baumgardner, and Brian C. Flynn of Boston University photographed on Jan 25 the giant sodium cloud produced by Io's volcanoes. And a giant it is, extending more than 52 million km (32 million mi) out from Jupiter. If we could observe it directly, it would cover a region the size of a dozen Moons placed side by side. Mendillo calls it "possibly the largest permanently observable feature of the solar system." The Boston University team was able to photograph it from the McDonald

Observatory in Fort Davis TX by using a 100-mm (4-in.) telescope equipped with a special filter and a light-intensifying detector.

What Voyager 2 saw on Neptune's moon Triton in Aug 1989 appeared to be the kind of plumes volcanoes emit, towering almost 8 km (5 mi) above the frozen surface of the large satellite. These plumes consist mainly of nitrogen gas. Initially, the plumes were described as geysers, and Triton was allowed by science to join the ranks of the volcanically active. The idea was that warmer subsurface nitrogen gas, heated perhaps by sunlight, breaks through the frozen nitrogen surface to form the "geysers." The dark color of the plumes could be caused by dust or ice carried upward by the transparent gas.

Only 4° C (7° F) warming is needed to turn the frozen nitrogen into a gas able to propel such a scenario. There must be regular weak spots in the crust, since the geysers reappear in the same place. There also must be some way to concentrate enough gas in one place to produce the plumes. In 1990 Randolph L. Kirk of the US Geological Service in Flagstaff AZ suggested that underground cracks could collect warm gas over a large region. When enough gas was in one place beneath some thin crust, it could break through. This would also explain why the plumes reappear in the same place.

In more recent times, at least two of the scientists involved in studying the Tritonian plumes have changed their minds about the correct interpretation of the apparent geysers. Andrew P. Ingersoll and Kimberly A. Tryka of Caltech have proclaimed that the plumes are giant dust devils, not geysers at all. They have been led to the conclusion by the undoubted phenomenon that the upper atmosphere of Triton is warmer than the lower atmosphere. Ingersoll thinks that whirlwinds would be just the medium to move heat from a surface warmed by the Sun to the upper atmosphere. Along the way, the whirlwinds would pick up water ice that had turned to ice dust on the surface. One of the main problems with this theory is that the atmosphere is rather thin for picking up any kind of dust. Most astronomers still think the plumes are geysers.

(Periodical References and Additional Reading: *Astronomy* 1-90, p 10; *Science News* 3-24-90, p 191; *Science News* 6-9-90, p 359; *Science News* 11-10-90, p 302; *Discover* 1-91, p 38; *Discover* 3-91, p 12; *Science* 9-20-91, pp 1331 & 1394; *Science* 10-4-91, p 89)

NEWS FROM THE OUTER PLANETS

With Galileo on the way to Jupiter, we can expect new information to begin to arrive in abundance in 1995, provided its antenna problems are solved (see "Problems and Some Resolutions," p 113). Meanwhile, an earlier traveler, Voyager 2, is still a source of previously unknown material. In addition, improving ground-based operations are also bringing us news from the planets beyond Jupiter.

New Moon at Saturn

Scientists trying to explain the presence of rings about the gas giant planets, notably the showy rings of Saturn, have invoked the concept of "shepherd moons." The shepherd moons' gravitational force would keep the dust particles that form the rings in alignment. The theory seems to have merit, but much of the evidence has been lacking. Specifically, very few of the proposed shepherd moons have been found. In 1990, however, a careful search through old data produced another one, the eighteenth observed moon about Saturn.

In 1985 Jeffrey N. Cuzzi and Jeffrey D. Scargle of NASA's Ames Research Center in Mountain View CA noticed evidence that there might be an unknown Saturnian moon hidden among the rings. Mark Showalter of Ames, Cuzzi, and coworkers began a search for it in 1986, but found nothing. In 1990 Showalter used a new computer program and Voyager 2 images to extract likely regions in which to look for the moon. The program searched through over 30,000 images and picked out twenty that had the right characteristics, including good enough resolution for its human partners to be able to find a moon if one were there. The humans were able to find eleven of the images that had a moon in them—always the same moon, but not one previously known.

The new moon is about 20 km (12 mi) across and is probably made of ice. It was temporarily named 1981s13, meaning that it was the thirteenth satellite found during 1981, even though its "finding" was not noticed for nine years. It soon was renamed Pan by the International Astronomical Union, a name proposed by Showalter based on the Greek deity Pan having been the god of shepherds.

Pan is found in a 320-km (200-mi) gap in the rings known as the Encke Gap, a gap the moon probably creates in the process of keeping nearby rings stable. As such, it is the only moon known to be within the ring system. Showalter has used the images to locate Pan's position with great accuracy, to about 1 km (½ mi). It is 133,583 km (82,821 mi) from Saturn's center. Showalter also was able to observe a tiny ring of dust that follows the new moon's orbit in the middle of the Encke Gap.

It is not expected that Pan will be observed again in the near future, although there is a slight chance that the Hubble Space Telescope will be able to pick it up during a favorable viewing opportunity in 1995. If that fails, the next most likely chance is during the Cassini space probe visit scheduled to begin in 2002. Cassini may also be the first to find Saturn's nineteenth satellite, believed to be causing ripples in the F ring.

The Aug 2 1991 session of the International Astronomical Union's executive committee that named Pan also named six small moons of Neptune that were discovered by Voyager 2 in 1989. Although the planet uses the Roman name of Neptune for the god of the seas, the satellite names were based on the god's Greek identity of Poseidon and various names of the god's Greek associates and family members. The new names are Naiad and Galatea (after Greek water nymphs), Thalassa and Larissa (among Poseidon's lovers), Proteus (a son), and Despina (a daughter).

Climate and Weather on Saturn

In some ways, climate and weather on Saturn are like climate and weather on Earth—there are seasons, winds, and storms, for instance. There are seasons because Saturn's axis tilts 26° 44′, just as seasons on Earth are caused by a 23° 27′ tilt. When it is late summer for Saturn's northern hemisphere, there are apt to be big storms. But because Saturn's year is 29.5 Earth years long, a Saturn summer from our point of view comes around infrequently and lasts more than 7 years at a time. Big, late summer storms in the northern hemisphere were observed on Saturn during the past five Saturnian summers, in 1876, 1903, 1933, 1960, and 1990.

In other ways the climate and weather on Saturn are very different from that on Earth. Despite there being a Saturnian summer, temperatures during it are not noticeably different from temperatures during the rest of Saturn's year. Saturn is far from the Sun, and most of the atmospheric heat comes from inside the planet. Winds whip around Saturn from east to west at speeds averaging about 1600 km (1000 mi) an hour, faster than the planet rotates. No one knows what causes the winds to blow in this direction or to be so very fast. Like Jupiter and Neptune, which also have high-speed winds, Saturn has an internal heat source that no doubt helps power the winds.

Also, when storms form on a giant gas planet they are much larger and more persistent than storms on Earth. Our largest late-summer storms, the hurricanes, may cover regions a few hundred kilometers in diameter and last for three weeks at most. The Great Red Spot on Jupiter is 32,000 km (20,000 mi) long and 13,000 km (8000 mi) wide and has persisted for hundreds of years.

On Sep 24 1990 two amateur astronomers, Alberto Montalvo in Los Angeles and Stuart Wilber at Las Cruces NM, observed the start of Saturn's summer storm of 1990, quickly dubbed (in imitation of Jupiter's storm) the Great White Spot. Wilber was the first to report his observation to the

Smithsonian Astrophysical Observatory in Cambridge MA, so the storm is sometimes called Wilber's Spot. It was the biggest such storm since 1933, about 21,000 km (13,000 mi) long and covering about a sixth of the visible surface of Saturn. For unknown reasons, much larger storms occur every other summer, or about 60 years apart.

Like most very large storms on Saturn, the Great White Spot appeared just north of the equator about 3 years into summer. Within weeks Saturn's swift winds had spread it from a spot to a band, stretching all the way around the planet by Oct 26, about a month after it was first observed. Shortly after that, Saturn moved behind the Sun, so observations from Earth ceased. The Hubble Space Telescope obtained some detailed photographs of the Spot before Saturn was lost from view.

Like hurricanes on Earth, Saturnian storms are powered by heat from the Sun, little though that heat might be. The whiteness of the Spot is probably a result of crystals of new ammonia formed as warm gas in the storm cools (just as water vapor in a hurricane cools to liquid water). Older ammonia crystals are yellow, the predominant color of Saturn. Ground-based observations by Agustin Sanches-Lavega and coworkers from the University of Pais Vasco in Bilbao, Spain, support this explanation as do observations from the Hubble Space Telescope.

According to theory and computer simulations, even larger spots should form 30° south of Saturn's equator, but the view from Earth of any such spots is blocked by the rings. The only way to check this prediction would entail using a space probe to observe from a different angle.

Interpreting Voyager Data from Uranus and Neptune

On Aug 24 1990, Bill R. Sandel and Floyd L. Herbert of the University of Arizona reported that a long search through Voyager 2 data had revealed the locations of the north and south auroras of Uranus. These are mostly centered around the Uranian magnetic poles, which are strangely tipped about 60° away from the 180° tipped axis of rotation. They also found what appeared to be a small aurora some distance away from the north magnetic pole—not where standard theory would allow one to be. More study may resolve this mystery and also may clarify why the northern aurora has three lobes instead of the expected two and is nearly circular instead of elliptical.

Scientists are also still trying to use the Voyager 2 data from 1989 to understand the very different weather on Neptune from that of its twin in size, Uranus. Where Uranus is calm, Neptune has the swiftest winds of any planet and has weather that in general is more like that of Jupiter or Saturn. Like them it has a "spot," called the Great Dark Spot. Unlike the transitory Great White Spot of Saturn, Neptune's Great Dark Spot looks to be at least semipermanent, much like the Great Red Spot of Jupiter. Yet it is very different from the Great Red Spot in many ways. Neptune also has a second semipermanent spot nearby, known as Dark Spot 2, or DS2 for short.

The most likely cause of Neptune's violent weather is that more heat is rising from its interior than it receives from the Sun. As a result, the gas giant is simmering like a stockpot, with hot gas continually rising toward the upper atmosphere, gas that may start fairly deep in Neptune's interior.

(Periodical References and Additional Reading: *Science News* 7-14-90, p 30; *Science News* 8-4-90, p 69; *Science News* 10-13-90, p 228; *Science News* 10-20-90, p 248; *Science News* 11-17-90, p 325; *Astronomy* 12-90, p 24; *Astronomy* 2-91, pp 22 & 34; *Astronomy* 3-91, p 36; *Science* 6-14-91, p 1489; *Science News* 7-27-91, p 63; *Astronomy* 8-91, p 38; *New York Times* 10-1-91, p C2; *Science News* 10-5-91, p 212)

COMET, METEOROID, ASTEROID UPDATE

Throughout the solar system there are small bodies that orbit the Sun on their own. If they are small enough, they are called meteoroids.

Bigger bodies orbiting the Sun by themselves are called asteroids or comets (or planets, if they are sufficiently big). About 5000 asteroids have had their precise orbits determined. This allows them to be designated with a number signifying their order of entry into this group, such as 1 Ceres (discovered first) or 3200 Phaëthon, although the number is usually dropped after the first reference. Some use just the number in parentheses to refer to the asteroid, using (1) for Ceres or (4789) for Sprattia, one of the most recent asteroids to have been assigned a number.

Another 13,000 asteroids have been found, but their orbits are yet to be calculated. They get temporary names consisting of the year found followed by a letter code that indicates the date they were first observed, such as 1991 BA or 1989 PB. Theory suggests that there are about a million more asteroids with a diameter greater than 1 km (0.62 mi) that have yet to be discovered.

Asteroids are mostly rocky or metallic. Comets are dusty frozen gases. Seen through a telescope, asteroids are pointlike, like stars (hence their name), while comets are fuzzy. If an asteroid or a comet explosively strikes Earth, it is sometimes called a bolide. Some astronomers do not make distinctions between medium-sized bodies in space. They call both comets and asteroids either planetoids or planetesimals.

A meteoroid that burns up in Earth's atmosphere is a meteor. Until recently, it was assumed that all meteors are made of rock or iron, like the meteoroids that do not burn up and reach the surface to become meteorites; there is good reason, however, to suspect that some, if not most, meteors are really little comets.

The Curious Case of Chiron

Sometimes it is not easy to tell whether a body in space is an asteroid or a comet. It was once thought that the asteroids all are in orbit between Mars and Jupiter, so any largish object with an orbit beyond Jupiter might be expected to be a comet. Now we know that asteroids can be found in various parts of the solar system, although there are more between Mars and Jupiter; so if an object is found far beyond Jupiter, further evidence is needed to identify it.

In 1977 Charles T. Kowal of Lowell Observatory discovered 2060 Chiron, a large (but not planet-sized) body orbiting the Sun between Saturn and Neptune. This was the first such body ever observed beyond the orbit of Saturn. From the beginning no one was quite sure what it was—it was an odd place for an asteroid to be, but not an impossible place; and it was too big to be a comet. If it was an asteroid, it was the most distant from the Sun then known; its orbit of 50.7 years would take it outside the orbit of Uranus at its farthest point. When discovered, however, it was heading for its perihelion, the closest point to the Sun, 1257 million km (2053 million mi), just inside the orbit of Saturn. It will make perihelion in 1996.

On Feb 20 1988 Chiron was found to be getting brighter. Comets brighten as they approach the Sun, before they look fuzzy and develop a tail. Checking earlier records showed that Chiron had started getting brighter—much brighter than expected from just being closer to the Sun—as early as mid-1987. Astronomers began keeping a close watch to see what was going to happen next.

Chiron got dimmer. At first this seemed surprising, but it is actually fairly common behavior for comets. From time to time, often far away from the Sun, heated vapor and dust escape from a comet, causing it to brighten for a few days because the escaped material reflects additional sunlight. Then, as the material disburses or—more likely—falls back on the comet, everything returns to normal. (See below for an account of Halley's comet's similar behavior.)

Some astronomers, such as William K. Hartmann of the Planetary Science Institute in Tucson AZ, a member of the team that first observed the sudden brightening of Chiron on Feb 20, think that science has until now misunderstood the difference between comets and asteroids. Hartmann argues that there are two classes of planetesimals, but they should be judged on their surface reflectivity, not on their behavior when near the Sun. The bright ones, most of which orbit the Sun at less than 400 million km (230 mi), are the rocky, metal-rich bodies that conform to the classic picture of an asteroid. Farther from the Sun, most bodies are dark and, Hartmann avers, consist more of dust and ice that can vaporize in bursts or produce gases steadily when close enough to the Sun. Chiron is a large example of one of the latter class. Other familiar examples include Halley's comet, Mars's dark moon Phobos, and Ceres—generally thought of as the largest known asteroid and the first discovered (in 1801). Ceres makes its way onto the list because it is black and because Michael A'Hearn of the University of Maryland discovered in 1990 a thin cloud of hydroxl ions surrounding the planetesimal, suggesting the fuzziness we associate with comets.

Chiron proceeded after its dim spell to continue on its way to turning into a recognizable comet. In Apr 1989 Michael Belton and Karen Meech of the University of Hawaii observed a faint cloud of dust surrounding Chiron. By Feb 1991 Schelte J. Bus, David Schleicher, and Edward Bowell of Lowell University, and Michael A'Hearn, could report that they had found a coma, or cloud, of cyanogen radicals (CN) around Chiron, a common feature of comets, although usually found closer to the Sun. The development of Chiron into a comet so far from the Sun suggests that it is a recent visitor to the inner solar system, probably making its first pass at the Sun.

Also in Feb, Mark Sykes of the University of Arizona and Russell Walker of Jamieson Science and Engineering reported on their calculations of Chiron's size, which they made to be 372 km (230 mi) in diameter—vastly bigger than any other comet known. Halley's comet, for example, is only about 16 km (10 mi) long.

An Early Return Act for Comet Halley

Halley's comet, or Comet Halley as it is more properly known, is by far the best known and in many ways the most familiar comet in the solar system. Not only does it make a fairly bright return to the vicinity of the Sun every 76 years or so, it is also the only comet so far to have its neighborhood visited by spacecraft—in 1986 by a half dozen at a time. Thus, while scientists certainly expected to learn more about Comet Halley, although probably not until its next visit in 2061, they did not expect any surprises.

On Feb 12 1991, however, Belgian astronomers Olivier Hainaut and Alain Smette, working at the European Southern Observatory in La Silla, Chile, decided they would take a glance at Comet Halley. To their astonishment, the comet was about 300 times brighter than it had been a year previously. A few days later other astronomers found it about a thousand times brighter than expected. At a distance of more than 2.1 km (1.3 billion mi) from the Sun, something big had erupted. This set a new record by hundreds of millions of kilometers for comet eruptions away from the Sun.

At that distance, the ice that causes a comet to erupt by vaporizing ought to be at a temperature of about -200° C (-330° F), an unlikely temperature to vaporize anything that was not already lifted into a gaseous state when temperatures were higher as the comet was nearer the Sun. One possibility, however, is that frozen carbon monoxide, which could turn to gas at -200°C, somehow broke through the skin of the comet. Two other ideas are that Comet Halley ran into something, or that a particularly energetic breeze from the solar wind set off the outburst. There was an increase in solar activity just before the brightening.

Asteroids in the News

A lot of people were startled when Clark R. Chapman of the Planetary Science Institute and David Morrison of NASA's Ames Research Center calculated that a US citizen is six times as likely to be killed by an asteroid impact as by traveling on a US airline. Their argument was based on statistics, not on calculations of the paths of asteroids. But some asteroids come uncomfortably close to Earth; they cross its orbit. These are called the Apollo asteroids (after the asteroid Apollo, the first of its class to be discovered—by Karl Reinmuth in 1932). So far the Palomar Planet-Crossing Asteroid Survey (PPCAS), a project headed by Eleanor "Glo" Helin, has found more than forty Apollo asteroids, which together with those found by other groups make a total of more than sixty so far. But there are others. These are part of a group of about 140 known asteroids collectively called "Earth-approaching asteroids." These include the Atens, inside Earth's orbit, and the Amors, just outside, as well as the Apollo orbit-crossers.

How close do they come? Since the PPCAS project started, one of the closer approaches came on Jul 10 1990. An asteroid named 1990 MF whizzed by only 4.83 million km (3 million mi) from Earth at a speed of 35,000 km (22,000 mi) per hour. Since it is only 100 m (300 ft) by 300 m (1000 ft) or so, 1990 MF is not a very big asteroid. It could do a lot of damage if it hit Earth, however. In Mar of the previous year, 1989 FC came within 800,000 km (495,000 mi) of Earth. But the closest recognized approach came in Jan 1991, when 1991 BA passed only 170,000 km (105,000 mi) from Earth, about half the distance between Earth and the Moon. But 1991 BA is even smaller than 1990 MF, only about 10 m (33 ft) in diameter. Many would call the meteor estimated at about 80 m (260 ft) in diameter that sliced through the top of the atmosphere in 1972 a small asteroid. Some think that the mysterious 1908 event on Jul 30 near the Tunguska River in Siberia was a collision between Earth and an asteroid that caused widespread devastation. And well-identified craters and other evidence indicate many collisions in the past.

Based on craters on the Moon and Earth, one estimate is that an asteroid strikes our planet or the Moon once on the average of every 65,000 to 70,000 years. Other estimates range from one every 250,000 years to one every million years for a sizable asteroid more than a kilometer (0.62 mi) in diameter. The largest known Apollo asteroid is Ganymed, with a diameter of 38 km (24 mi). Some of the efforts now underway to detect Earth-crossing asteroids and to determine their future positions are intended to provide advance warning against disastrous collisions with such asteroids.

What is the farthest away an asteroid has ever been found? The object 1991 DA, like Chiron, looks like an asteroid, but Robert McNaught of the University of Adelaide has calculated that its orbit extends beyond the orbit of Uranus, although it then swings toward the Sun and travels inside the orbit of Mars. If it is an asteroid, 1991 DA is 3.3 billion km (2 billion mi) from the Sun at the far point of its orbit, and holds the record. The problem is that, although no fuzziness has been spotted, its orbit looks suspiciously like that of a comet instead of that of an asteroid. Further observation may reveal which 1991 DA really is.

On Aug 25 1989 asteroid 1989 PB (discovered by Eleanor Helin and subsequently catalogued as 4769 Castalia) came 4,050,000 km (2,500,000 mi) from Earth, fortunately right in the proper place to be observed by the Arecibo Observatory's fixed radio telescope. The Goldstone radio telescope, busy communicating with Voyager 2 at Neptune during the asteroid's closest approach, also took a look at Castalia after Voyager headed on out to space. Analysis of the radio images revealed that Castalia is either two close objects or one object with two lobes, either possibility a surprising conclusion. Studies of how such an object or pair of objects could be stable suggest that Castalia was originally two similar-sized asteroids that had a low-impact collision and stuck. It is possible that such pairs are relatively common, as several craters on the Moon are twinned.

Another way an asteroid made the news came when NASA astronomers located a near-Earth asteroid made from metal. The newsworthy aspect was the calculation that, in addition to billions of tons of iron and nickel, the asteroid probably contains 10,000 tons of gold and 100,000 tons of platinum. Its nearest approach to Earth is 32 million km (20 million mi), suggesting a new reason for people to venture into space. Timothy D. Swindle and John S. Lewis of the University of Arizona and Lucy-Ann A. McFadden of the University of California, San Diego, La Jolla, argued in Oct 1991 that returning materials from such asteroids would be much easier than obtaining them from the Moon or other bodies in the solar system. They also compare the costs of obtaining materials for construction in low-earth orbit by lifting the materials from Earth with those of mining the materials from asteroids, with the advantage being in favor of asteroids.

Many readers were startled by the close-up photograph of asteroid 951 Gaspra that appeared in newspapers around the country on Nov 15 1991 (see "Successes from Space," p 118).

Star Specks in Meteorites

For a long time scientists did not believe meteorites existed, although a fall of about 200 in L'Aigle, France, in 1803 convinced most of them. Then scientists came to believe that meteorites are remnants of smashed planets or, later, part of the cloud of dust and gas that formed the planets. In 1983 researchers learned that at least some meteorites come from the Moon or from Mars. Five years later Edward Anders of the University of Chicago proposed that specks of dust embedded in some meteorites come from the stars.

The basic idea is that certain kinds of stars toss out a lot of material when they explode. This condenses into tiny grains that drift through space until pulled in by the gravitational force of some object. Suppose that the object is the cloud of dust and gas that is believed to have formed the Sun and the rest of the solar system. As the cloud condensed into bigger chunks, some of the smaller bigger chunks, the meteoroids, incorporated these star specks without melting them (big chunks seem to have melted at one time or another). When one of the small chunks studded with star specks bumps into Earth, if it is big enough but not too big, only the outside melts. The inside remains pristine, as do the tiny grains of extra-solar material within it.

How can one recognize that a speck of diamond, graphite, or silicon carbide comes from *outer* space? From analysis of the spectra of the Sun and planets, analysis of the bulk composition of meteorites, and theoretical considerations, we think we know what the isotopic and elemental composition of the solar system is. It is not quite the same as that of the crust of Earth. In the case of the suspected star material, the isotopic composition of every element looked odd. But xenon, for which all solar system samples are alike, held the key. For most of these specks, the xenon isotopes looked like those from a type of red giant star known by the awkward name of Asymptotic Giant Branch stars, a group of stars thought to bring up carbon from deep within their interiors and toss the carbon into their stellar winds. Some other specks, analyzed by Ernst Zinner of Washington University in St. Louis MO, were proclaimed in 1991 to be the result of supernova explosions. Also, analysis of some of the grains' radioactive materials gives dates that are before the origin of the solar system. If these dates hold up, the star specks will be the oldest material of any kind found on Earth or in the solar system.

(Periodical References and Additional Reading: *Astronomy* 1-90, p 10; *Science News* 4-21-90, p 248; *Science* 6-22-90, pp 1467 & 1523; *Astronomy* 8- 90, p 44; *Astronomy* 11-90, p 24; *Science News* 11-3-90, p 286; *Science* 2-15-91, pp 719, 774 & 777; *Science News* 3-2-91, p 133; *New York Times* 3-7-91, p A20; *Science* 3-15-91, p 1311; *Astronomy* 5-91, p 22; *Astronomy* 6-91, p 24; *Science* 6-7-91, pp 1351 & 1399; *New York Times* 6-11-91, p C5; *Science News* 6-15-91, p 373; *Planetary Report* 7-8-91, p 23; *Science* 7-26-91, p 381; *Astronomy* 9-91, p 50; *Planetary Report* 9-10-91, p 12; *Scientific American* 10-91, p 88; *Science News* 10-12-91, p 239; *EOS* 10-29-91, p 473; *Discover* 11-91, p 58; *Scientific American* 11-91, p 30; *New York Times* 11-15-91, p D19)

STARS AND OTHER INTERESTING BODIES

LOOKING FOR PLANETS

The history of science is the story of how human beings have been forced to realize that they are not as special as they thought. Nonetheless, we are still a major factor on the only planet known to harbor life in the only set of planets known to orbit a star. Scientists are, however, making major efforts to find other systems of planets around other stars. Early in 1991 the US National Research Council recommended that NASA undertake "a major scientific effort" to find planets around stars other than the Sun.

The Pulsar Planet Appears and Disappears

In 1991 the first apparent good evidence of a planet-sized body orbiting a star other than the sun was announced. Many astronomers found it to be bizarre, however. The problem is not the evidence, which consists of periodic shifts in the path of the star, shifts of the kind caused by the gravitational attraction of an orbiting body. The problem is the type of star—a neutron star, or pulsar. A neutron star is simply not a good candidate to be a sun harboring a planetary system.

A neutron star is the ash of a supernova; that is, it is what is left after a fairly large star violently explodes. A supernova explosion would be expected to blast away any planets that had formed around the original star. Thus, when Andrew G. Lyne, Matthew Bailes, and Setnam Shemar at the Nuffield Radio Astronomy Laboratories at the University of Manchester, UK, reported in the Jul 25 *Nature* that they had for 5 years observed what appeared to be the effects of a planet orbiting pulsar PSR1829-10, other astronomers were astonished and a bit skeptical.

The signals from a pulsar are normally among the steadiest ticks of the cosmic clock, although they slow down slightly over time as the neutron star gradually slows its rotation. When PSR1829-10 was discovered in 1985, some 33,000 light-years from Earth, its pulses were assumed to be as regular as any others. But by 1990, it became apparent that there was a 6-month variation in the signal. The signal, roughly a third of a second long, gradually got 8 milliseconds shorter for about 3 months and then slowly returned to its maximum value. The most probable cause of such a variation, Lyne and coworkers suggested, is a planet about ten to fifteen times the size of Earth orbiting the star in a nearly circular path at about the distance that Venus is from the Sun, about 105 million km (65 million mi).

Background

Neutron stars were first proposed as theoretically possible by Walter Baade [German-US: 1893-1960] and Fritz Zwicky [Swiss: 1898-1974] in 1934, just two years after the discovery of neutrons, in a paper on supernova explosions. It was not until 1968, however, that a neutron star was observed (in the debris of the explosion of the supernova of 1054, now known as the Crab nebula).

The original concept of a neutron star was that the force of a supernova explosion fuses the protons and electrons of atoms together to form neutrons. This idea does not make sense in terms of modern ideas of physics, although it is often still given as an explanation of how a neutron star forms. Instead, it is more nearly correct to say that gravitational forces squeeze the electrons out of the atoms, leaving just the protons and neutrons—collectively, nucleons—behind. The electrons do not really go away, however. Overcome by gravity, they blend into each other. The protons and neutrons cannot collapse further unless gravitational strength is much greater. (A great enough gravitational force blends nucleons together and a black hole results instead of a neutron star.) A better name than neutron star would be "nucleon star" or "nucleus star," since the resulting object is essentially the same as an atomic nucleus, except that it is electronically neutral instead of positive. The object has essentially the same density as an atomic nucleus. Thus, even though a neutron star's diameter is thought to be on the order of 10 km (6 mi), its mass may be nearly as great as three times that of the Sun.

The supernova explosion leaves most of the angular momentum of a large star in an object that is very much smaller in diameter. Therefore, neutron stars rotate rapidly when they are first formed. They continually emit radio waves as they rotate. The radio signals emerge from the magnetic poles of the stars in beams. We can detect neutron stars because such beams appear as rapid pulses of radio waves, thus, the other name for neutron star, <u>pulsar.</u>

When a planet orbits a star, as Earth orbits the Sun, the paths of both planet and star focus on the center of gravity of the system. Thus, the star travels in a small orbit of its own around that point. If the distance that signals travel from the pulsar to Earth varies continuously over a 6-month period, the timing of the signals also will be affected. The effect is subtle, but pulsar signals occur at such regular intervals that even a tiny change can be detected.

Since such a planet could not survive the blast of a supernova, it has been suggested that the pulsar formed after the explosion, coalescing out of the debris scattered about the star. This situation also seems unlikely since debris from supernova explosions have been observed traveling away from the explosions at speeds near that of light. Nevertheless, Douglas N.C. Lin, Peter Bodenheimer, and Stanford E. Woosley of the University of California at Santa Cruz published calculations in the Oct 31 1991 *Nature* showing that an origin from supernova debris is at least possible.

Two additional ideas were first suggested by Andrew Lyne and his team. The pulsar could have formed in the absence of a supernova (but no one knows how that could happen); or, after the pulsar formed, the rapidly rotating neutron star could have produced a surrounding disk of material that could in time form planets, just as planets are thought to have formed in our own solar system. Yet a fourth idea came from Julian H. Krolik of Johns Hopkins University, who proposed that a companion star to the one that went supernova could have had its outer parts stripped in the explosion, leaving a planet-sized core.

Assume then that the suspected planet got into the vicinity of pulsar 1829-10 by one of these methods or even some other way—gravitational capture of a wandering object, for example. There was still the problem of the orbit. Tidal effects should force a planet that close to a neutron star out of a circular orbit.

In Jan 1992, however, this original pulsar planet was found to be an artifact of Earth's motion. Meanwhile other astronomers discovered evidence of planets orbiting a different pulsar.

Disks of Dust

The Sun did not always have planets around it. According to the theories that have been most accepted for almost 200 years, the planets formed out of a cloud of dust that once surrounded the star. In late 1983 and early 1984 the Infrared Astronomical Satellite (IRAS) detected just such clouds around several nearby stars—including Vega, Formalhaut, HL Tau 500, R Mon, and about seventy other candidates for these dust clouds.

At the end of 1984, Bradford A. Smith of the Jet Propulsion Laboratory and Richard J. Terrile of the University of Arizona made an even more spectacular discovery while following up on the IRAS candidates. They were able to photograph a disk of gas and dust surrounding Beta Pictoris, a bright star about twice the mass of the sun that is visible from the southern hemisphere. The disk extends some 160 billion km (100 billion mi) from the star, about thirty times the average distance between the Sun and Pluto. The photograph they produced appears to be dramatic evidence of planet formation in a very early state.

Shortly after it began operation, the Hubble Space Telescope (HST) was able to observe Beta Pictoris using its Goddard High Resolution Spectrograph. Observing in the ultraviolet region, where the flaw in the HST mirror is not much of a problem, the spectrograph found what appear to be large clumps of matter in the disk, clumps that spiral in toward the star.

Brown Dwarfs

Finding a new star is no surprise; the number of stars is legendary. Finding a new planet is major news. Celestial bodies that are in between are hard to explain, but exciting for astronomers. If it is bigger than a planet yet not quite a star, an astronomical body is classed as a brown dwarf, a name coined in 1975 by Jill Tarter. It is "brown" because it does not have enough mass to ignite the fusion process that lights up a star. Instead, gravitational contraction supplies enough energy for the body to radiate at a low level of luminosity. Thus, although a dwarf when compared to a star, the body has to be much larger than a planet, or gravitational contraction would not provide enough energy to radiate in the visible region. (Early in life it may temporarily fuse heavy hydrogen, but the supply runs out fast and fusion soon halts.)

These considerations put limits on what can be called a brown dwarf. The maximum size is about eighty times the mass of Jupiter, which is about 8 percent the mass of the Sun. Any larger and hydrogen would fuse. It is less clear what the minimum size might be, but most astronomers would say the body has to be at least ten times Jupiter's mass to qualify. However, a body that size could also be a large planet. The current thinking is that a planet forms when small bodies—planetesimals—crash into each other, so a planet grows from a small body to a large one. A brown dwarf, on the other hand, starts out as a big gas cloud and just collapses. So, to distinguish a big planet from a small brown dwarf, you must know how the body formed.

In recent years about ten astronomical bodies have been touted as brown dwarfs. Most of these are companion bodies in binary systems, observed or inferred. None has ever been firmly identified. The tenth candidate is unusual in that it is an isolated body first observed in 1988 by Michael R.S. Hawkins of the Royal Observatory in Edinburgh, Scotland. Since then, Philip Ianna of the University of Virginia and Michael S. Bessell of Mount Stromlo and Siding Spring observatories in Australia have studied Hawkins's discovery and have concluded that it is most likely a brown dwarf with a mass of about 5 percent that of the Sun. It is certain that the object is the faintest one ever observed outside the solar system, with a luminosity about 1/4000 that of the Sun.

(Periodical References and Additional Reading: *Science News* 4-7-90, p 213; *Astronomy* 8-90, p 22; *Science News* 1-12-91, p 21; *Science* 2-22-91, p 870;; *Discover* 4-91, p 40; *New York Times* 5-18-91, p I8; *New York Times* 7-25-91, p A1; *Science* 7-26-91, p 385; *Science News* 7-27-91, p 53; *New York Times* 10-31-91, p B8)

BLACK HOLES—BIG AND ENORMOUS

Brown dwarfs are the smallest astronomical bodies that shine, however feebly, on their own. It is suspected that the largest astronomical bodies in terms of mass do not shine at all. These are the black holes that have become such a familiar part of people's thinking in recent times. In fact, the term "black hole" is often used for anything that absorbs something— money or love, for example—and fails to give it back. Even casual acquaintance with the idea of a black hole suggests why it is difficult to observe one. The name comes from the property a black hole has of gravitationally attracting light, so that no light can be emitted from it at all.

Despite the difficulty of observing something that emits no electromagnetic radiation or any particles except for the as-yet unobserved graviton, astronomers frequently announce that they have located a black hole. Since these discoveries are always on the basis of circumstantial evidence, no one is absolutely certain that any of the black hole "sightings" are real.

Ordinary Black Hole

Perhaps the best candidate so far for a black hole was reported by Carole Haswell of the University of Texas and Allen Shafter of San Diego State University in the Aug 20 1990 *Astrophysical Journal Letters*. They studied a black-hole candidate in what appears to be a binary system with the ordinary star V616 Monocerotis.

A binary system is a good place to look for a black hole. Even though the black hole may not be visible, the other member of the pair, if an ordinary star, will be. Thus, the ordinary star will appear to be orbiting a mutual center of gravity with something unseen. If the unseen something is large enough, it could be a black hole.

Another clue that a binary might include a black hole would be if it were to emit energetic X rays. Ordinary stars do not emit much in the way of X rays, but binary pairs that include either a neutron star or a black hole pour them out. The reason is that the gravitational pull of the neutron star or black hole pulls matter from the ordinary star. The matter moves so quickly on its way to destruction that it emits X rays along the way.

Thus, if a binary is emitting X rays, knowing the mass of the invisible body in the binary will tell whether it is a black hole or a neutron star. Various attempts in the past have tried to determine the mass of the unseen component of an X-ray binary. Perhaps the most famous attempt was made in connection with Cygnus X-1, which has been calculated to have a mass of six times that of the Sun. That is twice the three solar masses required for it to be a black hole—but there is a disagreement over the calculation.

Haswell and Shafter observed that the binary known by its X-ray cata-

---

USING SPECTRA TO RECOGNIZE BINARIES

Most stars are not isolated from each other, but appear in groups, of which the most common is the group of two stars that is called a binary system. A typical binary consists of two stars of somewhat different characteristics that orbit a mutual center of gravity. Sometimes it is possible to see the two stars with a good telescope, but more often one has to infer that there are two stars because of two different types of spectra. Ordinary stars can be classed by the types of spectra they show. The classes are assigned letters representing a range from hottest to coolest: O B A F G K M—memorized by the mnemonic "Oh, Be A Fine Girl (Gent), Kiss Me." The presence of an O spectrum and a K spectrum in what appears to be a single star is a sure sign of a binary. Even if the two stars are of the same spectral type, the binary nature of the system can often be detected by the Doppler shifts in their spectra as they orbit the mutual center of gravity.

---

log number as A0620-00 consists of a star of spectral type K (V616 Monocerotis) and what appears to be an accretion disk around the presumed black hole. With these elements observable, Haswell and Shafter could compute the ratio of their velocities to a certainty. The computation showed that the ratio of the mass of the presumed black hole to V616 Monocerotis is 10.6:1. From there, one has to make an informed guess. Most K-type stars have a mass that is at least 0.36 that of the Sun. Thus, at minimum, the companion must have a mass that is 3.8 times that of the Sun. Since this is over the three solar mass limit, the companion must be a black hole, or, at least, the best candidate for now.

Giant Black Hole

A study published on Apr 10 1991 in the *Astrophysical Journal Letters* claims not only the location of a black hole but the largest ever located. Joss Bland-Hawthorn of Rice University, Andrew Wilson of the University of Maryland, and Brent Tully of the University of Hawaii analyzed a strange object in the galaxy NGC 6240 with a new interferometer attached to the 224-cm (88-in.) telescope the University of Hawaii runs on Mauna Kea. With it they were able to detect shifts of a spectral line of hydrogen gas that they interpreted as being caused by two rotating disks of gas with different centers. One of these is the ordinary disk that pervades the galaxy. The other is a disk about 4000 light-years across that rotates almost as a unit. That second disk is presumed to surround the giant black hole.

One theory is that the object within the disk was once a quasar. If the quasar were a black hole, it would offer a possible mechanism for the emission of the immense amount of energy given off by a quasar, energy produced when stars, gas, and dust are sucked into a black hole. When most of the available matter in the neighborhood has been consumed, the quasar dies down. But the black hole, having grown to an enormous mass by "eating" everything about, remains behind.

Background

In locating what is perhaps an extraordinarily large black hole, the actual light analyzed is that caused when an electron in a hydrogen atom drops from one possible orbit, designated 3, to another, designated 2. The amount of energy released when the electron drops is known precisely. As radiation, one way the energy manifests itself is as light of a specific wavelength, and hence color. Passed though a spectroscope, the light falls as a single line in a known location.

Suppose that the source of the light is a hydrogen cloud that is moving with respect to Earth. As a result of the Doppler effect the emitted light will have its wavelength lowered if the cloud is moving away from Earth and raised if it is moving toward Earth. Thus, observing the position of the hydrogen line can give the general movement of the cloud.

When you know the overall motion of the cloud of hydrogen with respect to Earth, you can compensate for that and look for another motion. If the cloud is rotating in any way except face on to Earth, part of it will be moving away from us at the same time that another part is moving toward us. While this motion may be much less than the overall movement of the cloud, it can be detected and its speed measured by an interferometer, which magnifies differences in wavelengths.

With this sort of analysis, the astronomers were able to determine that there are two rotating clouds of gas, one that appears to be coextensive with the visible galaxy and one that is off to one side. The former rotates as if the main gravitational influence is the cloud and the stars embedded in it, with the outer regions moving much slower than the inner ones. The latter disk rotates in a different fashion. Within 4000 light-years of its center, it rotates very rapidly, at about 400 km (650 mi) per second, behaving more like a solid disk than a cloud. Beyond that radius the rotation rate drops off sharply.

Both clouds are assumed to be obeying the laws of gravity. But for the second cloud, these laws imply that a massive concentration of matter is controlling the rotation speed of the disk—somewhere between 40 and 200 billion solar masses. This could be, say, 100 billion stars; but stars would emit light, and there is not much light coming from the disk. Another possibility would be even a greater number of brown dwarfs, all crowded together.

The remaining possibility is a giant black hole. It would supply the mass and it would not emit large amounts of light.

(Periodical References and Additional Reading: *Astronomy* 2-91, p 22; *Science* 2-16-90, pp 775 & 817; *New York Times* 4-9-91, p C10; *Science* 4-19- 91, p 377; *Astronomy* 7-91, p 22; *Scientific American* 7-91, p 32)

THE STATE OF THE GALAXY

UNDERSTANDING THE MILKY WAY

Nicolaus Copernicus [Polish: 1473-1543] is justly famed for teaching humans that Earth and other planets go around the Sun, forming the solar system. But William Herschel [English: 1738-1822], in the period between 1784 to 1818, and Edwin Hubble [US: 1889-1953], in 1925, between them may have accomplished as much by demonstrating the general nature and shape of our galaxy, the Milky Way (Herschel), and its relation to the rest of the universe (Hubble). Today we know that the Milky Way galaxy is a spiral assembly of over 200 billion stars floating amid an uncounted number of other such galaxies.

Since the solar system is a part of the Milky Way, all we see of the galaxy is a band of stars across the night sky, too many to resolve with the naked eye. But various instruments and methods have enabled astronomers to study the Milky Way, a remarkable neighborhood to live in. Stephen M. Kent and coworkers at the Harvard-Smithsonian Center for Astrophysics in Cambridge MA were able to combine infrared views of the galaxy taken from the space shuttle in 1985 with those of spiral galaxies that can be seen face on to produce a more detailed picture of the overall structure of the Milky Way than ever seen before. This was confirmed by a similar, but more direct, infrared image obtained in 1990 by the Cosmic Background Explorer satellite. Both images show the Milky Way edge on, a view that looks like a central egg or watermelon surrounded on all sides by a wide disk.

Seeing the Center

The center of the Milky Way as observed from Earth is in the constellation Sagittarius, a bit below the not very bright star (fifth magnitude) X Sagittarii. The center is about 28,000 light-years from Earth. The center is virtually impossible to see because of the intervening dust and gas. Radio waves, however, are not affected by dust and gas and reveal that there is considerable activity at the center. Most of this is from Sagittarius A*, the strongest and most compact radio wavelength source of energy in the region.

Despite the dust and gas, astronomers Hans Zinnecker, Michael Rosa, and Andrea Moneti have been able to obtain an optical image of the center, including the mysterious Sagittarius A*. They combined the good viewing ability of the NTT at the European Southern Observatory with a clever strategy.

Since Sagittarius A* is a high-energy emitter, most of its radiation is toward shorter wavelengths. In optical terms, this means blue light. But blue light is more affected by dust and gas than any other light (scattering by dust and gas of blue light makes the sky blue, for example). Looking in the blue-light part of the spectrum would not reveal an image even if most of the light generated were blue. Infrared light would pass through the dust and gas more easily, but a high-energy object should not be emitting a large proportion of its light in the infrared. Therefore, the astronomers looked in the reddest part of the visible spectrum to see if they could image anything. That part of the spectrum might get through the dust and gas and might also include the lower limit of emitted light. Then the team combined five images into one, subtracted a star, and further processed the resulting image.

The astronomers located two starlike objects near the galactic center. The closest one to the center appears to be almost completely blue, with no infrared radiation at all. The team thinks that this object, denoted GZ-A, is most likely one and the same as Sagittarius A*. The other one, GZ-B, is a known infrared source a little bit farther from the galactic center.

The question of what GZ-A, or Sagittarius A*, is remains open. It could be a compact cluster of very blue stars, the kind astronomers call type O supergiants. Or, it could be a ring of dust, gas, and pieces of stars swirling around a black hole as the black hole consumes them, a ring known as the accretion disk of a black hole. Such accretion disks become very hot as they are pulled together by the black hole, which is why they are expected to emit blue light as well as other radiation. This radiation knocks electrons out of the gas farther from the black hole, producing radio waves.

Something New Near the Center

On Dec 28 1990 a new radio source suddenly appeared near the center of the Milky Way. It is close to the center but clearly resolved by the Very Large Array radio telescope as a separate point source about 4 light-years southeast of Sagittarius A*. Its strength has varied considerably each time it has been observed.

The only definite conclusion the radio observations have produced is that the waves are generated by high-speed electrons that have been pushed into curved paths by a strong magnetic field in a process called synchrotron radiation. The easy explanation, then, that the energy comes from some kind of explosion, as in a supernova, cannot account for the new radio source.

(Periodical References and Additional Reading: *Science News* 1-13-90, p 21; *Science News* 6-2-90, p 340; *Science News* 6-23-90, p 388; *Astronomy* 10-90, p 39; *Science News* 11-17-90, p 310; *Astronomy* 3-91, p 22; *Astronomy* 4-91, p 46; *Astronomy* 8-91, p 18)

THE GREAT ANNIHILATOR

Sometimes history in astronomy is made simply by looking in the right direction at the right time. On Oct 13 and 14 1990, a French gamma-ray telescope mounted in a Soviet satellite was the lucky one. It observed a tremendous 20-hour burst of gamma rays from an otherwise ordinary X-ray source. During this period the source was emitting 50,000 times as much energy as the Sun. Furthermore, the energy of the photons that make up the gamma rays appeared to be exactly the amount that would be caused by the mutual annihilation of about 10^{44} electrons and their antiparticles, positrons, every second. Thus, Marvin Leventhal of Bell Laboratories suggested that the X-ray source, known formally as 1E1740.7-2942 and informally as the Einstein source, be called "the Great Annihilator" (its nickname "Einstein source" derives from the discovery of this X-ray emitter, among others, by the Einstein satellite in 1979).

The Soviet satellite is called GRANAT and was launched into a high, elliptical, 4-day orbit from Kazakhstan in Dec 1989. The French telescope aboard it, called the Sigma Telescope, is the product of a collaboration between Jacques Paul of Saclay and Pierre Mandrou of the *Centre d'Etude Spatiale des Rayonnements* in Toulouse. It is based on an idea of Robert Dicke of Princeton to use what amounts to a random array of pinhole cameras in front of a gamma-ray detector. A computer program takes the pattern of images on the detector and computes the direction of the gamma rays. Unlike charged particles, which are buffeted about on their journey through space by stray magnetic fields, gamma rays mostly travel in straight lines, just as light does. As Dicke's idea is implemented in the Sigma Telescope, this arrangement has an excellent spatial resolution—it can tell almost exactly where the gamma rays are coming from—but not such a good energy resolution—the amount of energy of the gamma rays is somewhat blurred. At the lower wavelengths of light this would be described as having the ability to tell where the light is coming from but not having the ability to recognize its exact color. This energy blurring results from Sigma's sodium-iodide detector. Future plans are to combine a high-resolution germanium detector with a Dicke pinhole array, to obtain a gamma-ray telescope that can both identify where the "light" is coming from and what color it is.

Germanium detectors that can recognize energy levels almost exactly but that have a low spatial resolution have been observing the sky since 1977, when a balloon-based detector observed what appeared to be electron-positron annihilations near the Galactic center. Subsequent balloon flights sometimes observed repeats of these annihilations and sometimes did not. It appeared to be a phenomenon that waxed and waned, although not on a regular schedule.

Background

Albert Einstein's famous 1905 equation $E = mc^2$ (energy equals mass times the square of the speed of light) implies that mass and energy are two different ways to think about the same thing. It is convenient to measure the very small masses of subatomic particles in terms of their energy equivalents, which are much greater than the masses. Even so, a very small unit, the electron volt (abbreviated eV), and its multiples, such as the kilo-, mega-, or giga-electron volt, are needed to quantify the energy of the particles. For example, a kilo-electron volt is a thousand times as great as an electron volt. (A further discussion of this topic can be found in "Subatomic Particles," p 558.)

The gamma rays found coming from the Great Annihilator can be recognized because most of the rays have an energy of 511 keV (kilo-electron volts). The energy of 511 keV is one of the great unexplained numbers of modern physics because it is precisely the rest mass of every electron ever observed. (Physicists say "rest mass" because moving objects acquire additional mass from their motion, an effect that is more noticeable at speeds near the speed of light.) The antiparticle for the electron, the positron, also has a rest mass of 511 keV.

Particles and antiparticles that meet annihilate each other. To imagine this process, think of the antiparticle as a precisely shaped hole that the particle just fits. When particle and antiparticle meet, the particle falls in the hole and both hole and particle completely disappear with a thud. The "thud" is all the energy that both the particle and the hole together contained. Even though the particle and the hole disappear, the energy cannot vanish. In the particular type of annihilation suspected for the Great Annihilator—one in which the electron and positron first orbit each other, a combination known as positronium—the energy appears as a gamma ray for the particle's mass and another for the hole's mass. Each of these masses is 511 keV when measured in terms of energy, so the appearance in 1 second of 10^{44} gamma rays that have that energy suggests that half of 10^{44} such positronium annihilations have occurred during that second, although some astrophysicists think that some related annihilation processes may also be involved.

These sporadic observations excited astronomers, however, because they seemed to be coming from the center of the Milky Way galaxy. Observations of other galaxies have suggested that many may contain massive black holes at their centers; one popular idea about quasars is that quasars are black holes at the centers of invisible galaxies. Furthermore, a number of interesting and weird phenomena have been observed near our Galactic center, even though we can just barely image it at visible wavelengths because of intervening dust. But radio and infrared observations show gas clouds and stars near the center behaving as if they were orbiting some object that has several million times the mass of Earth's sun—an excellent candidate for a supermassive black hole.

Thus, one surprise when the lucky observation was reported by Jacques Paul of Saclay and Evgeny Churazov of the Moscow Space Research Institute was that the Great Annihilator is *not* at the Galactic center—it is 45 arc minutes away. Depending on how big you think the Milky Way galaxy is and the angles between the Great Annihilator, the center, and us, the gamma-ray source could be 300 light-years away from the center. A more conservative estimate would put it at 100 light-years away.

Furthermore, the best explanation for the gamma-ray burst is that the Great Annihilator is itself a black hole and that the burst came as a gas cloud or star fell into it, never to be seen in this universe again. It would seem odd that there could be a supermassive black hole at the center and another modest black hole hanging around 100 to 300 light years away. Also, the Sigma Telescope has never seen any electron-positron annihilations coming from the center, so if the center is a supermassive black hole, why doesn't it annihilate?

GRANAT took another look at the plane of the Milky Way in Apr and May 1991, finding what appear to be many black holes the size of stars that also seem to be annihilators.

Observations from the Gamma Ray Observatory launched Apr 6 1991 may help to resolve these questions, although its primary mission is to map all the gamma ray sources in the sky. This project will not be completed until late in 1992. Meanwhile, Leventhal is trying to get Russia to orbit a satellite containing one of his detectors.

(Periodical References and Additional Reading: *Science* 1-11-91, p 251; *Physics Today* 3-91, p 17; *Science News* 5-11-91, p 294; *Science News* 5-25-91, p 333; *Scientific American* 7-91, p 29)

BEYOND THE GALAXY

CLOSING IN ON THE MISSING MASS

As a result of theoretical work published in 1974 and later, astronomers and physicists worried about mass missing from the universe throughout the 1980s. Put simply, the most accepted theories of universe formation and structure all predicted far more matter than could be detected by any of the means used. If this matter exists, it is invisible because it does not reveal itself by giving off electromagnetic waves at any wavelength. The missing mass is often called "dark matter" for that reason. The only force to which it is known to respond is gravity. Dark matter must be affected by gravity because its gravitational force is what is apparent, even though dark matter is not apparent in any other way.

Various notions of what the missing mass might be have been put forward and discounted, mostly because any dark-matter theory has to postulate something that is undetectable by the means we ordinary use. As a result, the hypothetical source of the missing mass cannot be found, so no one believes it is really there.

Locating Missing Mass

But astronomers are clever. If the only way that dark matter can be detected is through gravity, then they will find a way to use gravity to locate it. Not the long-sought gravity waves, which are expected to come mostly from violent events, but the observable effects of gravity in the universe. In the early 1980s Vera Rubin of the Carnegie Institution used the rotational velocities of different parts of galaxies to show that each galaxy must be embedded in an envelope of unseen matter. This accounted for about 10 percent of the presumed mass that was missing. It did not, however, offer any clues as to what the dark matter is.

More recently, astronomers have found a way to observe some of the dark matter almost directly. The secret has been to use an artifact of gravity itself. Since 1984 it has been observed that galaxies can act as lenses, as predicted by R.W. Mandl and Albert Einstein in 1936. In 1990 J. Anthony Tyson and Richard Wenk of AT&T Bell Laboratories and Francisco Valdes of the US National Optical Observatories figured a way to use gravitational lenses to detect dark matter.

The basic concept of a gravitational lens is that gravity can bend electromagnetic radiation. It was the accurate prediction of the amount of this gravitational bending that made Einstein world famous in 1919. Light from a distant object is noticeably bent by an intervening mass provided the mass is heavy enough. Refraction of light from quasars by intervening galaxies was the first experimental observation of such bending. One quasar would appear to be two or three almost identical quasars, looking somewhat like multiple images in angled mirrors.

Another effect of a gravity lens is to distort an image. A point can be smeared into an arc, for example. Tyson, Wenk, and Valdes used a computer program to help them look for such distortions in apparent point sources that are very far away. Instead of quasars, however, they used the much more numerous faint blue galaxies. These are just barely observable, but virtually cover the sky. They are found if one focuses on an apparently dark bit of sky and gathers light for long enough.

Because of their faintness and also their red shifts (a good indicator of distance when measurable), the faint blue galaxies are thought to be very far away. Distortion in the image of one of these galaxies implies that some intervening massive material is causing the effect. Seeking galaxies with such distortions can be used as a way to map the dark matter, since the distant faint blue galaxies cover so much of the sky.

The team sought the missing mass in clusters of galaxies, since that is where most of it is thought to be. Missing mass is thought to have had a hand in causing the galaxies to cluster in the first place. The regions around a couple of clusters of galaxies were examined to see if apparently empty spaces act as gravitational lenses. Sure enough, the computer located about thirty-odd distorted faint blue galaxies in the two clusters examined. Some of these are seen as refracted from points into definite arcs.

The implication is that there is dark matter within the relatively near galactic clusters that is causing the distortions in the images of the very distant background galaxies. A full-scale examination of the complete sky for such distortions could be used to find the location of almost all the missing mass. It would not, of course, tell what the mass is.

In a reversal of this procedure, Glenn I. Langston and coworkers at the US National Radio Astronomy Observatory in Charlottesville VA have used a gravitational lens to estimate the mass of the lens. They based their estimate on distortion of a quasar's light by an intervening galaxy, which serves as the lens. The galaxy is too far away to be studied by conventional means. The gravitational lens experiment shows that this distant galaxy, like those in our neighborhood, contains more mass than can be observed using electromagnetic waves.

Cold and Hot Dark Matter

There are a number of possibilities as to what the missing mass might be. Astronomers have categorized missing-mass candidates into two large groups, cold matter and hot matter. Temperature in this context refers to the motion of the dark matter. If it is moving around the same speed as visible matter, then it is cold. Hot dark matter is streaming along at nearly the speed of light.

Cold dark matter could consist simply of moderately large bodies of ordinary matter that are too cool to radiate energy and too far away to see. Such bodies include large planets and small brown dwarfs. The latter could be orbiting ordinary stars, each other, or wandering in galaxies. (It is also possible that they are wandering between galaxies, but this is thought to be much less likely.) Another possibility for cold dark matter is that it consists of clouds of cool gas that would be undetectable unless energized by radiation from other forms of matter. Yet another possibility would be that it is made up of black holes that do not have any matter near enough them to form accretion disks. Theorists have also postulated a host of slow-moving Weakly Interacting Massive Particles (WIMPS, an acronym attributed to M. Turner), although no WIMPS have ever been observed. Recently, however, many writers have used the expression "cold dark matter" to refer primarily to slow-moving WIMPS.

It has also been argued that cold dark matter consists of Massive Compact Astrophysical Halo Objects (MACHOS, an acronym attributed to K. Griest), which are neutron stars and long-lived low-mass stars.

Cold dark matter was prominent in the news early in 1991. The spark was a new analysis of data from the 1983 Infrared Astronomical Satellite (IRAS) concerning the distribution of galaxies, clusters of galaxies, and superclusters of clusters. Will Saunders of Oxford University and a team of Canadian and British astronomers wrote in *Nature* that their analysis showed "There is more structure on large scales than is predicted by the standard cold dark matter theory of galaxy formation." Somehow this was interpreted in some of the press to mean that the big bang theory itself was in question, although this was not the case. Furthermore, an analysis by Changbon Park at Princeton the previous year had shown that cold dark matter, in a computer simulation, could explain the structure of the universe. There is further discussion of this question in the next article, which is on a closely related issue.

Hot dark matter would also consist of subatomic particles. Candidates can be produced by endowing the neutrino, a particle known to exist, with mass or by imagining other WIMPS, this time fast-moving ones. An experiment aboard ASTRO-1 in early 1991 was designed to observe the decay of tau neutrinos, following a prediction of Dennis Sciama. Had the decay been observed, it would have implied a mass for the tau neutrino of about 28 eV, which is not very much, but it would have helped supply some or all of the missing mass. No decay was detected. A mysterious 17 keV neutrino was detected in 1991 (see "What's Going On at 17 keV?" p 542); if its existence is confirmed, the neutrino might be another contributor to hot dark matter.

There is also an intermediate class of matter that has been proposed, grains of strange matter. Strange matter contains the strange quark as well as the up and down quarks that constitute ordinary matter. Such grains, if they exist, would be larger than individual particles but smaller than planets or gas clouds. They might be termed "lukewarm dark matter."

(Periodical References and Additional Reading: *Science News* 1-27-90, p 52; *Science* 2-9-90, p 247; *Science News* 3-3-90, p 133; *Astronomy* 5-90, p 10)

THE UNIVERSE IS TOO MUCH ALIKE

The argument over how the galaxies came to be, how they came to cluster, and how the clusters also came to cluster involves much more than the question of the missing mass. It is central to understanding the origin of the universe. Even if the dark matter could have provided the gravitational shove to bring the galaxies into being and together, that would simply move the question of smoothness from the visible matter to the dark matter. The problem is that very good evidence exists to show that the universe was very smooth when it was born; almost as good evidence exists to show that it is not very smooth today.

The Issues

The most likely way for the big bang to have happened is smoothly in all directions. The most popular version of the big bang, called the new inflationary universe theory, can either be interpreted with such a smooth universe or it can be assigned small fluctuations in temperature or other properties. The small fluctuations seem to be necessary to produce the universe as we observe it, with matter clumped into stars, galaxies, clusters of galaxies, and superclusters of clusters. Also, the uncertainty principle of quantum theory suggests that small fluctuations should exist. As the universe expands, the original small fluctuations expand with it and result in the large observable disparities of today.

The organization of the universe is generally assumed to have resulted from gravitational attraction, although there is a competing theory of organization by electromagnetic fields that is supported by a loud minority. Thus, whatever fluctuated must have had mass. One possibility would be that the differences were in the number of baryons—familiar particles, such as protons and neutrons, as well as more massive versions of the same particles. More baryons in some places than others would result in variations in the amount hydrogen or helium from place to place, which is what we observe.

Background

The big bang, according to current theories, really was a kind of explosion, complete with the release of a vast amount of energy into all the space that then existed. The easiest way to describe the amount of energy is in terms of temperature. The temperature of the big bang itself is not very meaningful because not much in the way of matter or ordinary forms of energy existed at the very beginning, although some scientists will state that the temperature 1 second after the big bang was something like 10,000,000,000 K (essentially the same for very high temperatures as Celsius and a little less than half what a Fahrenheit estimate would be), and that 100 seconds later, it would have dropped to about 1,000,000,000 K. By about 100,000 to 300,000 years after the big bang, however, matter as we know it had begun to condense and theorists are able to calculate its temperature: 30,000 K (54,000° F). At that time, the electromagnetic radiation would have been gamma rays.

The universe was expanding then as it is now. Just as lowering the pressure of a gas lowers its temperature, the expansion of the universe lowered this primordial temperature. The energy had nowhere to go, so it could not change; but the amount of space it occupied was increasing, so it got spread out and diminished. When the universe doubles in size, its temperature drops to half what it was. The gamma rays became longer and longer, moving down through the electromagnetic spectrum until today they are in the microwave region, corresponding to temperature of only 2.735 K (-270.415° C or -454.747° F). We detect this temperature by the electromagnetic waves it produces (just as a hot stove warms in part by radiation). Low temperatures produce long waves, while high temperatures make short waves. The exact measurement of this temperature in agreement with theory is what convinces most astronomers that the big bang happened.

If the big bang was smooth, looking in any one direction we should find electromagnetic waves from that direction registering the same temperature as waves from any other direction. In practice, however, waves from one specific direction have a higher frequency than waves from the opposite direction. This is caused by the Doppler effect—as we move toward waves they reach us more frequently and as we move away the come less frequently. This effect can be, and is, used to tell us for the first time exactly in what direction and how fast we are moving through the universe.

Simple as this idea is, baryon variations do not appear easily out of the equations that describe the early universe. The problem is that baryons interact too strongly with electromagnetic radiation. The radiation present during the early universe would tend to smooth out differences in baryon density. As a result, most cosmologists—astronomers or physicists who study the universe as a whole—postulate some other kind of massive particle in which the fluctuation takes place. The other particle, most commmonly one called an axion, is a member of the class of weakly interacting massive particles, or WIMPS, which have never been observed.

Neither baryons nor WIMPS would arrive in the universe with great speed. Since the average speed of particles in a large group is what we experience as temperature, the production of fluctuation of this type is explained by the cold dark matter hypothesis (see previous article). The cold dark matter hypothesis is not the only way that the origins of the diversity of the universe can be explained, but it has been the most popular theory.

If fluctuations exist, they should be reflected in small variations in the radiation that was produced when matter precipitated out of the universe somewhat after the big bang. Specifically, denser regions then should appear as warmer regions of the cosmic background radiation now. Finding and measuring those warmer regions could help explain the evolution of the universe.

A totally different issue is to check the details of the background radiation to see if it matches the big bang theory. Earth-based measurements have suggested very good agreement of observation and theory, but measurements from space can be much more precise. For one thing, not all of the background radiation is at frequencies that can penetrate the atmosphere.

In 1987 researchers from the University of California at Berkeley and Nagoya University in Japan used a rocket to sample the frequencies that cannot be detected in the lower atmosphere. Surprisingly, they measured much higher temperatures at those wavelengths than would be predicted by big bang theory. The temperatures were too high for most physicists and astronomers to believe, but still they cast doubt on the theory. A second rocket experiment failed, so people eagerly awaited additional data that might confirm either the theory or data from the original experiment.

The Instrument

The satellite Cosmic Background Explorer (COBE) was launched on Nov 18 1989 largely to resolve these issues. COBE's mission was to examine the part of the spectrum we usually think of in the kitchen, the microwaves. It is in this region, specifically in wavelengths around 0.5 mm to 5 mm (0.02 in. to 0.2 in.), that a remnant of the big bang can be found. COBE was to examine this remnant, called the cosmic background radiation, with unprecedented thoroughness. During the lifetime of the satellite it will map the cosmic background radiation with several instruments. The longer the mapping continues, the more precise the results, with final results expected to be about ten times as precise as those obtained in 1991.

COBE looks at the microwave radiation in several ways. The most dra-

matic is with an instrument called the Far Infrared Absolute Spectropho-
tometer (FIRAS), which measures the background radiation at a hundred
different wavelengths in each of a thousand different directions. Another
instrument, the Differential Microwave Radiometer (DMR), maps the
brightness of the cosmic background radiation over the entire sky at three
different frequencies. The third instrument, the Diffuse Infrared Background Experiment (DIRBE), maps the sky at three different infrared wavelengths, looking for infrared radiation that could come from point sources.
Such radiation might indicate early stars or galaxies formed about 500,000
years after the big bang. DIRBE is designed to locate sources that are only
1 percent brighter than the sky around them, an unprecedented sensitivity.

Both the infrared detectors have to be kept very cool or their own heat
will radiate so much energy at infrared wavelengths as to overwhelm
incoming energy. As a result, the detectors are cooled by immersion in liquid helium, which keeps them at 2 K (-271° C or -457° F). Even though it
is cold in space, the liquid helium slowly evaporates. The 96 kg (211 lb) of
liquid helium was expected to last about a year, defining the workable
lifetime of the satellite's infrared capability, although the DMR, which does
not have to be cooled, should last an additional year.

The Results

COBE's FIRAS showed that the first sixty-seven data points it measured
match up exactly with the hypothesis that the early universe behaved like a
perfect black body, disagreeing with the results of the rocket probe in 1987.
This implies that during the period when the background radiation was
formed, all the matter present had exactly the same temperature as the radiation in the universe. Thus, there were no places where baryonic matter or
anything else that could be described as matter was forming at faster or
slower speeds than anywhere else. John Mather of NASA's Goddard Space
Flight Center, principal investigator for FIRAS, described the result this way:
"COBE's latest cosmic background data show that the universe is smooth at
very early time scales. This is strong evidence against the so-called cold
dark matter theory that is a component of the big bang scenario." It is also
strong evidence that the big bang itself actually occurred. Furthermore, the
absence of deviation from black body radiation implies that there is no hidden source of energy, such as cosmic strings, in the universe.

The DMR confirmed these conclusions in a different way. It observed
no difference in brightness that could be accounted for in any way except
for the known Doppler effect. Thus, the cosmic background is the same
brightness in all directions.

Finally, DIRBE located known sources of infrared radiation, such as dust
in the solar system and elsewhere as well as known stars, but it found no
evidence that any clumps of matter were forming early in the history of

the universe. This result, combined with the overall smoothness COBE found, rules out another theory that says the structure of the universe was formed by explosions of massive structures early in the history of the cosmos. DIRBE says there was nothing there to explode.

The Beryllium Puzzle

In addition to a correct prediction of the cosmic background radiation and explanation of the expanding universe, the other main observational underpinning of big bang theory has been that it results in correct calculations for the amounts of elements lighter than iron that are observed in the universe, especially ratios of hydrogen to helium. Recently, an unexpected abundance of one of those light elements, beryllium, in certain older stars has suggested that the big bang was more uneven than the smooth cosmic background radiation would indicate.

Gerard Gilmore of Cambridge University, UK, and coworkers described their findings in the Sep 1 1991 *Astrophysical Journal*. Using spectroscopy and the Anglo-Australian Telescope in Coonabarabran, Australia, they compared the amount of beryllium to the amount of hydrogen in a 15-billion-year-old star called HD 140283. The team found the amount of beryllium to be about a thousand times as great as expected. According to big bang theory, there should be about 10^{16} times as many hydrogen atoms as beryllium atoms. The team also found excess beryllium in three other somewhat younger stars. All these stars are metal-poor according to their spectrum, meaning that they have not acquired much of the iron and other metals created by supernovas since the big bang.

More beryllium would be created in some regions of space if the big bang happened with different densities in different regions of the early universe. High-density regions would produce excess neutrons that would, in turn, fuse with other nucleons to make heavier atomic nuclei. Beryllium, with four protons and five neutrons, is the fourth lightest element, heavier than hydrogen, helium, and lithium. The next heavier element, boron, has five protons and six neutrons in its most common form.

Cosmic rays, which include protons and nuclei of light atoms, could account for the increased beryllium content of the metal-poor stars, but they would also produce ten times as much boron as is observed. Although further observations are needed, at present the only way to account for the excess beryllium is a lumpy big bang.

(Periodical References and Additional Reading: *Science* 1-26-90, p 411; *Astronomy* 2-90, p 16; *Astronomy* 4-90, p 10; *Science News* 4-21-90, p 245; *Science News* 4-28-90, p 262; *Astronomy* 6-90, p 20; *Science News* 6-20-90, p 36; *Science News* 10-27-90, p 260; *Science News* 11-10-90, p 301; *Astronomy* 1-91, p 24; *Discover* 1-91, p 34; *New York Times* 1-3-91, p A1; *Astronomy* 4-91, p 24; *Science* 11-22-91, p 1106)

COSMOLOGY UPDATE

Both the missing mass problem and the excessive smoothness of the big bang are main players in the relatively new science of cosmology. Cosmology's effective birth as a theoretical science came in 1915 with Einstein's general relativity theory and as an observational science in 1929 with Edwin Hubble's work demonstrating an expanding universe.

Although Einstein's equations predicted an expanding universe, no one paid much attention to the possibility until the expansion was observed. Since then, the pattern of theoretical results preceding observation has been common, but not absolute. For example, neutron stars and the cosmic background radiation were predicted and paid little attention to until they were observed. Once in a while an entity—the black hole is the obvious example—is predicted and becomes famous without ever being observed. Rarely, a quasar comes along—observed, unpredicted, and still not explained by theory.

With that situation in mind, where does cosmology stand today?

Predicted But Not Yet Observed

Probably no unobserved entity has the full faith and credit of the astrophysics community as much as gravity waves (although the top quark may be a close second). Gravity waves are caused when anything with mass moves or interacts gravitationally with any other mass. The catch is that both masses must be huge before the waves are detectable. A good gravity wave detector, or telescope if you will, would open a new window on such events as black holes, supernovas, and binary neutron stars. There is even a chance that there exists a gravity wave background left over from the big bang that could be used to determine very early events in the life of the universe.

Four gravity wave detectors are planned for the late 1990s, two in Europe and two in the United States. The US effort, the Laser Interferometer Gravitational Wave Observatory (LIGO), will use both American detectors, so that one can verify the other.

The way to observe a gravity wave is to notice the ripple in space-time as the wave passes. This can be observed in several ways, but LIGO plans to measure the change in two perpendicular masses at the ends of 4-km (2.5-mi) tubes—one will lengthen and the other will contract when the wave passes. A laser beam will be used to measure the distances with interferometry.

A prototype of LIGO has been operating with 40-m (130-ft) tubes, but it has not definitely detected gravity waves. The full-scale version will be a hundred times more sensitive, but it will cost $192 million to build, an amount the US Congress has been reluctant to fund.

Observed But Not Predicted or Understood

The large-scale structure of the universe, alluded to in the previous two articles, was mapped extensively for the first time in the 1980s, with major new projects for improving the maps continuing into the 1990s. Although there had been evidence of clusters of galaxies and superclusters of clusters, the discovery of great structures stretching as far as has been mapped (the "great wall") was totally unexpected. Furthermore, the theories put forward to explain the formation of galaxies and other large features did not seem adequate for these great structures. Finally, as noted above, the COBE results knocked out the cold dark matter hypothesis, which was the leading explanation.

On close heels of the new map of the cosmos based on IRAS observations, which was announced on Jan 3 1991, were early results from ROSAT that showed clusters of quasars. Like all quasars observed so far, these clusters are very far away, which implies that they are also very old (in an expanding universe, the older something is, the farther away it must be; the converse of this, although not necessarily true, is generally assumed). In these experiments, ROSAT collected X rays from apparently blank pieces of sky for as long as half a day at a time. Everywhere this kind of observation was made, ROSAT found dim X-ray sources that Guenther Hasinger and coworkers from the Max Planck Institute for Extraterrestrial Physics in Garching, Germany, interpreted as clusters of quasars. The ROSAT findings were viewed by some as showing that clumping began too early in the universe to be caused by small fluctuations in cold dark matter or any other known mechanisms. Later in 1991, Roger G. Clewes of the Royal Observatory in Edinburgh, Scotland, and Luis E. Campusano of the University of Chile in Santiago announced that they had discovered a very large cluster of quasars in an optical survey.

ROSAT also located an unexplained new class of stars that are visible only as ultraviolet sources.

Alain Picard of Caltech reported in Aug 1991 that he had found a phenomenon that goes beyond clustering. He located a wide discrepancy in the number of galaxies on each side of the sky using a device called COSMOS at the Royal Observatory in Edinburgh. With it, he eliminated the stars from two parts of the Palomar Sky Survey II, one from the photographic atlas of the northern sky and one from the southern sky. The result is something like two digitized views of the sky as it might be seen from outside the Milky Way, with only galaxies remaining. No actual image was produced, however, only galaxy counts. Surprisingly, galaxies are from 30 to 40 percent more in one section of the sky than the other, although each section was the same size, 15° on a side. This is far more clumpiness than found by any other method, so it will be even more difficult for cosmologists to explain.

How Old Is the Universe?

Careful reports of the distance to quasars or faraway galaxies often do not indicate in light-years or other measures of length just how far away the objects are. Instead, they rely on such criteria as how much the spectrum has been shifted to the red by the expansion of the universe. The reason for maintaining a low profile on distance is that there is considerable disagreement about how old the universe is. Many astronomers recently have assumed that the universe is about 10 to 12 billion years old, or perhaps as much as 15 billion years old; but some of the most respected cosmologists maintain that it is 18 to 20 billion years old.

In tables in *Current Science,* the assumption is made that the universe is 20 billion years old.

These discrepancies result from difficulty in measuring precisely how fast the universe is expanding. If the expansion rate were known exactly, then it would be a simple matter to calculate an age that everyone could agree on. In turn, this would make distances to faraway objects realistic in terms of light-years, parsecs, or other common units of measure.

The expansion rate of the universe is a number that has been named the Hubble constant, designated H, after Edwin Hubble. A value of H that is near 100 km/sec^2 (60 mi/sec^2) implies that the big bang happened a brief 10 billion years ago. The lower H is, however, the older the universe.

George Rhee of New Mexico State University in Las Cruces used a new method to determine H, which he reported in the Mar 21 1991 *Nature*. He studied two different images of the same quasar that were produced by a gravitational lens. Quasars often have slight variations in their output. Both images of the quasar, known as Q0957 561, used by Rhee have been observed for more than a decade. The variations in one image showed up as identical variations in the other image 415 days later. Comparison of the two paths helped to determine a more exact distance for the quasar. Distance is the hard part in determining H; the other component of H, velocity due to expansion, can be directly inferred from the red shift of a distant object.

Rhee's calculations show a value for H of 50 km/sec^2 with some uncertainty; H could be as low as 33 or as high as 67. A separate study of the same quasar using radio waves detected by the Very Large Array was conducted by a team led by David Roberts of Brandeis University. They obtained a value that could be as low as 46 or as high as 69 km/sec^2, with theory favoring the higher value. Astronomers using other methods have calculated H as between 75 and 100. Combining all these results, astronomer Sidney van den Bergh commented, "Almost any value between 50 and 80 is still possible at this time." Astronomers expect that eventually

THE HUBBLE "CONSTANT"

The Hubble constant, *H*, is the ratio of the expansion velocity of any observed object to its distance from the observer. For a nearby object, the velocity from something other than expansion may swamp the expansion velocity, but for objects far enough away the reverse is true. *H* is measured with the expansion velocity in kilometers per second and the distance in millions of parsecs (a parsec is 3.258 light-years). Technically, *H* is not a constant, since it has varied over the history of the universe. Careful astronomers call *H* the Hubble parameter.

The age of the universe is 1/*H* hundred billion years. Here is a conversion table.

H (km/sec^2)	Age of the universe
33	30 billion years
50	20 billion years
67	15 billion years
75	13 billion years
80	12.5 billion years
100	10 billion years

a repaired Hubble Space Telescope will provide a definite value on which everyone can agree.

(Periodical References and Additional Reading: *Science News* 2-3-90, p 67; *Science News* 3-24-90, p 184; *Science News* 4-28-90, p 262; *Astronomy* 6-90, p 10; *Science News* 6-9-90, p 358; *Science News* 6-16-90, p 373; *Science News* 6-23-90, p 389; *Science News* 7-20-90, p 45; *Astronomy* 8-90, p 28; *Science News* 9-1-90, p 133; *Science* 10-5-90, p 32; *New York Times* 11-21- 90, p D21; *Science News* 1-12-91, p 22; *Science* 1-18-91, p 272; *New York Times* 1-22-91, p C2; *Science News* 1-26-91, p 52; *Science* 2-1-91, pp 495 & 537; *Astronomy* 3-91, p 44; *Scientific American* 3-91, p 27; *New York Times* 3-21-91, p B10; *Astronomy* 5-91, p 24; *Discover* 5-91, p 24; *Science News* 6- 1-91, p 343; *Science News* 6-15-91, pp 375 & 381; *Science News* 6-22-91, p 396; *Astronomy* 7-91; *Science* 8-16-91, p 743; *Time* 9-2-91, p 62; *Astronomy* 11-91, p 25; *Discover* 11-91, p 20)

BASIC FACTS ABOUT THE PLANETS

The planets are of two main types, which are grouped separately below. The inner planets are somewhat like Earth, and are called the *terrestrial planets*. The outer planets, except for Pluto, are *gas giants,* similar to Jupiter. Pluto is not very much like the planets in either group as far as we know, but it is so far away that its exact nature is not known.

The numbers used below are sometimes given in *scientific notation* because of their size. A large number, such as 1,840,000,000, might be written as 1.84×10^9. The exponent 9 signifies the number of places after the first digit. A small number such as 0.000000000000001 is written as 10^{-15}, where the exponent -15 tells the number of zeroes, counting the one before the decimal place, before the 1.

Some of the words and phrases used in the table are briefly defined below:

Bar is a measure of pressure that is slightly less than Earth's air pressure at sea level under normal conditions, or about 73.825 cm (29.53 in.) of mercury, as the weather report on television would state it.

Eccentricity is a number that measures the shape of certain curves, including ellipital orbits; the smaller the number, the more an ellipse is like a circle, which has an eccentricity of 0.

Ecliptic refers to the apparent path of the Sun through the stars as viewed from Earth, which is in a plane inclined 23.5° to Earth's equator.

Escape velocity is the speed needed for an object to be propelled from the surface of a planet and not fall back.

Inclination of axis is the angle that the line about which a planet rotates makes with the plane defined by its path around the Sun.

Inclination of orbit to the ecliptic is the angle that the plane defined by a planet's path around the Sun makes with the plane defined by the apparent path of the Sun among the stars as seen from Earth.

Orbital velocity is the speed of a planet in its path around the Sun.

Retrograde means in the opposite direction of other planets.

Revolution is the trip a planet makes about the Sun.

Rotation is the turning of a planet about a line through its center.

Terrestrial Planets

	Mercury	*Venus*	*Earth*	*Mars*
Average distance from the Sun in kilometers	57,900,000	108,200,000	149,600,000	227,900,000
(in miles)	35,900,000	67,200,000	92,960,000	141,600,000
Rotation period in Earth days	59	-243.01 (retrograde)	1	1.0004
Period of revolution in Earth days	88	224.7	365.26	687
Average orbital velocity in miles per second	29.7	21.75	18.46	14.98
Inclination of axis	2°	3°	23°27′	25°12′
Inclination of orbit to the ecliptic	7°	3.39°	0°	1.9°
Eccentricity of orbit	0.206	0.007	0.017	0.093
Equatorial diameter	3030 mi	7521 mi	7926 mi	4217 mi
	(4880 km)	(12,104 km)	(12,756 km)	(6787 km)
Diameter relative to Earth	0.382 times	0.949 times	—	0.532 times
Mass	7.283×10^{23} lb	1.07×10^{25} lb	1.32×10^{25} lb	1.42×10^{24} lb
	3.303×10^{23} kg	4.87×10^{24} kg	5.98×10^{24} kg	6.42×10^{23} kg
Mass relative to Earth	0.0558 times	0.815 times	—	0.1074 times
Mass of Sun relative to planet mass (with atmosphere and satellites)	5,972,000	408,520	328,900	3,098,710
Average density	3.13 oz/in.^{3}	3.03 oz/in.^{3}	3.19 oz/in.^{3}	2.27 oz/in.^{3}
Gravity (at equator surface)	12.4 ft/sec^2	28.2 ft/sec^2	32.1 ft/sec^2	12.2 ft/sec^2
Gravity (relative to Earth)	0.38	0.88	1	0.38
Escape velocity at equator	4.3 km/sec	10.3 km/sec	11.2 km/sec	5 km/sec
	2.7 mi/sec	6.40 mi/sec	6.96 mi/sec	3.1 mi/sec
Average surface temperature	167° C 332° F	457° C 854° F	15° C 59° F	-55° C -67° F
Atmospheric pressure at surface	10^{-15} bars	90 bars	1.013 bar	0.006 bar

	Mercury	Venus	Earth	Mars
Atmosphere (main components)	Virtually none (but traces of sodium, potassium, oxygen, helium, hydrogen)	Carbon dioxide 96%; nitrogen 3.5%	Nitrogen 77%; oxygen 21%; water 1%; argon .93%	Carbon dioxide 95%; nitrogen 2.7%; argon 1.6%
Planetary rings	None	None	None	None
Planetary satellites	None	None	1 moon	2 moons

Outer Planets

	Jupiter	Saturn	Uranus	Neptune	Pluto
Position among planets	Fifth	Sixth	Seventh	Eighth	Ninth
Average distance from Sun	778,300,000 km 483,600,000 mi	1,472,000,000 km 914,000,000 mi	2,869,000,000 km 1,783,000,000 mi	4,496,000,000 km 2,794,000,000 mi	5,900,000,000 km 3,666,000,000 mi
Rotation period	9 hr, 55 min, 30 sec	10 hr, 39 min, 20 sec	-23.9 hr (retrograde)	22 hr (or less)	6 days, 9 hr, 18 min (retrograde)
Period of revolution	4332.6 Earth days (11.86 yr)	10,759.2 Earth days (29.46 yr)	30,685.4 Earth days (84.01 yr)	60,189 Earth days (164.1 yr)	90,465 Earth days (247.7 yr)
Average orbital velocity	13.06 km/sec 8.1 mi/sec	9.64 km/sec 5.99 mi/sec	6.8 km/sec 4.2 mi/sec	5.4 km/sec 3.35 mi/sec	4.7 km/sec 2.9 mi/sec
Inclination of axis	3°5′	26°7′	97°55′	28°41′	60°(?)
Inclination of orbit to ecliptic	1.3°	2.5°	0.8°	1.8°	17.2°
Eccentricity of orbit	0.048	0.056	0.047	0.009	0.25
Equatorial diameter	142,800 km 88,700 mi	120,400 km 74,800 mi	51,800 km 32,200 mi	49,500 km 30,800 mi	2290 km 1423 mi
Diameter relative to Earth	11.21 times	9.41 times	4.1 times	3.88 times	0.18 times
Mass	1.899×10^{27} kg 4.187×10^{27} lb	5.686×10^{26} kg 1.2538×10^{27} lb	8.66×10^{25} kg 1.909×10^{26} lb	1.030×10^{26} kg 2.271×10^{26} lb	c 6.6×10^{21} kg c 1.45×10^{22} lb
Mass relative to Earth	317.9 times	95.2 times	14.6 times	17.23 times	c 0.0017 times

(Continued)

	Jupiter	Saturn	Uranus	Neptune	Pluto
Mass of Sun relative to planet mass (with atmosphere and satellites)	1047	3498	22,759	19,332	3,000,000 (?)
Average density	0.759 oz/in.^{3}	0.40 oz/in.^{3}	c 0.7 oz/in.^{3}	1.0 oz/in.^{3}	c 0.2 oz/in.^{3}
Gravity (at equator surface)	75.06 ft/sec^2	29.69 ft/sec^2	25.5 ft/sec^2	36 ft/sec^2	c 14.1 ft/sec^2
Gravity (relative to Earth)	2.34	0.92	0.79	1.12	0.43
Escape velocity at equator	59.5 km/sec 36.9 mi/sec	35.6 km/sec 22.1 mi/sec	21.2 km/sec 13.2 mi/sec	23.6 km/sec 14.66 mi/sec	5.3 km/sec 3.29 mi/sec
Average temperature in atmosphere	(at surface) -108° C -163° F	(at surface) -133° C -208° F	(cloud tops) -215° C -355° F	(cloud tops) -230° C -382° F	(at surface) -273° C c -460° F
Atmosphere (main components)	(Near cloud tops) hydrogen 90%; helium c 10%	Hydrogen 94%; helium c 6%	Hydrogen, helium, methane	Hydrogen, helium, methane	Tenuous; methane & possibly neon
Other atmospheric gases	Methane, water, ammonia, ethane, acetylene, phosphine, hydrogen cyanide carbon monoxide	Methane, ammonia			

THE CONSTELLATIONS

Constellations are small groups of stars that, from our vantage point on Earth, seem to form some particular shape. In actuality, however, the stars that form a constellation are usually at vastly different distances from the solar system. They only appear to form a particular figure from Earth.

Long ago, probably well before recorded history of any kind, people began naming these groups. By Sumerian times there were already stories being told about how particular constellations were formed. Most of our present knowledge of such stories comes from the Greeks, who reflected much of their mythology in the stars and planets.

One group of constellations has exerted a special influence on human thought, at least since 1500 BC. As the Sun, Moon, and planets move through the sky, they pass through a group of twelve constellations: the constellations of the zodiac. Chaldean astronomers believed that the presence of the Sun, a planet, or even the Moon in one of these constellations at the time of a person's birth or at other significant times influences happenings on Earth. We call this belief *astrology*. Because of the precession of the equinoxes, the traditional twelve constellations are no longer where they were 3500 years ago. Modern astrologers have divided the year into twelve "houses" based on where the signs of the zodiac used to be. Thus, when an astronomer and an astrologer refer to the zodiac, they mean quite different things.

Today astronomers use constellations for their own purposes, especially to map the sky. Each part of the sky is named by a particular constellation. These constellations, especially in the southern hemisphere, may not be traditional ones, but rather groups of stars that astronomers have named so that all of the sky is covered (such constellations are labeled "Of modern origin" in the table). The International Astronomical Union has decreed that each such constellation be bounded by straight north-south and east-west lines. As a result, many of the larger traditional constellations extend beyond the boundaries of the astronomical constellation.

While astronomers often use the traditional names of stars, most of which come to us from Latin or Arabic sources, they also use another system for naming objects in the sky (galaxies, radio sources, quasars, and so forth) that is based on constellations. Generally, the brightest star in a particular astronomical constellation is called alpha, the next brightest beta, and so on through several letters of the Greek alphabet. Thus Sirius is also known as alpha Canis Majoris, usually abbreviated to α CMa, which means it is the brightest star in the constellation Big Dog (Sirius has long been known as the dog star). Since Sirius is a binary star, the much brighter main star is officially CMa A. Bright radio sources were once designated by the name of a constellation followed by a letter of the Roman alphabet, such as Cassiopeia A (Cassiopeia is a character from Greek mythology) or Cygnus A (the Swan). Unfortunately, this system has been largely abandoned, so the same radio source may have several different names depending on the astronomer. X-ray sources are still designated by the name of the constellation, followed by an X hyphen number. The number 1 is the brightest X-ray source in a given constellation, so Scorpius X-1 is the brightest X-ray source in the constellation Scorpio.

All astronomer's constellations are named in Latin. When astronomers use a constellation to locate a star, the genitive case, meaning "of the thing," is used. Thus the constellation Big Dog is officially Canis Major, but Sirius is alpha Canis Majoris, or "alpha of Big Dog."

In the table, the twenty-five brightest stars as seen from Earth are listed according to the constellation in which they can be seen. The number following the official name indicates the rank. Thus, Sirius is listed under Canis Major as "Contains Sirius (α CMa A) - 1," which means it is the brightest star, while Canopus, the second brightest, listed under Carina, is followed by - 2.

Name	Genitive	Abbreviation	Translation	Remarks
Andromeda	Andromedae	And	Andromeda	Character in Greek myth
Antlia	Antliae	Ant	Pump	Of modern origin
Apus	Apodis	Aps	Bird of Paradise	Of modern origin
Aquarius	Aquarii	Aqr	Water Bearer	In zodiac
Aquila	Aquilae	Aql	Eagle	Contains Altair (α Aql) - 12
Ara	Arae	Ara	Altar	
Aries	Arietis	Ari	Ram	In zodiac
Auriga	Aurigae	Aur	Charioteer	Contains Capella (α Aur) - 6
Boötes	Boötis	Boo	Herdsman	Contains Arcturus (α Boo) - 3
Caelum	Caeli	Cae	Chisel	Of modern origin
Camelopardalis	Camelo-pardalis	Cam	Giraffe	Of modern origin
Cancer	Cancri	Cnc	Crab	In zodiac
Canes Venatici	Canum Venaticorum	CVn	Hunting Dogs	Of modern origin
Canis Major	Canis Majoris	CMa	Big Dog	Contains Sirius (α CMa A) - 1 and Adhara (ε CMa A) - 22

Name	Genitive	Abbreviation	Translation	Remarks
Canis Minor	Canis Minoris	CMi	Little Dog	Contains Procyon (α CMi A) - 8
Capricornus	Capricorni	Cap	Goat	In zodiac
Carina	Carinae	Car	Ship's Keel*	Contains Canopus (α Car) - 2; of modern origin
Cassiopeia	Cassiopeiae	Cas	Cassiopeia	Character in Greek myth
Centaurus	Centauri	Cen	Centaur	Character in Greek myth; contains Rigil Kentaurus (α Cen A) - 4 and Hadar (β Cen AB) - 11
Cepheus	Cephei	Cep	Cepheus	Character in Greek myth
Cetus	Ceti	Cet	Whale	
Chamaeleon	Chamaeleonis	Cha	Chameleon	Of modern origin
Circinus	Circini	Cir	Compass	Of modern origin
Columba	Columbae	Col	Dove	Of modern origin
Coma Berenices	Comae Berenices	Com	Berenice's Hair	
Corona Australis	Coronae Australis	CrA	Southern Crown	Of modern origin
Corona Borealis	Coronae Borealis	CrB	Northern Crown	
Corvus	Corvi	Crv	Crow	
Crater	Crateris	Crt	Cup	
Crux	Crucis	Cru	Southern Cross	Of modern origin; contains Beta Crucis (β Cru) - 19 and Acrux (α Cru A) - 21; smallest constellation
Cygnus	Cygni	Cyg	Swan	Contains Deneb (α Cyg) - 18
Delphinus	Delphini	Del	Dolphin	
Dorado	Doradus	Dor	Goldfish	Of modern origin
Draco	Draconis	Dra	Dragon	
Equuleus	Equulei	Equ	Little Horse	
Eridanus	Eridani	Eri	River Eridanus	Contains Achernar (α Eri) - 10
Fornax	Fornacis	For	Furnace	Of modern origin
Gemini	Geminorum	Gem	Twins	In zodiac; contains Pollux (β Gem) - 17

(Continued)

Name	Genitive	Abbreviation	Translation	Remarks
Grus	Gruis	Gru	Crane	Of modern origin
Hercules	Herculis	Her	Hercules	Character from Greek myth
Horologium	Horologii	Hor	Clock	Of modern origin
Hydra	Hydrae	Hya	Hydra (water monster)	Monster from Greek myth; largest constellation
Hydrus	Hydri	Hyi	Sea Serpent	Of modern origin
Indus	Indi	Ind	Indian	Of modern origin
Lacerta	Lacertae	Lac	Lizard	Of modern origin
Leo	Leonis	Leo	Lion	In zodiac; contains Regulus (α Leo A) - 20
Leo Minor	Leonis Minoris	LMi	Little Lion	Of modern origin
Lepus	Leporis	Lep	Hare	
Libra	Librae	Lib	Scales	In zodiac
Lupus	Lupi	Lup	Wolf	
Lynx	Lyncis	Lyn	Lynx	Of modern origin
Lyra	Lyrae	Lyr	Harp	Contains Vega (α Lyr) - 5
Mensa	Mensae	Men	Table (mountain)	Of modern origin
Microscopium	Microscopii	Mic	Microscope	Of modern origin
Monoceros	Monocerotis	Mon	Unicorn	Of modern origin
Musca	Muscae	Mus	Fly	Of modern origin
Norma	Normae	Nor	Level (square)	Of modern origin
Octans	Octanis	Oct	Octant	Of modern origin
Ophiuchus	Ophiuchi	Oph	Ophiuchus (serpent bearer)	Character in Greek myth
Orion	Orionis	Ori	Orion	The hunter, character in Greek myth; contains Rigel (β Ori A) - 7, Betelgeuse (α Ori) - 9, and Bellatrix (gamma Ori) - 24
Pavo	Pavonis	Pav	Peacock	
Pegasus	Pegasi	Peg	Pegasus	Winged horse in Greek myth
Perseus	Persei	Per	Perseus	Character in Greek myth
Phoenix	Phoenicis	Phe	Phoenix	Of modern origin
Pictor	Pictoris	Pic	Easel	Of modern origin

Name	Genitive	Abbreviation	Translation	Remarks
Pisces	Piscium	Psc	Fish	In zodiac
Piscis Austrinus	Piscis Austrini	PsA	Southern Fish	Contains Fomalhaut (α PsA) - 16
Puppis	Puppis	Pup	Ship's Stern*	Of modern origin
Pyxis	Pyxidis	Pyx	Ship's Compass*	Of modern origin
Reticulum	Reticuli	Ret	Net	Of modern origin
Sagitta	Sagittae	Sge	Arrow	
Sagittarius	Sagittarii	Sgr	Archer	In zodiac
Scorpius	Scorpii	Sco	Scorpion	In zodiac; contains Antares (α Sco A) - 15 and Shaula (lambda Sco) - 23
Sculptor	Sculptoris	Scl	Sculptor	Of modern origin
Scutum	Scuti	Sct	Shield	Of modern origin
Serpens	Serpentis	Ser	Serpent	
Sextans	Sextantis	Sex	Sextant	Of modern origin
Taurus	Tauri	Tau	Bull	In zodiac; contains Aldebran (α Tau A) - 13 and Elnath (β Tau) - 25
Telescopium	Telescopii	Tel	Telescope	Of modern origin
Triangulum	Trianguli	Tri	Triangle	
Triangulum Australe	Trianguli Australis	TrA	Southern Triangle	Of modern origin
Tucana	Tucanae	Tuc	Toucan	Of modern origin
Ursa Major	Ursae Majoris	UMa	Big Bear	Big Dipper
Ursa Minor	Ursae Minoris	UMi	Little Bear	Little Dipper
Vela	Velorum	Vel	Ship's Sails*	Of modern origin
Virgo	Virginis	Vir	Virgin	In zodiac; contains Spica (α Vir) - 14
Volans	Volantis	Vol	Flying Fish	Of modern origin
Vulpecula	Vulpeculae	Vul	Little Fox	Of modern origin

* Formerly part of the constellation Argo Navis, the Argonaut's Ship.

METEOR SHOWERS

Each year on certain dates meteors begin to appear in the sky; these meteors apparently come from particular constellations. The constellation actually has nothing to do with the shower of meteors; it merely appears in the portion of the sky from which the meteors come. The meteors are caused when Earth passes through the debris left by a comet that crosses Earth's orbit. Each year Earth passes through the debris on or about the same day, since a specific day corresponds to a given part of the orbit. The line of meteors is especially dense when the stream of debris is young, and it gets wider and less dense as the stream ages—rather like the contrail of a jet airplane. A particularly dense shower is called a storm; at the height of a meteor storm from the Leonid group in 1966 about 72,000 meteors an hour could be seen—but the storm only lasted about an hour.

The connection between comets and meteor showers has been observed since 1863, when it was first postulated by US astronomer Hubert Newton. Three years later further observations by Giovanni Schiaparelli clinched the connection.

If a comet crosses Earth's orbit twice, it can leave behind two meteor showers that come at different times of the year. In some cases, the associated comet has never been observed.

Peak Date	Duration	Constellation	Meteor Shower	Associated Comet
Jan 4	2 da	Boötes, Draco, and Hercules	Quadrantids	none
Apr 22	5 da	Lyra	Lyrids	Thatcher 1861 1
May 5	6 da	Aquarius	Eta Aquarids	Halley
Jul 28	14 da	Aquarius	Delta Aquarids	
Aug 12	9 da	Perseus	Perseids	Swift-Tuttle 1862 III
Oct 21	4 da	Orion	Orionids	Halley
Nov 8	30 da	Taurus	Taurids	Encke 1786 I
Nov 17-18		Leo	Leonids	Tempel-Tuttle 1866 I
Dec 13	5 da	Gemini	Geminids	none

(Periodical References and Additional Reading: *Astronomy* 11-91, p 44)

MAJOR TELESCOPES

Telescopes were first discovered in Holland (the Netherlands) about 400 years ago. The first telescopes used lenses, familiar at the time from lenses in spectacles, to gather light and focus it. Later in the seventeenth century, scientists realized that curved mirrors could also gather and focus light. Since the light did not need to pass through the mirror (as light passes through a lens), mirrors proved to be more efficient than lenses for large telescopes.

Optical Telescopes

Year	Type	Importance
1608	Lens	Hans Lippershey in Holland applies for the first patent on a telescope
1609	Lens	Galileo builds the first optical telescope, eventually reaching 30 power
1611	Lens	Johannes Kepler introduces the convex lens, producing even greater power
1663	Mirror	James Gregory is the first to think that a reflecting telescope can be made
1668	Mirror	Isaac Newton builds the first telescope to use a mirror, rather than a lens, to collect light
1723	Mirror	John Hadley invents a reflecting telescope based on the parabola, which concentrates light at a point
1789	Lens	William Herschel builds a telescope with a 122-cm (48-in.) lens, the largest for many years
1897	Lens	Alvan Clark builds what is still the world's largest telescope to use a lens instead of a mirror
1917	Mirror	The Hooker Telescope at Mt Wilson is put into operation; it proves to be the world's largest for about 30 years
1929	Combination	Bernard Schmidt's telescopes, which combine lenses and mirrors, are first made; the Schmidt telescope becomes the workhorse of astronomy
1948	Mirror	The 5-m (200-in.) Hale Telescope, located on Mt Palomar, becomes the largest and the best on Earth
1962	Mirror	The largest telescope devoted to observing the Sun is erected at Kitt Peak in Arizona
1976	Mirror	The Soviet Zelenchuksaya Telescope becomes the largest, but various problems limit its effectiveness

(Continued)

Year	Type	Importance
1979	Mirrors	The Multiple Mirror Telescope, or MMT, uses six mirrors to obtain the equivalent light-gathering power of a 4.5 meter (177-in.) reflector
1989	Mirror	The 2.4-m (94-in.) Hubble Space Telescope becomes the first optical telescope in space
1991	Mirror	The Keck Telescope on Mauna Kea in Hawaii uses the world's largest mirror, 10 m (400 in.) in diameter with four times the light-gathering power of Hale

Radio Telescopes

Before 1931, all telescopes were optical—that is, they gathered and focused electromagnetic radiation in the range people can sense with their eyes. However, stars, planets, and other objects in the universe also produce other wavelengths of radiation. A radio telescope gathers and focuses radiation at long wavelengths, the same kind of electromagnetic radiation used for transmission of radio signals.

Year	Type	Importance
1931	Ordinary antenna	Karl Jansky accidentally discovers that radio waves are coming from space as he tries to track down sources of static
1937	Parabolic dish	Grote Reber builds the first radio telescope, in Wheaton IL
1957	Steerable dish	The 75-m (250-ft) parabolic dish at Jodrell Bank in England is the first major radio telescope
1962	Steerable dish	The 90-m (300-ft) dish at Green Bank is the first used to search for extraterrestrial life; it collapses mysteriously on Nov 15 1988
1963	Fixed dish	The largest fixed-dish radio telescope, 305-m (1,000-ft) across, is built into a valley at Arecibo PR
1970	Steerable dish	The world's largest steerable dish, 328 ft (100 m) in diameter, is installed at Effelsberg, W. Germany
1977	Several antennas	The first Very Long Baseline Interferometry begins operating at Caltech's Owens Valley Radio Observatory
1980	27 antennas	The Very Long Array (VLA) is built in a 21-km (13-mi) Y near Socorro NM

Other Types

Since both short and long wavelengths coming from space had been studied by optical and radio telescopes, it seemed likely that other wavelengths also could be detected. The problem is that Earth's atmosphere,
relatively transparent to optical and radio waves, is almost opaque to other
wavelengths of electromagnetic radiation. The solution is to put telescopes
in satellites traveling above Earth's atmosphere to detect other wavelengths.

Year	Type	Importance
1961	Gamma rays	The first telescope in space observes gamma rays that do not penetrate Earth's atmosphere
1970	X rays	Uhuru ("freedom" in Swahili), the first telescope to detect X rays, is launched into space
1972	Ultraviolet radiation	The Copernicus spacecraft incorporates a telescope designed to collect ultraviolet radiation
1978	X rays	The Einstein Observatory, which detects X rays from space, becomes one of the most productive satellite-based telescopes
1983	Infrared radiation	The Infrared Astronomy Satellite (IRAS) becomes the most successful satellite-based telescope, detect-ing possible new planetary systems and the forma-tion of stars
1990	Gamma rays	The Gamma Ray Observatory contains a more sen-sitive telescope and a wider range than previous satellite telescopes

ABUNDANCE OF ELEMENTS IN THE UNIVERSE

Astronomers and physicists believe that the big bang produced a universe that contained 80 percent hydrogen, 20 percent helium, and probably no other elements at all. When clouds of hydrogen began to collapse into small spaces as a result of gravitational forces, the nuclei of hydrogen atoms were pressed so close together that they fused into heavier hydrogen (deuterium and tritium), which in turned fused to form helium. The process released energy and the balls of hydrogen and helium became stars. The energy released balanced the force of gravity, and the stars stabilized in size. This process provided the source of energy for the Sun. In larger stars, the helium and hydrogen continued to fuse, producing nitrogen, carbon, neon, and some oxygen, as well as more helium. Even larger stars went farther and were able to produce magnesium, silicon, and iron, which have more protons and neutrons in their nuclei, and are therefore heavier than the elements mentioned previously.

Iron is the end of the line for this process, because fusing iron nuclei takes more energy than the process produces. Gravitational energy, however, causes a star's core to contract with great speed when the fusion process begins to slacken. This contraction provides the necessary energy to fuse iron, but it provides so much energy that the star blows up, becoming a supernova. In the process of exploding, the elements heavier than iron are created. Furthermore, the explosion sends all the elements from the supernova into space, creating clouds that contain all elements.

Since the big bang, some hydrogen and helium has remained as interstellar gases and some has formed smaller stars in which the fusion process does not go beyond fusion of helium. As a result, hydrogen and helium remain the most abundant elements. Carbon, nitrogen, and oxygen are the main components of the medium-sized star's fusion cycle, so they are the next most abundant elements. After iron is produced, the amount of heavier elements present through supernova explosions goes down considerably, although nickel occurs in quantities near that of iron. With few exceptions, fusion produces elements with even atomic numbers more easily than those with odd atomic numbers. Therefore, when arranged by atomic number, the abundance of the elements tends to seesaw back and forth. This effect becomes more pronounced for elements heavier than carbon.

In the following table, hydrogen is assumed to have an abundance of 1,000,000,000,000 (one trillion) units, arbitrarily chosen as a large number so that the other elements will not all be very small. Then the other elements can be assigned the following amounts, based on data from astrophysical theories, astronomical measurements, elements found on Earth, and elements found in meteorites.

Atomic No.	Element	Abundance	Atomic No.	Element	Abundance
1	Hydrogen	1,000,000,000,000	47	Silver	6.61
2	Helium	162,000,000,000	48	Cadmium	28.2
3	Lithium	3,160	49	Indium	5.12
4	Beryllium	631	50	Tin	37.2
5	Boron	758	51	Antimony	8.92
6	Carbon	398,000,000	52	Tellurium	112
7	Nitrogen	112,000,000	53	Iodine	22.4
8	Oxygen	891,000,000	54	Xenon	115
9	Fluorine	1,000,000	55	Cesium	14.5
10	Neon	551,000,000	56	Barium	120
11	Sodium	2,000,000	57	Lanthanum	12.6
12	Magnesium	25,100,000	58	Cerium	19.5
13	Aluminum	1,560,000	59	Praseodymium	4.57
14	Silicon	31,700,000	60	Neodymium	22.9
15	Phosphorus	251,000	61	Promethium	trace
16	Sulfur	22,400,000	62	Samarium	7.76
17	Chlorine	355,000	63	Europium	3.02
18	Argon	4,880,000	64	Gadolinium	11.2
19	Potassium	66,100	65	Terbium	1.74
20	Calcium	1,550,000	66	Dysprosium	12.0
21	Scandium	708	67	Holmium	2.45
22	Titanium	77,400	68	Erbium	6.92
23	Vanadium	6,610	69	Thulium	0.120
24	Chromium	240,000	70	Ytterbium	6.03
25	Manganese	132,000	71	Lutetium	0.115
26	Iron	3,710,000	72	Hafnium	2.51
27	Cobalt	56,200	73	Tantalum	0.0562
28	Nickel	891,000	74	Tungsten	3.98
29	Copper	10,000	75	Rhenium	7.94
30	Zinc	19,100	76	Osmium	25.1
31	Gallium	282	77	Iridium	15.8
32	Germanium	1,590	78	Platinum	50.1
33	Arsenic	129	79	Gold	4.63
34	Selenium	2,140	80	Mercury	5.62
35	Bromine	446	81	Thallium	3.59
36	Krypton	1,530	82	Lead	31.6
37	Rubidium	224	83	Bismuth	3.16
38	Strontium	501	84	Polonium	0.0000000316
39	Yttrium	56.2	85	Astatine	trace
40	Zirconium	316	86	Radon	0.0000000001
41	Niobium	31.6	87	Francium	trace
42	Molybdenum	75.8	88	Radium	0.000126
43	Technetium	trace	89	Actinium	0.0000000631
44	Ruthenium	27.5	90	Thorium	1.00
45	Rhodium	6.31	91	Protactinium	0.00001
46	Palladium	18.2	92	Uranium	0.0501

A S T R O N O M I C A L
R E C O R D   H O L D E R S

By Size

Largest asteroid	Ceres	Diameter: 947 km (588 mi)—but Ceres may be a comet, in which case Pallas or Vesta is largest; Pallas, like Ceres, may turn out to be a comet, so Vesta at 582 km (349 mi) may be the largest
Largest comet nucleus	Chiron	Diameter: 372 km (230 mi), but Ceres or Pallas may displace Chiron by being comets
Largest natural satellite	Ganymede (satellite of Jupiter)	Diameter: 5274 km (3276 mi)
Largest known planet	Jupiter	Size: 317.9 times the mass of Earth
Largest known object	Combination of the Lynx-Ursa Major supercluster with the Pisces-Perseus supercluster of galaxies	Size: About 700 million light-years long
Smallest known natural satellite	Leda (satellite of Jupiter)	Diameter: 15 km (9.3 mi)
Smallest known planet	Pluto	Size: 0.0017 as big as Earth Diameter: about 2250 km (1400 mi)

By Distance

Distance records that are based on red shifts depend on what one thinks the Hubble constant is. In the following, a value for the constant is used that makes the universe 20 billion years old. Many astronomers would prefer values that result in ages of only 12 or 15 billion years (see "Cosmology UPDATE," p 83).

Most distant supernova	In the galaxy cluster AC118	Distance: About 5,000,000,000 light-years from Earth
Most distant galaxy	4C41.17	Distance: About 15,000,000,000 light-years from Earth
Most distant object	Quasar PC 1247 +3406 as reported on Aug 26 1991, found by Donald Schneider of the Institute for Advanced Studies, Maarten Schmidt of Caltech, and James Gunn of Princeton	Distance: About 18,000,000,000 light-years from Earth; it has a red shift of 4.897, the greatest ever measured
Most distant star in Milky Way galaxy	Unnamed star in Virgo	Distance: About 160,000 light-years from Earth (about as far as the Magellanic Clouds, but in a different direction)
Most distant star ever observed	Supernova in galaxy cluster AC118 observed Aug 9 1988 by Hans Ulrik Norgaard-Neilsen	Distance: 4,000,000,000 to 5,000,000,000 light-years from Earth

Nearest star	Proxima Centauri	Distance: 4.22 light-years, or 40,000,000,000,000 km (24,800,000,000,000 mi)
Closest known approach by an asteroid	1991 BA	Distance: within 170,000 km (105,000 mi), about half the distance to the Moon
Farthest asteroid from sun	1991 DA, found by Australian astronomers on Feb 18 1991	Distance at aphelion: 22 astronomical units, or 3.3 billion km (2046 million mi); beyond orbit of Uranus

By Brightness

Brightest object in universe	Probable gas cloud in Ursa Major found by Michael Rowan Robinson of Queen Mary and Westfield College in London	Brightness: More than 300 trillion times as luminous as the Sun and more than 20,000 times as luminous as the Milky Way galaxy
Brightest-appearing star at optical wavelengths	Sirius A	Brightness: Apparent magnitude −1.47; absolute magnitude +1.45, about 26 times as bright as the sun; other stars are actually brighter, but much farther away
Brightest star in Milky Way in reality at optical wavelengths	Cygnus OB2 #12	Absolute visual magnitude of −9.9; would be first magnitude visually from Earth if not for dust
Brightest star in Milky Way including all wavelengths	HD 93129A	Absolute bolometric magnitude (including all wavelengths of electro-magnetic radiation) of −12.0, about 5 million times as luminous as the Sun
Brightest object	Quasar BR 1202-07	Brightness: Absolute magnitude of −33 is about 1,000,000,000,000,000 times as bright as the Sun
Faintest object outside the solar system	Unnamed probable brown dwarf discovered in 1988 by Michael R.S. Hawkins	Brightness: about 1/4000 that of the Sun

By Other Measures

Oldest star in Milky Way	CS 22876.32, recognized by Timothy Beers and coworkers at Michigan State University in 1990	Age: Between 15 and 20 billion years old
Oldest observatory	Tomb at Newgrange, Ireland, aligned with the sun at the winter solstice	Age: About 5150 years old
Most humans in space at one time	Starting on Dec 2 1990 and lasting for 9 days	Total: 12 (7 aboard US space shuttle *Columbia*; 3 on *USSR Soyuz TM-11*; and 2 on USSR space station *Mir*)

(Periodical References and Additional Reading: *Science* 9-6-91, p 1094; *Astronomy* 11-91, p 28)

THE STATE OF EARTH'S SPACE PROGRAMS

During the 35 years that Earth people have been putting objects and themselves in the regions beyond Earth's atmosphere that we call space, an amazing amount has been accomplished. Some of our hardware has left or is about to leave the solar system (depending on how one defines the solar system). Machinery has gone to the neighborhood of every planet but Pluto and people have walked on Earth's Moon. Daily life all over our planet has been changed by communications, weather, and positioning satellites, while science has been vastly enriched by new ways to study the universe and Earth itself.

Space and Politics

From the advent of Sputnik in 1957 to the future of Mars exploration in the 1990s, space programs around the world have been inextricably tangled up with politics. After the Soviet Union launched the first artificial satellite, cold war politics demanded that the United States catch up and do something more impressive quickly. Great Britain, France, China, and Japan all wanted to get into the act, not so much because of actual need at first, but to show that they are first-class powers. India wanted to demonstrate that even a Third World nation could take part in the space effort.

Today the politics has shifted to more practical issues. In the United States, the president talks of Mars, but there is not enough money for some of the space programs already being implemented. The original space-faring nation, the Soviet Union, has crumbled into pieces, and even while intact, it was economically pressed. Japan, which has plenty of money, has finally encountered a technical field in which its wizardry fails; Japanese space efforts in the 1990s are reminiscent of US efforts in the 1950s, complete with rockets that fail to get very far from the pad. Gradually, the European Space Agency is coming to be a leading player, probably because it combines the resources of seven nations. If Russia and other former Soviet states survive as a space power, and if the United States is to accomplish some of its more ambitious goals, cooperation between the two is probably the only way to success.

Cooperation between the European Space Agency, individual European nations, Japan, the successors to the USSR, and the United States seems to be a major trend. Very few missions of the early 1990s and fewer still planned for the mid-1990s have only a single sponsor.

The space-faring nations have primarily been the Soviet Union, the United States, the nations of Western Europe, and Japan. A few additional players are Canada (with more than a half-dozen scientific satellites), Australia, China, and India. Because of its spread-out geography, Indonesia is a major user of communications satellites.

The US space program is analyzed largely in the individual articles that follow this one. Here is a summary of the state of the other major nations.

USSR or Whatever

At the end of 1991 it appeared certain that the central government of the USSR had lost all power and that a loose commonwealth or community of successor states had replaced it. What will happen to the space program is not one of the high-priority unanswered questions. The following is written in the assumption that the new confederation, or perhaps Russia, will continue some sort of program. For want of a definite name at this time, the entity responsible for that program is called by the familiar, if misleading, name USSR.

Although development of the US space station remains problematical, the USSR continues to maintain its *Mir* space station complete with cosmonauts and visitors from around the world. When it seemed in 1991 that all constituent republics of the Soviet Union were declaring their independence, people joked that only the two cosmonauts in *Mir* had not defected. There is talk of an interchange in which a US astronaut would spend some time on *Mir* while a USSR cosmonaut would try out the shuttle (the USSR equivalent to the shuttle, called *Buran* or "Buzzard," seems to be mothballed after initial tests without human pilots).

Since Jun 1990, when the Kristall module became the third addition to the core *Mir* station, *Mir* has been the largest and most sophisticated space station ever flown. Consisting of the core, the three attached modules, and two vehicles that travel from *Mir* back and forth to the ground, one for cosmonauts and another for supplies, *Mir* is 32.5 m (82.5 ft) wide, 44 m (112 ft) long, and has a mass of 88 metric tons (97 short tons). (Earlier stations were the US *Skylab* and the USSR Salyut series.) Plans for 1992 and 1993 include two additional modules, which will bring the space station's total mass to about 115 metric tons (about 125 short tons).

The cosmonauts live in the core module and work in the others. The first module to be attached, Kvant 1, contains mostly astronomical instruments; Kvant 2 is the setting for biological experiments; and Kristall is used for what its name suggests, crystal growing. The crystal growing has proved to have considerable commercial value, producing protein and gallium arsenide crystals for sale and providing a setting for foreign corporations to order their own crystals grown. Another source of income has been the purchase by foreigners of trips to and stays on *Mir*, reputedly for about $10 to $12 million a visit.

In a surprise, given the state of the Soviet economy, the Center Institute for Machinery Research of the USSR's Ministry of General Machine Building, the agency that builds almost all USSR spacecraft and launch vehicles, announced that they will replace *Mir* with an advanced *Mir 2* in 1994 and maintain the new space station through the end of the 1990s. In the meantime, an earlier USSR space station, *Salyut-7,* was in the news when it reentered the atmosphere and burned up in Feb 1991.

Despite the announced development of *Mir 2,* rumors suggest that the Soviet space budget was cut by 20 percent in the early 1990s. A noticeable effect is that the number of annual launchings in the 1990s is half what it was in the 1980s. Also, not only have tests of the Russian shuttle stopped, but also tests on the giant Energia rocket have been suspended.

A major plan for a USSR Mars mission ran into budget difficulties in 1991 and was reorganized into two separate operations. The first, scheduled for 1994, will include an orbiter that will launch some hard-landing probes at Mars. The second, two years later, will also include an orbiter, but the main feature will be a soft-landing vehicle carrying two major experiments, a Martian land-roving vehicle and a French balloon that will travel by day and land by night. Soviet and US scientists and engineering students from Stanford University proposed on Jun 26 1991 a detailed plan to follow this with a US-USSR mission to send people to Mars using Soviet rockets and American money.

In another unusual cooperative effort, the United States arranged to purchase a compact Soviet nuclear reactor, an advanced version of a type used in two experimental USSR satellites. The Topaz 2 reactor is designed to produce as much as 10,000 watts of power for as long as 5 years. The United States has said that its intention is to study the reactor rather than to use it as a power source. Because of concern over radioactivity in the event of an accident, nearly all US spacecraft are solar-powered. The Soviet Union has used nuclear reactors primarily on spy satellites, but stopped after a 1988 incident when it appeared for a time that such a satellite was out of control.

The reactor purchase may be the beginning of many such deals. Although the USSR has been actively trying to sell space-related devices and launches to the United States (and anyone else), it has had little success. The first major US satellite to be launched by a Soviet rocket, an ozone-measuring device called TOMS, was lifted into orbit on Aug 15 1991. After the collapse of the Soviet economy, both the need to make sales to keep the space program afloat and US willingness to make space-related purchases improved. According to *New York Times* articles on Sep 3 1991 and Nov 4 1991, everything, even the *Mir* space station, is for sale. Items of particular interest to the United States include special alloys, a magnetic rocket engine, and plutonium 238 for use in powering space probes.

Europe

In the early 1990s, Europe was a participant in virtually all scientific missions except for Magellan, COBE, and ASTRO. This pattern is expected to continue for the foreseeable future. Most of these missions, but not all, were launched by the United States or the USSR.

Europe participates in several ways. Both Germany and France often act apart from other European nations. For example, ROSAT was launched by Germany, while GRANAT uses French instruments aboard a USSR-launched satellite. Since 1980 much of the European effort has been channeled through the European Space Agency (ESA), a consortium of Austria, Belgium, Denmark, Finland, France, Germany, Ireland, Italy, the Netherlands, Norway, Spain, Sweden, Switzerland, and the United Kingdom. Sometimes several members of ESA participate in a project without the others, as is the case with the Mars Observer, which also includes participation by the United States and USSR. Of the imminent missions, only the Infrared Space Observatory, scheduled for 1993, will be a purely ESA project.

At the end of 1991, however, the economic belt-tightening that has hit the USSR and the United States appeared to also affect the European space effort. While no programs were stopped completely, some seem likely to be slowed down in an effort to improve the equivalent of cash flow.

Japan

Japan has long been determined to become the fourth major player in the space game. Although it launched its first satellite as early as Feb 11 1970 (after a string of failures), its first noticeable impact came in 1986 when it launched two spacecraft to study Comet Halley, with the United States conspicuous by its absence. Much of Japan's space program is directed by the Institute of Space and Aeronautical Sciences (ISAS), which produced the Mu rocket that Japan has been using for its launches.

The major effort in the early 1990s was a mission to the Moon as a feasibility study for future experiments. A Mu rocket on Jan 24 1991 carried a vehicle named Hiten into an elongated orbit about Earth. On Mar 19, while Hiten was fairly near the Moon, it released a small vehicle that used its own propellant to carry it into lunar orbit. This is a new method of getting a vehicle to the Moon, but one that the Japanese plan to use for future missions that will contain instruments. After the smaller device went into lunar orbit, the Japanese took a photograph of the Moon from it and named the new lunar satellite Hagoromo. The only experiment aboard Hiten is a German device to count micrometeorites, and this only because Japanese engineers found that they had a little extra room and did not want it to go to waste. After dropping off Hagoromo, Hiten changed orbit as a maneuver that will be needed for another Japanese experiment planned for 1992.

That experiment will be Geotail, to be launched by the United States and to go into a very large Earth orbit, one that will carry the satellite beyond the orbit of the moon. The intent of Geotail is to analyze Earth's magnetic tail. Japan is also planning to land penetrators on the Moon to study another comet's tail and to develop a Venus orbiter.

Japan, like other nations, is not going it alone in space. One of its major efforts is expected to be a $2-billion module for the US space station *Freedom*. When it briefly appeared in 1991 that *Freedom* was going to be abandoned by the US Congress, Japan complained directly to the US vice president and the secretary of state. Japan indicated that it would not participate in other joint scientific projects, such as the Superconducting Supercollider, if *Freedom* were to be cancelled. It is likely that this attitude affected the Senate when its relevant committee voted to restore funding for the space station.

Japan is having great difficulties in producing a more powerful launch vehicle, one that could even carry a miniature space shuttle up to its *Freedom* module. Three times the new rocket has blown up in tests. The explosion on Aug 9 1991 also resulted in the death of one of the technicians working on it. Japan is handicapped in this area by having no experience in developing rockets for ballistic missiles or even in developing large jet engines, which use some of the same technology. The first launch using the new vehicle is currently scheduled for 1993.

Rethinking the Basic Ideas

In the early years of space exploration, the Soviet Union succeeded with fairly simple systems launched from very powerful vehicles. The United States, instead, developed sophisticated miniature devices. For a time, it appeared that the US approach had pulled far ahead, but that is less clear today. Sophisticated equipment is breaking down at an alarming rate, partly because, as a Sep 23 1991 US Government Accounting Office report stated, the US National Aeronautics and Space Administration (NASA) has no uniform testing policy. The original plan for the US space station had to be cut back because, among other things, it would have been too complicated to maintain. Some desirable missions are beyond the size of the launch vehicles available to the United States.

Typical of the rethinking going on is the debate over the Earth Observing System (EOS). EOS was planned as "big science," which is to say that it will cost so much that scientists not directly involved in the program are complaining. The mission of EOS is to study Earth from space. The original concept was for six 16-ton "platforms" to orbit Earth. Each platform would contain twelve to sixteen sophisticated experiments. In addition to the 20-m (50-ft) platforms, smaller satellites would be used to obtain additional data. The $30-billion program would also include ground-based observations to complement data from space. In this vision, the first platform would be launched in 1998 and the project would continue for the next 20 years.

On Aug 14 1991, as EOS was about to move from planning into development, Lennard A. Fisk, associate administrator for space science at NASA, held a news conference to discuss a different plan altogether. Mindful of the problems of the Hubble Space Telescope and other scientific satellites of the 1990s, some scientists have called for replacing the six platforms with a flock of small single-purpose satellites. Fisk said that it would not make sense to put each experiment on a separate satellite, since it is often essential to make sure that data are being obtained from the same environment. It might be possible to replace the six platforms with perhaps eighteen smaller satellites that each would carry fewer experiments than the platforms. On Sep 23, a presidential review panel studying EOS noted that the US Air Force was planning to launch modified Atlas-Centaur rockets from its West Coast facilities. The panel recommended that the first two large platforms be replaced with smaller ones carrying from two to nine experiments each, with the small platforms launched by the Air Force rockets.

In Aug 1991, Fisk also replaced a planned Orbiting Solar Laboratory with two smaller satellites, one to study solar X rays and the other to study Earth's magnetosphere. In addition to providing insurance against failure, the new plan would cost about $400 million instead of $750 million for a single large satellite.

Clustering many experiments on a single platform not only opens the way for disaster if the platform fails, it also means that many experiments have to be designed for less than optimum conditions. For example, although an experiment would work best from a height of 600 km (100 mi), it might be on a platform with experiments that operate most successfully at twice that height, leading to various compromises.

In Oct 1991 the science advisory committee to NASA's Office of Space Science and Applications submitted a letter to NASA recommending "small innovative missions" that "provide frequent access to space by each discipline" as the place for NASA to spend its money first. Larger or intermediate-sized missions should be a second priority. The letter did propose, however, that the ongoing plans should be completed first, to be followed by the small missions, with larger projects after that.

One group that has based its entire approach on the new "small is beautiful" thinking is Surrey Satellite Technology (SST), an organization that is partly owned by the University of Surrey in the UK and partly privately funded. Its backpack-sized microsatellites weigh 50 kg (100 lb) or less and are small enough to hitch rides on Delta or Ariane rockets that are in the process of launching much larger spacecraft. So far SST has launched five such microsatellites with three more in development. SST satellites have a dual function: the education of graduate students at the University of Surrey and the pursuit of commercial endeavors.

Currently, each launch costs only $250,000 for a ride on an ESA Ariane. So far there has been room not only for microsatellites from SST, but also for those from commercial companies such as Orbital Science Corporation of Fairfax VA as well as those developed by academic institutions. At that price per microsatellite, ESA does not make money on the launches, but the agency feels that the launches are good public relations. If the current boom in microsatellites continues, however, ESA will run out of room (NASA already has stopped launching microsatellites) and may need to become selective or to charge higher prices or both.

Putting Life on Mars

It now appears that there is no life on Mars, but US President Bush has proposed changing that situation in the next century. In Jul 1989 he set two goals: a return of humans to the Moon and a human expedition to Mars by 2019, the fiftieth anniversary of landing astronauts on the Moon. This proposal came to be called the Space Exploration Initiative, with the Moon return set for 2005, the Mars expedition to occur by 2014, and $15 million in the federal budget for 1992 to start planning. In 1990 a committee of twenty-seven experts of all stripes was appointed to fill in some of the outlines. They reported on Jun 11 1991 that a more powerful launch vehicle and research on improving the ability of humans to live and work in space were needed. Although the space station *Freedom* was originally planned to perform research of this type, current scaled-down plans make it unlikely that much of the needed research can be undertaken. On the other hand, some of the research could be carried out in the USSR space station *Mir* or perhaps in the planned *Mir 2*.

In the meantime, a more modest Mars expedition, without humans, is forthcoming from the US-launched Mars Observer, planned for a Sep 1992 launch, only two years behind schedule. It will begin orbiting Mars in Aug 1993 if all goes well. Mars and Earth are in the right positions for travel between them every two years, so missing 1992 would mean aiming for 1994. Like many of the forthcoming space projects, it will feature international cooperation, including Germany, Austria, France, the USSR, and the UK.

NASA is also considering a group of Mars missions for the 1998 to 2003 launch windows. These would each be fairly small penetrators, something like the Vikings of the 1970s, which successfully broadcast data from the Martian surface for several years.

Although the future of the space program in the former USSR is clouded, that nation has had an active Mars exploration program since 1960 and has announced support for a manned expedition to Mars in the mid-1980s.

Unfortunately, a lot of the USSR missions to Mars have been failures, but the nation has kept trying. Just before pieces of the Soviet Union began to break off, the central government's space agency announced that its planned 1994 Mars mission would become two separate trips, an orbiter and hard-landing probe with a 1994 launch and a soft-lander 2 years later. Part of the USSR plans include a possible return of material from Mars by a robot expedition to be launched in 1998 or 2001.

A promising part of all of this is that there has been increasing cooperation between the USSR and US in several space projects—all the recent and planned Mars missions as well as the Magellan Venus mission. Indeed, USSR President Gorbachev has since 1987 frequently listed joint USSR-US exploration of Mars as one of the ten national goals of his country.

What seems to be an outside shot at Mars is the Japanese effort, but they may surprise everyone. There is some talk of buying launch time on USSR vehicles as early as 1996 to send Japanese astronauts to Mars. Less likely is the development of Japanese vehicles in time for a human mission two or four years later.

Launch Vehicles

A key problem in the space programs of the early 1990s has centered around the launch vehicle, or rocket. The USSR clearly has the most powerful rockets, especially the Energia, the current world champion. The Soviets seem to have abandoned their space shuttle, however. At one point the United States came to rely completely on the space shuttle, but now space officials and scientists wish they had the launch vehicles of the 1960s and 1970s back. The European Space Agency has the fairly reliable Ariane, but it too needs a larger vehicle. Japan is struggling to develop a bigger rocket engine. It hopes eventually to have a Titan-class rocket, called the H-2, that will be able to launch 2-ton satellites. Launch vehicles have not posed as many problems since the late 1950s.

The US military, according to reports in Apr 1991, is planning to improve its launch capability with Timberwind, a Star Wars program using nuclear reactors. The basic idea of any rocket is Newton's third law of motion. In a chemical rocket, a controlled explosion pushes hot gases out a nozzle, which in turn pushes the rocket in the opposite direction. A nuclear reactor could be used to heat gases to higher temperatures (about 2500° C or 4500° F), forcing the molecules to move faster out of the nozzle. Thus, the rocket would move faster in the opposite direction; or, a more massive rocket could move at the same speed. While various tests have been conducted on parts of Timberwind, it is unlikely that it will become a reality in the 1990s. For one thing, the military objective, which is to lift payloads quickly in response to a nuclear attack, seems less and less reasonable as the political situation changes.

However, there are so many advantages to nuclear propulsion that NASA, joined by both the US Departments of Defense and Energy, established a Nuclear Propulsion Office at the Lewis Research Center in Ohio late in 1991. It will study both the thermal propulsion system described above and an alternative system using nuclear reactors to develop electricity. The electricity could then use electromagnetic repulsive forces to propel ions through the nozzle. NASA is interested because the higher thrust of nuclear rockets could cut the time of interplanetary travel, saving money and improving conditions for human travel in space. A trip to Mars might be cut from as much as 500 days to as little as 150 to 300 days. NASA spokespersons said that a nuclear propulsion system might be available as soon as 2010.

In Summary

If there is enough money available, scientific programs will proceed at a quicker rate, similar to that of the 1970s rather than the slow pace of the 1980s. But, as the following pages detail, these programs will have to be better planned and produced to succeed. If the USSR drops out of space exploration and research, a major player will be off the field and there is little likelihood at present that the European Space Agency or Japan will take up the slack.

What no nation will give up are the practical satellites used for communications, weather prediction, and other purposes. These have become as essential as telephones and paved roads.

The debate over whether humans or robots are best suited for space will continue. Money is again an issue. If there is enough money available, almost certainly political issues will keep humans in space. When short of money, robots will do.

(Periodical Sources and Additional Reading: *Science* 7-13-90, p 115; *New York Times* 1-1-91, p I37; *New York Times* 1-3-91, p A1; *New York Times* 1-7-91, p B8; *Planetary Report* 7/8-91, pp 3, 4, & 24; *Discover* 8-91, p 66; *Science* 8-23-91, p 848; *New York Times* 9-3-91, p C1; *Science News,* 9-7-91, p 156; *New York Times* 9-16-91, p A1; *New York Times* 9-24-91, p A32; *Science* 9-27, p 1481; *Science News* 9-28-91, p 198; *Astronomy,* 10-91, p 26; *EOS* 10-1-91, p 425; *EOS* 10-8-91, p 441; *New York Times* 11-4-91, p A1)

SCIENCE IN SPACE

PROBLEMS AND SOME RESOLUTIONS

No one ever said that science from space would be easy. In the beginning, the first space probes aimed at the Moon and Mars missed, the first US probe to another planet landed at the bottom of the Atlantic, and more satellites were blown up than reached orbit. Then things stabilized for a long time. Today, most—but not all—satellites and space probes are aimed more-or-less correctly and make it into space. However, possibly because of the increasing sophistication of modern equipment used, all sorts of strange things can happen. Here is a brief report on the problems of the early 1990s (also see "Hubble Trouble," p 26)

Hipparcos

Launched by the European Space Agency in Aug 1989, Hipparcos was supposed to determine accurate positions and motions of stars from a circular orbit. Its booster engine failed, however, and it wound up in a highly elliptical orbit. It can still fulfill part of its original mission from that orbit and will study more than 100,000 stars during a 2½-year mission.

Scientists worried for another reason. The length of time that Hipparcos was eclipsed by Earth was much greater than planned. The satellite had been designed to operate on solar panels while in sunlight and to store some power for eclipses, but only enough for brief periods of darkness. Because of the relationship of Earth, Sun, and orbit, the satellite encountered a series of long eclipses in Mar 1990. When on Mar 16 Hipparcos survived a 105-minute eclipse—with barely 5 minutes of battery power left—it appeared that everything would be all right. That was about as long an eclipse as is expected.

Venus Briefly Unobserved

The Magellan space probe, launched May 4 1989, was the first interplanetary explorer in a dozen years. It was also the first ever to be launched from the space shuttle, and it marked the return of NASA to the study of the planets, with several other missions planned for the 1990s. Magellan entered Venus's orbit on Aug 10. It was supposed to make a detailed radar map of Venus starting Sep 1 1990, but its start was delayed by two mysterious losses of radio contact, one for 13 hours and the other for 21, after less than two orbits of the craft about Venus. In both cases, something caused the spacecraft to fix on the wrong guide stars or otherwise to point in the wrong direction. NASA officials were at a loss to explain the problem but seemed to have solved it nevertheless after a brief period of control from Earth. Contingency plans were instituted, and when a third episode of mispointing occurred on Nov 15 1990, the spacecraft automatically corrected itself in about 40 minutes. By the time of a fourth "walkabout," as NASA scientists took to calling the events, the operation was routine.

Finally, in Jun 1991, scientists accidentally discovered the root of the problem when a piece of hardware malfunctioned during a ground test of equipment. They learned that a software design error caused Magellan to be subject to loss of control if it received certain commands while it was switching from routine operations to specialized tasks. Not only did the computer suddenly go into an infinite loop, it also wiped out the memory of what had just happened. The hardware problem in the ground equipment prevented the memory loss, enabling scientists to understand the event. They proceeded to reprogram to correct the difficulty.

The next problem was a defect in a tape recorder that collects the data for broadcast to Earth. Earth-based commands moved all the data onto another tape recorder, but there would now be three small gaps in data in each orbit. It was hoped that even those gaps could be filled by using a computer algorithm to restore lost data.

The Galileo Antenna Problem

The most common problem faced in recent scientific experiments in space has been that an antenna has failed to point in the right direction or failed to open up properly. Over a period of 17 months in the early 1990s, six major spacecraft costing $5.3 billion had antenna malfunctions, degrading the flow of information back to Earth. Perhaps the most alarming has been the failure of the main antenna on the Galileo space probe to open. Still a problem as of this writing, engineers have until 1995, when Galileo is due at Jupiter, to deploy it. In the meantime, two smaller antennas can be used as limited data links.

The Galileo antenna problem may never have occurred were it not for various delays in development and launch plans that pushed back the mission from a 1982 start to a 1989 start. After the final delay, caused by the long shutdown of the US space program after the *Challenger* disaster, five trips eventually affected the antenna.

The first four were the trips between Florida and California's Jet Propulsion Laboratory in Pasadena, where Galileo waited out post-*Challenger* delays in the shuttle program. During those truck journeys, the 5-m (16-ft) main antenna lay on its side and bounced about a bit. It now appears that the jostling caused the lubricant to rub off from the three or four ribs that were at the bottom. As a result, small pins that keep the antenna furled until it is ready for use got stuck. On NASA communications satellites, such pins are regularly replaced and relubricated just before launch, but this step was not taken for Galileo.

The fifth trip was that of Galileo through space. Safety considerations after the *Challenger* disaster had resulted in a reduced power for Galileo's engine, making the direct 18-month trip to Jupiter no longer feasible. Scientists worked out a route that involved three gravity assists from planets, including one from Venus. But Galileo was not designed to travel as close to the Sun as the Venus station, so other changes had to be made to protect the craft. One of the changes was to keep the main antenna furled like an umbrella until after the Venus flyby. When an attempt to open the antenna was made after the first Earth gravity assist, however, the antenna stuck. It is not known whether the passage close to the Sun further complicated the problem caused by the stuck pins, but it apparently did not help.

Properly open, the antenna forms a paraboloid to direct data toward a specific target; partially open, it is worthless. Initial efforts to fix the antenna involved heating or cooling it by turning it toward or away from the Sun. In Aug a major maneuver to freeze the antenna by putting it in the shade for 50 hours was attempted, but it failed to get the antenna cold enough to unstick it. The problem was still not solved in time for the Oct 29 1991 flyby of asteroid 951 Gaspra. The next try at correcting the antenna problem was scheduled for Dec 1991, when the spacecraft would be farther from the Sun.

The plan is to continue to try more cooling maneuvers if the Dec attempt is also unsuccessful. A second try at freeing the antenna through heating may also be implemented in Dec 1992, when Galileo has returned to the vicinity of Earth. If none of these efforts prevail, then the spacecraft's handlers will try to use mechanical force to open the antenna. The main difficulty if the antenna cannot be fixed will be with data from the Jupiter orbiter, which is designed to make high-resolution images of Jupiter's moons. A probe into Jupiter's atmosphere will not be affected as it was originally designed to transmit data at a slower rate. In the event that the antenna cannot be opened, scientists are considering sending a second spacecraft to orbit Jupiter to act as a relay for data back to Earth. Such a space vehicle would have to be launched at least by early 1993 to get to Jupiter at the same time as Galileo.

The Gamma Ray Observatory Antenna Problem

In a troubled time for NASA and the US shuttle program, everything about the Apr 5 1991 countdown for the launch of the shuttle *Atlantis* with the Gamma Ray Observatory (GRO—renamed the Compton Gamma Ray Observatory on Sep 23 1991 in honor of Nobel Prize winner Arthur Holly Compton) aboard went almost perfectly. Three days later, astronauts tried to open the high-gain antenna, GRO's primary communications link. When the antenna would not open while GRO was still at the end of the shuttle's remote manipulator system arm, NASA and the astronauts did just what anyone would do. They rocked it, they shook it, and, when that failed, they went out and kicked it. Astronauts Jay Apt and Jerry Ross made an unplanned space walk and finally unstuck the antenna. After that, all was well.

The Ulysses Antenna Problem

Ulysses is designed by NASA and the European Space Agency to report back on the part of the Sun we cannot observe from Earth, the polar regions. Since Earth's orbit is roughly parallel to the solar equator, the poles can be seen from Earth obliquely at best. The Ulysses spacecraft is to head out to Jupiter, where careful planning should cause the Jovian gravity to whip it into a polar orbit over the Sun, something difficult to do from Earth. On the trip to Jupiter, for example, Ulysses started out with a push of nearly 30 km per sec (18 mi per sec) from the motion of Earth. But that motion is not in the proper direction to put a satellite into a nearly solar-polar orbit. Even with the gravity assist from Jupiter, Ulysses will orbit at an inclination of about 80° (90° would be perfectly polar).

Ulysses was launched Oct 6 1990, more than 30 years after it was first discussed and more than 10 years after the project received formal approval. It will conduct eleven investigations, of which the most important is of the Sun, its magnetic field, its corona, and the solar wind. One of several mysteries about the Sun is why the corona is hotter than the Sun's surface. One unusual experiment will be an attempt by B. Bertotti to use any sudden deviations from orbit to detect gravity waves.

About 4 hours after an antenna on the Ulysses satellite was raised, the spacecraft began to loop around like a child's top shortly before it falls over and skitters across the ground. This motion is called nutation although it is basically the same as precession. Perhaps astronomers decided to call this motion of the spacecraft *nutation* rather than *precession* because nutation of Earth is both more rapid and slighter than precession (for the planet, nutation is a motion caused by the gravitational attraction of the Moon, but precession is caused by a combination of the gravitational attractions of both the Sun and Moon). By any name, the slight nodding is not desirable in a scientific satellite since it interferes with observations.

For the spacecraft, the nutation was caused by heating and cooling of the antenna. When the motion became great enough, an on-board damper managed to stop it. Still, its maximum angle of 6° was enough to interfere with observations. Fortunately, as the distance from the Sun increases, the nutation decreases. Also, during periods when the offending antenna is shaded by the satellite, the nutation stops. Engineers are hoping to take steps to ensure that this last condition always prevails, but it is not entirely certain that their efforts will work.

Ulysses is not the only spacecraft to orbit the Sun. Several early space probes that missed targets such as the Moon or Venus went into solar orbits. The oldest working space probe, Pioneer 6, was deliberately launched into a solar orbit on Dec 16 1965 and continues to return useful data.

ROSAT Blinded in One Eye

ROSAT, a satellite mapping the X ray and extreme ultraviolet universe has four "telescopes" aboard. In late Jan 1991 charged particles from a solar flare apparently were the cause of problems in the attitude control system that led to the loss of one of these devices. When the satellite spun out of control, a device called a Proportional Counter, installed by German experts, accidentally aimed at the sun and was burned out. Fortunately, one of the other devices aboard the satellite is a near twin of the lost one, so scientific work will be only slightly impaired.

ASTRO Almost Foiled by Lint

The plan was to observe astronomical bodies for 10 days using a group of three ultraviolet telescopes and one X-ray telescope. Both ultraviolet and X-ray radiation are blocked from Earth by the atmosphere, so space-based observation is the only way to proceed. Therefore, the telescopes, a configuration called the ASTRO observatory, were placed aboard the space shuttle. Although ASTRO cost $150 million to develop and was intended originally for more than one flight, it turned out to have some built-in problems. The automatic guide system for the ultraviolet telescopes was too sensitive and had to be abandoned, so its functions were controlled by the astronauts manually at first, using computers to guide them. Then both computers failed as a result of clogging from lint from ASTRO that started floating around the shuttle *Columbia*. After that, ground-based computers radioed up the commands necessary to aim the ultraviolet telescopes (the X-ray telescope was out for a while as well, but was repaired in space). Some planned targets were also lost to launch delays. As a result, ASTRO was able to observe only 135 of 524 planned targets.

(Periodical References and Additional Reading: *Astronomy* 1-90, p 46; *Astronomy* 4-90, p 30; *Astronomy* 5-90, p 18; *Astronomy* 7-90, p 26; *Science News* 8-25-90, p 116; *Science* 8-31-90, p 977; *Science News* 9-1-90, p 135; *Science* 10-5-90, p 27; *New York Times* 11-30-90, p A20; *Science News* 12-1-90, p 340; *Science News* 12-8-90, p 356; *New York Times* 12-11-90, p C1; *Science News* 12-15-90, p 372; *Dis-*

cover 1-91, p 36; *Science News* 1-5-91, p 10; *Science News* 2-24-91, p 119; *Astronomy* 3-91, p 30; *Astronomy* 4-91, p 39; *Science* 4-12-91, pp 247 & 260; *New York Times* 4-17-91, p A16; *Science* 5-3-91, p 638; *New York Times* 5-21-91, p C9; *New York Times* 5-30-91, p B11; *Science* 6-7-91, p 1373; *Science News* 6-8-91, p 356; *EOS* 6-18-91, p 267; *Planetary Report* 7-8-91, p 25; *Astronomy* 8-91, p 24; *Scientific American* 8-91, p 24; *Science News* 8-3-91, p 79; *Science* 8-23-91, p 846; *Science News,* 9-7-91, p 156; *Science* 9-27-91, p 1492; *Science* 10-18-91, p 381; *New York Times* 10-27-91, p L18)

SUCCESSES FROM SPACE

Despite all the problems, a great deal of good science was accomplished from space-based experiments. Some of this is reported elsewhere (see "Hubble Trouble," p 26; "Venus Laid Bare," p 48; "Weird Volcanoes of the Solar System," p 51; "News from the Outer Planets," p 53; "The Great Annihilator," p 72; and "The Universe Is Too Much Alike," p 78).

Galileo's Pictures Along the Way

Just as a vacationer sometimes finds that there is more to see on the way to Serendip than one would expect, the convoluted path Galileo followed to Jupiter provided a lot of interesting material along the way. The Feb 10 1990 flyby of Venus was notable for its maps of the atmosphere below the clouds. Passing by Earth and its Moon on Dec 8 1990, Galileo made some excellent snapshots of both. Of most scientific interest were measurements of the magnetosphere and of the composition of the Moon. On its return to Earth's vicinity on Dec 8 1992, Galileo will be able to observe regions of the Moon that have never been seen previously.

About 150 photographs were collected at asteroid 951 Gaspra as Galileo passed 1600 km (1000 mi) from the planetoid at 29,000 km (18,000 mi) per hour on Oct 29 1991. The asteroid was 330 million km (205 million mi) from the Sun, in the asteroid belt between the orbits of Mars and Jupiter. A few photographs were returned to Earth for immediate processing and were shown starting in mid Nov. Unless the main antenna is fixed before then, most of the photographs will be stored for the transmission in Nov 1992, just before the Dec gravity assist from Earth. At that time the data can be transmitted by the two small antennas. The small antennas transmit data at a rate of 10 bits per second as compared with the planned 134,000 bits per second for the main antenna. Several thousand images of Gaspra were originally planned before the antenna problem developed.

The first photographs tended to confirm what had been deduced from observations from Earth. Gaspra is a small asteroid, no more than 20 km

(12 mi) long, 12 km (8 mi) across, and 11 km (7 mi) wide. It is shaped something like a potato or a lumpy banana. Its skin is covered with pockmarks from bombardment by smaller bodies, but its overall look is that of a survivor of numerous collisions with bodies of all sizes. The last major collision, judging from the number of craters on its surface, took place about 300 to 500 million years ago.

ASTRO a Success Despite Losing Almost Half Its Targets

The ASTRO mission certainly seemed like a hard-luck operation if ever there was one. It was originally planned to have been the next mission in 1986 after a *Challenger* flight. As a result of the *Challenger* blowing up and other shuttle problems (its launch was scrubbed four times during 1990), it did not reach space until Dec 1990. After many complications in space, it finally returned to Earth with 135 observations and the knowledge that the additional times the equipment was to have been lofted had been dropped from NASA's schedule. Indeed, the mission had been originally named ASTRO-1, but it seemed prudent to drop the "-1" with "-2," "-3," "-4," and "-5" cancelled. After analysis, however, the data looked so good that plans were made to put ASTRO up again in late 1993 or early 1994.

What kind of data was it? Much of it consisted of better measurements using wavelengths that cannot be observed from Earth. One whole new set of observations was on polarized ultraviolet light; these revealed that there are two different interstellar dust populations, one based on carbon and another based on silicon. There was some good X-ray evidence of a likely black hole and the possible detection of the pulsar thought to be in supernova 1987A. A single measurement of 1000 seconds from the Broad Band X Ray Telescope was sufficient to accurately determine that the distance to the Perseus cluster of galaxies is 350 million light-years. Other results are still being analyzed.

ROSAT Maps the X-Ray Universe

When Wilhelm Roentgen [German: 1845-1923] discovered X rays in 1895 he could never have guessed that one of the results would be an artificial satellite named after him, the *Röntgensatellit*, launched by the United States on Jun 1 1990 and sponsored by NASA, the US Air Force, Germany, and the United Kingdom. As the somewhat more pronounceable ROSAT, the orbiting X-ray telescope had already increased the number of known X-ray emitters from 1000 to 100,000 in its first pass at the universe, completed Feb 9 1991. ROSAT's specialty is less energetic, or "soft," X rays. "Hard" X rays will be observed in roughly the same time frame by the Gamma Ray Observatory. ROSAT also has an extreme ultraviolet telescope hitchhiking a ride on it. The telescope located a "thousand points of ultra-violet light" during the first all-sky survey.

Among the highlights of the all-sky survey are

LMC X-1 in the Large Magellanic Cloud (LMC), thought to be an ordinary star being torn apart by a black hole;

an unexplained and very large "glitch" in an X-ray pulsar in the LMC, or sudden change in the spin rate; and

what appear to be clusters of quasars that cause the overall background X-ray radiation that suffuses the sky.

For the next phase in its program, ROSAT is making a detailed study of the thousand most interesting objects found in its all-sky survey. That phase is expected to last about 2 years.

Gamma Ray Puzzle May Be Solved by GRO

In 1967 the US Defense Department orbited the experimental Vela satellites to see if they could be used to monitor compliance with the treaty that bans atmospheric nuclear explosions. One of the things the satellites looked for was sudden bursts of gamma rays. Causing some alarm, Vela soon found some; but it quickly became clear that the bursts came from outer space, not from Earth.

The cause of these irregular and exceptionally powerful bursts has never been determined, although scientists have proposed many scenarios that might result in such outpourings of energy. Three popular theories have been put forth. The first is that the bursts are caused by thermonuclear explosions after all, but explosions on neutron stars. Such explosions could be caused by pressure, magnetic confinement of hydrogen, and high temperatures, all produced in the vicinity of a neutron star. These conditions are similar to those humans are trying to achieve to produce controlled nuclear fusion on Earth. Another theory is that the bursts are high energy "bangs" associated with objects falling onto a neutron star after acceleration by the star's powerful gravitational force. A third idea is that neutron stars undergo something like earthquakes, but produce gamma rays during the process. The gamma-ray bursts are so variable that it is likely that several different mechanisms are involved. Furthermore, recent observations suggest that some mechanism not yet described may account for many gamma-ray bursts.

The best way to observe gamma rays is by their effects as they pass through a liquid or a crystal. But there are few individual photons at gamma-ray energy in space and gamma rays cannot be focused, so detectors have to be large, about the size of a subcompact car. A major goal of

ABOUT GAMMA RAYS

Gamma rays were named when the nature of radioactivity was poorly understood but experiments had revealed three very different types of energy produced by radioactive elements. The alpha rays are now known to be helium nuclei, the beta rays are electrons, and the gamma rays are high-energy photons. Viewed as waves, gamma rays are electromagnetic radiation with very short wavelengths, shorter even than X rays. Specifically, gamma rays are sometimes defined as photons with energies greater than 0.3 MeV (million electron volts), or as electromagnetic waves with wavelengths less than one ten millionth of a centimeter. Sometimes very high-energy gamma rays from space are called cosmic rays, although most cosmic rays are not gamma rays.

Because gamma rays interact easily with matter, they cannot penetrate the atmosphere. A few very high-energy gamma rays can be observed from Earth's surface by the effects they cause in the upper atmosphere. Astronomical gamma rays are produced by synchrotron radiation (acceleration of electrons), radioactive decay of elements, annihilation of particles by antiparticles, and decay of some elementary particles. Each process produces gamma rays of particular energies, so astronomers can learn a great deal from studying this radiation.

the heaviest scientific satellite ever launched by the United States, the 15.5-metric-ton (17-ton) Compton Gamma Ray Observatory (GRO), is to learn more about the gamma ray bursters. One of its four detectors, for example, is specially designed to detect gamma-ray bursts. The main strategy for GRO is the same as for ROSAT, an all-sky survey followed by concentration on the most interesting objects.

The Burst and Transient Source Experiment (BATSE), whose chief investigator is Gerald Fishman of NASA's Marshall Space Flight Center in Huntsville AL, picks up a powerful burst once or more each day on the average, more than any previous detector has recorded. That pattern is the chief clue so far to the nature of the bursts. Since no weak bursts have been detected, at first it was thought that the energy sources must be local, inside the Milky Way. Otherwise, the sensitive BATSE would be expected to detect about fifty weak bursts a day from other galaxies. Later, enough bursts were detected to show that they came uniformly from all regions of the sky. That suggests that the bursters are outside the Milky Way or near to the solar system. If the former is true, the gamma ray bursts are extremely powerful, with each burst releasing more energy than a supernova explosion. If near to the solar system, gamma ray bursters do not have to be very powerful, but it is unlikely that they are caused by neutron stars. One piece of evidence supporting the latter theory is that the source of the bursts seems to stop at a certain (if unknown in extent) distance, as if they are all in a region with an outside boundary, or edge.

This result contradicts a May 1991 analysis of gamma ray bursts detected by the USSR space probes Phobos and Venera 13 and 14. A French-Soviet team led by J.L. Atteia of the Center for the Study of Space Radiation in Toulouse announced that the data showed that the bursts were aligned with the plane of the Milky Way, implying that they are coming from our own galaxy.

The bursts vary a lot in duration and in type. Some are true bursts, finished in seconds, while others last a hundred times as long. Some flicker, while others do not. This variation suggests to Fishman that there are several different types of sources for the bursts. Only a few have ever been linked with optical sources.

Interestingly, GRO was expected to have serious interference from gamma rays from nuclear-powered USSR spy satellites when it was designed. But in 1988 the Soviet Union stopped launching such satellites and all of those previously in orbit have been decommissioned. So GRO's equipment for avoiding unwelcome effects from such satellites is just extra baggage.

GRO is also studying other sources of gamma rays that produce radiation steadily or that vary in output continuously, not just in bursts. These sources are also mysterious at this time.

Mission to Planet Earth Begins

NASA's multiyear study of Earth, especially of any changes in climate or the atmosphere caused by human activity, started officially with the launch from the space shuttle *Discovery*, on Sep 14 1991, of the $710-million Upper Atmosphere Research Satellite (UARS). The satellite began a 20-month mission of studying physical and chemical processes in Earth's upper atmosphere. Of most concern was an analysis of how quickly ozone depletion is occurring in the stratosphere. Data from satellites not planned expressly to study ozone suggest that such depletion is proceeding at twice the rate originally expected. With instruments committed to observing ozone-related gases—ozone itself and chlorine compounds thought to be mainly responsible for most ozone depletion, for example—UARS will make measurements that are more direct than the previous ones from space.

The ten instruments aboard the satellite will also measure gases thought to be related to global warming, including carbon dioxide, methane, and water vapor. Two of the instruments are specifically dedicated to determining the directions and speeds of winds in the region between 10 km (6 mi) and 80 km (49 mi) above the surface. Another four observe the amount and effects of energy from the Sun.

(Periodical Sources and Additional Reading: *Astronomy* 1-90, p 30; *Science News* 10-20-90, p 255; *New York Times* 12-16-90, p 132; *Astronomy* 1-91, p 24; *Astronomy* 3-91, p 24; *New York Times* 4-2-91, p C1; *New York Times* 4-7-91, p 126; *Science* 4-12-91, p 215; *Astronomy* 5-91, p 38; *EOS* 5-28-91, p 241; *Astronomy* 6-91, p 42; *Science* 6-7-91, p 1376; *Science News* 6-8-91, p 365; *Astronomy* 7-91, pp 26 & 44; *Discover* 8-91, p 12; *Astronomy* 9-91, p 22; *New York Times* 9-10-91, p C4; *New York Times* 9-19-91, p A24; *Science News* 9-21-91, p 181; *New York Times* 9-24-91, p C2; *Science* 9-20-91, p 1352; *Science News* 9-28-91, p 196; *Science* 10-4-91, p 34; *Astronomy* 10-91, p 48; *Astronomy* 11-91, p 26; *Discovery* 1-1-92, p 28)

WATCH THIS SPACE

It often takes years for a space probe to reach its destination, and yet the laws of motion enable scientists to be certain with uncanny accuracy when those encounters will occur. For space probes that are already launched, dates of major events are almost definite. Sometimes, however, equipment fails at or before the rendezvous. Here are some of the dates you might want to keep. (All craft are from the United States unless otherwise noted.)

May 1992   The latest space shuttle, *Endeavour*, is scheduled to make its maiden flight.

Jul 10 1992   Giotto is the European Space Agency space probe that was the leading player (after the comet itself) in the 1986 visit of Comet Halley. It flew within 600 km (375 mi) of the comet on Mar 13, suffering considerable damage from the dust particles in the coma. Four of nine experiments were swamped by the bombardment. But the spacecraft survived and now is planning another excursion in comet study. In Mar 1990 Giotto was redirected into a path that will take it 23,000 km (14,300 mi) from Earth on its way to Comet Grigg-Skjellerup. During most of this trip it will be dormant to save its remaining hydrazine fuel. A few months before the planned Jul 10 1992 encounter with the comet, the probe will be reactivated.

Jul 1992   The Geotail satellite is scheduled to be launched from a Delta II rocket.

Sep 1992   The Mars Observer space probe, a cooperative venture among the United States, Germany, Austria, France, the USSR, and the United Kingdom, will be launched. It is expected to map Mars from orbit during 1993 and 1994.

Dec 8 1992   Galileo will take its second and last gravity assist from Earth before swinging out toward its planned rendezvous with Jupiter. Shortly before that time, the data from the asteroid 951 Gaspra, captured in Oct 1991, will be beamed back to Earth, unless the main antenna has been previously fixed.

1993	A joint USSR-UK mission called Spectrum X, to map the cosmos at far ultraviolet and X-ray frequencies, is planned for this year, but may fall to political and financial disarray in the USSR and budget cutting in the United Kingdom.
May 1993	The European Space Agency plans to orbit the Infrared Space Observatory (ISO).
Aug 1993	Mars Observer is expected to enter orbit about Mars and begin returning data.
Aug 28 1993	If Galileo's trip has been going well, it will fly by the asteroid Ida.
1994	A Thermosphere-Ionosphere-Mesosphere Energetics and Dynamics (TIMED) satellite.
Jun-Nov 1994	Ulysses will pass over the Sun's south pole.
1995	The High-Energy Physics Space Program (HESP).
	Launch of the Space Infrared Telescope Facility (SIRTF).
Mar 13 1995	Ulysses will come the closest to the Sun in its path, a distance of 1.3 times the average distance of the Earth and Sun.
Jun-Sep 1995	Ulysses will pass over the Sun's north pole.
1996	Possibilities include a Pluto Flyby or a Neptune Orbiter.
	Launch of the Inner Magnetosphere Imager (IMI).
1997	The Mars Environmental Survey (MESUR).
	The Grand Tour Cluster (GTC).
1998	The Orbiting Solar Laboratory.
	The Sub-Millimeter-Intermediate Mission (SMIM).
1999	The Advanced X-Ray Astrophysics Facility (AXAF).

(Periodical Sources and Additional Reading: *Astronomy* 7-90, p 26; *Science* 2-15-91, p 740; *EOS* 10-1-91, p 425; *Science* 10-25-91, p 508; *New York Times* 10-27-91, p L18; *Science* 11-22-91, p 1109)

THE USES OF SPACE

GOES-NEXT MAY NOT GO NEXT (AND WHY)

Placing scientific satellites in orbit around Earth has been a major effort since 1957, but the practical side of space has also become increasingly important. One of the most practical programs, that for weather satellites, is in deep trouble in the United States.

The satellite weather report that people in the United States see on television comes from a system called the Geostationary Operational Environmental Satellites (GOES). The satellite GOES-7 is currently the only weather satellite in orbit over the United States, although other weather satellites monitor Europe and the Far East. GOES-7 was to have been replaced by the first of a technically superior group of satellites. The group is GOES-NEXT and the first weather satellite scheduled to lift off is GOES-I. Development of the GOES-NEXT satellites is 3 years behind schedule, however. In the meantime, GOES-7 is on its last legs. Its planned lifetime was originally scheduled to expire in Feb 1992. Even with conservation of fuel, it is still expected to drift away from its course a bit in Jul 1992 and to continue that drift over the following year. The result would be that the United States would have no weather satellite coverage.

The problem is largely with the design and execution of the five NEXT satellites. Instead of sticking with the same basic design as the first GOES series, "spin-scan" satellites that maintained direction and thermal equilibrium by rapidly rotating, the NEXT group was to stay with one face always toward Earth. Recent experience with scientific satellites designed this way suggests that this kind of orientation cannot be easily accomplished.

A complex design is only the beginning, however. Not only are there technical problems with some of the instruments, but also the wrong type of wiring was installed. The US General Accounting Office has said that the problems were caused by the design choice, poor management, and poor workmanship. Others attribute the problems to penny-pinching by NASA, which did not provide enough money for subassembly tests. There may also have been a lack of experienced or even trained engineers to design and build such a complex instrument. The best engineers for this kind of operation were off working for the US military on the Star Wars program. A general level of incompetence pervaded the program. One of the detectors was even lost.

In Sep 1991 the House Committee on Investigations and Oversight recommended that no attempt be made soon to launch GOES-I. Instead, they suggested that some other satellite should be used until the NEXT program is straightened out.

GEOSTATIONARY SATELLITES

Since 1974 the United States has had continuous coverage from geostationary (geosynchronous) weather satellites. Distance from Earth's center of mass is the most important factor in determining the speed of revolution of a satellite. A geostationary satellite is at an altitude (roughly 35,900 km or 22,300 mi) that makes one revolution of the satellite almost exactly the same as a single rotation of Earth. Such a satellite must be above the equator for this to work. As a result of gravitational effects from the Moon and Sun, which vary as the relative positions of Earth, Moon, and Sun change over time, and the irregular gravitational field of Earth, a perfectly geostationary orbit is impossible to achieve. Instead, a geostationary satellite moves in a figure-8 pattern that is predictable—and compensated for by changing the direction of antennas on Earth to point at the correct satellite position. But superimposed on this pattern is a gentle drift away from the geostationary orbit that must be counteracted by the use of small thrusters that expel compressed gas. When this gas is used up, the drift can no longer be corrected and the satellite moves into an orbit that is no longer geostationary. This is expected to be the fate of GOES-7 sometime in the early to mid-1990s.

Meanwhile, the US National Oceanic and Atmospheric Administration (NOAA) has negotiated with Eumetsat, the European weather satellite operation, to have the organization, in the event of the failure of GOES-7, move their Meteosat-3 farther west to cover North America and to turn over full control to the United States. The European satellite would be moved from 50° W longitude to 90° W. Meteosat-3 is a spin-scan satellite that is now above the east coast of South America, a position that is useful for watching storms in the Atlantic Ocean as well as all of South America. Prior to any move to farther west, the United States would need to build a control and communications station on Wallops Island VA. This would cost about $10 million as opposed to $100 million for GMS-5. Meteosat-3's planned life ended in Jun 1991, but it seems to have enough fuel to last until the end of 1993, when NOAA still hopes it can be replaced by GOES-I. NOAA is also discussing acquisition of other weather satellites from Eumetsat.

(Periodical Sources and Additional Reading: *New York Times* 7-11-91, p A16; *Science* 7-12-91, p 133; *New York Times* 7-14-91, p A14; *New York Times* 7-26-91, p A13; *Science News* 8-3-91, p 68; *EOS* 8-6-91, p 337; *EOS* 8-13-91, p 347; *Science News* 8-18-90, p 102; *New York Times* 9-9-91, p A1; *New York Times* 9-15-91, p L15; *EOS* 9-24-91, p 418; *Science News* 9-28-91, p 207; *Science* 10-4-91, p 28)

P E O P L E   I N   S P A C E

END OF THE SHUTTLE AGE AT NASA

The US National Aeronautics and Space Administration (NASA) was founded in 1958 amid a conspicuous string of US space failures. Only one of NASA's first ten missions was a success, with most failing even before entering orbit. By the 1960s, however, NASA began to enjoy a string of successes that culminated in the lunar missions of 1969-72. Throughout the 1970s and early 1980s, it was assumed that NASA was as close to perfection as an agency of the US government could get. The *Challenger* disaster of Jan 28 1986 changed that perception. Recovering after the *Challenger* explosion, the space shuttle returned to action in Sep of 1988, but nothing has been the same since. It has seemed increasingly difficult for NASA to get the shuttles off the ground, for example. And when shuttles did fly, they had to deal with many difficulties.

The Space Telescope was launched in Apr 1990, but the whole year posed problems. After a flight by *Columbia* ending on Jan 20, fibers—probably from sanding down a new launching pad—were found to have contaminated the engines. In trying to fix the problem, NASA scientists started a series of leaks. When leaks were also found in *Atlantis*, all three shuttles were grounded for four months. Even after the problems were declared fixed, the launch of a scientific mission was postponed for days as new leaks and other troubles surfaced. From May 29 to Sep 18 NASA tried to launch *Columbia,* but did not succeed. It then turned to the third shuttle, *Discovery,* which had never developed leaks, and finally got a mission off the ground in Oct, after nearly a 6-month gap.

A flight before the shutdown launched a military satellite that failed. Although this was not NASA's fault, it contributed to the feeling that developed in 1990 that all was not well with NASA. *Columbia* was finally launched again on Dec 2, after nearly a year on the ground. As *Columbia* orbited Earth that Dec, almost daily stories appeared in the press and on television detailing new problems with its scientific payload, the ASTRO Observatory.

Even as *Columbia* orbited, a twelve-member panel that had been appointed by Vice President Dan Quayle, the head of the US National Space Council, issued a report that was sharply critical of all aspects of the shuttle operation and of the way NASA functioned. Among other things, the panel recommended that NASA stop building shuttles and return to the type of large unmanned rockets that had been so successful in the 1960s.

In 1991 further problems developed. Two of the forty-four thrusters used to steer the ship in orbit developed serious leaks and a third had a bad weld. Originally it was thought that this problem would delay the mission from late Feb 1991 to early Mar, about a week. Still in Feb, after *Discovery* was positioned on the launch pad for a Mar military mission, cracks were found on the hinges of three of the four doors used for access to internal fuel-line connectors. Two of the cracks were as much as 5-cm (2-in.) long. Since the cause of the cracks was unknown, NASA also inspected similar door hinges on *Columbia* and *Atlantis*, finding small cracks in *Columbia's* hinges and tiny cracks in those of *Atlantis*. *Discovery* was taken off the launch pad and fitted with the hinges taken from the newest shuttle, *Endeavour*, which was still under construction and not scheduled for launch until spring 1992. One of these hinges also was found to have a tiny crack, but was used anyway. *Columbia's* hinges were removed and returned to the shop for repair. The hinges on *Atlantis* were deemed usable despite the tiny cracks. In Sep 1991, however, an entirely different set of cracks were discovered in *Atlantis*, cracks on the heat-resistant seals on the wings and on a panel between the nose cap and landing gear. The cause of the damage was not known.

As a result of the hinge-crack problem, the first shuttle mission of 1991 did not take place until Apr 5, nearly 4 months after the previous mission. The mission of *Atlantis* to launch the 17-ton Gamma Ray Observatory (GRO) went smoothly except for just before the actual deployment of the GRO. As has become common in the space program lately, the antenna failed to open properly. Fortunately, the GRO was still attached to the shuttle's mechanical arm, so astronaut Jerry L. Ross was able to don a space suit and release it. This was followed by a successful launch.

The delayed *Discovery* military mission got another delay when it again reached the launch pad. The problem on Apr 23 was a signal from the fuel system of one of the main engines that reported a rise in pressure that engineers were sure was not occurring. After nearly 2 months of delay, *Discovery* finally headed into orbit on Apr 28, but it immediately ran into instrument problems. After 4 hours of use, both of the mission's tape recorders failed. These were supposed to store data from the mission, which was to observe natural phenomena that might be mistaken for military strikes. Six experiments had to be cancelled outright, as three of the five instruments aboard could not function properly without the recorders. Near the end of the mission, however, the astronauts aboard *Discovery* were able to make jury-rigged repairs that enabled them to carry out some observations. Although there were other minor problems along the way, the mission did return safely with useful information.

With flights from both *Atlantis* and *Discovery* in Apr, NASA succeeded for the first time in 5 years in launching two shuttles in the same month. When shuttle flights first began, some NASA officials predicted regular launchings every 2 weeks throughout the year.

The next mission, using the aging *Columbia*, was to start on May 22, but a multitude of equipment failures became apparent on May 21. One of the five flight computers failed and a device for transmitting data to the ship's computers from the engines also went down. But the most potentially damaging problem was one that had already flown on five shuttle missions. A new version of the nine temperature sensors for the fuel lines was found to be faulty. The sensors had been one of the elements examined in the search for leaks in 1990. NASA was suspicious of them then and asked shuttle manufacturer Rockwell International to check them further, but Rockwell sent them to the wrong supplier by mistake. It was not until Jan 28 that RFD Corporation of Hudson NH finally got them. Then it took RFD until Apr to report to Rockwell that the sensors were dangerous. Rockwell tested some more, and finally told NASA on May 20 that the problem was a bad weld that could break, causing the sensor to fall into the main engine. This could produce a shutdown, an explosion, or at least a precarious moment or two.

The *Columbia* flight was rescheduled for May 31, then for Jun 1, then Jun 5. A new problem was the discovery that one of the measurement units for the internal guidance system was giving false and inconsistent reports, necessitating its replacement. *Columbia* finally was launched on Jun 5, despite a few last-minute mechanical problems. In flight, problems with various cooling systems required frequent repairs. A more serious problem was a loose seal observed on one of the cargo bay doors, a potentially dangerous situation since failure of the seal on reentry could cause the spacecraft to burn up. Ground engineers experimented with replicas of the door and deemed it usable, however. Fortunately, the NASA engineers were right and *Columbia* landed safely on Jun 14.

The next scheduled mission was for *Atlantis* on Jul 22. Small glitches delayed it until the 23rd, and then a major computer problem in an engine-control computer delayed it even further. On the same day as the longer delay was announced, Vice President Quayle announced that the United States is not going to build any more space shuttles. The new policy would be to focus on maintenance of the present fleet of three shuttles and the new *Endeavour*. Moreover, the policy continued the rule that the shuttle would be used only when absolutely necessary. Vehicles without people aboard would get preference.

The Age of the Shuttle, which began in 1981, was definitely over, even though the shuttle itself would continue.

(Periodical References and Additional Reading: *Science News* 7-7-90, p 11; *Science* 7-20-90, p 238; *Discover* 1-91, p 33; *Science* 1-25-91, p 357; *New York Times* 6-11-91, p C1; *New York Times* 7-25-91, p A18; *New York Times* 9-27-91, p A14)

SPACE STATION REDESIGN

Even projects only in the planning stage have not gone well for NASA. On Jul 20 1990, a NASA self-study headed by astronaut William F. Fisher and Charles R. Price of the Johnson Space Center in Houston (consequently known as the Fisher-Price report) revealed that the space station *Freedom,* scheduled for construction from 1995 to 1998 after numerous delays, needed reworking because the original design required more maintenance than seemed feasible. According to the study, astronauts would have to spend 3276 astronaut hours each year in extravehicular activity (EVA) just to perform routine maintenance. That's about 9 astronaut hours a day, and almost double previous estimates. Fisher had had personal experience with such work, having spent a dozen hours in EVA in 1985 repairing a communications satellite. It should be noted that maintenance on the USSR space station *Mir* also occupies much of the cosmonaut's daily duties.

At the same time as the Fisher-Price report, NASA released a set of recommendations that it claimed would reduce the amount of space walking needed to just 485 hours a year. Among the means proposed were structural modifications that would result in fewer repair needs, automated repair for some parts, removal of nonessential components, and increased efficiency in preparing for EVA.

The Fisher-Price document was not a complete surprise. An earlier version of their report—with less drastic conclusions—had been leaked to *The New York Times,* resulting in a US House committee investigation in Mar 1990, complete with testimony from both Fisher and Price. At that point NASA did what it could to discredit the report and to defend the space station. After the detailed report was released, however, that position became indefensible. Also, Congress mandated a redesign of the station after learning of the complete report.

At the beginning of 1991, Fisher resigned from the astronaut program, saying that NASA did not appear to be planning the amount of change that was needed for the space station.

That perception changed as NASA's revised plans gradually became known in detail. As NASA simplified the station by cutting down its size and capabilities, prospective users of the facility began to protest. Among the first to make its views known was the Space Studies Board of the National Research Council, which complained in early Mar that the proposed station "does not meet the basic research requirements of the two principal scientific disciplines for which it is intended." Specifically, the new station would not have enough power to do microgravity research nor enough astronauts and equipment to do space medicine research. Among the missing equipment in the plan was a centrifuge for producing artificial gravity. The White House struck back with a statement by Vice

President Quayle that seemed to say America needed to build a space station just to prove it was the leader in space (the Soviet Union, of course, has had a space station, *Mir*, since Feb 1986).

The actual redesign plan was announced on Mar 21, the day after Quayle's defense of it. The new station would cost only $30 billion instead of $38.3 billion to build, would only take 2 years longer to build, would be 72 percent the size of the original, generate three-quarters as much power, and house half as many astronauts in 61 percent of the space previously planned for living quarters. NASA claimed that the planned EVA for maintenance would be less than 300 hours per year. The station would not have a centrifuge. The centrifuge was reinstated, however, in Apr, albeit as the last part of the station to be built—sometime after 2000. NASA said that the redesigned station would average about $2 billion a year to operate during its planned lifetime of nearly 30 years, making the total cost in 1991 dollars $84 billion. In May the US congressional General Accounting Office refigured the amount as $118 billion.

Congress responded to these estimates. On May 15 the House Appropriations Subcommittee for the VA, HUD, and Independent Agencies, which controls NASA funding, voted to cut off all appropriations for *Freedom,* with the money saved to go instead to various NASA scientific projects. The subcommittee suggested replacing the station with two small orbiting laboratories devoted to research in microgravity and life science in space. But when the budget came before the full House on Jun 6, the space station won and other science projects lost. Full funding for *Freedom* was restored after extensive lobbying by NASA and the White House and complaints from Japan, Canada, and the European Space Agency, which were planning to have modules of their own on the station.

On Jul 9 the presidents of fourteen professional scientific societies, ranging from the Society of Rheology to the main bodies in physics, chemistry, mathematics, and earth science, wrote a joint letter to all members of the US Senate expressing their concerns about the cost of the space station. Some societies, such as the American Geophysical Union, were already on record elsewhere as being more than concerned: they opposed it outright. Engineering societies and the American Astronomical Society did not participate. A few days later the Senate subcommittee that handles NASA appropriations showed that it had failed to be swayed, recommending full funding.

(Periodical References and Additional Reading: *Science* 4-6-90, p 26; *Science News* 7-28-90, p 53; *New York Times* 1-9-91, p A18; *Science* 3-8-91, p 1167; *Science* 3-29-91, p 1556; *Planetary Report* 5-6-91, p 17; *EOS* 5-21-91, p 233; 17; *Science News* 6-15-91, p 375; *EOS* 6-18-91, p 265; *New York Times* 7-10-91, p A13; *Science* 7-19-91, p 256; *EOS* 7-22-91, p 313; *Discover* 8-91, p 66; *Astronomy* 9-91, p 28; *Science* 10-11-91, p 191)

SPACE FLIGHTS
CARRYING PEOPLE

The space programs of the Soviet Union and the United States both have had dramatic flights by human pilots as an important component, although many scientists feel that most goals of the space program can be achieved without risking lives.

The Vostok, Voskhod, and Soyuz missions are part of the Soviet program; all other flights are part of the US program.

Craft	Date	Duration	Crew	Remarks
Proving That People Can Venture into Space				
Vostok 1	4/12/61	1hr 48 min	Yuri Gagarin	First space flight by a human (1 orbit)
Mercury 3	5/5/61	5 min	Alan B. Shepard, Jr.	*Freedom 7* (suborbital)
Mercury 4	7/21/61	16 min	Virgil I. Grissom	*Liberty Bell 7* (suborbital)
Vostok 2	8/6/61	25 hr 18 min	Gherman S. Titov	First multi-orbit flight; 17 orbits
Mercury 6	2/20/62	4 hr 55 min	John H. Glenn, Jr.	*Friendship 7*; first orbital flight by American; 3 orbits
Mercury 7	5/24/62	4 hr 56 min	M. Scott Carpenter	*Aurora 7*; 3 orbits
Vostok 3	8/11/62	94 hr 24 min	Andrian G. Nikolayev	Landing by parachute; 64 orbits
Vostok 4	8/12/62	70 hr 57 min	Pavel R. Popovitch	Dual launch with Vostok 3; 48 orbits
Mercury 8	10/3/62	9 hr 13 min	Walter M. Schirra	*Sigma 7*; 6 orbits
Mercury 9	5/15/63	34 hr 20 min	L.Gordon Cooper	*Faith 7*; 22 orbits
Vostok 5	6/14/63	119 hr 6 min	Valery F.Bikovsky	81 orbits
Vostok 6	6/16/63	70 hr 50 min	Valentina Tereshkova	First woman cosmonaut; dual launch with Vostok 5; 48 orbits
Voskhod 1	0/12/64	24 hr 17 min	Vladimir M. Komarov Konstantin P. Feotistov Boris B. Yegorov	First multi-person crew; 16 orbits
Planning for Operations in Space				
Voskhod 2	3/18/65	26 hr	Aleksei A. Leonov Pavel I. Belyayev	First extravehicular activity (EVA) by Leonov (20 min); 17 orbits
Gemini 3	3/23/65	4 hr 53 min	Virgil I. Grissom John W. Young	First American multi-person crew; 3 orbits
Gemini 4	6/3/65	97 hr 56 min	James A. McDivitt Edward H. White II	First American EVA; first use of personal propulsion unit; 62 orbits

Craft	Date	Duration	Crew	Remarks
Gemini 5	8/21/65	190 hr 56 min	L. Gordon Cooper Charles Conrad, Jr.	Demonstrates feasibility of lunar mission; simulated rendezvous; 120 orbits
Gemini 7	12/4/65	330 hr 35 min	Frank Borman James A. Lovell, Jr.	Extends testing and perfomance; target for first rendezvous; 206 orbits
Gemini 6A	12/15/65	25 hr 51 min	Walter M. Schirra Thomas P. Stafford	First rendezvous (with Gemini 7); 15 orbits
Gemini 8c	3/16/66	10 hr 42 min	Neil A. Armstrong David R. Scott	First dual launch and docking; first Pacific landing; 6.5 orbits
Gemini 9A	6/3/66	72 hr 21 min	Thomas P. Stafford Eugene A. Cernan	Unable to dock with target vehicle; 2 hr 7 min of EVA; 44 orbits
Gemini 10	7/18/66	70 hr 47 min	John W. Young Michael Collins	First dual rendezvous; docked vehicle maneuvers; umbilical EVA; 43 orbits
Gemini 11	9/12/66	71 hr 17 min	Charles Conrad, Jr. Richard F. Gordon, Jr.	Rendezvous and docking; 44 orbits
Gemini 12	11/11/66	94 hr 34 min	James A. Lovell, Jr. Edwin A. Aldrin	Final Gemini mission; 5 hr of EVA; 59 orbits

To the Moon and Experiments in Space

Craft	Date	Duration	Crew	Remarks
Soyuz 1	4/23/67	26 hr 48 min	Vladimir M. Komarov	Komarov is killed when parachute fails, first fatality of space program; 18 orbits
Apollo 7	10/11/68	260 hr 8 min	Walter M. Schirra Donn F. Eisele R. Walter Cunningham	Eight service propulsion firings; 7 live TV sessions with crew; rendezvous with S-IVB stage performed
Soyuz 3	10/26/68	94 hr 51 min	Georgi T. Beregovoi	Approachs the unpiloted Soyuz 2 to a distance of 650 ft; 64 orbits
Apollo 8	12/21/68	147 hr	Frank Borman James A. Lovell, Jr. William A. Anders	First Saturn V propelled flight; first lunar orbital mission; returns good lunar photography; 10 orbits
Soyuz 4	1/14/69	71 hr 14 min	Vladimir A. Shatalov	Docks with Soyuz 5 in first linkup of two space vehicles that both carry people; 48 orbits
Soyuz 5	1/15/69	72 hr 46 min	Boris V. Volynov Alexei S. Yeliseyev Yevgeni V. Khrunov	Three cosmonauts perform EVA and are transferred to Soyuz 4 in rescue rehearsal
Apollo 9	3/3/69	241 hr 1 min	James McDivitt David R. Scott Russell Schweickart	First flight of all lunar hardware in Earth orbit, including lunar module (LM)

(Continued)

Craft	Date	Duration	Crew	Remarks
Apollo 10	5/18/69	192 hr 3 min	Eugene A. Cernan John W. Young Thomas P. Stafford	Lunar mission development flight to evaluate LM performance in lunar environment; descent to within 15,240 m (50,000 ft) of Moon
Apollo 11	7/16/69	165 hr 18 min	Neil A. Armstrong Michael Collins Edwin E. Aldrin, Jr.	First lunar landing; limited inspection, photography, evaluation, and sampling of lunar soil; touchdown Jul 20
Soyuz 6	10/11/69	5 da	Georgi S. Shonin Valery N. Kubasov	First triple launch (with Soyuz 7 and 8)
Soyuz 7	10/12/69	5 da	Anatoly V. Filipchenko Vladislav N. Volkov Viktor V. Gorbatko	Conducts experiments in navigation and photography with Soyuz 7 and 8
Soyuz 8	10/13/69	5 da	Vladimir A. Shatalov Aleksei S. Yeliseyev	80 orbits
Apollo 12	11/14/69	244 hr 36 min	Charles Conrad, Jr. Richard F. Gordon, Jr. Alan L. Bean	Second lunar landing; demonstrates point landing capability; samples more area; total EVA time: 15 hr 32 min
Apollo 13	4/11/70	142 hr 55 min	James A. Lovell, Jr. Fred W. Haise, Jr. John L. Swigert, Jr.	Third lunar landing attempt aborted due to loss of pressure in liquid oxygen in service module and fuel cell failure
Soyuz 9	6/2/70	17 da 16 hr	Andrian G. Nikolayev Vitaly I. Sevastianov	Longest space flight to date
Apollo 14	1/31/71	216 hr 42 min	Alan B. Shepard, Jr. Stuart A. Roosa Edgar D. Mitchell	Third lunar landing; returns 98 pounds of lunar material
Soyuz 10	4/23/71	2 da	Vladimir A. Shatalov Alexei S. Yeiseyev Nikolai N. Rukavishnikov	Docks with *Salyut 1*, the first space station
Soyuz 11	6/6/71	24 da	Georgi Dobrovolsky Viktor I. Patsayev Vladislav N. Volkov	All three cosmonauts killed during reentry
Apollo 15	7/26/71	295 hr 12 min	David R. Scott Alfred M. Worden James B. Irwin	Fourth lunar landing; first to carry Lunar Roving Vehicle (LRV); total EVA time: 18 hr 46 min; returns 173 lb of material
Apollo 16	4/16/72	265 hr 51 min	John W. Young Thomas Mattingly II Charles M. Duke	Fifth lunar landing; second to carry LRV; total EVA time: 20 hr 14 min; returns 213 lb of material
Apollo 17	12/7/72	301 hr 52 min	Eugene A. Cernan Ronald E. Evans Harrison Schmitt	Last manned lunar landing; third with LRV; total EVA time: 44 hr 8 min; returns 243 lb of material

Craft	Date	Duration	Crew	Remarks
First Stations in Space				
Skylab 2	5/25/73	28 da 49 min	Charles Conrad, Jr. Joseph P. Kerwin Paul J. Weitz	First Skylab launch; establishes Skylab Orbital Assembly in Earth orbit; conducts medical and other experiments
Skylab 3	7/29/73	59 da 11 hr	Alan L. Bean Owen K. Garriott Jack R. Lousma	Second Skylab; crew performs systems and operational tests, experiments, and thermal shield deployment
Soyuz 12	9/27/73	2 da	Vasily G. Lazarev Oleg G. Makarov	First Soviet spaceflight to carry humans since Soyuz 11 tragedy
Skylab 4	11/16/73	84 da 1 hr	Gerald P. Carr Edward G. Gibson William R. Pogue	Third Skylab; crew performs unmanned Saturn workshop operations; obtains medical data for extending spaceflights
Soyuz 13	12/18/73	8 da	Pytor I. Klimuk Valentin Lebedev	Performs astrophysical and biological experiments
Soyuz 14	7/3/74	16 da	Pavel R. Popovich Yuri P. Artyukhin	Crew occupies *Salyut 3* space station; studies Earth resources
Soyuz 15	8/26/74	2 da	Gennady Sarafanov Lev Demin	Makes unsuccessful attempt to dock with *Salyut 3*
Soyuz 16	12/2/74	6 da	Anatoly V. Filipchenko Nikolai N. Rukavishnikov	Taken to check modifications in Salyut system
Soyuz 17	1/10/75	30 da	Alexei A. Gubarev Georgi M. Grechko	Docks with *Salyut 4*; sets Soviet endurance record to date
Soyuz 18A	4/5/75	22 min	Vasily G. Lazarev Oleg G. Makarov	Separation from booster fails and craft does not reach orbit, but crew lands successfully in western Siberia
Soyuz 18B	5/24/75	63 da	Pyotr I. Klimuk Vitaly I. Sevastyanov	Docks with *Salyut 4*
ASTP	7/15/75	9 da 1 hr	Thomas P. Stafford Vance D. Brand Donald K. Slayton	Apollo-Soyuz Test Project, cooperative US-USSR mission
Soyuz 19	7/15/75	5 da 23 hr	Alexei A. Leonov Valery N. Kubasov	Docks with ASTP, the US Apollo capsule

The Soviets Study Human Biology in Space

Craft	Date	Duration	Crew	Remarks
Soyuz 20	11/17/75	90 da	No crew	Biological mission; docks with *Salyut 4*
Soyuz 21	7/6/76	49 da	Boris V. Volynov Vitaly Zholobov	Docks with *Salyut 5* and performs Earth resource work
Soyuz 22	9/15/76	8 da	Valery F. Bykovsky Vladimir Aksenov	Takes Earth resource photographs

(Continued)

Craft	Date	Duration	Crew	Remarks
Soyuz 23	10/14/76	2 da	Vyacheslav Zudov Valery Rozhdestvensky	Unsuccessfully attempts to dock with *Salyut 5*
Soyuz 24	2/7/77	18 da	Viktor V. Gorbatko Yuri N. Glazkov	Docks with *Salyut 5* for 18 days of experiments
Soyuz 25	10/9/77	6 da	Vladimir Kovalyonok Valery Ryumin	Unsuccessfully attempts to dock with *Salyut 6*
Soyuz 26	12/10/77	96 da	Yuri V. Romanenko Georgi M. Grechko	Docks with *Salyut 6;* crew sets endurance record for this time
Soyuz 27	1/10/78	6 da	Vladimir Dzhanibekov Oleg G. Makarov	Carries second crew to dock with *Salyut 6* space station
Soyuz 28	3/2/78	8 da	Vladimir Remek Alexi A. Gubarev	Carries third crew to board *Salyut 6;* first non-Russian, non-American in space (the Czech Remek)
Soyuz 29	6/15/78	140 da	Vladimir Kovalyonok Aleksander S. Ivanchenkov	Docks with *Salyut 6;* crew sets new space endurance record for this time
Soyuz 30	6/27/78	8 da	Pyotr I. Klimuk Miroslaw Hermaszewski	Carries second international crew to *Salyut 6;* first Polish cosmonaut (Hermaszewski)
Soyuz 31	8/25/78	8 da	Valery F. Bykovsky Sigmund Jahn	Carries third international crew to *Salyut 6;* first East German in space (Jahn)
Soyuz 32	2/25/79	175 da	Vladimir Lyakhov Valery Ryumin	Carries crew to *Salyut 6;* new endurance record set
Soyuz 33	4/10/79	2 da	Nikolai N. Rukavishnikov Georgi Ivanov	Engine failure prior to docking forces early termination; first Bulgarian in space (Ivanov)
Soyuz 34	6/6/79	74 da	No crew	Launched with no crew; returns with crew from *Salyut 6*
Soyuz 35	4/9/80	185 da	Valery Ryumin Leonid Popov	Carries 2 crew members to *Salyut 6*
Soyuz 36	5/26/80	8 da	Valery N. Kubasov Bertalan Farkas	Carries 2 crew members to *Salyut 6;* crew returns in Soyuz 35; first Hungarian (Farkas)
Soyuz T 2	6/5/80	4 da	Yuri Malyshev Vladimir Aksenov	Test of modified Soyuz craft; docks with *Salyut 6*
Soyuz 37	7/23/80	8 da	Viktor V. Gorbatko Pham Tuan	Exchanges cosmonauts in *Salyut 6;* returns *Soyuz 35* crew after 185 days in orbit
Soyuz 38	9/18/80	8 da	Yuri V. Romanenko Arnaldo Tomayo-Mendez	Ferry to *Salyut 6;* First Cuban in space (Tomayo-Mendez)
Soyuz T 3	11/27/80	13 da	Leonid Kizim Oleg G. Makarov Gennadi M. Strekalov	Ferry to *Salyut 6;* first 3-person crew since *Soyuz 11*

Craft	Date	Duration	Crew	Remarks
Soyuz T 4	3/12/81	75 da	Vladimir Kovalyonok Viktor Savinykh	Mission to *Salyut 6*
Soyuz 39	3/22/81	8 da	Vladimir Dzhanibekov Jugderdemuduyn Gurragcha	Docks with *Salyut 6*; first Mongolian (Gurragcha)

The Space Shuttle: The US Reenters Space

Craft	Date	Duration	Crew	Remarks
Columbia	4/12/81	2 da 6 hr	John W. Young Robert L. Crippen	First flight of reusable space shuttle proves concept; first landing of US spacecraft on land
Soyuz 40	5/14/81	8 da	Leonid I.Popov Dumitru Prunariu	First Romanian (Prunariu) in space
Columbia	11/12/81	2 da 6 hr	Joe H. Engle Richard H. Truly	First reuse of space shuttle; ends early due to loss of fuel cell
Columbia	3/22/82	8 da	Jack R. Lousma C. Gordon Fullerton	Third shuttle flight; payload includes space science experiments
Soyuz T 5	5/13/82	211 da	Anatoly Berezovoy Valentin Lebedev	First flight to *Salyut 7*; space station equipped to measure body functions
Soyuz T 6	6/24/82	8 da	Vladimir Dzhanibekov Jean-Loup Chrétien Aleksandr Ivanchenkov	Mission to *Salyut 7*; Soviet/French team
Columbia	6/27/82	7 da 1 hr	Thomas K. Mattingly II Henry Hartsfield, Jr.	Fourth shuttle mission; first landing on hard surface
Soyuz T 7	8/16/82	8 da	Leonid I. Popov Sventlana Savitskaya Alexander Serebrov	Mission to *Salyut 7*; second Soviet woman in space, Savitskaya
Columbia	11/11/82	5 da 2 hr	Vance D. Brand Robert F. Overmyer Joseph P. Allen William B. Lenoir	First operational mission; first 4-man crew; first deployment of satellites from shuttle
Challenger	4/4/83	5 da	Paul J. Weitz Karol J. Bobko Donald H. Peterson F. Story Musgrave	Second shuttle joins fleet; deploys TDRS tracking satellite; first shuttle EVA
Soyuz T 8	4/20/83	2 da	Vladimir G. Titov Gennadi M. Strekalov Aleksandr A. Serebrov	Cosmonauts fail in planned rendevous with *Salyut 7*
Challenger	6/18/83	6 da 2 hr	Robert L. Crippen Frederick H. Hauck John M. Fabian Sally K. Ride Norman E. Thagard	First 5-person crew; first American woman in space (Ride); first use of Remote Manipulating Structure ("Arm") to deploy and retrieve satellite

(Continued)

Craft	Date	Duration	Crew	Remarks
Soyuz T 9	6/27/83	150 da	Vladimir Lyakhov Aleksandr Alexandrov	Crew spends 49 days in *Salyut* 7 after *Soyuz 10* fails in relief mission
Challenger	8/30/83	6 da	Richard H. Truly Daniel Brandenstein William Thornton Guion S. Bluford, Jr. Dale Gardner	First night launch; first African-American (Bluford) in space; launches weather/ communications satellite for India
Columbia	11/28/83	10 da	John Young Brewster Shaw, Jr. Robert Parker Owen Garriott Byron Lichtenberg Ulf Merbold	Carries Spacelab; 6-man crew performs numerous experiments in astronomy and medicine
Challenger	2/3/84	8 da	Vance Brand Bruce McCandless, II Robert Stewart Ronald McNair Robert Gibson	Jet-propelled backpacks carry 2 astronauts on first untethered space walks; 2 satellites (Western Union and Indonesian) lost; first landing at Kennedy Space Center
Soyuz T 10B	2/8/84	237 da	Leonid Kizim Vladimir Solovyov Oleg Atkov	Mission to *Salyut* 7 to repair propulsion system; sets new duration in space record for crew
Soyuz T 11	4/2/84	8 da	Yuri Malyshev Gennadi M. Strekalov Rakesh Sharma	Docks with *Salyut* 7; first Indian cosmonaut (Sharma)
Challenger	4/7/84	8 da	Robert L. Crippen Richard Scobee Terry Hart George Nelson James van Hoften	Deploys Long Duration Exposure Facility for experiments in space durability; snares Solar Max satellite and repairs attitude control system
Soyuz T 12	7/18/84	12 da	Svetlana Savitskaya Vladimir Dzhanibekov Igor Volk	Savitskaya becomes the first woman to walk in space
Discovery	8/30/84	6 da	Henry W. Hartsfield, Jr. Michael L. Coates Steven A. Hawley Judith Resnik Richard M. Mullane Charles D. Walker	Third shuttle in fleet deploys 3 satellites and tests a solar sail
Challenger	10/5/84	7 da	Robert L. Crippen Jon A. McBride Kathryn D. Sullivan Sally K. Ride Marc Gameau David C. Leestma Paul D. Scully-Power	Carries first Canadian astronaut (Gameau); deploys Earth Radiation Budget Satellite and monitors land formations, ocean currents, and wind patterns; uses Sir-B radar system to see beneath surface of sand

Craft	Date	Duration	Crew	Remarks
Discovery	11/8/84	7 da	Frederick H. Hauck David M. Walker Anna L. Fisher Joseph P. Allen Dale A. Gardner	Salvages two inoperative satellites and returns them to Earth for repair
Discovery	1/24/85	2 da	Thomas K. Mattingly, II Loren J. Schriver James F. Buchli Ellison S. Onizuka Gary E. Payton	"Secret" military mission
Discovery	4/12/85	6 da	Karol J. Bobko Donald E. Williams Jake Garn Charles D. Walker Jeffrey A. Hoffman S. David Griggs M. Rhea Seddon	First US senator in space (Garn)
Challenger	4/29/85	7 da	Robert F. Overmyer Frederick D. Gregory Don L. Lind Taylor G. Wang Lodewijk van den Berg Norman E. Thagard William Thornton	Carries European Spacelab module to conduct 15 experiments in space
Soyuz T 13	6/6/85	112 da	Vladimir Dzhanibekov Viktor Savinykh	Successfully repairs damage to *Salut 7*, which had suffered power failure
Discovery	6/17/85	6 da	John O. Creighton Shannon W. Lucid Steven R. Nagel Daniel C. Brandenstein John W. Fabian Salman al-Saud Patrick Baudry	First Arab in space (Prince Sultan Salman al-Saud); successfully launches 4 satellites
Challenger	7/29/85	7 da	Roy D. Bridges, Jr. Anthony W. England Karl G. Henize F. Story Musgrave C. Gordon Fullerton Loren W. Acton John-David F. Bartoe	Carries Spacelab 2, a group of scientific experiments
Discovery	8/27/85	7 da	John M. Lounge James D. van Hoften William F. Fisher Joe H. Engle Richard O. Covey	Repairs satellite Syncom 3
Soyuz T 14	9/17/85	65 da	Vladimir Vasyutin Aleksandr N. Volkov Georgi M. Grechko	Takes supplies to *Salut 7*; terminates early to return Vasyutin to Earth because he is ill

(Continued)

Craft	Date	Duration	Crew	Remarks
Atlantis	10/4/85	2 da	Karol J. Bobko Ronald J. Grabe David C. Hilmers William A. Pailes Robert C. Stewart	Fourth shuttle brings fleet up to planned size
Challenger	10/30/85	7 da	Henry W. Hartsfield, Jr. Steven R. Nagel Bonnie J. Dunbar Guion S. Bluford, Jr. Ernst Messerschmid Reinhard Furrer Wubbo J. Ockels	Carries Spacelab 1-D scientific experiments conducted by Germans
Atlantis	11/26/85	7 da	Brewster H. Shaw, Jr. Bryan D. O'Conner Charles Walker Rudolfo Neri Vela Jerry L. Ross Sherwood C. Spring Mary L. Cleave	Works on techniques for assembling structures in space; first Mexican astronaut (Vela)
Columbia	1/12/86	5 da	Robert L. Gibson Charles F. Bolden, Jr. George D. Nelson Franklin R. Chang-Diaz Steven A. Hawley Robert J. Cenker	First US congressman in space (Nelson)
Challenger	1/28/86	73 sec	Francis R. Scobee Michael J. Smith Robert E. McNair Ellison S. Onizuka Judith A. Resnik Gregory B. Jarvis Christa McAuliffe	When O-rings in the solid-fuel boosters wear through, the entire fuel supply explodes, killing all 6 regular astronauts and elementary-school teacher Christa McAuliffe

After The *Challenger* Disaster; The First Real Space Station

Craft	Date	Duration	Crew	Remarks
Mir	2/20/86	Still in space	Variable	Soviet space station is launched without a crew
Soyuz T 15	5/5/86	125 da	Vladimir Solovyev Leonid Kizim	First cosmonauts board *Mir* space station
Soyuz TM 2	2/7/87	326 da	Yuri Romanenko Aleksandr Laveykin	Both cosmonauts begin long tours in space
Soyuz TM 3	7/23/87	8 da	Aleksandr Alexandrov Aleksandr Viktorenko Muhammad Faris	First Syrian in space (Faris)
Soyuz TM 4	12/20/87	366 da	Vladimir Titov Musa Manarov Anatoly Levchenko	Cosmonauts set new record of year in space about the *Mir* space station—366 days

Craft	Date	Duration	Crew	Remarks
Soyuz TM 5	6/7/88	10 da	Aleksandr Alexandrov Viktor P. Savinykh Anatoly Y. Solovyov	Alexandrov becomes the first Bulgarian in space; he is not the same Alexandrov as on Soyuz TM 3 flight
Soyuz TM 6	8/29/88	9 da	Vladimir Lyakhov Valery Polyakov (Abdul) Ahad (Mohmand)	Ahad first Afghan in space; on 9/6/88 Lyakhov and Ahad are stranded 24 hr as they attempt to return in Soyuz TM 5; they land safely on 9/7/88
Discovery	9/29/88	4 da	John M. Lounge David Hilmers Frederick H. Hauck George D. Nelson Richard O. Covey	Redesigned shuttle makes first flight since Challenger disaster
Soyuz TM 7	11/26/88	152 da	Aleksandr Volkov Sergei Krikalev Jean-Loup Chrétien	Mir is temporarily abandoned for first time when cosmonauts return to Earth
Atlantis	12/2/88	4 da	Robert L. Gibson Jerry L. Ross William M. Shepherd Guy S. Gardner Richard M. Mullane	"Secret" military mission deploys a radar spy satellite
Discovery	3/13/89	4 da	Michael L. Coats John E. Blanha James F. Buchli James P. Bagian Robert C. Springer	Deploys NASA's third relay satellite and tests thermal control system for proposed US space station
Atlantis	5/4/89	4 da	David M. Walker Ronald J. Grabe Mary L. Cleave Norman E. Thagard Mark C. Lee	Launches space probe Magellan on its way to map Venus with radar
Columbia	8/9/89	5 da	Brewster H. Shaw, Jr. Richard N. Richards David C. Leestma James C. Adamson Mark N. Brown	"Secret" military mission launches spy satellite
Atlantis	10/18/89	5 da	Donald E. Williams Michael J. McCulley Shannon W. Lucid Franklyn R. Chang-Diaz Ellen S. Baker	Launches Galileo space probe, which travels to Venus, returns twice to vicinity of Earth, and then will take up long-term orbit about Jupiter
Discovery	11/22/89	5 da	Frederick D. Gregory John E. Blaha F. Story Musgrave Kathryn C. Thornton Manley Lanier Carter, Jr.	"Secret" military mission

(Continued)

Craft	Date	Duration	Crew	Remarks
Columbia	1/9/90	10 da 21 hr	Daniel C. Brandenstein Bonnie J. Dunbar Marsha S. Ivens G. David Low James D. Wetherbee	Launches communications satellite Syncom IV and retrieves the Long Duration Exposure Facility, which has been in orbit since 4/7/84; longest shuttle flight to date; after return, problems found with fibers in working parts of shuttle
Soyuz TM 9	2/13/90	6 da	Anatoly Solovyev Aleksander Balandin	To relieve Alexander S. Vitorenko and Alexander A. Serebrov; on Jul 18, Solovyev and Balandin are briefly trapped outside *Mir* by a faulty hatch
Atlantis	2/28/90	6 da	John O. Creighton John H. Caspar David C. Hilmers Richard M. Mullane Pierre J. Thuot	"Secret" military mission launches spy satellite that fails and is soon burned up in atmosphere
Discovery	4/24/90	5 da	Loren J. Shriver Charles F. Bolden, Jr. Bruce McCandless 2nd Steven H. Hawley Kathryn D. Sullivan	Launches Hubble Space Telescope
Soyuz TM 10	8/4/90	6 da	Gennadi Manakov Gennadi Strekalov	Cosmonauts take their turn in *Mir*
Discovery	10/6/90	4 da	Richard N. Richards Robert D. Cabana Bruce E. Melnick William M. Shepherd Thomas D. Akers	Launches Ulysses space probe into orbit about Sun
Atlantis	11/15/90	6 da	Richard O. Covey Frank L. Culbertson, Jr. Charles D. Gemar Carl J. Meade Robert C. Springer	"Secret" military mission proclaimed as last that will be intended to be kept from public
Columbia	12/2/90	9 da	Vincent D. Brand Guy S. Gardner Jeffrey A. Hoffman John M. (Mike) Lounge Robert A.R. Parker Samuel T. Durrance Ronald A. Parise	After long delays caused by leaking hydrogen, shuttle carries a set of 3 ultraviolet and 1 X-ray telescopes, an instrument called ASTRO-1
Soyuz TM 11	12/2/90	9 da	Viktor Afansev Musa Manarov Toyohiro Akiyama	Japanese journalist Akiyama visits *Mir* on an outing sponsored by Japanese corporations and returns; other 2 replace Gennadi Manakov and Gennadi Strekalov in *Mir*

Craft	Date	Duration	Crew	Remarks
Atlantis	4/5/91	6 da	Steven R. Nagel Kenneth D. Cameron Linda M. Godwin Jerry L. Ross Jerome Apt	Launches 17-ton Gammar Ray Observatory; an unscheduled spacewalk is required to get the satellite's antenna to open properly
Discovery	4/28/91	8 da	Michael L. Coats L. Blaine Hammond, Jr. Gregory J. Harbaugh Charles Lacy Veach Guion S. Bluford, Jr. Richard J. Hieb Donald R. McMonagle	First military mission under new nonsecret policy; tests detection devices developed for space use
Columbia	6/5/91	10 da	Bryan D. O'Connor Sidney M. Guttierrez James P. Bagian Margaret Rhea Seddon Francis A. Gaffney Millie Hughes-Fulford Tamara E. Jernigan	Performs experiments to test human and animal adaptation to space
Atlantis	8/3/91	8 da	John E. Blaha Michael A. Baker Shannon W. Lucid G. David Low James C. Adamson	Performs 22 experiments and launches communications satellite
Discovery	9/12/91	5 da	John O. Creighton Kenneth S. Reightler, Jr. Mark N. Brown James F. Buchli Charles D. Gemar	Launches Upper Atmospere Research satellite
Atlantis	11/24/91	7 da	Frederick D. Gregory Terence T. Henricks James S. Voss Mario Runco Jr. F. Story Musgrave Thomas J. Hennen	Delayed 5 days, this is first completely nonsecret military flight; studies how well military installations can be seen from space, which turns out to be very well indeed; deploys satellite as well; lands early as a result of failure of a navigation unit (second time ever that early landing is forced by equipment failure)

MAJOR ACCOMPLISHMENTS OF SATELLITES AND SPACE PROBES

While much attention is focused on human beings in space, most of the serious scientific progress has been made by satellites or probes—the general name for space vehicles not carrying humans and not orbiting Earth—that are directed internally or from Earth. Since 1957 there have been hundreds of such satellites and probes. Here are some of their most important achievements.

Launch Date	Name	Accomplishment
10/4/57	Sputnik 1	First satellite to orbit Earth (USSR)
1/31/58	Explorer 1	First satellite to detect Van Allen radiation belts; first US satellite
1/2/59	Mechta	First space probe to go into orbit around Sun, passing 8050 km (5000 mi) from Moon (at which it was aimed) (USSR)
3/3/59	Pioneer 4	First US probe aimed at Moon; like Mechta, it misses and goes into orbit about sun
3/17/59	Vanguard	Demonstrates that Earth is pear-shaped with slight bulge in southern hemisphere (US)
9/12/59	Lunik II	First space probe to reach Moon, where it crash-lands (USSR)
10/4/59	Lunik III	First space probe to return photographs of far side of Moon (USSR)
4/1/60	Tiros I	First weather satellite (US)
6/22/60	Transit I-B	First navigational satellite (US)
8/12/60	Echo I	First communications satellite—actually a large balloon off which radio signals can be bounced (US)
8/18/60	Corona (Discoverer 14)	First US spy satellite
12/12/61	Venera 1	First space probe intended to reach another planet—Venus (USSR)
3/7/62	OSO I	Orbiting Solar Observatory, first major astronomical satellite (US)
4/23/62	Ranger IV	First US space probe to reach Moon
4/26/62	Cosmos	First Soviet spy satellite
7/10/62	Telstar	First active communications satellite, allowing direct television between Europe and United States (US)

Launch Date	Name	Accomplishment
8/27/62	Mariner 2	First space probe to reach vicinity of another planet (Venus) and return scientific information (US)
10/31/62	Anna I-B	First satellite intended for accurately measuring shape of Earth (US)
11/1/62	Mars I	First space probe aimed at Mars; contact lost about 106 million km (66 million mi) from Earth (USSR)
6/26/63	Syncom II	First communications satellite to go into synchronous orbit with Earth
7/28/64	Ranger VII	Returns close-up photographs of Moon before crashing into it (US)
8/28/64	Nimbus I	First weather satellite to be stabilized so that its cameras always point toward Earth (US)
11/28/64	Mariner 4	Flies by Mars and takes twenty-one pictures of its surface, successfully transmitting them back to Earth; its closest approach is 9850 km (6118 mi) (US)
4/6/65	Early Bird	First commercial satellite (US)
4/23/65	Molniya I	First Soviet communications satellite
7/16/65	Proton I	At 12,200 kg (26,896 lb), it is largest Earth satellite to date (USSR)
11/16/65	Venera 3	Crash-lands on Venus; first space probe to make physical contact with another planet; radio contact is lost before it reaches immediate vicinity of planet (USSR)
11/26/65	A-1	First satellite to be lauched by nation other than Soviet Union or United States (France)
12/16/65	Pioneer 6	Launched into solar orbit, it continues to function to this day (US)
1/31/66	Luna 9	Although main vehicle crash-lands, ejected capsule lands nondestructively and transmits photographs to Earth (USSR)
3/31/66	Luna 10	First space vehicle to go into orbit about Moon (USSR)
5/30/66	Surveyor 1	First soft landing of complete vehicle on Moon (US)
8/10/66	Lunar Orbiter 1	First US space vehicle to go into orbit about Moon
6/12/67	Venera 4	Ejects instrument package into atmosphere of Venus; package parachutes toward surface; contact lost before reaching surface (USSR)
6/14/67	Mariner 5	Second US satellite to reach vicinity of Venus
9/15/68	Zond 5	First Soviet satellite to return to Earth from vicinity of Moon

(Continued)

Launch Date	Name	Accomplishment
2/11/70	Ohsumi	First satellite launched by Japan
4/24/70	Mao I	First satellite launched by China; it broadcasts song "The East Is Red" once a minute from orbit, pausing at the end to send other signals
8/17/70	Venera 7	First Venus probe to return signals from surface of planet (USSR)
9/12/70	Luna 16	First space probe to land on Moon without humans aboard; scoops up samples and returns them to Earth (USSR)
11/10/70	Luna 17	Carries roving vehicle to Moon's surface; vehicle roams for 2 weeks at a time (during daylight), then "sleeps" during darkness; returns photos and other data to Earth (USSR)
12/12/70	Uhuru	First X-ray satellite telescope (US)
5/28/71	Mars 3	First space probe to soft-land on Mars, although it quickly ceases functioning (USSR)
5/30/71	Mariner 9	First space probe to orbit another planet (Mars); returns 7329 photographs of planet (US)
3/2/72	Pioneer 10	First space probe to study Jupiter and, on Jun 13 1983, first to leave solar system (US)
7/23/72	Landsat I	First Earth resources satellite (US)
3/6/73	Pioneer 11	First space probe to reach vicinity of Saturn (US)
11/3/73	Mariner 10	First space probe to observe two planets, Venus and Mercury, and only probe ever to observe Mercury (US)
12/10/74	Helios	First German space probe
6/8/75	Venera 9	Returns first photographs from surface of Venus (USSR)
8/20/75	Viking 1	First US space probe to soft-land on Mars; continues to return data until May 1983
9/20/75	Viking 2	Successfully soft-lands on Mars (US)
8/20/77	Voyager 2	After studying Jupiter and Saturn, becomes first space probe to reach vicinities of Uranus and Neptune (US)
9/5/77	Voyager 1	After studying Jupiter, becomes first space probe to reach vicinity of Saturn (US)
1/26/78	IUE	International Ultraviolet Explorer—the only astronomical satellite to be placed in geosynchronous orbit; it is still sending back data
5/20/78	Pioneer Venus 1	First space probe to go into orbit about Venus (US)

Launch Date	Name	Accomplishment
6/26/78	Seasat	Analyzes ocean currents and ice flow (US)
8/12/78	ISEE-3 (ICE)	Originally the third International Sun-Earth Explorer, space probe is renamed International Cometary Explorer (ICE) when it is redirected to study tail of comet Giacobini-Zinner in 1983 (US)
12/13/78	HEAO-2 (Einstein)	High-Energy Astronomy Observatory makes high-resolution X-ray images of universe (US)
2/24/79	P78-1	Studies solar radiation until purposely shot down by US Air Force Sep 13 1985; still working at time of its destruction, satellite is deemed by many scientists to be too valuable to be used as target (US)
2/14/80	Solar Max	Studies solar radiation; after failure in Nov 1980, it is repaired and relaunched from space shuttle in Apr 1984; is finally pushed to its destruction by massive solar flare, Dec 2 1989 (US)
1/25/83	IRAS	Infrared Astronomical Satellite studies galactic and extragalactic infrared sources and discovers new stars forming as well as possible planet formation (US)
12/15/84	Vega 1	First Soviet mission to study Halley's comet; along the way it drops balloon probe into atmosphere of Venus
12/21/84	Vega 2	Second Soviet mission to Halley's comet, it also releases balloon probe at Venus
1/7/85	Sakigake	First Japanese mission to study Halley's comet (this one from far away)
7/2/85	Giotto	Joint European mission to Halley's comet; passes closest to comet, 600 km (375 mi)
8/18/85	Suisei	Japanese mission to Halley's comet
2/21/86	SPOT	French satellite designed to photograph surface details of Earth as small as 9 m (30 ft) across
5/4/89	Magellan	US probe now orbiting Venus and mapping it in detail with radar
10/18/89	Galileo	US-German probe that, after passing near Venus and Earth (twice), will orbit Jupiter, report on Jovian Moons, and drop probe into Jupiter's atmosphere
11/18/89	Cosmic Background Explorer (COBE)	Studies cosmic background radiation caused by big bang
4/24/90	Hubble Space Telescope	An optical telescope on satellite orbiting Earth

(Continued)

Launch Date	Name	Accomplishment
6/1/90	ROSAT	German-British-American satellite that, after surveying the X-ray sky, will focus on 1000 most interesting targets
10/6/90	Ulysses	US-European Space Agency mission to study previously unobserved north and south poles of Sun
4/5/91	Gamma Ray Observatory	A 15.5-metric-ton (17-ton) telescope for observing universe at very short wavelengths
9/12/91	Upper Atmosphere Research Satellite	A 6.6-metric-ton (7.25-ton) satellite designed to observe ozone, greenhouse gases, and winds in the upper atmosphere
early 1992 (sched)	Extreme Ultraviolet Explorer (EUVE)	US satellite that will study high range of ultraviolet radiation in universe
7/92 (sched)	Geotail	US-Japanese satellite with elongated Earth orbit to study the tail of Earth's magnetosphere
9/92 (sched)	Mars Observer	US space probe that will orbit Mars

Chemistry

THE STATE OF CHEMISTRY

Chemists are excited in the early 1990s. After a period when, to many in the field, chemistry seemed to have become an area in which everything interesting had been discovered, new developments have stirred things up. The most dramatic of these are outlined on the following pages, but the excitement extends beyond the big stories.

Chemists Take Over the Universe

A lingering confusion persists in the public mind as to where chemistry stops and physics or biology begins. Chemists have no such difficulty: "Chemistry is the study of matter" is an old definition, and chemists take it more seriously than ever. If a scientist is studying matter in any way, he or she is doing chemistry. The American Chemical Society makes the point that "everything is a chemical." Some would add that chemistry, then, is the science of everything.

Some physicists fail to recognize this. They have taken to calling themselves materials scientists, although a lot of what they do is simply chemistry.

On the other side of the equation are the biochemists. While biochemistry is recognized today as a discipline in its own right, some of the synthesis of new molecules that have been inspired by biology is really the basic chemistry of bonds of various types. Are synthetic molecules that are not direct imitations of naturally occurring products of living creatures really part of biochemistry even if they are designed to do some of the same things biological molecules do? The answer to this question is not obvious.

It might be said that a chemist is any materials scientist who cares to call himself or herself a chemist. Increasingly, chemistry merges seamlessly into physics and biology. This is far from saying that chemistry is disappearing. Just because there is a continuous spectrum from red to violet is no sign that one cannot identify green. And as there is green, there is also red and violet.

BASIC DEFINITIONS

Fundamental understanding of the basic concepts of any science requires more detailed description of interlocking concepts than can be offered here, but the following brief definitions may be used either as a reminder or as an introduction. Further details on electrons, neutrons, and protons can be located in "Subatomic Particles," p 558).

Atom An electrically neutral composite particle that forms the basic building block of ordinary matter; it consists of one or more protons bound to neutrons (except for the hydrogen atom, which has a single proton and no neutrons) surrounded by the same number of electrons as protons.

Atomic number The number of protons in an atom of an element; each element is characterized by its atomic number, which ranges from 1 for hydrogen to 109 for an as-yet unnamed element.

Bond Any of several types of attachment that can form between atoms to produce a molecule.

Catalyst A chemical that increases the rate of a reaction between two other chemicals without itself being changed. (Also see Background material on p 168.)

Electron A small negatively charged fundamental particle.

Element A chemical that cannot be broken down into simpler substances because it is formed entirely of one kind of atom.

Ion An atom that has become electrically charged because it has gained or lost an electron.

Matter Ordinary matter is any chemical made from atoms or ions; some subatomic particles, including but not limited to electrons, neutrons, and protons, are also considered to be matter.

Molecule A particle consisting of two or more atoms connected to each other by bonds.

Neutron A neutral particle that is found in most atoms; it resembles a proton closely and often decays into a proton, releasing its charge and other energy in the form of smaller particles to produce one kind of radioactivity.

Proton A relatively heavy positively charged particle found in all atoms.

What Can One Look for from Chemistry?

First, one can expect progress in the fields discussed in this section (except possibly cold fusion—see "Cold Fusion?." p 162). Most of the topics covered are exciting to chemists because these studies are near the beginning of major developments.

New tools are making a difference. The diamond anvil (see "The State of Earth Science," p 191) allows chemists to study materials under very high pressures. Such studies are interesting for reasons that go beyond the local concerns of earth scientists studying Earth's interior or astronomers wondering about Jupiter's core. The scanning tunneling microscope (see "Scanning Tunneling Microscope Finds New Uses," p 596) that allows the examination and manipulation of individual atoms, molecules, and small clusters may have been invented by a pair of physicists, but chemists soon found out about it, as well as the various devices inspired by it. Pushing pairs of individual atoms or molecules together in ways that they normally do not fit looks like it will turn out to be an important activity.

Lasers have been a vital tool in chemistry for various purposes for some time, but recently they have become increasingly important in new techniques for slowing down atoms and molecules. At room temperature, atoms and molecules are constantly moving, although in solids they do not move very far in any one direction. Lasers have been used to slow individual atoms or molecules almost to stopping, allowing them to participate in reactions that could not occur at their usual speed. Lasers have also been vital in studying small clusters of atoms or molecules. The only other known way to slow atoms and molecules to near zero velocity has been bulk cooling to temperatures near absolute zero. This has not been especially useful in recent chemistry, although it remains extremely important to materials scientists.

There is a Monty Python television routine about a Society for Putting One Thing on Top of Another, the point of which is that its members are engaged in a perfectly useless activity. Chemists could organize a similar Society for Putting One Thing Inside Another. The work of such a society would represent not only a way to solve some theoretical problems, it would also have the potential to produce materials of considerable practical value. Putting an atom (or ion or molecule) inside a molecule or other small cage both alters the properties of the cage and protects the entity inside it from the outside. The latter property has been mainly exploited in biological delivery systems, but it could have many far-reaching uses.

Chemists have always been specialists in finding new ways to assemble molecules. The recent work with self-replicating and self-assembling molecules suggests that some chemists are abdicating their traditional role to the molecule itself. In this case, the successes of the late 1980s and early 1990s are clearly opening thrusts. Progress in self-replication could lead to many practical possibilities and also could help people to understand how life evolved.

The chemistry of short-lived elements, all of which are artificial and highly radioactive, is becoming gradually known. While this is largely of theoretical importance so far, there is some reason to believe that elements just a few protons beyond the current high-proton champion of atomic number 109 may be more stable. If that is the case, then chemistry on the fly may also turn out to pave the way for practical applications as well.

Another field that is recognized as important in the early 1990s is the analysis of biological substances. Although biologists or medical researchers usually discover the actions of specific substances in plants or other organisms, chemists are needed to find out what the substances really are and, if possible, how to synthesize them. This program is urgent because of the possible nearness to extinction of many species (see "Is This a Mass Extinction?," p 390).

Applied Chemistry Is Still Important

Most of the developments referred to in the previous paragraphs pertain to phenomena that are being studied in the laboratory. They are a long way from being the polyvinyls, the high-octane gasolines, the synthetic rubbers, the 12-hour cold capsules, the water-based house paints, or similar advances that have changed the way we live. Practical progress is being made, however. There is still a big future in new plastics, for example. While the only true artificial enzyme took 2 years to design and 2 months to manufacture in laboratory-sized amounts, inorganic catalysts of great promise are moving quickly from the laboratory into the factory (see "Zeolites," p 171).

Timetable of Chemistry to 1990

450 BC	Greek philosopher Leucippus [b. 490 BC] of Miletus is the first to introduce the concept of an atom, expanded on about 430 BC by his pupil Democritus [Greek: 470–380? BC]
1662	Robert Boyle [English: 1627–1691] announces Boyle's law: for a gas kept at a constant temperature, pressure and volume vary inversely
1670	Boyle discovers hydrogen
1755	Joseph Black [Scottish: 1728–1799] discovers carbon dioxide
1772	Joseph Priestley [English–US: 1733–1804] notes that burning hydrogen produces water
	Daniel Rutherford [Scottish: 1749–1819] and several other chemists discover nitrogen
	Karl Wilhelm Scheele [Swedish: 1742–1786] discovers oxygen, but does not announce his discovery until after the independent discovery by Joseph Priestley in 1774
1778	Antoine-Laurent Lavoisier [French: 1743–1794] discovers that air is mostly a mixture of nitrogen and oxygen
1781	Lavoisier states the law of conservation of matter
1784	Henry Cavendish [English: 1731–1810] announces that water is a compound of hydrogen and oxygen

1791	Jeremias Richter [German: 1762–1807] shows that acids and bases always neutralize each other in the same proportion
1803	John Dalton [English: 1766–1844] develops the atomic theory of matter
1807	Amedeo Avogadro [Italian: 1776–1856] proposes that equal volumes of gas at the same temperature and pressure contain the same number of molecules (Avogadro's law)
1824	Joseph-Louis Gay-Lussac [French: 1778–1850] discovers chemical isomers, chemicals with the same formula but different structures
1828	Friedrich Wöhler [German: 1800–1882] prepares an organic compound from inorganic chemicals, showing that living matter is basically the same as other matter
1859	Gustav Kirchhoff [German: 1824–1887] and Robert Bunsen [German: 1811–1899] introduce the use of the spectroscope to identify elements from the light they give off when heated or burned
1868	Pierre-Jules-César Janssen [French: 1824–1907] and Sir Joseph Lockyer [English: 1836–1920] discover helium by observing the spectrum of the sun
1869	Dimitri Mendeléev [Russian: 1834–1907] publishes his first version of the periodic table of the elements
1875	Paul-Emile Lecoq de Boisbaudran [French: 1838–1912] discovers gallium, the first discovery of a predicted element (predicted by Mendeléev on the basis of his periodic table)
1906	Mikhail Tsvett [Russian: 1872–1919] develops paper chromatography, the beginning of modern methods of chemical analysis
1908	Fritz Haber [German: 1868–1934] develops a cheap process for making ammonia from nitrogen in the air
1943	Albert Hofmann [Swiss: 1906–] discovers that LSD is hallucinogenic
1962	Neil Bartlett [English: 1932–] shows that the noble gases can form compounds by creating a compound of xenon
1974	Frank Sherwood Rowland [US: 1927–] and Mario Molina [US: 1943–] warn that chlorofluorocarbons (Freons) are destroying the ozone layer in the atmosphere that protects life on Earth from harmful ultraviolet radiation
1982	German scientists produce one atom of element 109
1985	Richard E. Smalley (US: 1943–] and Harry W. Kroto (English: 1939–] discover fullerenes, such as buckminsterfullerene, forms of carbon in the shape of geodesic domes and spheres
1987	Scientists confirm that a hole in the ozone that forms in Antarctica in Aug and Sep is caused by chlorofluorocarbons
1989	On Mar 23 B. Stanley Pons and Martin Fleischmann [English: 1927–] announce that they have discovered how to fuse hydrogen nuclei at room temperature, a claim hotly disputed to this day

NEW FORMS OF MATTER

CLUSTERS OF ATOMS

A paradox from the Greek philosopher Eublides asks how there can be a heap of sand. Surely one grain of sand is not a heap. Neither are two grains. Or three. So adding one grain does not make a heap, unless you already have a heap. Yet if you continue the process, you do get a heap.

A cluster in chemistry is a heap of atoms or molecules. The reason for discussing clusters is that they do not behave like single atoms or molecules. Nor do they behave like familiar solids, most of which are crystals (it is hard to remember that metals such as iron are crystals; today one is likely to picture New Age pieces of quartz when encountering the word *crystal*). Unlike the sand heap of the Eublides paradox, clusters can be strictly defined. A cluster is more than a molecule but less than a piece of a crystal. If you keep adding the appropriate atom or molecule to a cluster, eventually the substance will start to acquire crystalline structure. At that point, it is no longer a cluster. Sometimes bulk properties appear for very few atoms. For example, a cluster of eight sodium atoms has electrical and optical properties similar to solid sodium.

Although clusters have been studied for almost 20 years, only in the late 1980s and early 1990s has the subject commanded extensive attention. One reason is that new methods have been developed recently to produce clusters with sizes from 3 to 20,000 atoms. Another is that newly developed types of lasers can be used to analyze the clusters after they are created.

Clusters are interesting because they do not behave like single atoms or like molecules or crystals. Some have called clusters a fifth state of matter because their properties differ from other kinds of matter. For example, solid cadmium sulfide melts at 1477° C (2691° F), but clusters of the same material have a melting point of only 848° C (1559° F). Clusters of metal atoms act like giant atoms electrically, a construct known as jellium. Because of their jellium structure, metal clusters have so-called "magic numbers" of atoms. Clusters with such magic numbers of atoms form easily, while those with other numbers of atoms are difficult to form. (Large atoms also have similar magic numbers of protons and neutrons.)

Clusters are often sensitive to size or to the exact numbers of atoms or molecules per cluster. Cadmium selenium clusters change color from red to orange as they go from about 3000 to about 1000 atoms, while other sizes produce black or even light green. A cobalt cluster with 10 atoms reacts easily with either hydrogen or nitrogen, while one with 9 atoms does not react with either.

Because clusters have much more surface for their volume than solids do, they are more reactive than solids. Most chemical reactions take place

STATES OF MATTER

In elementary school one used to be taught that there are three states of matter—solid, liquid, and gas. As with much in life, it is more complicated than that. To begin with, the solid state might more properly be called the crystalline state for most solids; solids that are not crystals, such as glasses, are more like liquids in some ways than they are like solids. The classic three states are identified by how ordered they are; a crystal forms in a definite pattern, a liquid has atoms or molecules that slide over each other, while a gas has atoms or molecules that are not much attracted to each other so that they tend to move apart. This results in the familiar idea that a solid holds its shape, a liquid flows to match the shape of its container but stays flat on top, while a gas completely takes the shape of its container and disperses if not contained.

These three states are composed of atoms, ions, or molecules. Atoms and molecules are comprised of subatomic particles called protons, neutrons, and electrons. The protons and neutrons tend to stick together to form an entity called the nucleus. When heated sufficiently, the electrons, which stay closely associated with the nucleus in atoms and molecules, go off on their own. The resulting material is so different in many properties from the other states that it is given its own name, a plasma. Plasma is the fourth state of matter.

---

at the surface. Therefore, some clusters can be used as catalysts. Not all clusters can be used this way, however. For instance, experiments show that silicon clusters with fewer than 70 atoms are less reactive than silicon crystals, not more. The most likely reason for this is that the clusters form structures that are more compact—atoms get closer to each other—than silicon crystals.

Because of their sensitivity, chemists think that clusters can be finely tuned to specific reactions. Their optical properties can also be useful in optical computing (see "Optical Computers," p 580) or in communications. Also, they may have structural uses in new forms of metals or ceramics. A ceramic is typically the result of heating a powder, often one contained in another medium that is driven off by heat. The archetypical ceramic is formed by heating a powder of fine clay particles in water to form pottery. Changing from an ordinary powder to a powder made from clusters results in changed properties for the resulting ceramic. Titanium oxide, like most ceramics, is quite brittle when made the ordinary way. But forming it from specifically sized clusters can make it as ductile as some metal alloys. Similarly, palladium made from appropriately sized clusters is four times harder and stronger than ordinary bulk palladium.

In one cluster experiment in 1990, the result was the opposite of what was expected. Clusters of 6 to 12 atoms of indium phosphide, studied by Kirk D. Kolenbrander and Mary L. Mandich of AT&T Bell Laboratories, behave very much like the bulk counterpart. In this case, the result may be useful, since many scientists and engineers are trying to build very small electronic devices. It is possible that indium phosphide will be able to replace silicon in such tiny devices.

(Periodical References and Additional Reading: *Science* 6-8-90, p 1186; *Science News* 11-3-90, p 279; *Physics Today* 12-90, p 42; *LBL Research Review* spring 91, p 2; *Science* 5-24-91, p 1085)

QUEST FOR METALLIC HYDROGEN

Hydrogen is a transparent gas. Metals are shiny solids. In the periodic table of the elements, metals are shown at the left and nonmetals at the right. But a quick glance at the most common layout for the periodic table shows hydrogen far to the left, riding high over the metals. Less commonly, periodic tables move hydrogen out of the metals to put it at the top of the column that includes fluorine and chlorine, very active gases. A way around this problem puts hydrogen off the table altogether, in a class by itself.

It would be intellectually satisfying to know for sure whether hydrogen should be classed as a metal that shares properties with sodium and potassium or a gas that shares properties with fluorine and chlorine. For the past 56 years, most scientists have believed that hydrogen would be observed as a shiny metal at sufficiently high a pressure. Under that theory, the core of the gas giant planet Jupiter, which is mostly hydrogen, might be metallic hydrogen.

Furthermore, there might be some practical benefits if metallic hydrogen exists. There is some reason to think that metallic hydrogen might persist as a metal once the pressure that created it was removed. If so, the metal would have unusual properties, among them high-temperature superconductivity (see "Superconductivity UPDATE," p 586). Hot fusion (see "Cold Fusion," p 162 and "Progress in Hot Fusion," p 546) might be accomplished more easily with metallic hydrogen as a fuel.

The advent of the diamond anvil (see "The State of Earth Science," p 191) has made it possible to achieve very high pressures, raising the hopes for metallic hydrogen. Indeed, several researchers have reported it, but others generally think their evidence is weak. Perhaps the most noticeable claim came from Ho-Kwang "Dave" Mao and Russell J. Hemley of the Carnegie Institution's Geophysical Laboratory in Jun 1989. After putting hydrogen under a pressure of above 2 million atmospheres in a diamond anvil, they observed the result by bouncing infrared radiation off the sample. They reported that the reflected radiation showed something opaque or even shiny. As usual, others quickly rushed in to dispute the claim. Mao and

THE SHELL GAME

The position in which an element appears in the periodic table is controlled by the configuration of electrons in its shells. The laws of quantum mechanics decree that the total number of electrons in each shell can be only some small fixed number, and that the outer shell can contain at maximum eight electrons. In general, elements in the same column have the same number of electrons in the outer shell, so hydrogen would seem to belong in column IA, where every element has one electron in the outer shell, and not in column VIIA, where every element has seven electrons in the outer shell. If hydrogen is in column IA, it must be a metal.

But there is another way of looking at this. Although the outer shells of most elements are complete when they contain eight electrons, the smallest outer shell, important only for helium and hydrogen, is complete when it contains two electrons. The elements in column VIIA are those that are one electron short of completing their eight-electron shell. Helium is put with the inert gases because its two-electron shell is just as complete as their eight-electron shells. By this reasoning, hydrogen, one shy of completing its two-electron shell, belongs in VIIA; in that case, it is not a metal.

Hemley, nevertheless, continued into the 1990s to investigate the transition to a metal. Also, Isaac F. Silvera claimed in Mar 1990 that his team at Harvard University found evidence of metallic hydrogen under a pressure of 1.5 million atmospheres and under temperatures as low as 4.5 K (–268.65° C or –451.6° F). Mao and Hemley also note a transition of some kind at that pressure, but observe that the material is still transparent after the transition, not a very reasonable appearance for a metal.

No one has reported a form of hydrogen that remains a metal at ordinary pressures. Also, no one has performed what some scientists say would be the acid test, showing that material believed to be metallic hydrogen conducts electricity. Do not expect to drive a car with hydrogen bumpers or rewire your house with hydrogen wire in the near future.

(Periodical References and Additional Reading: *Science News* 3-17-90, p 164; *Science* 3-30-90, p 1545; *Science* 7-27-90, pp 339 & 391; *New York Times* 3-26-91, p C1)

BUCKYBALL FEVER

Buckyballs are molecules that look like soccer balls, are ubiquitous and superconducting, and may have various other tricks. They were unknown until 1985. Certainly, no single molecule has had the impact of buckminsterfullerene and its close relatives since scientists first learned the importance of deoxyribonucleic acid (DNA—see "The State of Life Science," p 421). While it is unlikely that the fullerenes will come close to DNA in true significance, they have brought a lot of excitement to the usually staid discipline of chemistry.

Fullerenes were discovered in 1985 by Rick Smalley and Harry Kroto, but at first, there was considerable controversy over whether or not Smalley and Kroto had found anything new. Their claim was that under various conditions, including ordinary combustion, carbon atoms tended to form soccer-ball-shaped clusters of sixty atoms—in chemist's shorthand, C_{60}. Since Smalley and Kroto believe that the sixty-atom size of the typical cluster is caused by a shape identical to that of a geodesic sphere, they named the molecule and its cousins after the famous architect, inventor, and philosopher Buckminster Fuller [US: 1895-1983], who developed the half-spheres called geodesic domes.

In May 1990, Donald Huffman and Lowell Lamb of the University of Arizona and Wolfgang Krätschmer and Konstantin Fostiropoulos of the Max Planck Institute for Nuclear Physics in Heidelberg, Germany, found a

Background

Carbon is unusual among elements in the way it can bond with itself. Not only can carbon form as many as four bonds with atoms of other elements, it can form single, double, or triple bonds with itself. It has long been known that carbon can form long chains in which each carbon atom is bonded to another carbon atom—sometimes for as many as seventy carbon atoms in a row. A second result of carbon's versatility is that two carbon-based chemicals (known as organic chemicals) with the same number of atoms of the same types can have very different properties. The geometric arrangement of atoms in a molecule often determines what it does, and in three dimensions a carbon-based molecule can be left-handed or right-handed or differ in other ways. For example, both glucose and fructose, two different sugars, share the formula $C_6H_{12}O_6$, but there is a key difference in how the atoms are arranged.

Carbon atoms even by themselves can be arranged in varying geometric configurations that have vastly different properties. The example of the black greasy graphite that is used for pencil leads and the transparent hard diamond is well known. In diamond, carbon atoms bond very closely to each other in the pyramidal shape that mathematicians call a tetrahedron, so that each carbon atom is bonded to exactly four other carbon atoms in a rigid structure. In graphite, each carbon atom is bonded to three other atoms in a plane, a structure that consists of hexagons in layers that are loosely bound to each other. It was widely believed that this was the whole story, although studies showing that soot, charcoal, and other forms of carbon are basically the same as graphite were not very convincing. Nevertheless, textbooks felt comfortable in stating, "All the way to make large amounts of nearly pure samples of the material (the C_{60} buckyballs are mixed with C_{70} fullerenes). Actually, they had found the basic step of vaporizing a graphite rod in helium at one seventh of an atmosphere pressure as early as 1983. But because that was before Smalley and Kroto had discovered buckminsterfullerene, they did not know what it was they were producing. The two teams announced their discovery in Sep. By early 1991 buckyballs were commercially available at $1250 a gram and buckyball fever was in full swing. Large numbers of chemists dropped their previous research to work on aspects of fullerenes that were close to their specialties—whether electrochemistry, polymers, hydrocarbons, materials, or another.

Chemists quickly learned that buckyballs are tough, able to survive collisions with targets at 32,000 km (20,000 mi) an hour, impacts that would rip other molecules to pieces. They learned that C_{60} reacts vigorously with various organic molecules, such as ethylene diamine (resulting in a structure of a buckyball with hair). Joel M. Hawkins of the University of California, Berkeley, and coworkers were able to make X-ray diffraction studies of buckminsterfullerene that confirmed for the first time that the proposed

---

apparently amorphous forms of carbon are found to possess the graphite structure."

However, in 1985 Richard E. Smalley of Rice University in Houston TX and Harold W. Kroto of the University of Sussex UK discovered that some of the "apparently amorphous" forms of carbon possess a radically different structure. Pentagons are capable of inserting themselves into the basic hexagonal form of graphite. This causes the flat layers found in graphite to warp. If exactly a dozen pentagons are present, it is possible for the warping to be sufficient to produce a closed surface—essentially a sphere, although technically a truncated icosahedron.

The main clue Smalley and Kroto had at first was not evidence of pentagons, however. What they observed was that carbon that was encouraged to form clusters (see above) tended very vigorously to form in clusters of sixty atoms. In other words, sixty is a magic number for carbon clusters. When Smalley and Kroto tried to determine what sort of structure might cause this, they recognized that a geodesic sphere, the most stable structure made from pentagons and hexagons, has sixty vertices. The properties and possibilities of such structures had been worked out and popularized by Buckminster Fuller, so Smalley and Kroto named the sixty-atom carbon structure buckminsterfullerene.

Leonhard Euler established in the eighteenth century that all closed three-dimensional structures of pentagons and hexagons must have exactly twelve pentagons. Since buckminsterfullerene is the smallest such molecule, its structure is the most symmetric possible in ordinary three-dimensional space.

geodesic shape was correct. Most spectacular of all, Arthur F. Hebard and coworkers at AT&T Bell Laboratories announced at the Apr 1991 Atlanta GA meeting of the American Chemical Society that buckyballs doped with potassium ions are superconducting at temperatures below 18 K (–255° C or –428° F).

Chemists were not the only ones excited by the molecule—the month before the chemistry meeting, the crowded buckyball session at the American Physical Society ran from 7:30 p.m. to 1 a.m. And the interest didn't stop in the spring. At the Chemical Congress of North America in the fall of 1991, a talk on fullerenes by Smalley was stopped as he began so that a wall could be removed to accommodate the crowd that had gathered.

Forms of buckminsterfullerene were soon developed that showed superconducting properties at higher temperatures, but not so high as the so-called high-temperature superconductors (see "Superconductivity UPDATE," p 586). As experimental results were announced or rumored, however, many came to believe that large fullerenes—perhaps with as many as 540 carbon atoms—offer the most hope for the long-sought room-temperature superconductor. Surprisingly, the temperature of the transition to superconductivity lowers when metallofullerites are put under high pressures. In most superconductors, it works the other way; high pressure increases the temperature at which superconductivity begins. Because buckyball superconductors conduct equally well in all three directions, they may turn out to be more useful than the high-temperature superconductors, which are strongly two-dimensional.

As 1991 continued, it became almost impossible to keep up with the new discoveries about the fullerenes and related materials. By Jul there were reports that one doped form acted as a magnet; other forms were announced in Aug as nonlinear optical materials, ones for which light passing through can be changed in direction by altering the intensity of the light. Such materials might have applications in optical computers (see "Optical Computers," p 580). In both cases, the buckyball-related materials seem to offer advantages when compared with previous organic materials that exhibited these properties.

The practical benefits of buckyballs, if any, may still be a long way off. Although it is easy to produce enough of the materials for study, it is not yet possible to make fullerenes and related compounds on the scale needed for mass markets. Even though buckyballs combined with other chemicals seem promising for various applications, it may be difficult to get the other chemicals to lodge on or in the buckyballs in precisely the places needed. Plans to use buckyballs as lubricants are affected by both problems, for example.

Smalley has continued to be an active researcher on fullerenes, specializing in trying to substitute other atoms, such as boron or potassium, for some of the carbon atoms in a buckyball. He thinks he may have suc-

NEW NAMES

buckminsterfullerene The most stable fullerene, with sixty carbon atoms, or C_{60}.

buckyballs The soccer ball of buckminsterfullerene and the rugby balls of C_{70} as well as larger fullerenes; fullerenes smaller than C_{60} are sometimes called baby buckyballs.

fullerenes Any carbon structure with exactly twelve pentagons and enough hexagons to form a closed geodesic dome with carbon atoms at the vertices.

fullerites Solid forms of carbon in which the atoms are arranged to form fullerenes.

metallofullerites Solid materials built from fullerenes that each enclose an atom of some metal, such as potassium; these have not yet been made, although individual fullerene structures enclosing potassium and cesium atoms have been produced.

---

ceeded, but has not been able to prove it yet. Smalley and others are also occupied with putting foreign atoms into the interiors of buckyballs. This was first achieved with helium, making what Smalley calls "the world's smallest helium balloon." Smalley himself has trapped a lanthanum atom inside a buckyball, Martin M. Ross and coworkers at the US Naval Research Laboratory have succeeded with yttrium, and others have used various other metals. This process could lead to a whole new form of chemistry. One of the outstanding unresolved issues at the end of 1991, however, was how to separate the filled molecules from the empty ones. Smalley is also trying to put a buckyball inside a larger fullerene, a construct he calls a Russian doll, after the well-known nested sequence of wooden dolls.

Perhaps the most exciting thing about the fullerenes in 1991 was that no one knew for sure what might emerge. In addition to the various known possibilities, there is still likely to be a surprise or two on the way.

(Periodical References and Additional Reading: *Science* 3-23-90, p 1468; *Science* 8-24-90 pp 835, 895, & 897; *Discover* 9-90, p 53; *Science News* 10-13-90, p 238; *Science News* 12-8-90, p 357; *New York Times* 12-25-90, p I37; *Science* 2-1-91, p 516; *Science* 4-5-91, p 29; *Science* 4-12-91, pp 187 & 312; *Science* 4-26-91, pp 483, 547, & 548; *Science* 5-3-91, p 646; *New York Times* 5-21-91, p C2; *Science* 5-24-91, pp 1043, 1154, & 1160; *Science* 5-31-91, p 1288; *Science* 6-7-91, pp 1351, 1417, & 1419; *Science News* 6-8-91, p 358; *Science* 6-28-91, pp 1763, 1785, & 1829; *Science* 7-12-91, p 171; *Science* 7-19-91, pp 247, 301, & 330; *Science* 7-26-91, p 429; *Science* 8-9-91, pp 599 & 646; *Science News* 8-10-91, p 84; *Science* 8-23-91, pp 884 & 886; *Science News* 8-24-91, pp 120 & 127; *New York Times* 9-3-91, p C1; *Science* 9-27, p 1476; *Scientific American* 10-91, p 54; *Science* 10-4-91, p 30; *Science News* 12-14-91, p 391)

CHEMICAL REACTIONS

COLD FUSION?

On Mar 23 1989, electrochemists B. Stanley Pons of the University of Utah at Salt Lake City and Martin Fleischmann of the University of Southhampton in England announced that they had discovered a way to use simple equipment to make nuclear fusion occur. Most previous efforts to achieve nuclear fusion involved giant magnets, powerful lasers, or other very expensive, high-tech devices. The Pons-Fleischmann experiment used a jar filled with heavy water (water in which some of the ordinary hydrogen has been replaced with deuterium, a form of hydrogen that contains a neutron), a platinum electrode, a palladium electrode, and a source of electric current. Some nicknamed it "fusion in a jar," but the name that has stuck is "cold fusion."

Since heavy water is abundant and cheap—the technology for making it has been known for about 50 years and it is used in certain types of nuclear reactors—and indications are that less-expensive metals, such as titanium, could be substituted for palladium, if necessary. Cold fusion would be an important energy boon if it works. Pons and Fleischmann claim to have produced four times the energy that was put into their device. No one, however, has convinced the larger scientific community that they have consistently succeeded in obtaining such a result. Furthermore, most physicists who study fusion think that the reaction is not likely to occur in amounts sufficient to be useful as an energy source.

Pons and Fleischmann think that in the palladium electrode, palladium absorbs the deuterium and hydrogen that electrolysis releases from heavy water. As the deuterium and hydrogen concentrate in the palladium, atoms of the gases are squeezed close enough for fusion to occur. It is not necessary for molecules to be moving very fast for this to happen—which is the same as saying that the gas does not have to be hot; hence cold fusion.

There are several pathways by which hydrogen and heavy hydrogen can fuse. All the likely ones produce large numbers of neutrons. They also create radioactive elements. Pons and Fleischmann reported that their experiments produced a few neutrons, but only about a billionth as many as known fusion pathways make. As with the other results reported by Pons and Fleischmann, this relative absence of neutrons would make cold fusion a more attractive energy source.

The likely fusion pathways should also result in the production of either helium or tritium, the radioactive form of hydrogen that has two neutrons. Pons and Fleischmann initially did not try to detect these by-products in their early experiments.

Background

All nuclear fusion, whether cold or hot, involves the blending of two atomic nuclei into one. This releases energy because the amount of energy necessary to hold the single new nucleus together is less than the sum of the amounts needed to hold the two previous nuclei together. Fusion is the process that produces most of the energy released by the Sun and stars and hydrogen bombs.

If cold fusion occurs, the process would be similar in some ways to the fusion process in young stars and in hydrogen bombs. In stars, gravity pushes nuclei of hydrogen together, forming either heavier types of hydrogen (with additional neutrons) or forms of helium (with one more proton and perhaps additional neutrons). This process is accelerated by heat, which causes nuclei to move faster and collide more often.

The key to understanding fusion is understanding the structure of the atomic nucleus. The simplest nucleus is that of hydrogen, which consists of a single proton. Adding one neutron to that nucleus produces deuterium and adding two makes tritium. Some energy is required as glue to hold the heavier particles together. This energy can be thought of as particles called pions that pass between protons and neutrons, changing protons to neutrons and neutrons to protons. Because of this unusual glue, a nucleus cannot consist of two protons or two neutrons by themselves. Therefore, the smallest nucleus with two protons also has one neutron; it is known as helium-3. The more common form of helium (because its nucleus is glued together better) has two protons and two neutrons and is therefore helium-4.

Both splitting nuclei (fission) and gluing them together (fusion) produce energy. In fission, elements heavier than iron are broken apart. So much energy is needed as glue to keep these heavy nuclei together, that fission releases it. In fusion, light elements are combined to produce nuclei that can be no heavier than iron. For these elements, the heavier the combination, the less energy needed to hold them together. When they combine they release some of this energy, either as photons (gamma rays) or as kinetic energy (motion of subatomic particles).

Putting all this together, one could expect the following typical pathways for fusion, whether cold or hot:

hydrogen + deuterium yields helium-3 + gamma rays
deuterium + deuterium yields helium-4 + gamma rays
deuterium + deuterium yields tritium + neutron + gamma rays

Other reactions are possible when tritium is one of the founding nuclei. Furthermore, neutrons can release electrons and antineutrinos to change into protons, providing additional pathways. But in every case, one expects at least gamma rays and tritium or helium as a result, with neutrons appearing also in what physicists calculate as the most likely pathways. Most scientists would therefore accept proof that gamma rays, neutrons, tritium, or helium were produced from hydrogen and deuterium as good evidence of fusion, but would not accept heat alone.

The University of Utah had already decided in 1989 to invest $5 million in a Center for Cold Fusion Research, to be headed by Pons, and in 1989 also a cold-fusion newsletter was being marketed for $345 for twelve annual issues.

On Jun 15 1989, the American Association for the Advancement of Science (AAAS) magazine *Science* reported direct allegations of fraud—not by Pons or Fleischmann, but by scientists at Texas A&M University, where early reports seemed to confirm cold fusion. It was claimed that someone in the laboratory of chemist John O'Malley Bockris had deliberately added tritium, the radioactive form of hydrogen, to Bockris's apparatus ("spiked" the experiment). Bockris then claimed to detect nuclear activity because the equipment produced tritium. Texas A&M investigated the charges and reported on Nov 18 1990 that they are probably false. Bockris results could not have been produced by spiking. The investigating panel consisted of John Poston, Edward S. Fry, and Joseph B. Natowitz. Bockris still believes he is getting real results and that others are as well: "One hundred laboratories in some twelve countries have replicated phenomena that are similar to those reported." Despite Bockris's continuing enthusiasm, his own team apparently ceased obtaining tritium in Nov 1989.

Many other experiments seemed to confirm those of Pons and Fleischmann at first, but they often were later withdrawn because of flaws. At the First Annual Conference of Cold Fusion held in Salt Lake City UT in 1990, approximately a year after the initial announcement of the Pons-Fleischmann results, at least sixteen laboratories reported excess heat in their versions of cold-fusion experiments, while about twenty reported detection of tritium and a few found neutrons. At the same time, physicists writing in *Nature* claimed that no one, not even Pons and Fleischmann, had found no signs of fusion; editorial commentaries in the magazine denounced the whole field as worthless.

One exception is a series of experiments conducted independent of and concurrent with the original Pons and Fleischmann research. These experiments, conducted by Steven E. Jones of Brigham Young University in Provo UT, show that cold fusion can take place, but only at levels of energy so small as to have no practical value. Unlike the Pons-Fleischmann results, which many scientists feel are not valid, the Jones experiments are generally accepted.

Cold fusion began to make a comeback in 1991. More than 250 papers on the subject were produced, mostly ones showing some success in producing the effect. For example, nuclear processes coupled with heat production in Pons-Fleischmann jars were reported by chemist Melvin H. Miles and coworkers at the Naval Weapons Center at China Lake CA and at the University of Texas. Their report claimed that the level of nuclear processes and the amount of heat go up and down together, suggesting strongly that the two phenomena are related. Various other researchers obtained apparent cold fusion by using high pressure to pack deuterium gas into metal, or employing some other version of the technique, instead of using the electrochemical method pioneered by Pons and Fleischmann.

From the beginning, chemists wanted to believe in cold fusion, and many still do, while physicists claimed that it is theoretically impossible. It remains to be seen whether this controversy is a battle over turf or a genuine scientific disagreement. In Apr and May of 1991, however, some physicists broke ranks. First, Frederick J. Mayer, a consultant in Ann Arbor MI, and John R. Reitz proposed a new theory to account for cold fusion, basing it on the hypothetical existence under certain circumstances of a combined proton and electron, a particle that in their theory accomplishes the action of fusion. The following month, two books on the subject were published. Although one book on cold fusion, by physicist Frank Close of Oak Ridge, took the conventional physicists' negative point of view, a competing book that was sympathetic received an endorsement by physicist Julian Schwinger of the University of California at Los Angeles.

Through it all, the main problem has been irreproducible results. Problems centered on the delicacy of the experiments (greater than first suggested by Pons and Fleischmann), the length of time it takes to get any results (weeks or even months), and the sudden disappearance of the effect for no apparent reason. Getting results seems to depend largely on how high a concentration of deuterium nuclei one can pack into a metal—but a high density of nuclei does not guarantee success. Furthermore, the rate of packing seems to affect the process, with both high and low rates coming up empty.

By 1991 hundreds of investigators had found that some things happened some of the time, but not on a regular basis. These irregular results took three forms:

1. More energy (in the form of heat) coming from the experiment than the energy put in.

2. A low level of neutrons appearing around the apparatus.

3. Small amounts of tritium that were not present before the experiment started.

While there are ways to explain the excess heat without invoking fusion (chemical reactions often produce heat, for example), production of neutrons and tritium cannot be explained without some form of nuclear process. The amounts of these nuclear by-products are too low, however, for the amount of heat produced. One way to explain this discrepancy is to say that perhaps more than one process is going on.

(Periodical References and Additional Reading: *Science News* 4-7-90, p 212; *Science* 6-15-90, p 1299; *Science News* 6-16-90, p 374; *Science* 8-3-90, p 463; *Science* 12-14-90, p 1507; *Science* 2-1-91, p 499; *Science* 3-22-91, p 1415; *New York Times* 4-14-91, p IV1; *New York Times* 4-26-91, p A18; *Science* 6-14-91, p 1479; *Science News* 6-22-91, p 392)

RE-CREATING NATURE

SELF-REPLICATING MOLECULES THAT ARE NOT ALIVE

Julius Rebeck, Jr., Tjama Tjivkua, and Pablo Ballester of MIT announced in the Jan 30 1990 *Journal of the American Chemical Society* that they had created a molecule that could, in a suitable medium, reproduce using means similar to those employed by living systems. While theirs is not the first self-replicating molecule, it is considered more sophisticated than the small self-replicating molecules that were first produced from nucleic acids in the mid-1980s by Günter von Kiedrowsky of the University of Göttingen in Germany and Leslie Orgel of the Salk Institute in San Diego CA.

The molecule developed by Rebeck's team is officially amino adenosine triad ester, or AATE, although it is sometimes called the J-molecule from its shape. Placed in a chloroform solution of its two components, an amine and an ester, AATE grabs one of each, bonding an amine to its own ester end and an ester to its own amine end. The ester and amine are positioned to form what is called an amide bond between them. This is the same kind of bond that forms between amino acids to make proteins. The result of this process is two J-molecules that are loosely bonded to each other by hydrogen bonds. Thermal agitation is enough to break the hydrogen bonds, so the two J-molecules can float apart and seek partners to make new copies. Chemists think of it this way:

AATE + amine + ester → 2 AATE

Because there is a geometric progression in the number of copies, AATE can duplicate itself a million times a second if it has the proper "food." If there is no AATE added to the solution to be a template, the same reaction occurs, but a hundred times slower. A major breakthrough would be to produce a self-replicating molecule that assembles two or more components in a particular sequence that happens only with the template present.

Rebeck has been working on this problem for several years. The principal challenge has been to fine-tune the amount of attraction in the weak bonds. Early versions tended to be too strong. One version caused the template to grab its own ends and effectively stop the reaction before it could make a new molecule. Another held onto the first new molecule it produced, stopping the reaction at that point.

The practical importance of a self-replicating molecule is that it might have as one of its components a useful chemical that is difficult to manufacture by ordinary means. After the molecule had been allowed to duplicate itself enough times, it should be easy to break off the useful part and separate it from the rest—with the separated parts being returned to solution as part of the "food" for the larger molecule. This is somewhat like the way that living cells make difficult-to-produce substances. It is also somewhat like catalysis in that part of the chemicals involved could be recycled without using them up.

The interesting scientific question related to this issue is how closely creating such self-replicating molecules resembles the origin of life. A prominent theory of the origin of life involves self-replicating molecules of ribonucleic acid (RNA) floating in a solution that contains suitable bases. The RNA in this hypothesis uses the bases to make copies of itself.

Self-Assembly

In a related development, two groups of chemists announced that they had developed chemicals that, when the right ingredients are combined in the right order, assemble into large molecules. There is no template, such as the J-molecule. Instead, the large molecules put themselves together by using electrostatic interactions and hydrophobic (water avoiding) regions coupled with the proper shapes for self-assembly.

Think of the hydrophobic regions as being directed by forces comparable to those between similar poles of a bar magnet (likes repel) and the electrostatic bonds as directed by forces comparable to those of the opposite poles of magnets (opposites attract). If you put a cylinder and a long piece that could thread through it together in a jar and shake them, you can use such forces to cause the long piece to stick with its middle in the cylinder and the two ends outside. Then, if you put in a couple of wing nuts that had opposite magnetic orientation to the ends of the long piece and shook it some more, you would end up with a large piece in which the long piece is held in place both by the internal situation with the cylinder and the wing-nut caps at each end.

In effect, this is what David Lawrence and Tata Venkata S. Rao of the State University of New York at Buffalo did at the molecular level, as they reported in the Apr 25 1990 *Journal of the American Chemical Society.* Their cylinder was a starchlike molecule called a cyclodextrin. The long piece was a diammonium salt that had a middle part that avoids contact with water—so it wants to hide where it is protected by the cyclodextrin. The ends of the diammonium are electrostatically attracted to molecules that resemble four-bladed wing nuts.

A more complex self-assembly based on similar ideas was reported by Lawrence and John S. Manka in the Mar 14 1990 issue of the same journal. They used some of the same basic pieces coupled with others to make an eleven-piece self-assembler. Lawrence is working toward developing self-assembling molecules that bond and release oxygen so that they could serve as artificial blood.

(Periodical References and Additional Reading: *Science News* 2-3-90, p 69; *Science News* 5-19-90, p 318; *Science* 6-29-90, p 1609; *Discover* 12-90, p 28)

NEW ENZYME SYNTHESIZED

Many of the most important functions in living beings are carried out by protein catalysts called enzymes that both assemble and disassemble other organic molecules. Genetic engineers can synthesize some naturally occurring enzymes, but one goal of biochemistry is to create new enzymes that can be tailored to specific needs. New enzymes might be more powerful at accomplishing the same tasks naturally occurring enzymes do, or they might do something totally different. Since 1986 researchers have used antibodies to create something like those enzymes, but such catalytic antibodies are only semisynthetic. The real trick is to start with a bunch of

Background

A catalyst is any substance that speeds up a reaction without changing itself. For example, molecules of chemical A and chemical B may react with each other in a solution as a result of random collisions. If permanent bonding forms, the result over a long period of time might be new chemicals combination, which can be labeled AB. If, instead, the time it takes AB to break back into A and B as a result of random shaking of the molecule by heat is about the same as the time span for the reaction forming it, the solution will always contain A and B with perhaps a trace amount of AB. Even though a reaction is occurring slowly, the opposite reaction is undoing it at about the same speed.

Now enter the catalyst, which can be called C. Adding C makes the reaction that combines A and B into AB happen faster. In general, the method C uses to accomplish this is by binding loosely with A and B to form an unstable ABC, a combination that very quickly breaks apart into AB and C. Since this speeds up the production of AB so that it is faster than AB's breakdown, the solution soon consists almost entirely of AB and C, with perhaps a trace of A and B in uncombined form. The catalyst may then be removed to be used again.

Often the reaction that is desired is the opposite one. If AB forms faster than it disassociates into A and B, AB by itself is considered a stable compound. If there is a catalyst D that causes AB to disassociate into A and B faster, then AB will no longer be stable in the presence of D.

An enzyme is just a catalyst that is also a protein. Considering all living things, there are vast numbers of different enzymes. Indeed, there are vast numbers in just us. Perhaps the best understood enzyme is chymotrypsin, one of the enzymes in pancreatic juice, a substance produced by the pancreas and released into the small intestine. Chymotrypsin breaks large proteins found in food into smaller pieces; these are then broken down further by a succession of other enzymes into amino acids. Later, after the amino acids have traveled through the bloodstream into cells, other enzymes assemble from them proteins that the body needs.

amino acids, the twenty-some molecules from which proteins are made, and produce a brand new enzyme.

A version of this type of enzyme synthesis was finally accomplished in 1990 by John Stewart, Karl W. Hahn, and Wieslaw A. Klis of the University of Colorado Medical School. For two years Hahn worked with a computer to design the new enzyme. He started with a computer model of the natural pancreatic enzyme chymotrypsin, one of the most studied enzymes because it is easy to crystallize. The computer showed in a three-dimensional display the exact position of each amino acid component of the molecule. Then Hahn removed all the amino acid components of chymotrypsin from the computer display except for the three that constitute

Chymotrypsin itself contains all twenty amino acids in various combinations, resulting in a total of 245 amino acids that contain about 4000 atoms. Technically, when amino acids are combined, bits and pieces get lost, so the protein components are called amino acid residues. The specific order of the residues and the interactions among parts of the amino acid residues result in a particular shape for the enzyme. A consequence of this structure is formation of an active site on the enzyme, the place that accomplishes the work of breaking down large proteins by splitting a peptide bond, a bond between carbon and nitrogen. The process involves replacing one bond in the protein with a bond between carbon and a hydroxl group, OH, which is found in water, familiar as H_2O, but better thought of in this context as H–O–H. The remaining H of H–O–H attaches itself to the nitrogen, splitting the peptide bond at that point. This happens naturally, but very slowly, for proteins dissolved in water without the presence of the enzyme. The enzyme speeds up the process by a billion times.

In 1970 David Blow and coworkers at the University of Cambridge, UK, worked out how chymotrypsin accomplishes its task. Three amino acid residues are located in the enzyme molecule in such a way that they can pull apart peptide bonds. The other 242 amino acid residues not only hold the three key ones in the correct position, they also influence such factors as the electrical forces on different parts of the key amino acid residues (and have the ability to break peptide bonds by themselves, but not quickly). As a result, the three can act sequentially as an acid, producing the first half of the break, and then as a base, producing the second half. In the immediate vicinity of the three key amino acid residues, the result is an environment with a lot of free hydroxl ions and free positive hydrogen ions. The active site also stabilizes the transition state for the bond that exists during the process of breaking it, enabling the chemical to pass over an energy barrier. This has the same effect on the transition as raising the temperature would. In combination, these processes speed up the reaction.

the main active site of the enzyme. These were left on the display in their original positions in space. Over a 2-year period, Hahn worked out the positions of seventy additional amino acids that could be attached to the three so that they would retain their positions and so that electrical activity with neighboring amino acids would maintain the catalytic ability of the new molecule.

Once Hahn's design was complete, Klis could use techniques developed by Stewart to synthesize the new seventy-three-amino-acid molecule. They named the new peptide chymohelizyme-1 (CHZ-1). (A small molecule made of amino acid components is called a peptide; bigger ones are polypeptides; still bigger ones are proteins.)

Essentially, CHZ-1 does the same thing as chymotrypsin does, although about a thousandth as fast. Chymotrypsin gets additional speed from catalysis that occurs in parts of the molecule away from the main active site. The Stewart team performed extensive tests showing that CHZ-1 behaves like a real enzyme and that it specifically does what it was planned to do.

Nobel laureate biochemist Bruce Merrifield of Rockefeller University in New York City comments, "If others can reproduce and expand on this work, it will be one of the most important achievements in biology or chemistry."

(Periodical References and Additional Reading: *Science* 6-22-90, p 1544; *Science News* 6-23-90, p 388)

APPLICATIONS AND MATERIALS

ZEOLITES

Most of the excitement about buckyballs comes from their open structure and the possibilities of combining them with various other atoms, molecules, or ions to accomplish specific tasks. Chemists have good reason to be excited about this combination because some of the major successes in applied chemistry have come from exactly these characteristics in a group of minerals and synthetic materials called zeolites. Zeolites have many uses—water softeners and oil-cracking catalysts are the best known, but they also are used to separate molecules of various sizes and to absorb specific gases or vapors.

There are a few more than forty known zeolites. The late 1980s and early 1990s were graced with the discovery of two new ones of exceptional promise. Also, in 1991, an unusual zeolite was produced by growing its crystals on the US space shuttle *Atlantis* during a 6-day mission that began on Apr 5 1991.

Boggsite and Joe Smith

In the late 1980s a group of eighteen amateur rock hounds went looking for samples of a recently discovered zeolite called tschernischite in a hole in the ground near Goble WA. Instead they discovered little white balls of a mystery mineral. X-ray analysis showed that it was a previously undiscovered substance with a structure that was difficult to unravel. The new mineral was named boggsite.

The problem was turned over to Joseph V. Smith and Joseph Pluth of the University of Chicago. Working with sets of simple models representing the five regular, or Platonic, solids (tetrahedron, cube, octahedron, dodecahedron, and icosahedron) and the thirteen Archimedian solids (semiregular solids that have two and only two regular polygons as faces, such as the truncated icosahedron that is a geodesic dome or the shape of a buckyball), Smith and Pluth deduced that boggsite is a new zeolite with the commercially interesting combination of ten-member rings and twelve-member rings, similar to that of various useful natural and artificial zeolites. Boggsite also has two channels of unusually large pores for a zeolite—5500 nanometers (5.5 angstroms) and 7000 nanometers (7 angstroms). The combination of these features might make boggsite an unusual catalyst.

Background

Zeolites are natural or artificial alumininosilicates in which the atoms of aluminum, silicon, and oxygen are arranged in regular or semiregular polyhedra that connect to form channels. Natural zeolites have these channels filled with water, but the water is easily driven out by heating. The result is a mineral that is largely empty space, but empty space that has a definite geometric structure.

By themselves zeolites can be used as size- or size-and-shape-specific sieves. It is also easy to produce active sites inside the channels by detaching protons from the framework. Such active sites can be used as catalysts. In addition, it is also possible to insert specific atoms, molecules, or ions into the spaces in the structure for this purpose.

There is, however, probably not enough boggsite in nature for large-scale applications. Scientists are working to develop an artificial zeolite with the same characteristics.

A Four-Leaf Zeolite

In 1991 French and Swiss chemists led by Henri Kessler of the National Superior School of Chemistry in Mulhouse, France, created a new artificial zeolite with a pore structure that has a cross section resembling a four-leaf clover. Although the French word for clover is *trèfle,* they named their creation cloverite. The researchers reported in the Jul 25 *Nature* that cloverite "provides new possibilities for shape-selective sorption," in other words, separating molecules according to their shapes. Cloverite is based on the use of gallium and phosphate, and each of its clover rings contains twenty atoms.

Zeolites from Space

Most of the action on the Apr 5 1991 *Atlantis* mission that was reported in the press at the time concerned space walks and the launch of the Gamma Ray Observatory. At the same time, strange crystals were quietly growing in a gel in a test chamber aboard the shuttle.

Crystal growing in space has been a major activity on the shuttle, as various scientists have tried to grow more perfect crystals without the interfering force of gravity. For the most part, not much has come of it. Most of the crystals tried were proteins, and the results were about the same as those on Earth, although purer and more regular. Some lead iodide crystals grown on a *Discovery* flight in 1988 were different in both growth pattern and purity, however, suggesting that commercially important molecular sieves might be grown in space. One of the properties of a zeolite is that it is a molecular sieve.

Robin Stewart and coworkers from the US National Institute of Standards and Technology in Boulder CO, in collaboration with Instrumentation Technology Associates, Inc. of Exton PA, followed up on the lead iodide experiment by growing a zeolite in space on the *Atlantis* flight. This time the results were even more surprising. Instead of the cubic or octahedral crystals that were expected, the crystals were shaped like rods. It is not clear why this happened or whether or not the finding will be useful. Stewart is now planning another experiment in hopes of obtaining larger crystals that could provide further insights.

(Periodical References and Additional Reading: *Science* 3-23-90, p 1413; *Science* 6-8-90, p 1190; *Science News* 7-13-91, p 22; *Science News* 8-3-91, p 77)

PROPERTIES, ABUNDANCE, AND DISCOVERY OF THE ELEMENTS

All ordinary matter is made from one or more substances called *elements* (because they cannot be changed by chemical means). Ninety elements are found in nature, and people have created others, for a current total of 109. In this table each of the elements is listed in alphabetic order along with several of its important properties. The chemical symbol and the atomic number can be used to locate other information about the elements in the Periodic Table on page 183. The relative abundance of the elements is given as parts per million in Earth's crust—83,600 parts per million for aluminum means that of a million atoms chosen at random from the crust, on the average 83,600 atoms would be aluminum atoms and the others would be atoms of other elements. Some elements have so few parts per million that they are simply listed as rare, while others are "synthetic"—for artificial elements that are not found in the crust at all. Many elements, known from ancient times, are labeled "prehistoric." Others are given with their first discovery, although many elements were independently rediscovered.

Element	Symbol/ Atomic No.	Type*	Melting Point*	Boiling Point*	Parts per Million in Crust	Year Disc.& by Whom	Derivation of Name
Actinium	Ac 89	Radioactive metal	1920°F 1050°C	5790°F 3200°C	Rare	1899 André-Louis Debierne	Greek *aktis*, a ray
Aluminum	Al 13	Metal	1220°F 660°C	4473°F 2467°C	83,600	1825 Hans Christian Oersted	Latin *alumen*, a substance having an astringent taste
Americium	Am 95	Radioactive metal	1821°F 994°C	4725°F 2607°C	Synthetic	1944 Glenn T. Seaborg & coworkers	*America*
Antimony	Sb 51	Metal	1167°F 631°C	3180°F 1750°C	0.2	c 900 Rhazes	Greek *antimonos*, opposed to solitude; symbol Sb is from Greek name *stibi*
Argon	A 18	Gas	-308.6°F -189.2°C	-302.3°F -185.7°C	Rare	1892 Sir William Ramsay	Greek *argus*, neutral, inactive

Element	Symbol/ Atomic No.	Type*	Melting Point*	Boiling Point*	Parts per Million in Crust	Year Disc.& by Whom	Derivation of Name
Arsenic	As 33	Nonmetal	1502°F** 817°C	1135°F* 613°C	1.8	1649 J. Schroder & N. Lémery	Greek *arsenicos*, valiant or bold; from its action on other metals
Astatine	At 85	Radioactive nonmetal	576°F 302°C	639°F 337°C	Synthetic	1940 Emilio Segrè & coworkers	Greek *astatos*, unstable
Barium	Ba 56	Metal	1337°F 725°C	2980°F 1640°C	390	1774 Karl Wilhelm Scheele	Greek *baros*, heavy; because its compounds are dense
Berkelium	Bk 97	Radioactive metal	N.A.	N.A.	Synthetic	1949 Glenn T. Seaborg & coworkers	First made at U. of California at Berkeley
Beryllium	Be 4	Metal	2332°F 1278°C	5380°F 2970°C	2	1798 Louis-Nicolas Vauquelin	Latin *beryllus*, Greek *beryllos*, gem
Bismuth	Bi 83	Metal	520°F 271°C	2840°F 1560°C	0.008	1450 Basil Valentine	German *weisse masse*, white mass; changed to *bismat*
Boron	B 5	Nonmetal	4170°F 2300°C	4620°F 2550°C	9	1808 Joseph-Louis Gay-Lussac & Louis-Jacques Thénard	Aryan *borak*, white
Bromine	Br 35	Liquid nonmetal	19°F -7.2°C	137.8°F 58.8°C	2.5	1825 Carl Löwig	Greek *bromos*, a stench; because of odor of its vapor
Cadmium	Cd 48	Metal	609.6°F 320.9°C	1409°F 765°C	0.16	1817 Friedrich Strohmeyer	Greek *cadmia*, earthy
Calcium	Ca 20	Metal	1542°F 839°C	2703°F 1484°C	46,600	1808 Humphry Davy	Latin *calx, calcis*, lime
Californium	Cf 98	Radioactive metal	N.A.	N.A.	Synthetic	1950 Glenn T. Seaborg	First made at U. of California
Carbon	C 6	Nonmetal	6420°F 3550°C	8721°F 4827°C	180	Prehistoric	Latin *carbo*, coal
Cerium	Ce 58	Rare earth	1468°F 798°C	5895°F 3257°C	66.4	1803 Martin Klaproth	From asteroid *Ceres*, discovered in 1801
Cesium	Cs 55	Metal	83.1°F 28.4°C	1253.1°F 678.4°C	2.6	1860 Gustav Kirchhoff & Robert Bunsen	Latin *caesius*, bluish gray

(Continued)

Element	Symbol/ Atomic No.	Type*	Melting Point*	Boiling Point*	Parts per Million in Crust	Year Disc.& by Whom	Derivation of Name
Chlorine	Cl 17	Gas	-150°F -101°C	-30.3°F -34.6°C	126	1774 Karl Wilhelm Scheele	Greek *chloros*, grass green; from the color of the gas
Chromium	Cr 24	Metal	3375°F 1857°C	4842°F 2672°C	122	1797 Louis-Nicolas Vauquelin	Greek *chroma*, color; because many of its com- pounds are colored
Cobalt	Co 27	Metal	2723°F 1495°C	5200°F 2870°C	29	1735 George Brandt	Greek *kobolis*, a goblin
Copper	Cu 29	Metal	1981°F 1083°C	4653°F 2567°C	68	Prehistoric	Latin *cuprum*, for island of Cyprus
Curium	Cm 96	Radioactive metal	2444°F 1340°C	N.A.	Synthetic	1944 Glenn T. Seaborg	After Pierre and Marie Curie
Dysprosium	Dy 66	Rare earth	2568°F 1409°C	4235°F 2335°C	Rare	1886 Paul-Emile Lecoq de Boisbaudran	Greek *dysprosi- tos*, difficult of access
Einsteinium	Es 99	Radioactive metal	N.A.	N.A.	Synthetic	1952 Albert Ghiorso & coworkers	After Albert Einstein
Element 106	N.A. 106	Radioactive metal	N.A.	N.A.	Synthetic	1974 N.A.	Claimed by USSR and US
Element 107	N.A. 107	Radioactive metal	N.A.	N.A.	Synthetic	1981 N.A.	Identified in Germany follow- ing earlier dis- puted claim by USSR
Element 108	N.A. 108	Radioactive metal	N.A.	N.A.	Synthetic	1984 N.A.	Created in Germany by bombarding lead with iron ions
Element 109	N.A. 109	Radioactive metal	N.A.	N.A.	Synthetic	1982 N.A.	Created in W. Germany by bombarding bis- muth with iron ions
Erbium	Er 68	Rare earth	2772°F 1522°C	4550°F 2510°C	3.46	1843 Carl Gustav Mosander	From *Ytterby,* vil- lage in Sweden
Europium	Eu 63	Rare earth	1512°F 822°C	2907°F 1597°C	2.1	1896 Eugène-Anatole Demarçay	*Europe*
Fermium	Fm 100	Radioactive metal	N.A.	N.A.	Synthetic	1953 Albert Ghioroso & coworkers	After Enrico Fermi, Italian physicist

Element	Symbol/ Atomic No.	Type*	Melting Point*	Boiling Point*	Parts per Million in Crust	Year Disc.& by Whom	Derivation of Name
Fluorine	F 9	Gas	-363.3°F -219.6°C	-306.7°F -188.1°C	544	1886 Ferdinand-Frèdèric-Henri Moissan	Latin *fluere*, to flow
Francium	Fr 87	Radioactive metal	80.6°F 27°C	1256°F 677°C	Rare	1939 Marguerite Perey	*France*
Gadolinium	Gd 64	Rare earth	2392°F 1311°C	5851°F 3233°C	6.1	1886 Jean-Charles Marignac	After Johan Gadolin, Finnish chemist
Gallium	Ga 31	Liquid metal	86.6°F 29.8°C	4357°F 2403°C	19	1875 Paul-Emile Lecoq de Boisbaudran	Latin *Gallia*, France; also Latin *gallus*, a cock, pun on Lecoq de Boisbaudran
Germanium	Ge 32	Metal	1719°F 937°C	5126°F 2830°C	1.5	1886 Clemens Winkler	*Germany*
Gold	Au 79	Metal	1947°F 1064°C	5085°F 2807°C	0.002	Prehistoric	Anglo-Saxon *gold* ;Sanskrit *juel*, to shine; symbol is from Latin *aurum*, shining down
Hafnium	Hf 72	Metal	4041°F 2227°C	8316°F 4602°C	2.8	1923 Dirk Coster & György Hevesy	*Hafnia*, ancient name of Copenhagen
Hahnium (niels-bohrium)	Ha 105	Radioactive metal	N.A.	N.A.	Synthetic	1970 Albert Ghiorso & coworkers	Hahnium, the name used in the US, is after Otto Hahn; nielsbohrium, used in the USSR, is after Niels Bohr
Helium	He 2	Gas	-458°F -272°C	-452°F -269°C	Rare	1868 Pierre-Jules-Cèsar Janssen & Sir Joseph Norman Lockyer	Greek *helios*, the sun; first observed in sun's atmosphere
Holmium	Ho 67	Rare earth	2678°F 1470°C	4928°F 2720°C	1.26	1879 Per Teodor Cleve	*Holmia*, Latinized form of *Stockholm*
Hydrogen	H 1	Gas	-434.6°F -259.1°C	-423.2°F -252.9°C	1520	1766 Henry Cavendish	Greek *hydro*, water, plus *gen*, forming

(Continued)

Element	Symbol/ Atomic No.	Type*	Melting Point*	Boiling Point*	Parts per Million in Crust	Year Disc.& by Whom	Derivation of Name
Indium	In 49	Metal	313.9°F 156.6°C	3776°F 2080°C	0.24	1863 Ferdinand Reich & Hieronymus Theodor Richter	Latin *indicum,* indigo
Iodine	I 53	Nonmetal	236.3°F 113.5°C	363.9°F 184.4°C	0.46	1811 Bernard Courtois	Greek *iodes,* violet; from the color of its vapor
Iridium	Ir 77	Metal	4370°F 2410°C	7466°F 4130°C	0.001	1803 Smithson Tennant	Greek *iris,* a rainbow; from the changing color of its salts
Iron	Fe 26	Metal	2795°F 1535°C	4982°F 2750°	62,200	Prehistoric	Anglo-Saxon *iren;* Fe is from Latin *ferrum*
Krypton	Kr 36	Gas	-249.9°F -156.6°C	-242.1°F -153.3°C	Rare	1898 Alexander Ramsay & Morris William Travers	Greek *kryptos,* hidden
Lanthanum	La 57	Rare earth	1688°F 920°C	6249°F 3454°C	34.6	1839 Carl Mosander	Greek *lanthanein,* to be concealed
Lawrencium	Lr 103	Radioactive metal	N.A.	N.A.	Synthetic	1961 Albert Ghioroso & coworkers	After Ernest Lawrence, American physicist
Lead	Pb 82	Metal	621.5°F 327.5°C	3164°F 1740°C	13.0	Prehistoric	Anglo-Saxon *lead;* Pb is from Latin name, *plumbum*
Lithium	Li 3	Metal	356.9°F 180.5°C	2457°F 1347°C	18	1817 Johan August Arfvedson	Greek *lithos,* stony
Lutetium	Lu 71	Rare earth	3013°F 1656°C	5999°F 3315°C	Rare	1907 Georges Urbain	Latin *Lutetia,* ancient name for Paris
Magnesium	Mg 12	Metal	1200°F 649°C	1994°F 1090°C	27,640	1808 Humphry Davy	Latin *Magnesia,* a district in Asia Minor
Manganese	Mn 25	Metal	2271°F 1244°C	3564°F 1962°C	1060	1774 Karl Wilhelm Scheele	Latin *magnes,* magnet; because of confusion with magnetic iron ores
Mendelevium	Md 101	Radioactive metal	N.A.	N.A.	Synthetic	1955 Albert Ghiorso & coworkers	After Dmitri Mendeléev, Russian chemist
Mercury	Hg 80	Liquid metal	-38.0°F -38.9°C	673.9°F 356.6°C	0.08	Prehistoric	For Roman god *Mercurius;* symbol from Latin *ydrogyrum*

Element	Symbol/ Atomic No.	Type*	Melting Point*	Boiling Point*	Parts per Million in Crust	Year Disc.& by Whom	Derivation of Name
Molybde-num	Mo 42	Metal	4743°F 2617°C	8334°F 4612°C	1.2	1778 Karl Wilhelm Scheele	Greek *molybdaina*, galena (lead ore)
Neodymium	Nd 60	Rare earth	1850°F 1010°C	5661°F 3127°C	39.6	1885 Karl Auer	Greek *neo*, new, plus *didymon*, twin (with praseodymium)
Neon	Ne 10	Gas	-416.7°F -248.7°C	-411°F -246°C	Rare	1898 Alexander Ramsay & Morris William Travers	Greek *neo*, new
Neptunium	Np 93	Radioactive metal	1184°F 640°C	7056°F 3902°C	Synthetic	1940 Edwin McMillan & Philip Abelson	For planet *Neptune*
Nickel	Ni 28	Metal	2647°F 1453°C	4950°F 2732°C	99	1751 Axel Cronstedt	German *Nickel*, Satan (Old Nick)
Niobium	Nb 41	Metal	4474°F 2468°C	8568°F 4742°C	20	1801 Charles Hachett	Latin *Niobe*, daughter of Tantalus
Nitrogen	N 7	Gas	-345.8°F -209.9°C	-320.4°F -195.8°C	19	1772 Daniel Rutherford	Latin, *nitrum* plus *gen*, forming niter, a compound of nitrogen
Nobelium	No 102	Radioactive metal	N.A.	N.A.	Synthetic	1957 P.R. Fields & coworkers	After Alfred Nobel; made at Nobel Institute
Osmium	Os 76	Metal	5513°F 3045°C	9081°F 5027°C	0.005	1803 Smithson Tennant	Greek *osme*, smell; for mal-odorousness
Oxygen	O 8	Gas	-361°F -218.4°C	-297°F -183°C	456,000	1774 Joseph Priestley	Greek *oxyx*, sharp, plus *gen*, form-ing; from the incorrect belief that oxygen forms acids
Palladium	Pd 46	Metal	2826°F 1552°C	5684°F 3140°	0.015	1803 William Hyde Wollaston	Greek goddess *Pallas;* from asteroid Pallas
Phosphorus	P 15	Nonmetal	111.4°F 44.1°C	536°F 280°C	1120	1669 Hennig Brand	Greek *phospho-ros*, light-bringer; glows because of rapid oxidation
Platinum	Pt 78	Metal	3222°F 1772°C	6921°F 3827°C	0.01	1735 Antonio de Ulloa	Spanish *plata*, silver; from the color of the metal

(Continued)

Element	Symbol/ Atomic No.	Type*	Melting Point*	Boiling Point*	Parts per Million in Crust	Year Disc.& by Whom	Derivation of Name
Plutonium	Pu 94	Radioactive metal	1186°F 641°C	5850°F 3232°C	Synthetic	1940 Glenn T. Seaborg & coworkers	For planet *Pluto*
Polonium	Po 84	Radioactive metal	489°F 254°C	1764°F 962°C	Rare	1898 Marie & Pierre Curie	Named by Marie Curie for her native Poland
Potassium	K 19	Metal	146.7°F 63.7°C	1425°F 774°C	18,400	1807 Humphry Davy	From *potash*, a compound of potassium; symbol is from Latin *kalium*
Praseo- dymium	Pr 59	Rare earth	1708°F 931°C	5814°F 3212°C	9.1	1885 Karl Auer	Greek *prasios*, green, plus *didymos*, twin
Promethium	Pm 61	Radioactive rare earth	1976°F 1080°C	4460°F 2460°C	Rare	1947 J. A. Marinsky, L.E. Glendenin, & C. D. Coryell	For Greek god *Prometheus*, who stole fire from heaven
Protactinium	Pa 91	Radioactive metal	2912°F 1600°C	N.A. N.A.	Rare	1917 Otto Hahn & Lise Meitner	Latin *proto*, first, plus actinium
Radium	Ra 88	Radioactive metal	1292°F 700°C	2084°F 1140°C	Rare	1898 Marie & Pierre Curie	Latin *radius*, ray
Radon	Rn 85	Radioactive gas	-96°F -71°C	-79°F -61.8°C	Rare	1900 Friedrich Ernst Dorn	*Radium* plus on, as in *neon*
Rhenium	Re 75	Metal	5756°F 3180°C	10,161°F 5,627°C	0.0007	1925 Walter Noddack, Ida Tacke, & Otto Berg	Latin *Rhenus*, Rhine
Rhodium	Rh 45	Metal	3571°F 1966°C	6741°F 3727°C	Rare	1803 William Hyde Wollaston	Greek *rhodios*, roselike; for red color of its salts
Rubidium	Rb 37	Metal	102°F 38.9°C	1270°F 688°C	78	1861 Gustav Kirchhoff & Robert Bunsen	Latin *rubidus*, red; from red lines in its spectrum
Ruthenium	Ru 44	Metal	4190°F 2310°C	7052°F 3900°C	Rare	1844 Carl Claus	From *Ruthenia* in Urals, where ore was first found
Rutherfordium	Rf 104	Radioactive metal	N.A.	N.A.	Synthetic	1969 Albert Ghiorso & coworkers	After Ernest Rutherford, British physicist
Samarium	Sm 62	Rare earth	1962°F 1072°C	3232°F 1778°C	7.0	1879 Paul-Emile Lecoq de Boisbaudran	For Scandinavian mineral samarskite

Element	Symbol/ Atomic No.	Type*	Melting Point*	Boiling Point*	Parts per Million in Crust	Year Disc.& by Whom	Derivation of Name
Scandium	Sc 21	Metal	2802°F 1539°C	5130°F 2832°C	25	1879 Lars Fredrik Nilson	Scandinavia
Selenium	Se 34	Nonmetal	423°F 217°C	1265°F 685°C	0.05	1817 Jöns Jakob Berzelius	Greek *selene*, the Moon
Silicon	Si 14	Nonmetal	2570°F 1410°C	4271°F 2355°C	273,000	1823 Jöns Jakob Berzelius	Latin *silex*, flint
Silver	Ag 47	Metal	1763.4°F 961.9°C	4014°F 2212°C	0.08	Prehistoric	Assyrian *sarpu*; Anglo-Saxon *soelfor*; the symbol is from Latin *argentum*
Sodium	Na 11	Metal	208.0°F 97.8°C	1621.2°F 882.9°C	22,700	1807 Humphry Davy	English *soda*, a compound of sodium; the symbol is from Latin name, *natrium*
Strontium	Sr 38	Metal	1416°F 769°C	2523°F 1384°C	384	1808 Humphry Davy	For *Strontian*, a town in Scotland
Sulfur	S 16	Nonmetal	235.0°F 112.8°C	832.5°F 444.7°C	340	Prehistoric	Sanskrit *solvere*
Tantalum	Ta 73	Metal	5425°F 2996°C	9797°F 5425°C	1.7	1802 Anders Ekeberg	For mythical king Tantalus, condemned to thirst; because of its insolubility
Technetium	Tc 43	Radioactive metal	3942°F 2172°C	8811°F 4877°C	Synthetic	1937 Emilio Segrè	Greek *technetos*, artificial; first artificial element
Tellurium	Te 52	Metal	841.1°F 449.5°C	1814°F 990°C	Rare	1782 Franz Joseph Müller	Latin *tellus*, the earth
Terbium	Tb 65	Rare earth	2480°F 1360°C	5506°F 3041°C	1.18	1843 Carl Gustav Mosander	For *Ytterby*, village in Sweden
Thallium	Tl 81	Metal	578.3°F 303.5°C	2655°F 1457°C	0.7	1861 William Crookes	Greek *thallos*, a young, or green, twig (after color of its spectrum)
Thorium	Th 90	Radioactive metal	3182°F 1750°C	8654°F 4790°C	8.1	1829 Jöns Jakob Berzelius	For Norse god *Thor*
Thulium	Tm 69	Rare earth	2813°F 1545°C	3141°F 1727°C	0.5	1879 Per Teodor Cleve	Greek *Thoule*, northernmost region of the world

(Continued)

Element	Symbol/ Atomic No.	Type*	Melting Point*	Boiling Point*	Parts per Million in Crust	Year Disc.& by Whom	Derivation of Name
Tin	Sn 50	Metal	450°F 232°C	4118°F 2270°C	2.1	Prehistoric	Anglo-Saxon *tin*; symbol is from Latin name, *stannum*
Titanium	Ti 22	Metal	3020°F 1660°C	5949°F 3287°C	6320	1791 William Gregor	For *Titans* of Classical mythology
Tungsten	W 74	Metal	6170°F 3410°C	10,220°F 5,660°C	1.2	1783 Don Fausto d'Elhuyar	Swedish *tung-sten*, heavy stone; symbol is from German name, *Wolfram*
Uranium	U 92	Radioactive metal	2070°F 1132°C	6904°F 3818°C	2.3	1789 Martin Klaproth	For planet *Uranus*
Vanadium	V 23	Metal	3434°F 1890°C	6116°F 3380°C	136	1801 Andrès del Rio	For Scandinavian goddess *Vanadin*
Xenon	Xe 54	Gas	-169.4°F -111.9°C	-161°F -107°C	Rare	1898 Alexander Ramsay & Morris William Travers	Greek *xenon*, stranger
Ytterbium	Yb 70	Rare earth	1516.1°F 824.5°C	2179°F 1193°C	3.1	1907 George Urbain	From *Ytterby*, village in Sweden
Yttrium	Y 39	Rare earth	2773°F 1523°C	6039°F 3337°C	31	1794 Johan Gadolin	From *Ytterby*, village in Sweden
Zinc	Zn 30	Metal	787.3°F 419.6°C	1665°F 907°C	76	Prehistoric	German *zink*
Zirconium	Zr 40	Metal	3366°F 1852°C	7911°F 4377°C	162	1789 Martin Klaproth	Arabic *zargun*, gold color

*At a pressure of one atmosphere and, for type, at room temperature.
** At a pressure of 28 atmospheres.
*** Instead of melting, this element turns into a gas at this temperature (sublimes).

THE PERIODIC TABLE

In the nineteenth century, chemists began to determine atomic weight—how much one atom of an element weighs relative to one atom of a different element (now known as the *atomic mass* and measured in *atomic mass units,* or *amu,* a mass equal to one-twelfth the mass of the most common form of a carbon atom). The first really good list was prepared by Jöns Jakob Berzelius [Swedish: 1779-1848] in 1828.

When chemists made lists of elements in the order of atomic weights, they noticed that every seventh or eighth element in the list had similar properties. In 1869 Dmitri Mendeléev [Russian: 1834-1907] went further and boldly interchanged some elements in the list and left blanks for others to make sure the properties matched for every "period" of eight elements. This was the first periodic table. Mendeléev had only sixty-three elements to work with, but he correctly predicted three more that would make his list more complete. Today, there are 109 elements in the periodic table.

Early in the twentieth century, atoms were discovered to consist of protons and electrons. In 1932 it was discovered that neutrons are also found in atoms. Normally, the number of protons and electrons is equal. This is the atomic number, which is a different counting number for every element from hydrogen (atomic number 1) to the unnamed element numbered 109. When the concept of atomic number was discovered, it was possible to improve the periodic table by arranging the elements in order of atomic number instead of atomic weight. This did not require the rearrangements Mendeléev had to make, and it clearly showed where the blanks were—all of which have been filled in since 1940. Any other newly discovered or created elements must go to the end of the table.

What It Reveals

Each column of the periodic table includes elements with similar properties, although hydrogen in the first column is less typical in this respect (see "Quest for Metallic Hydrogen," p 156). The other elements in the first column are all soft metals that react vigorously. Similarly, the last column of the table contains only gases that hardly react at all. In general, elements on the left side of the table are metals (except for hydrogen); they become mostly nonmetals in the last six columns. However, these columns include some elements that are metals, such as aluminum. (A broken heavy line separates the metals from the nonmetals.)

The table is shown on pages 184-185.

THE PERIODIC TABLE OF THE ELEMENTS

6	— atomic number
C	— chemical symbol
12.01	— atomic mass
Carbon	— name of element

alkali metals
IA

| Period 1 | 1 H 1.01 Hydrogen |

alkaline earth metals
II A

| Period 2 | 3 Li 6.94 Lithium | 4 Be 9.01 Beryllium |

| Period 3 | 11 Na 23.00 Sodium | 12 Mg 24.31 Magnesium |

transition metals

			III B	IV B	V B	VI B	VII B			VIII
Period 4	19 K 39.10 Potassium	20 Ca 40.08 Calcium	21 Sc 44.96 Scandium	22 Ti 47.90 Titanium	23 V 50.94 Vanadium	24 Cr 52.00 Chromium	25 Mn 54.94 Manganese	26 Fe 55.85 Iron	27 Co 58.93 Cobalt
Period 5	37 Rb 85.47 Rubidium	38 Sr 87.62 Strontium	39 Y 88.91 Yttrium	40 Zr 91.22 Zirconium	41 Nb 92.91 Niobium	42 Mo 95.94 Molybdenum	43 Tc 98.91 Technetium	44 Ru 101.07 Ruthenium	45 Rh 102.91 Rhodium
Period 6	55 Cs 132.91 Cesium	56 Ba 137.34 Barium		72 Hf 178.49 Hafnium	73 Ta 180.95 Tantalum	74 W 183.85 Tungsten	75 Re 186.2 Rhenium	76 Os 190.2 Osmium	77 Ir 192.22 Iridium
Period 7	87 Fr (223) Francium	88 Ra (226) Radium		104 Rf (261) Rutherfordium	105 Ha (262) Hahnium	106 (263)	107 (262)	108 (265)	109 (266)

rare earth elements — Lanthanide series

| 57 La 138.91 Lanthanum | 58 Ce 140.12 Cerium | 59 Pr 140.91 Praseodymium | 60 Nd 144.24 Neodymium | 61 Pm (145) Promethium | 62 Sm 150.4 Samarium |

Actinide series

| 89 Ac (227) Actinium | 90 Th 232.04 Thorium | 91 Pa 231.04 Protactinium | 92 U 238.03 Uranium | 93 Np 237.05 Neptunium | 94 Pu (244) Plutonium |

noble
gases
O

				nonmetals			He
							2
							He
							4.00
							Helium

	III A	IV A	V A	VI A	VII A	

5	6	7	8	9	10
B	**C**	**N**	**O**	**F**	**Ne**
10.81	12.01	14.01	16.00	19.00	20.18
Boron	Carbon	Nitrogen	Oxygen	Fluorine	Neon

13	14	15	16	17	18
Al	**Si**	**P**	**S**	**Cl**	**Ar**
26.98	28.09	30.97	32.06	35.45	39.95
Aluminum	Silicon	Phosphorus	Sulfur	Chlorine	Argon

I B   II B

28	29	30	31	32	33	34	35	36
Ni	**Cu**	**Zn**	**Ga**	**Ge**	**As**	**Se**	**Br**	**Kr**
58.71	63.55	65.37	69.72	72.59	74.92	78.96	79.90	83.80
Nickel	Copper	Zinc	Gallium	Germanium	Arsenic	Selenium	Bromine	Krypton

46	47	48	49	50	51	52	53	54
Pd	**Ag**	**Cd**	**In**	**Sn**	**Sb**	**Te**	**I**	**Xe**
106.4	107.87	112.40	114.82	118.69	121.75	127.60	126.90	131.30
Palladium	Silver	Cadmium	Indium	Tin	Antimony	Tellurium	Iodine	Xenon

78	79	80	81	82	83	84	85	86
Pt	**Au**	**Hg**	**Ti**	**Pb**	**Bi**	**Po**	**At**	**Rn**
195.09	196.97	200.59	204.37	207.2	208.98	(209)	(210)	(222)
Platinum	Gold	Mercury	Thallium	Lead	Bismuth	Polonium	Astatine	Radon

other metals

63	64	65	66	67	68	69	70	71
Eu	**Gd**	**Tb**	**Dy**	**Ho**	**Er**	**Tm**	**Yb**	**Lu**
151.96	157.25	158.93	162.50	164.93	167.26	168.93	173.04	174.97
Europium	Gadolinium	Terbium	Dysprosium	Holmium	Erbium	Thulium	Ytterbium	Lutetium

95	96	97	98	99	100	101	102	103
Am	**Cm**	**Bk**	**Cf**	**Es**	**Fm**	**Md**	**No**	**Lw**
(243)	(247)	(247)	(251)	(254)	(257)	(258)	(255)	(256)
Americium	Curium	Berkelium	Californium	Einsteinium	Fermium	Mendelevium	Nobelium	Lawrencium

Two rows of elements do not fit neatly into the rest of the table. Elements from atomic number 57 to 71 are all similar to lanthanum, and are known as the rare earths; while elements from atomic number 89 to 103 are similar to actinium, and are called actinides. The rare earths generally are not rare and do not resemble soil. They are moderately common metals that, because of atomic structure, are very similar chemically.

The periodic table also includes the atomic mass as well as the atomic number. The atomic mass is essentially the sum of the protons and neutrons in an atom of an element, although different standards have been used at various times to measure this. As protons and neutrons join to form an atomic nucleus, a little of their energy becomes mass, the amount of which depends on how many protons and neutrons there are (this effect is exploited in nuclear fission, when the reverse process, splitting the nucleus of heavy elements, releases the energy). Consequently, a particular atom is chosen upon which to base the amu. Today, the atomic mass is adjusted to make the most common form of carbon have an atomic mass of exactly 12 (six protons and six neutrons).

Most elements occur with several different atomic masses. The number of protons is always equal to the atomic number for a given element, but the number of neutrons varies. In addition to carbon-12, for example, both carbon-13 and carbon-14 are relatively common, while carbons from 10 to 16 are all possible. Carbon-14 has six protons and eight neutrons and is radioactive. These different forms are called *isotopes*. Therefore, in the periodic table, the atomic mass given for most elements is the one that would be found by averaging the different isotopes in the amounts that they naturally occur. Carbon is given at the atomic mass of 12.01 because there is so much more carbon-12 than there is carbon-13 or carbon-14 in an ordinary sample of carbon. For some radioactive elements, natural abundance is meaningless since there is no stable form. For these, the atomic mass of the most stable form is given, indicated by putting the atomic mass in parentheses.

THE NOBEL PRIZE FOR CHEMISTRY

Date	Name [Nationality]	Achievement
1991	Richard R. Ernst [German: 1933–]	Improvements in nuclear magnetic resonance techniques
1990	Elias James Corey [US: 1928–]	New ways to synthesize organic molecules
1989	Sidney Altman [US: 1939–] Thomas Cech [US: 1947–]	Discovery of catalytic properties of RNA
1988	Johann Deisenhofer [German: 1943–] Robert Huber [German: 1937– Hartmut Michel [German: 1948–]	Determination of the structure of molecules involved in photosynthesis

Date	Name [Nationality]	Achievement
1987	Charles J. Pedersen [US: 1904–1989] Donald J. Cram [US: 1919–] Jean-Marie Lehn [French: 1939–]	Making artificial molecules per- form the same functions as natu- ral proteins
1986	Dudley R. Herschbach [US: 1932– Yuan T. Lee [US: 1936–] John C. Polanyi [Canadian: 1929–]	Herschbach and Lee for the crossed- beam molecular technique; Polanyi for chemiluminescence for studying chemical reactions
1985	Herbert A. Hauptman [US: 1917–] Jerome Karle [US: 1918–]	Work on equations to determin the structure of molecules
1984	R. Bruce Merrifield [US: 1921–]	Method for creating peptides and proteins
1983	Henry Taube [US: 1915–]	New discoveries in basic mecha- nism of chemical reactions
1982	Aaron Klug [South African: 1926–]	Developments in electron microscopy and study of acid- protein complexes
1981	Kenichi Fukui [Japanese: 1918–] Roald Hoffmann [US: 1937–	Application of laws of quantum mechanics to chemical reactions
1980	Paul Berg [US: 1926–] Walter Gilbert [US: 1932–] Frederick Sanger [English: 1918–]	Berg for development of recom- binant DNA; Gilbert and Sanger for methods to map the struc- ture of DNA
1979	Herbert C. Brown [English–US: 1912–] Georg Wittig [German: 1897–1987]	Brown for study of boron-con- taining organic compounds and for phosphorus-containing compounds
1978	Peter Mitchell [English: 1920–]	Study of biological energy trans- fer by mitochondria
1977	Ilya Prigogine [Russian–Belgian: 1917–]	Nonequilibrium theories in thermodynamics
1976	William N. Lipscomb, Jr. [US: 1919–]	Study of bonding in boranes
1975	John W. Cornforth [Australian: 1917–] Vladimir Prelog [Yugoslavian– Swiss: 1906–]	Cornforth for work on structure of enzyme-substrate combina- tions and Prelog for study of symmetric compounds
1974	Paul J. Flory [US: 1910–1985]	Study of long-chain molecules
1973	Ernst Otto Fischer [German: 1918–] Geoffrey Wilkinson [English: 1921–]	Work on the structure of ferocene
1972	Christian B. Anfinsen [US: 1916–] Stanford Moore [US: 1913–1982] William H. Stein [US: 1911–1980]	Pioneering research in enzyme chemistry
1971	Gerhard Herzberg [German–Canadian: 1904–]	Study of geometry of molecules in gases
1970	Luis F. Leloir [Argentinian: 1906–1987]	Discovery of sugar nucleotides and their biosynthesis of carbo- hydrates

(Continued)

Date	Name [Nationality]	Achievement
1969	Derek H.R. Barton [English: 1918–] Odd Hassel [Norwegian: 1897–1981]	Determination of three-dimensional shape of organic compounds
1968	Lars Onsager [US: 1903–1976]	Theoretical basis of diffusion of isotopes in a gas
1967	Manfred Eigen [German: 1927–] Ronald G.W. Norrish [English: 1897–1978] George Porter [English: 1920–]	Study of high-speed chemical reactions
1966	Robert Mulliken [US: 1896–1986]	Study of atomic bonds in molecules
1965	Robert B. Woodward [US: 1917–1979]	Synthesis of organic compounds, including quinine, cholesterol, cortisone, reserpine, and chlorophyll
1964	Dorothy Crowfoot Hodgkin [English: 1910–]	Analysis of the structure of vitamin B_{12}
1963	Giulio Natta [Italian: 1903–1979] Karl Ziegler [German: 1898–1973]	Synthesis of polymers for plastics
1962	John C. Kendrew [English: 1917–] Max F. Perutz [English: 1914–]	Kendrew for location of the position of the atoms in myoglobin and Perutz for hemoglobin using X-ray diffraction
1961	Melvin Calvin [US: 1911–]	Work on chemistry of photosynthesis
1960	Willard F. Libby [US: 1908–1980]	Invention of radiocarbon dating
1959	Jaroslav Heyrovsky [Czech: 1890–1967]	Polarography for electrochemical analysis
1958	Frederick Sanger [English: 1918–]	Discovery of structure of insulin
1957	Sir Alexander Todd [English: 1907–]	Study of chemistry of nucleic acids
1956	Sir Cyril Hinshelwood [English: 1897–1967] Nikolai Semenov [Russian: 1896–1986]	Parallel work on kinetics of chemical chain reactions
1955	Vincent Du Vigneaud [US: 1901–1978]	Synthesis of polypeptide hormone oxytocin
1954	Linus C. Pauling [US: 1901–]	Explanation of chemical bonds
1953	Hermann Staudinger [German: 1881–1965]	Study of polymers
1952	Archer J.P. Martin [English: 1910–] Richard L.M. Synge [English: 1914–]	Separation of elements by paper chromatography
1951	Edwin M. McMillan [US: 1907–] Glenn T. Seaborg [US: 1912–]	Discovery of plutonium and research on transuranium elements
1950	Otto Diels [German: 1876–1954]	Synthesis of organic compounds of the diene group

Date	Name [Nationality]	Achievement
1949	William Francis Giauque [US: 1895–1982]	Methods of obtaining temperatures very close to absolute zero
1948	Arne Tiselius [Swedish: 1902–1971]	Research on blood serum proteins using electrophoresis
1947	Sir Robert Robinson [English: 1886–1975]	Studies of plant alkaloids such as morphine and strychnine
1946	James B. Sumner [US: 1887–1955] John H. Northrop [US: 1891–1987] Wendell Stanley [US: 1904–1971]	Sumner's first crystallization of an enzyme; Northrop's crystallization of enzymes; Stanley's crystallization of tobacco-mosaic virus
1945	Artturi Virtanen [Finnish: 1895–1973]	Improvement of fodder preservation
1944	Otto Hahn [German: 1879–1968]	Discovery of atomic fission
1943	György Hevesy [Hungarian–Swedish: 1885–1966]	Use of isotopes as tracers
1942	No award	————
1941	No award	————
1940	No award	————
1939	Adolf Butenandt [German: 1903–] Leopold Ružička [Croatian–Swiss: 1887–1976]	Butenandt for study of sexual hormones (declined award at Hitler's direction); Ružička for work with atomic structures and terpenes
1938	Richard Kuhn [Austrian–German: 1900–1967]	Carotenoid and vitamin research (declined award at Hitler's direction)
1937	Walter N. Haworth [English: 1883–1950] Paul Karrer [Swiss: 1889–1971]	Haworth for work on carbohydrates and vitamin C and Karrer on carotenoids, flavins, and vitamins
1936	Peter J.W. Debye [Dutch–US: 1884–1966]	Study of dipolar moments of ions in solution
1935	Fédéric Joliot-Curie [French: 1900-1958] Irène Joliot-Curie [French: 1897–1956]	Synthesis of new radioactive elements
1934	Harold C. Urey [US: 1893–1981]	Discovery of heavy hydrogen
1933	No award	————
1932	Irving Langmuir [US: 1881–1957]	Study of suface chemistry of monomolecular films
1931	Karl Bosch [German: 1874–1940] Friedrich Bergius [German: 1884–1949]	Invention of high-pressure methods for producing ammonia
1930	Hans Fischer [German: 1881–1945]	Analysis of the structure of heme
1929	Arthur Harden [English: 1865–1940] Hans von Euler-Chelpin [German–Swedish: 1873–1964]	Harden for the first known coenzyme and Euler-Chelpin for analysis of its structure
1928	Adolf Windaus [German: 1876–1959]	Study of cholesterol

(Continued)

Date	Name [Nationality]	Achievement
1927	Heinrich O. Wieland [German: 1877–1957]	Analysis of the structure of steroids
1926	Theodor Svedberg [Swedish: 1884–1971]	Development of ultracentrifuge
1925	Richard Zsigmondy [German: 1865–1929]	Study of colloid solutions
1924	No award	———————
1923	Fritz Pregl [Austrian: 1859–1930]	Microanalysis of organics
1922	Francis W. Aston [English: 1877–1945]	Development of the mass spectrograph and the whole-number rule of atomic weights
1921	Frederick Soddy [English: 1877–1956]	Discovery of isotopes
1920	Walther Nernst [German: 1864–1941]	Discovery of the third law of therodynamics, the impossibility of obtaining absolute zero
1919	No award	———————
1918	Fritz Haber [German: 1868–1934]	Synthesis of ammonia from atmospheric nitrogen
1917	No award	———————
1916	No award	———————
1915	Richard Willstätter [German: 1872–1942]	Research on chlorophyll
1914	Theodore Richards [US: 1868–1928]	Determination of atomic weights
1913	Alfred Werner [German–Swiss: 1866–1919]	Discovery of coordination bonds (secondary valence)
1912	Victor Grignard [French: 1871–1935] Paul Sabatier [French: 1854–1941]	Grignard for discovery of reagents and Sabatier for hydrogenated compounds
1911	Marie Curie [Polish–French: 1867–1934]	Discovery of radium and polonium
1910	Otto Wallach [German: 1847–1931]	Work with terpenes
1909	Wilhelm Ostwald [Russian–German: 1853-1932]	Work on catalysis, chemical equilibrium, and reaction rates
1908	Ernest Rutherford [English: 1871-1937]	Study of transmutation of elements
1907	Eduard Buchner [German: 1860–1917]	Noncellular fermentation
1906	Henri Moissan [French: 1852–1907]	Isolation of fluorine and introduction of electric furnace
1905	Adolf von Baeyer [German: 1835–1917]	Work on organic dyes
1904	Sir William Ramsay [English: 1852–1916]	Discovery of inert gas elements and placement in periodic table
1903	Svante Arrhenius [Swedish: 1859–1927]	Theory of electrolytic dissociation
1902	Emil Fischer [German: 1852–1919]	Sugar and purine synthesis
1901	Jacobus van't Hoff [Dutch: 1852–1911]	Laws of chemical dynamics and osmotic pressure

Earth Science

THE STATE OF
EARTH SCIENCE

Earth science is really a group of sciences that centers on a single object: our planet and how it works. As a result of that focus, which is different in many ways from that of other broad areas of science, earth science partakes of a bit of virtually all the major branches of science:

Since Earth is a planet, astronomy and space are both involved with earth science; indeed, some earth scientists prefer to be called planetary scientists; like astronomy, earth science requires an understanding of "deep time";

Earth is made from minerals, some of which (from our surface-dwelling point of view) are under unusual stresses; mineralogy is allied to chemistry, and many earth scientists can be considered geochemists;

For now at least, Earth is our environment, and it is often difficult to determine the boundary between earth science and environmental science (for example, where exactly does the study of soil erosion fit?);

Paleontologists are classed as earth scientists even though they specialize in studying the remains of living creatures, the subject of evolutionary biology;

Increasingly an earth scientist or two is required for any anthropological or archaeological expedition;

Earth scientists who study earthquakes or meteorology use some of the most sophisticated mathematical analyses in science;

Geophysics is becoming the heart of earth science, although the term is used very broadly by earth scientists;

The largest source of employment for earth scientists around the world is technology—the petroleum industry.

Major Trends in Physical Geology

Understanding Earth's Interior Earth science is comparatively young when compared with biology and physics. It was not until the eighteenth century that important results were achieved, and these were mostly connected with easily visible processes—weathering and erosion, deposition, crystallization, fossilization in sedimentary rock, and so on. Great progress was made in understanding these surface phenomena throughout the nineteenth century.

Understanding what goes on beneath the surface did not start until the twentieth century. Large structures, such as the mantle and core, were observed with earthquake waves. Studies of meteorites and materials ejected by volcanoes also helped scientists understand the chemistry of Earth's interior.

At the same time, a combination of factors, with residual magnetism on the ocean floor as the main key, led earth scientists to the theory of plate tectonics (see "Advances in Plate Tectonics," p 244). This theory took shape in the 1950s and 1960s. The mechanism that moves tectonic plates is an important part of the theory. Convection currents caused by differences in temperature drive the plates. If a material is all of the same composition, heating the bottom makes it less dense than the cooler upper part. If it is fluid, the cool part from the top flows downward under gravity, pushing the warmer stuff up. A continuous movement of the material in a circular pattern called a convection current results.

Convection currents require a heat source at the bottom, a fairly uniform material, and some fluidity. Plate tectonics theory requires that the interior of Earth contain a region with the appropriate characteristics. Originally it was thought that the large region between Earth's crust and core, known as the mantle, meets the criteria. As more data were collected, however, problems began to develop. During the 1980s, earth scientists began to take sides among the rival theories that tried to reconcile the data with the concept of convection currents as the driving force for tectonic activity on the planet. As the 1990s began, opinions were sharply divided, but new tools and new studies look to resolve the problem over the decade.

CT Scans of Earth In 1755 John Winthrop, observing an aftershock of the Boston earthquake of that year, noted that the shock produced an actual wave in the bricks as it passed through his fireplace. Earth scientists have been using such waves to deduce facts about the interior of Earth since 1906, when Richard D. Oldham speculated that patterns in earthquake waves reveal that Earth has a partially liquid core; the amount of information that can be obtained from earthquake waves has increased dramatically in recent years. Mathematics has been one key and the computer revolution the other.

The first sophisticated mathematical analysis of earthquake waves was conducted by Beno Gutenberg [German-US: 1889-1960], who refined the speculations of Oldham and others and correctly located the boundary between the core and mantle, called the Gutenberg discontinuity. But such calculations could reveal only the general outlines of regions hundreds or thousands of kilometers beneath Earth's crust.

In 1956, however, mathematician Ronald N. Bracewell developed the projection-slice theorem, the essential tool for recovering the shape of an object from various projections—shadows—of the object. Initially, the mathematics required was too daunting for the theorem to have many practical applications. However, cheaper and faster computers changed the picture, and by 1972 the first application of the theorem to medicine, computerized axial tomography (familiar as a CT scan or CAT scan), was in use in the United Kingdom. Earth scientists put the idea into practice about 10 years later. Since earthquake waves are used, this method is called seismic tomography. At first seismic tomographers did not have records from enough seismographs or enough earthquakes to obtain very good resolution of layers in Earth's interior. As new seismograph networks were set up and new techniques for producing waves other than earthquake waves were introduced, the ratio of computer power to cost was skyrocketing, enabling more and more geophysicists to use seismic tomography and refine it. Today, with ultrapowerful supercomputers, seismograph networks around the globe, modern communications, and Bracewell's indispensable theorem, geophysicists can not only analyze evidence for actual shapes within the Earth, they can also simulate how these shapes change over time according to various models. This means that for the first time geophysicists can test their ideas using a simulated Earth.

It should be noted that a few earth scientists, such as Thomas Jordan of MIT, think that seismic tomography "has run out of steam," and will not reveal much more. Certainly, different groups continue to develop conflicting pictures using the technique. In the meantime, agreement among seismic tomographers may become easier with the 1991 publication of new seismic wave travel timetables developed by the International Association of Seismology and Physics of the Earth's Interior. These 1991 tables are expected to replace the previous official 1940 tables of the British Association Seismological Committee, known as the Jeffreys & Bullen Tables, as well as various tables for specific wave types developed around 1970 by individual scientists.

Chemistry Under Pressure It has long been known that rock under great pressure and extremely high temperatures does not have the same properties as rock under what people at the surface view as normal conditions. It has been difficult to speculate about what these differences caused by pressure and temperature mean to our interpretation of events in Earth's interior, where the principal information is the behavior of earthquake waves.

The development of the diamond anvil by Charles Weir, Alvin van

Valkenburg, and coworkers at the National Bureau of Standards in the 1970s revolutionized the way earth scientists study the interior of Earth. A diamond anvil applies pressure to a small sample held between the surfaces of two specially cut, gem-quality diamonds. A fundamental law of physics states that pressure equals force divided by area of application. Since the diamonds are tiny, and the anvil's hydraulics, levers, and screws multiply force to a high level, pressure between points is great. As diamond is very hard, the points do not break or bend. Using this device, scientists have been able to produce pressures in the laboratory comparable to those in Earth's interior. Diamond anvils routinely develop pressures 2 million times air pressure at sea level (2 megabars). The record for a diamond anvil is 4.16 megabars, set by Arthur L. Ruoff and coworkers at Cornell University. The maximum theoretical strength for a diamond anvil is about 8 to 10 megabars, the pressure at which diamond becomes a metal and too malleable to be useful. Theoretical pressure might be increased by replacing diamond with buckminsterfullerene, since buckminsterfullerene is expected to become twice as stiff as diamond under high pressures (see "Buckyball Fever," p 157).

A few other devices produce even higher pressures than diamond anvils. The two-stage, light-gas gun is a 32-m (105-ft) cannon whose first stage is a gunpowder explosion that pushes a piston that compresses a light gas, almost always hydrogen. At high pressure, the hydrogen bursts a barrier and propels a plastic projectile. The plastic projectile travels about 60 cm (23 in.) through a vacuum to its target. When the plastic projectile hits a target, it transfers its momentum to the target, producing pressures as great as 50 megabars.

Hydrogen is often used in this device because molecules of light gases accelerate faster than those of heavier gases. Although momentum is a product of both mass and velocity, it is easier to increase momentum by accelerating particles of low mass to a high velocity than by accelerating heavy particles to a low velocity. Since acceleration equals force divided by mass, for a given force, lower mass means higher acceleration. After a short period of acceleration, the greater velocity of the light gas molecules contributes more to the momentum than the low mass of the molecules takes away. In practice, only about ten of the lightest molecules are useful in gas guns (or for rocket fuels). Plastic projectiles are used to reduce the loss of momentum—and therefore pressure—that would be produced by bouncing off the target. The projectiles do not bounce and plastic does not compress, also contributing to the transfer of momentum.

Even higher pressures have been produced by nuclear explosions (158 megabars) and implosion of a capsule using lasers (about 1000 megabars).

The advantage of the diamond anvil over devices that have produced higher pressures is that impact and implosion methods create pressures that last for very short times, while diamond anvils increase pressure slowly. All methods for producing high pressures suffer from the small size of the samples that can be used and the difficulty of measuring physical properties while maintaining the pressure. In some ways, the ability to measure what they have achieved, often by bouncing laser light off the samples, is as remarkable as the pressures themselves.

So far, high-pressure chemistry is just beginning to reveal important results about Earth's interior (see "Understanding Earth's Mantle and Core," p 256).

Comparative Planetology Increasing knowledge of other planets from the work of space probes is providing a wealth of new insights for earth-bound scientists who want to understand how their own planet works. For example, the Magellan space probe studies of Venus suggest that Venus is like a young Earth (see "Venus Laid Bare," p 48). Similarly, studies of impact craters on other planets and their satellites, as well as on the Moon, have been essential in developing ideas about the impact history of Earth. Exploration of Jupiter by the Galileo space probe is expected to contribute to the understanding of Earth's meteorology, while studies of its moons may help explain a number of subjects, ranging from core formation to volcanology.

Solar system studies will also be enriched by the first close-up looks at asteroids and additional close encounters with comets. These smaller bodies are thought to contain the main clues to the formation and early history of Earth.

Major Trends in Paleontology

Understanding the Dinosaurs and Their Fate Reevaluation of the place of dinosaurs in the history of life continues, as does the argument about what caused them to become extinct. Increasingly, young paleontologists are espousing some or all of the new controversial ideas:

Dinosaurs were not really reptiles at all.

They were warm-blooded creatures that did not become extinct, but evolved into birds.

They did not live in tropical climes only, but ranged far to the north and south.

Some did not abandon their offspring, the way most modern reptiles do, but cared for them in colonies that resemble seabird enclaves.

They ran fairly quickly and carried their tails high.

All in all, as this wave of reinterpretation continues, we may expect to enter the next millennium with a completely different concept of these familiar and yet unfamiliar animals.

Understanding the Cambrian Revolution About 600 million years ago (mya), the fossil record begins in earnest—animals had evolved hard shells to protect themselves from each other, hard shells that made good fossils. Traces of earlier animals that had soft parts only are rare. What is worse, until recently, many of those rare soft-body fossils, especially of the famous Burgess Shale formation in British Columbia, were wrongly interpreted. A major reevaluation has produced not only a new feeling for early forms of life, but also a new feeling for the process of evolution itself. Chief popularizer of the new view is Stephen Jay Gould of Harvard University, although he would be the first to point out that he is not doing the hard work. Most of that has been accomplished by Derek E.G. Briggs of the University of Bristol, UK, Simon Conway Morris of Cambridge University, UK, Adolph Seilacher, and Harry B. Whittington, who have undertaken the task of reanalyzing the record of soft-bodied fossils.

Major progress began in 1984, but it did not become news in the West until 1991. In China, Hou Xianguang discovered at a site called Chengjiang a fossil field that is as rich in soft-bodied fossils as, and even older than, the Burgess Shale, the fossil bed that has prompted the main reevaluation of early evolution. Details concerning this discovery are discussed on page 203.

Sources of soft-bodied fossils	Organisms preserved	Geological time period
Ediacara, Australia	"Quilted" creatures that have no mouths and may have absorbed nutrients through their skins	Vendian period (610-570 mya)
Burgess Shale, British Columbia, Canada Chengjiang, Yunnan Province, China House Range UT Kangaroo Island, Australia Lancaster PA Peary Land, Greenland	Sponges, brachiopods, polychaete worms, priapulid worms, trilobites and other arthropods, mollusks, echinoderms, chordates, and onychphorans, along with many creatures of no known phylum, including *Anomalocarus, Opabinia,* and *Microdictyon*	Cambrian period (570-510 mya)
Beecher's Bed NY	Trilobite (limbs)	Ordovician period (510-439 mya)
Lesmahagow, Scotland Waukesha WI	Early marine uniramian that resembles a centipede; possible leech; *Ainiktozoon,* of no known phylum	Silurian period (439-409 mya)
Gilboa NY Hunsrück, Germany	Trigonotarbids, mites, centipedes, spiders, and trilobite (limbs)	Devonian period (409-363 mya)

Sources of soft-bodied fossils	Organisms preserved	Geological time period
Granton Shrimp Bed, Edinburgh, Scotland	Conodont animal, a primitive chordate(?)	Mississippian subperiod of Carboniferous (363-323 mya)
Mazon Creek IL	Priapulid worms and *Tullimonstrum* (the Tully Monster) of no known phylum, along with hundreds of animals and plants	Pennsylvanian subperiod of Carboniferous (323-290 mya)
Solnhofen, Bavaria, Germany Christian Malford, England	*Archaeopteryx lithographica,* an early dinosaurlike bird; squid	Jurassic period (208-146 mya)
Santana, Brazil	Fish with muscles, guts, and gut contents	Cretaceous period (146-65 mya)
Green River WY	Freshwater fish, insects, and plants	Paleocene epoch of Tertiary period (65-56.5 mya)
Messel River, Germany		Eocene epoch of Tertiary period (56.5-35.5 mya)

Taphonomy The word *taphonomy* may not be in your dictionary, but if you are interested in paleontology you are encountering it and its derivatives increasingly often. The word was coined in 1940 by the Russian geologist Efremov from the Greek *taphos* for "burial" and *nomos* for "law." From the point of view of paleontologists, taphonomy is the study of the transfer of forms from the biosphere to the lithosphere—what happens when living organisms are preserved in stone. Studies of the early stages of the process have been most influential so far in helping paleontologists interpret the fossil record.

For example, the preservation of soft bodies that is the subject of the preceding section has long been assumed to be caused by burial of the bodies in sediments where there was no oxygen. A 1990 study by Susan M. Kidwell of the University of Chicago and Tomasz Baumiller, however, tested this idea by treating dead sea urchins to conditions that simulated burial on beaches by waves, varying such quantities as temperature, oxygen, and time. Surprisingly, they found that temperature rather than oxygen controlled the preservation of the sea urchin—the ligaments that hold its body plates together decay more slowly at lower temperatures. Oxygen has nothing to do with it, since microbes that live without oxygen are as willing to consume sea urchin ligaments as those that depend on the gas.

Similar studies, based more often on observations than on experiments, have been conducted for at least the past 25 years. A 1991 survey by Kidwell and Daniel Bosence of the University of London in Egham of the studies of preservation of marine shells showed that the fossil record is better than previously thought, but it also pointed to problems. For example, using rare species to distinguish between two communities probably does not work.

Taphonomy will continue in the 1990s to be a major new tool in understanding the fossil record.

Major Trends in Oceanography

Ocean Acoustic Tomography Just as tomography has been applied to studies of the interior of the Earth, it has also become a significant new tool in tracking the movements of water in the ocean. Pressure, temperature, and composition affect the speed of seismic waves through the Earth, while temperature and pressure affect the speed of sound waves through the ocean (composition contributes also, but the slight variations in composition of seawater do not make much difference). Small changes in the speed of sound waves bends them, just as small changes in the speed of light between air and water cause a pencil sticking out of a glass of water to appear bent. Low-frequency sound waves are used in ocean acoustic tomography because these frequencies travel great distances through the water.

The first experimental work with the technique was conducted in 1981 by Robert C. Spindel of the University of Washington, Peter F. Worcester of the Scripps Institute of Oceanography, and colleagues. Since then much experimentation has followed, with steady improvement in the technique. Primarily, ocean acoustic tomography has demonstrated that most of the variability in the ocean is not caused by large slow currents, but rather by much smaller patterns that can persist for months. These are the ocean equivalents of weather in the atmosphere. That is to say, with previous techniques scientists could study only the ocean's "climate," but with tomography they can study its "weather."

Satellite Studies Ocean acoustic tomography is concerned with action below the surface. Satellites have already shown their effectiveness in studying the surface of the ocean, observing temperature, height of sea level, and even currents. A major new tool planned for launch in 1992 is the Ocean Topography Experiment satellite, which will use tomographic techniques to study the surface of the ocean. It is hoped that these satellite observations can be combined with those from underwater ocean acoustic tomography.

(Periodical References and Additional Reading: *Science News* 8-25-90, p 128; *Discover* 10-90, p 96; *Scientific American* 10-90, p 94; *American Scientist* 3/4-91, p 130; *Science* 4-5-91, p 32; *EOS* 5-7-91, p 216; *Science News* 5-18-91, p 310; *EOS* 5-21-91, p 236; *New York Times* 5-21-91, p C2; *Scientific American* 6-91, p 101)

Timetable of Earth Science to 1990

BC

c 15,000	Map on bone is made of region around what is now Mezhrich, USSR
c 1000	Duke of Chou [Chinese] builds "south-pointing carriage," which may be early magnetic device
	Map on clay table shows Earth with Babylon at center
c 575	Anaximander of Miletus [Greek: c 610-c 546] in Ionia (Turkey) states that fossil fish are remains of early life
c 530	Pythagoras of Samos [Greek: c 560-c 480] argues on philosophical grounds that Earth is a sphere
c 500	Hecataeus of Miletus [Greek: c 550-c 476] makes map showing Europe and Asia as semicircular regions surrounded by ocean
c 350	Aristotle [Greek: c 384-322] lists observations that support Pythagorean idea that Earth is a sphere
	Explorer Pytheas of Massalia (Marseilles, France) [Greek: c 330-?] suggests that Moon causes tides
c 300	Dicaearchus of Messina (Sicily) [Hellenic: 355-285] develops map of Earth on a sphere using lines of latitude
	Chinese *Book of the Devil* contains first clear reference to magnetic properties of lodestone
271	In China, first form of magnetic compass is used for locating south
c 240	Erathosthenes of Cyrene (Shahat, Libya) [Hellenic: c 276-c 196] correctly calculates diameter of Earth
c 79	Explorer Hippalus [Hellenic] discovers the regularity of the monsoon

CE

79	Pliny the Elder [Roman: 23-79] is killed by eruption of Mt Vesuvius that buries Pompeii and destroys Herculaneum; nephew Pliny the Younger writes detailed account of disaster
132	Inventor Zhang Heng [Chinese: 78-139] develops first seismograph
c 1110	Chinese sailors begin to use magnetic compass for navigation
c 1180	*De naturis rerum* by Alexander Neckham [English: 1157-1217] contains first known Western reference to magnetic compass
1600	William Gilbert [English: 1544-1603] suggests that Earth is a giant magnet
1643	Evangelista Torricelli [Italian: 1608-1647] invents barometer
1667	Robert Hooke [English: 1635-1703] invents anemometer for measuring wind speed
1669	Nicolaus Steno [Danish: 1638-1686] correctly explains origin of fossils as once-living organisms preserved in stone
c 1672	Isaac Newton [English: 1642-1727] shows that rotation of Earth causes bulge at equator and flattening at poles
1701	Edmond Halley [English: 1656-1742] publishes chart of variations in Earth's magnetic field

1735	George Hadley [English] describes how temperature controls the major air movements around the globe
1736	Pierre de Maupertuis [French: 1698-1759] proves experimentally that Newton's theory that Earth is flattened at poles is correct
1761	Chronometer Number Four, designed by John Harrison [English: 1693-1776], successfully demonstrates method of finding longitude at sea
1774	Abraham Gottlob Werner [German: 1750-1817] introduces standard ways to classify minerals
1777	Nicolas Desmarest [French: 1725-1815] proposes that basalt is formed from lava
1785	James Hutton [Scottish: 1726-1797] proposes that geologic features of Earth result from tiny changes taking place over very long periods of time
1795	Georges Cuvier [French: 1769-1832] shows that giant bones found in Meuse River are remains of extinct giant reptile
1797	James Hall [Scottish: 1761-1832] shows that melted rocks form crystals on cooling
1803	Luke Howard [English: 1744-1864] classifies cloud types
1816	William Smith [English: 1769-1839] publishes first geological map in which rock strata are identified by fossils
1821	Ignatz Venetz [Swiss: 1788-1859] proposes that glaciers once covered Europe
1822	According to Gideon Mantell [English: 1790-1852], he and his wife Mary Ann are first to discover and recognize dinosaur bones this year; external evidence suggests Mantell may have found first bones as early as 1818
	Friedrich Mohs [German: 1773-1839] introduces Mohs' scale of hardness for minerals
1830	Charles Lyell [Scottish: 1797-1875] begins to publish *The Principles of Geology,* the work that convinces geologists that Earth is at least several hundred million years old
1835	Gustave Coriolis [French: 1792-1843] shows that rotation of a sphere causes displacement of paths of objects moving on its surface
1837	Louis Agassiz [Swiss-US: 1807-1873] introduces term *ice age*
1839	Karl Friederich Gauss [German: 1777-1855] uses data gathered by Paul Erman [German: 1764-1851] and his son Georg Adolf [German: 1806-1877] to develop mathematical theory of Earth's magnetic field
1842	Charles Darwin [English: 1809-1882] classifies coral reefs and explains how atolls are formed
1850	Matthew Maury [US: 1806-1873] charts the Atlantic Ocean
1855	Luigi Palmieri [Italian: 1807-1896] builds first Western seismometer
1859	Edwin Drake [US: 1819-1880] drills first oil well
1866	Gabriel-August Daubrée [French: 1814-1896] suggests that Earth has a core of iron and nickel

1880   John Milne [English: 1850-1913] invents modern seismograph

1896   Svante Arrhenius [Swedish: 1859-1927] discovers greenhouse effect of carbon dioxide in the atmosphere; global temperatures rise for higher levels of carbon dioxide

1899   Clarence Dutton [US: 1841-1912] describes vertical movements of Earth's crust in terms of isostasy, the equilibrium between the crust's weight and forces that cause parts of the crust to rise

1902   Léon-Phillipe Teisserenc de Bort [French: 1855-1913] discovers the stratosphere

1906   Robert D. Oldham [English: 1858-1936] uses earthquake waves to show that Earth's core exists

Bernard Brunhes shows that Earth's magnetic field can be recorded by cooling rock

1907   Bertram B. Boltwood [US: 1870-1927] shows that age of rocks containing uranium can be determined by measuring the ratio of uranium to lead

1909   Andrija Mohorovičič [Croatian: 1857-1936] discovers boundary between Earth's crust and mantle, now known as the Mohorovičič discontinuity, or "Moho"

1912   Alfred Wegener [German: 1880-1930] proposes theory of continental drift

1913   Charles Fabry [French: 1867-1945] discovers ozone layer in atmosphere

1925   The German *Meteor* expedition discovers mid-Atlantic ridge

1929   Motonori Matuyama [Japanese: 1884-1958] shows that Earth's magnetic field reverses polarity every few hundred million years

1935   Maurice Ewing [US: 1906-1974] starts study of ocean floor using refraction of waves caused by explosions

Charles Richter [US: 1900-1985] develops scale for measuring energy of earthquakes

1943   Mexican farmer discovers volcano (later named Mt Parícutan) growing in his cornfield

1944   US B-29 bombers attacking Tokyo on Nov 24 find they are flying faster than theoretical top speed of aircraft, resulting in discovery of jet stream

1946   Vincent Schaefer [US: 1906-] discovers dry ice can be used to cause clouds to release rain

1950   Hannes Alfvén [Swedish: 1908-] shows that solar wind interacting with Earth's magnetic field causes auroras

1953   Maurice Ewing discovers rift that runs down the middle of mid-Atlantic ridge

1958   James Van Allen [US: 1914-] discovers belts of radiation that surround Earth in space, now known as Van Allen belts

1960   Harry Hess [US: 1906-1969] develops theory of seafloor spreading

1968   The US National Science Foundation's Deep Sea Drilling Project, using the ship *Glomar Challenger,* begins; between this year and Nov 1983, it takes 20,000 core samples from 624 sites

1977   On Feb 19, John Corliss [US: 1936-] and two crew mates aboard research submersible *Alvin* discover living organisms near undersea volcanic vents; food pyramid for bacteria, worms, clams, and crabs is based on sulfur, not sunlight

1980   Walter Alvarez [US: 1940-] and coworkers discover layer of iridium at K/T boundary identified with demise of dinosaurs; he attributes both iridium and extinction to impact of large comet or meteorite

1985   The British Antarctic Expedition detects hole that forms annually in the ozone layer over Antarctica

1986   Using theories developed in wake of very strong 1982-1983 El Niño, US National Weather Service successfully predicts 1986-1987 El Niño

1987   US National Aeronautics and Space Administration (NASA) observes that continents move in ways predicted by theory of plate tectonics

1989   Eirik J. Krogstad and Gilbert N. Hanson [US: 1936-] of the State University of New York, Stony Brook; S. Balakrichnan and V. Rajamani of Jawaharlal Nehru University, India; and D. K. Mukhopadhyay of Roorkee University, India, find evidence that tectonic plates of 2.5 billion years ago clashed in what is now the Kolar schist belt in India, the earliest indication that current geologic processes were under way

Wade E. Miller [US: 1932-] and coworkers from Brigham Young University in Provo UT discover that the ilium of a dinosaur known as Supersaurus (possibly a large member of the Diplodocidae family) is hollow

A mapping project by scientists from the University of Washington in Seattle and Woods Hole (MA) Oceanographic Institution reveals a string of sixteen spreading centers that make up the mid-Atlantic ridge; each center is over a bullseye-shaped region of low gravity that may be caused by hot mantle rising toward the center

William F. Ruddiman of the Lamont-Doherty Geological Laboratory and John E. Ketzbach of the University of Wisconsin-Madison use a computer model to demonstrate that tectonic uplift of the Tibetan plateau and the Rocky Mountains causes global cooling and weather patterns that could have set off ice ages

Quarry workers uncover the most extensive set of dinosaur tracks in North America in Culpepper VA; about a thousand footprints of a carnasaur, a coelurosaur, and an unknown quadruped dating from 210 million years ago are well preserved

THE EVOLUTION OF LIFE

EARLY COURSE OF EVOLUTION

In 1984 Hou Xianguang of the Nanjing Institute of Geology and Paleontology made what may turn out to be one of the pivotal fossil discoveries of the century. It was not the bones of some spectacular giant dinosaur or mammal—in fact, there was not a bone to be found and the largest creatures known so far from the site were about half a meter (almost 2 ft) tall. The importance of this site comes from its 570-mya age, right at the beginning of the Cambrian revolution, its excellent state of preservation, and its extent. The fossil bed extends as far as 50 km (30 mi) in one direction and 20 km (12 mi) in the other. Its reach suggests that it formed when violent storms stirred up mud that settled on, killed, and preserved the organisms.

This remarkable find is in south central China's Yunnan Province at Chengjiang. The Chengjiang bed is being excavated by Hou in conjunction with Chen Junyuan, also of the Nanjing Institute, and studied by Hou and Chen and by Jan Bergstrom, Maurits Lindstrom, and others at the Swedish Museum of Natural History. The first thorough Western account appeared in the Apr 1991 issue of *Research and Exploration,* a scholarly journal of the National Geographic Society. In it, Bergstrom and colleagues wrote, "Evolution of these creatures seems to have been a sudden and widespread phenomenon."

This account was soon followed by a major report leading to a new interpretation of early Cambrian life. An article by Hou and Lars Ramsköld of the Swedish Museum of Natural History in the May 16 issue of *Nature* reported on an animal that Stefan Bengtson of Uppsala University in Sweden describes as a cross between a centipede and the Michelin Man. Its eleven pairs of legs are balloon-shaped like the legs of the famous Michelin trademark, although these legs are tipped with two-pronged claws. Such balloon legs are typical of modern onychophorans, a phylum of eighty or so species that includes the velvet worms, which live in moist litter in tropical rain forests. Many think onychophorans are the link between the annelid worms and the arthropods, which would suggest that some should be present around the beginning of the Cambrian, when both phyla are first encountered. The unnamed marine creature is about 5-6 cm (about 2-in.) long, large for an animal from the early Cambrian. Although some have previously identified onychophorans in middle Cambrian marine fossils (*Aysheaia pedunculata* from the Burgess Shale and *Aysheaia prolata* from the House Range UT), these identifications have been controversial.

This discovery could also result in a new reinterpretation of the mysterious *Hallucigenia.* This Burgess Shale beast was recently reclassified in the

1970s by Simon Conway Morris of Cambridge University, UK, as not in any modern phylum. Ramsköld and Hou think it is actually an onychophoran. They turn *Hallucigenia* upside down from Conway Morris's reconstruction and conjecture a missing set of legs. Conway Morris thinks they may be correct, but he wants another look at the original *Hallucigenia* specimens from the Burgess Shale before making up his mind.

Background

As long ago as 600 BC the Ionian scholar Xenophanes reported that he had seen fossils of sea creatures high on mountaintops. Other scientists continued to report ancient shells and bones—fossils—with varying interpretations until well into the seventeenth century, when finally Athanasius Kircher suggested that some fossils might be remains of animals that had missed Noah's Ark and therefore become extinct. By the nineteenth century it was generally recognized that most fossils were the bones, shells, and hard parts of extinct animals or plants. Although a few fossils with preserved soft parts, such as the feathers of Archaeopteryx, were found as early as 1861, it was generally believed that early fossils that revealed soft body parts were impossible.

This belief was used to explain the "Cambrian revolution," a name given to the sudden appearance of life in the geologic record about 570 mya. Before this time, there was nothing, (although recent work has revealed traces of bacteria and single-celled alga). After this "revolution," there was practically everything, with many representatives of the different animal phyla that still survive. It was thought that there had been a long period of evolution of multicellular life prior to 570 mya, but scientists could not find traces of it because there were no hard parts to record this development.

In 1909 the Burgess Shale was found in British Columbia by Charles Doolittle Walcott. This fossil bed preserved the soft parts of animals, probably because it was caused by mud slides that buried complete undersea communities so that the organisms were protected from oxygen, and therefore from decay, for the long period of time it took the mud to turn to rock. Although recognized as important, the pivotal place of these fossils in understanding early evolution was not fully grasped until the 1980s (see "The State of Earth Science," p 191).

One implication of the Burgess Shale discoveries is that the old idea that pre-Cambrian life evolved slowly (but unseen because of lack of hard parts) is very likely wrong. This ancient community looks as if evolution was proceeding very fast 550 mya. The new discovery in Chengjiang demonstrates that extremely rapid evolution started earlier and was even faster than previously expected. Most of the thirty-some phyla known today were already in evidence then, as well as half as many phyla that have since become extinct.

(Periodical References and Additional Reading: *Science News* 8-25-90, p 120; *New York Times*, 4-23-91, p C1; *New York Times* 4-28-91, p E7; *Science News* 5-18-91, p 310)

ORIGINS OF ORGANISMS

When people think of very early organisms, most envision the animals of the early Cambrian sea, but further reflection suggests that animals cannot exist alone. Today most animals are dependent either directly or indirectly on plants because green plants produce energy directly from sunlight by photosynthesis. In the distant past, however, animals may have preceded plants. At first, this seems to be a paradox; but think of monerans and protists! Both kingdoms contain organisms that photosynthesize (and monerans also include organisms that subsist on inorganic chemicals), and both are likely to have preceded plants and animals. Thus, in looking for truly early organisms, one looks for—and finds—monerans and protists.

Early Algae

While monerans are single-celled or colonial, some protists, such as some algae, are true multicellular organisms. The origin of multicellularity in algae precedes the development of known multicellular animals by a very long time. The earliest known animals date from about 600 mya. Some suspected multicellular algae date from 1400 mya. In 1990 Nicholas J. Butterfield and Andrew H. Knoll of Harvard along with Keene Swett of the University of Iowa reported on an early multicellular red algae, similar to but smaller than related modern red algae. Its age is between about 1270 mya and 720 mya, but probably closer to the earlier date. The fossils, found on Somerset Island in Canada's Northwest Territory, are extremely well preserved and show that algal evolution was far along at that time.

Algae cannot live on land because they lack any kind of stiff framework. Without water to buoy them up, they would just collapse into a blob. Nevertheless, some photosynthetic organism must have moved onto land before animals did, or there would be no reason for animals to leave the water. In 1989 algae were found to contain a chemical very similar to lignin, the substance that stiffens plants and enables them to live on land. While this chemical is not used for stiffening by the algae, its closeness to lignin suggests that it would be an easy evolutionary step to make it into lignin. Then the algaelike plant or the plantlike algae could invade the land.

Animals Move onto Land

Scientists have long searched for evidence that would show what kinds of animals first left the sea and when they did. In 1990 the search advanced significantly. Andrew J. Jeram of the Ulster Museum in Belfast, Paul A. Selden of the University of Manchester, and Dianne Edwards of the University of Wales announced the discovery of the earliest known land animals—centipedes and spiderlike arachnids called trigonotarbids.

Both centipedes and trigonotarbids are predators, not herbivores. Thus, some as yet unknown animal must have preceded them onto land. Selden speculates that such an animal might be like today's ubiquitous and nearly invisible mites that live on decaying vegetable matter (including carpets, and anything lodged in them, in people's homes).

Early Spider and Earlier Fliers

In 1982 Patricia Bonamo of the State University of New York at Binghamton, looking for plant fossils in the famous Gilboa NY shale, observed some remains that she thought were animals, but she could not identify them positively. Seven years later, continuing her study of the fossils, she observed a telltale organ, no bigger than a pinhole. This was identified by William Shear of Hampden-Sydney College in Virginia as the spinneret of a spider, the organ a spider uses to manufacture silk. The fossil is now the earliest known remains of a spider, dating from about 380 mya.

Insects, the most common prey of spiders today, had evolved by 380 mya, so the spider had something to eat. But at that time the only known insects did not fly. Since one of the main uses of a spider's silk is to build webs for catching flying insects, why build webs to catch flying insects if none were around? The strong possibility is that insect flight evolved earlier than expected, and that spider silk *was* used to capture flying insects. Shear thinks that continued study of the Gilboa shale will reveal remains of insects that flew some 80 million years earlier than previously thought.

Early Birds

If insects learned to fly at least 380 mya, when did birds join them in the air? Although archaeopteryx is considered the earliest known bird at about 145 mya (there is a competing claim for *Protoavis* from 225 mya—see below), archaeopteryx could not fly very well and perhaps could not even fly at all. The first birds that appear to have anything like the modern capability of flight are now thought to have appeared about 135 mya, based on a fossil found by a 10-year-old boy in northeast China. It was analyzed by Paul C. Sereno of the University of Chicago and Cheng-gan Rao of the Beijing Natural History Museum.

In addition to adaptations for flying, the unnamed Chinese fossil has feet better adapted to perching on tree limbs, also considered a sign of frequent flight. Another change from archaeopteryx is a much shorter tail. The long tail of archaeopteryx is similar to the long tail that bipedal dinosaurs used to keep their center of gravity over their legs, useful in walking. A short tail moves the center of gravity to the shoulders, more useful in flight. Despite such adaptations, the Chinese bird lacks many of the features pervasive in modern birds, such as fused hand bones and the absence of stomach ribs.

Only slightly younger birds are known from Spanish fossils, but they have already progressed farther toward modern birds. For example, the Spanish fossils have lost their stomach ribs. Because the Spanish fossils are close chronologically to the Chinese find, but distant geographically, birds were probably all over Earth at the time, and evolving rapidly. Fossils are hard to come by, however; the life-style of the birds and their hollow skeletons are not amenable to fossilization.

About *Protoavis:* The fossils for this creature were found in west Texas in 1986 by Sankar Chatterjee of Texas Tech University in Lubbock. Formal description of the head was published in Jun 1991 by Chatterjee along with his claim that it is the head of a bird, not a dinosaur. Most paleontologists think that Chatterjee is probably wrong, but also that whatever *Protoavis* is, it is an extremely interesting branch on the early reptile/bird tree. No one disputes that it is about 225 million years old, making it contemporaneous with the earliest dinosaurs known.

A major problem is that the fossils are not articulated, so there may be several different animals whose bones have been mixed. The type of rock in which the fossils were found did not preserve feathers, as the rock holding Archaeopteryx did. Also, individual bones are broken or bent, so there is considerable dispute about the correct way to reconstruct the living creature. A later publication will detail Chatterjee's reconstruction of the body, providing more fuel for the debate.

(Periodical References and Additional Reading: *Discover* 4-90, p 21; *Science* 10-5-90, pp 7 & 104; *Science News* 10-20-90, p 246; *Science* 11-2-90, pp 607 & 667; *Science News* 11-10-90, p 292)

EVOLUTION OF ARTHROPODS

In 1990 Michael J. Emerson and Frederick R. Schram reported on their reconstruction of a fossil that they believe is at least a partial "missing link" between the uniramians and the crustaceans. Newly discovered fossils in better condition helped. Called *Tesnusocaris goldichi,* the "missing link" is related to an extant group of primitive crustaceans called the remipedes. Remipedes are tiny (to 3 cm or about 1 in.) creatures found a dozen or so years ago in caves that connect to the ocean. They look something like upside down shrimp that have many extra segments, but their anatomy is much more primitive. Each remipede segment has a pair of biramous appendages on it. *T. goldichi,* as reconstructed, is similar to a remipede, but it has two sets of uniramous limbs on each segment. Thus, *Tesnusocaris* appears to be the "missing link," a uniramous crustacean.

Background

Arthropods are animals whose segmented bodies are covered with a distinctive hardened exterior, termed an exoskeleton or a cuticle, and jointed limbs. They include, among other groups, centipedes and millipedes (myriapods); spiders, horseshoe crabs, scorpions, mites, and ticks (chelicerates); crustaceans; and insects. Arthropods might be termed the dominant phylum of animals on Earth. Their claim comes from ubiquity, diversity, and longevity. There are now and have been since early Cambrian times more arthropods and more species of arthropods than of any other animal. They occupy every habitat known on Earth. However, their exoskeleton, which requires growth by molting, and their primitive lungless breathing apparatuses limit their size. Most arthropods are small, although the largest, Japanese spider crabs, have leg spans up to 4 m (13 ft).

The evolution of arthropods has presented several puzzles. Hundreds of books have been written trying to explain where arthropods originated and developed, with many speculative and different ideas. The close relation of arthropods to annelids, or segmented worms, such as earthworms and leeches, has been accepted until recently. In this view, either the arthropods arose from an annelid ancestor or both phyla came from a common ancestor. Recent evidence, however, suggests that repeated segments, which are the main link between arthropods and annelids (and between vertebrates and, possibly, annelids) arose separately in the phyla.

Within the arthropod phylum, there have been widely diverse views of relationships. In 1947, for example, O.W. Teigs argued that there are two unrelated groups (except insofar as all animals are distantly related) that are artificially classed as arthropods. One of Teigs's groups includes myriapods and insects, while the other consists of chelicerates, crustaceans, and the now extinct trilobites. Later, Sidnie M. Manton and D.T. Anderson broke these two groups into four different phyla that they felt did not share common annelid or annelidlike ancestors. Despite these theories, most taxonomists continue to put all arthropods into a single phylum, although within that phylum there is considerable variation as to how classes should be defined. Different characteristics are seen as basic by different taxonomists.

One set of differences that has frequently been the focus of taxonomists is the characteristics of the jointed appendages—legs in insects, myriapods, and chelicerates; and swimmerets, legs, or various specialized appendages in crustaceans. The distinctive difference is that the appendages of crustaceans are composed of two branches, or are <u>biramous</u>, while those of the other arthropods are single, or <u>uniramous</u>. The chelicerates have many common features that are not shared with other arthropods, so they are considered as a separate subphylum. Currently, many taxonomists group the insects and myriapods into a subphylum they call Uniramia, for their uniramous limbs. In this scheme, the crustaceans are a third subphylum of arthropods, while now extinct trilobites form either a separate subphylum or are grouped with the Crustacea.

The crustaceans and uniramians are enough alike that many taxonomists have proposed that one of the two subphyla emerged from the other. Although there are various characteristics in which individuals differ, the major difference to be explained is why the uniramians are uniramous and the crustaceans are biramous. Popular theories have been that the uniramian limb somehow reorganized itself to become branched or that the crustacean appendage was simplified to become unbranched. Exactly how such a transition could take place is difficult to explain. Furthermore, no "missing link" fossils have been found. In fact, no one can say for sure what such a missing link fossil would resemble.

What does this mean for the evolutionary history of the arthropods? Emerson and Schram put forth a compelling idea: Uniramian segments fused to become typical crustacean segments with biramous appendages. They view *Tesnusocaris* as partway along the path to such a fusion, since the segments seem to have fused, but the appendages have not. Later in the evolutionary process, the limbs also fused, changing therefore from uniramous to biramous. Thus, the crustaceans, in this view, have an ancestry in the uniramians, most likely from a centipedelike creature. Emerson and Schram, in their article in the Nov 2 *Science*, back up this theory with various arguments relating to the details of the body plans of uniramians and crustaceans.

Some of the most compelling evidence they offer comes from seemingly far afield. Studies of fruit fly mutations have been basic to the understanding of the genetics of all living things. But fruit flies—*Drosophila*—are uniramians, and their mutations affect the segments. These mutations show that in fruit flies and in other uniramians, segments are treated genetically as pairs. In crustacea, however, segments are genetically and developmentally distinct. Thus, it seems possible that the paired segments of uniramians could become the unpaired segments of crustacea.

(Periodical References and Additional Reading: *Science* 11-2-90, pp 607, 632 & 658)

NEW VIEWS OF THE DINOSAURS

During the past 20 years or so, our view of the dinosaurs has changed dramatically, partly as a result of new fossils but largely because of new interpretations of existing fossils and an improved understanding of evolution. While the tendency has been to see the dinosaurs and their life-styles in terms of the animals and behavior of today, there have also been many small changes in viewpoint and new information that affect the interpretation of individual fossils. Also, the 1980s saw serious debate about the cause of the extinction of the dinosaurs; that controversy continues into the 1990s.

One change was in the understanding of the role of the forelimbs of *Tyrannosaurus rex,* one of the most familiar dinosaurs. Because the forelimbs are very small in comparison with the hind limbs, and because of *T. rex*'s impressive daggerlike teeth, people have generally assumed that the forelimbs were weak and not involved in capturing prey. Not so, say Matt B. Smith of the Museum of the Rockies in Bozeman MT and Kenneth Carpenter of the Denver Museum of Natural History. Their 1990 reconstruction gives the forelimbs muscles that would be the envy of any bodybuilder, as well as opposing claws that could hold prey like the barb on a fish hook.

The idea that dinosaurs and birds are members of the same class got a boost in 1990 when David D. Gillette, state paleontologist of Utah, and

Background

In the 1820s, Gideon Mantell, aided by his wife Mary Ann, was collecting fossil bone fragments and teeth from English gravel beds. He concluded that he had found the remains of a giant plant-eating reptile. These fossils are now recognized as the first evidence for dinosaurs to be uncovered. Georges Cuvier and Richard Owen were among the paleontologists who studied these and other fossils of large reptilelike creatures. In 1841 Owen grouped many of the fossils into a new taxon called Dinosauria, a name that in its anglicized form, dinosaur, has stuck until today. By the late nineteenth century, however, paleontologists had determined that there were two groups of "dinosaurs" that were no more closely related to each other than they were to crocodilians or to pterosaurs (such as Pterodactyl). These groups are the Saurischia and the Ornithischia. Despite this, members of both groups are still generally known as dinosaurs. However, both groups differ from other reptiles: reptiles sprawl and dinosaurs stood upright (see below for a discussion of whether dinosaurs actually are reptiles.) To further complicate matters, in 1990 Rolf E. Johnston of the Milwaukee Public Museum and John H. Ostrom of Yale University made the controversial claim that Triceratops and related dinosaurs sprawled. Dinosaurs used their upright stance to walk the Earth from Late Triassic times until the end of the Cretaceous period, about 140 million years.

Traditional views of the dinosaurs as slow-moving, stupid creatures with life-styles similar to those of crocodilians began to be attacked in 1968, when Robert T. Bakker proclaimed their upright stance made dinosaurs swift-moving, not sluggish. The following year John Ostrom suggested that dinosaurs, or at least some of them, might have been warm-blooded (technically, endothermic). Bakker took up the cause, first at Yale (where he was Ostrom's student) and then at Harvard, going far beyond Ostrom's suggestion to argue throughout the 1970s in favor of dinosaur endothermy. Despite this, most paleontologists think today that few, if any, dinosaurs were endothermic. Indeed, another set of revisionist ideas concerns structures such as the plates along the back of stegosaurus coworkers discovered evidence that seismosaurus, the world's largest known dinosaur, had a crop and a gizzard as modern birds do. The evidence consists of about 180 small stones and a larger one found in or near the fossil. Gillette thinks that the stones were digestive aids, similar to the gravel that chickens consume to help grind their food. Most of the stones were in locations in the body that correspond to where birds have their crop and their gizzard. The large stone may be an error on the seismosaurus's part: swallowing a rock the size of a grapefruit may have caused the creature's death.

Symbolic of the new theories is the closing for the early 1990s of The Halls of Dinosaurs in New York City's American Museum of Natural History, the preeminent US display of dinosaurs for the general public. When

and the frills found on Triceratops and related dinosaurs. These have been described by some as devices used to cool large dinosaurs, dinosaurs whose immense size would make their blood warm without the need for any mechanism such as those found in endotherms.

Another recent issue concerning dinosaurs has been their relationship to birds. As early as 1867, Thomas Henry Huxley proposed that birds are descended from some branch of the dinosaurs. Until the 1970s, however, most scientists believed the dinosaurs and birds shared a common ancestor that was neither one nor the other. Starting in 1973, Ostrom argued, with considerable success, that Huxley had been right. He based his claim largely on a careful reexamination of archaeopteryx, which, aside from the feathers, he found almost identical to small dinosaurs of the time. In 1975, Bakker took the idea one step further and proposed that dinosaurs and birds together formed one class of vertebrates, while reptiles were a separate class from which the dinosaur/bird class had evolved.

Another idea about dinosaurs that emerged in the 1970s was that dinosaurs were sometimes social animals that lived in herds and even cared for their young, traits not possessed by reptiles, although a few snakes show signs of parental care. This view was sparked by the discovery in 1978 and 1979 of apparent nesting colonies of duck-billed dinosaurs by John R. Horner, then of Princeton University, and Montana high school teacher Robert Makela.

The cause of the extinction of the dinosaurs around the end of the Cretaceous period has long been debated, with many ingenious theories put forth. When Walter Alvarez and coworkers discovered evidence in 1980 for a giant impact on Earth by a large object from space, they proposed that the climatic effects from that impact had killed all the dinosaurs (see "Impact Craters Located," p 235). Bakker and others who support the idea that dinosaurs and birds are of the same class would revise this to say that the impact killed only those dinosaurs that had failed to evolve feathers.

it reopens, the entrance to the dinosaur exhibits will be dominated by a 17-m-high (55-ft-high) cast of a barosaur protecting its baby from an allosaur, suggesting that dinosaurs behaved more like birds than like crocodiles. Another much noted change will be that the 45-m (150-ft) apatosaur (commonly called a brontosaur) fossil will have its head turned. The present skull, that of a camarasaur, was mounted on the giant apatosaur fossil by mistake. The head will be removed and an aptosaur's skull will replace it, correcting a reconstruction that has been on exhibit at the museum since May 21 1953.

New Distribution Evidence

We usually picture dinosaurs in tropical settings. Although dinosaurs lived at a time in Earth's history when the average global temperature was higher, not all dinosaurs were in especially warm regions. Evidence uncovered in 1987 places some populations fairly near the north pole, in what is now Alaska. William Clemens of the University of California at Berkeley found dinosaur fossils in Alaska's North Slope, a region that would have been inside the arctic circle even during the dinosaur's time. Although the climate then was warmer, a habitat with long periods of darkness seemed like no place for a dinosaur. Clemens thinks the dinosaurs must have migrated into sunlit regions during the part of the year that the sun does not rise in that region.

Dinosaur fossils were also found in Antarctic regions in the 1980s and the early 1990s. In Dec 1990 David Elliot of Ohio State University discovered the first dinosaur fossils known from the Antarctic continent itself. The remains were about 3800 m (12,500 ft) up the side of Mt Kirkpatrick, near the Ross Ice Shelf. The fossils were first studied by William Hammer of Augustana College in Rock Island IL. Hammer determined that the bones belonged to a plant-eating dinosaur from about 200 mya. There was also the tooth of a carnivorous dinosaur at the same site. This discovery rolled back the record of dinosaurs in Antarctica by about 125 million years, as the 1986 find on an island near Antarctica was of a species from about 75 to 80 mya.

The Antarctic dinosaurs did not have a problem with long periods of darkness like the ones from Alaska. Antarctica 200 mya was where the southwestern Pacific Ocean is today, not below the antarctic circle. At the time, it was also joined to South American and Africa in the supercontinent, Gondwanaland.

New Extinction Theories

In the early 1990s, scientists speculated on many possible reasons for the extinction of dinosaurs. Most theories were based on the idea that a giant impact by an extraterrestrial body or bodies was responsible for their demise. The supposition of the first impact theory of dinosaur demise was that the primary causes of death were cold (supported by a 1990 analysis of fossil leaves made by Jack Wolfe from the US Geological Survey in Denver) and darkness, both the result of material tossed into the stratosphere by the impact that blocked sunlight. More recent ideas suggest other ways an impact could result in dinosaur extinction. The new theories include a rain of extremely hot particles the size of sand grains over all of Earth's surface (shown in a computer simulation from the University of Arizona) and a rain of nitric acid (espoused by Ronald G. Prinn of MIT).

(Periodical References and Additional Reading: *Science News* 1-20-90, p 40; *Science News* 2-3-90, p 79; *Science News* 2-10-90, p 85; *Discover* 5-90, p 12; *Science News* 7-14-90, p 31;*Science News* 9-29-90, p 212; *Discover* 10-90, p 76; *Science News* 10-6-90, p 212; *Science News* 10-20-90, p 255; *Science News* 10-27-90, p 270; *New York Times* 11-28-90; *New York Times* 12-10-90; *Discover* 1-91, p 43; *Science* 1-11-91, p 160; *Discover* 2-91, 52; *New York Times* 3-13-91, A20; *Discover* 4-91, p 14; *Natural History* 4-91, p 33; *Discover* 5-91, p 6; *Science* 6-14-91, p 1496; *Science* 9-6-91, p 1089)

WHALE WITH FEET

The transition of mammals from land to full-time existence in water has been compared to the transition of the first fish/amphibians from the sea to land, or early reptiles from land to sea and from land to air. Other vital passages took insects, birds, and mammals into the air. Finding organisms part way along in such a transition is one of the keys to understanding evolution. In 1990 the world learned of a major discovery that cast new light on how whales moved back into the water from the land.

In Dec 1989 husband and wife team Philip D. Gingerich and B. Holly Smith of the University of Michigan and Elwyn L. Simons of Duke University unearthed a fossil of a whale that still possessed hind limbs complete with feet, the first ever found on a species of whale. The whale, *Basilosaurus isis*, is a long, slender sea creature that has been known since before 1835, although no complete skeleton has ever been assembled. The main purpose of the 1989 expedition to the place paleontologists call Zeuglodon Valley ("Zeuglodon" is a previous name for early whales) was to locate more *B. isis* bones in the hope of reconstructing still unknown parts, notably the front flipper. Five days before the end of the expedition, Gingerich located a bone showing the place where the pelvis attached to the vertebrae, indicating the position along the 15-m (50-ft) length where leg bones might be found. Some other early whales are known to have small remnants of former hind limbs, so it seemed worthwhile to look for those of *B. isis*. Returning to other partially exposed fossils, Gingerich, Smith, and Simons started searching in that region of the fossils and in the remaining days of the trip located the legs, from the femur down to the toes. After studying the bones

---

WHALE FAMILY TREE

The ancestors of whales were land mammals until about 60 million years ago (mya). The early whale ancestors, the mesonychids, were carnivorous relatives of the ungulates, the herbivores who became today's horses, cattle, pigs, and so forth. Although the mesonychids became extinct on land about 40 mya, the ones that moved into the water gradually evolved into today's whales and dolphins. The oldest known whale, *Pakicetus,* discovered by Gingerich in 1978, lived about 50 mya and may still have ventured onto land from time to time, perhaps like a modern sea otter.

---

back in Ann Arbor, Gingerich reported that the legs were about 50 cm (20 in.) long, not very big for a 15-m (50-ft) whale, but much longer than expected. Furthermore, these were clearly functional legs, not rudiments.

The whale, which lived 40 million years ago in a shallow sea where Egypt is today, apparently did not use the legs in swimming. Speculation about their use has ranged from helping the whale work its way out of mud (by Lawrence G. Barnes of the Natural History Museum of Los Angeles County) to clasping the opposite sex during mating (by Gingerich). Gingerich has pointed out that since the pelvis is 11 m (35 ft) from the brain along a highly flexible body, legs may have been needed to keep the proper alignment for copulation. Barnes notes that sexual organs are often different sizes in males and females; if more limbs are found, and if they differ greatly in size, he will concede the case to Gingerich.

The four toes found for *B. isis* suggest that it is more closely related to the even-toed ungulates, or Artiodactyla, than to the odd-toed ungulates. In other words, whales descended from the ancestors of cattle, deer, pigs, and so forth.

B. isis is not the only fossil whale found in Zeuglodon Valley. Gingerich and Smith also found what they now believe are leg bones from a small whale called *Dorudon,* the probable ancestor of modern whales. It seems increasingly likely that all sufficiently early whales had legs, probably for the first 10 million years after they began to live full time in the water.

(Periodical References and Additional Reading: *Science* 7-13-90, p 154; *Science News* 7-14-90, p 21; *Discover* 5-91, p 44)

RECOVERING ANCIENT LIFE

11,000 Years Ago

Peat bogs have been known to preserve organic remains for a long time. Bodies of humans hundreds of years old found in such bogs sometimes touch off modern murder investigations because the police assume that such fresh flesh must be recent. On May 3 1990, an even more startling example of peat preservation was announced. A team led by archaeologist Bradley Lepper of the Ohio Historical Society had excavated the remains of an 11,000-year-old mastodon found in 1990 in a peat bog on a golf course in Newark OH. The amazing news was not the mastodon, since mastodon remains are fairly common in North America. It was that the fossil's preserved intestines contained recognizable remains of its last meal and living bacteria, identified by microbiologist Gerald Goldstein of Ohio Wesleyan University as *Enterobacter cloacae.*

The ancient bog was capped by a meter (3 ft) of clay, which helped keep it at a chilly 7° C (45° F) year-round.

Goldstein found *E. cloacae* in a smelly reddish brown deposit near the mastodon's ribs, but none in a dozen samples of surrounding soil. A second blind analysis from another laboratory confirmed these results. Other scientists were skeptical of the claim regarding the bacteria, announced at a press conference on the golf course. It is difficult to prove that contamination by present-day bacteria had not occurred.

Nevertheless, everyone was enthusiastic about the rest of the discovery. Although soft parts of mammoths—like mastodons, members of the elephant family—from the ice ages have been preserved in permafrost (see below), no soft parts of a mastodon had been found previously.

A sample of the "last meal" was carbon-14 dated at about 11,000 years ago. These food remains suggest that the mastodon diet was very different from what scientists had previously believed. Although the common American mastodon of the last retreat of ice age glaciers is supposed to have dined mostly on branches and needles of evergreen trees, browsing as deer and giraffes do, this mastodon's last meal was identified by Dee Anne Wymer of Bloomsburg University in Pennsylvania as swamp grass, moss, leaves, seeds, and a bit of water lily, a diet similar to the summer diet of the moose. This analysis was confirmed by J. Gordon Ogden of Dalhousie University in Nova Scotia, who found pollen stuck in the fossil's teeth that matched the food found in the intestines. Since mastodon fossils are regularly associated with bogs and swamps, it may be the case that this habitat, not the surrounding forest, supplied a major portion of the diet. If so, the theory that the mastodon's extinction was caused by the retreat of coniferous forests to the north along with the retreat of the glaciers may need revision. Or, like modern moose, the mastodons may have eaten highland browse in the winter and lowland water vegetation in the summer.

Paleontologist Daniel C. Fisher of the University of Michigan claimed that the discovery vindicated his contention that early Native Americans preserved kills of mastodons by anchoring the remains under water. He believes this find to be a cache that had been abandoned.

Other members of the project were Tod Frolking of Denison University,

---

MASTODON DIET

The mastodons were originally given their name by Georges Cuvier [French: 1769-1832] to distinguish them from their relatives the mammoths, ancestors of today's elephants. The name refers to their teeth, which were identified by Cuvier as the type used for browsing—eating mainly leaves and small branches of bushes and trees—as opposed to the type used for grazing of the mammoths. Since remains of mastodons were associated with swamps and bogs of the type often found in forests, it was assumed that mastodons were forest creatures who browsed on forest branches. Ice age forests in North America were composed mostly of conifers, such as spruce, so the mastodon of the end of the most recent advance and retreat of the ice, the one whose fossils are so common, had been thought to eat spruce needles and branches. Mammoths, on the other hand, were evidently creatures of the steppes and plains, where grass, not trees, was the main vegetation.

In 1990 Soviet mammoth expert Nicolai Vereschagin brought to the US Smithsonian Institution the freeze-dried stomach contents of a 30,000-year-old mammoth. Results of an investigation by Smithsonian archaeologists have not yet been announced.

Jon Sanger of Ohio Wesleyan, and Paul Hooge of the Licking County Archaeology and Landmarks Society.

Goldstein commented on the discovery of the living bacteria that "these could be the oldest living organisms ever found." The remainder of this article suggests some other possibilities.

3 Million Years Ago

In northern Siberia, a half meter to a meter (2 to 3 ft) under the surface, the soil never thaws. This is the region of permafrost. According to scientists from the USSR, the permafrost north of Cherskiy, close to the Arctic Ocean, has been frozen for 3 million years. The pattern of geomagnetic reversals (see page 241) shows that the sediments frozen here are older than at least 2.4 million years. It is believed that these sediments have not been above -12° C (-10° F) during this whole time.

Mammoths have been found trapped in more recent permafrost layers, frozen like a steak in a freezer. In fact, adventurous souls have dined on frozen mammoth flesh from over 10,000 years ago. What kind of remains might be found in 3 mya permafrost?

A joint expedition to answer this question included a US team led by E. Imre Friedmann, director of the Polar Desert Research Center of Florida State University, along with Christopher P. McKay of the NASA Ames Research Center in California and Michael A. Meyer of the Desert Research Institute in Reno NV; a USSR team led by David A. Gilichinsky of the USSR

Academy of Science's Institute of Soil Science and Photosynthesis joined them. The team traveled to the region north of Cherskiy during the period between Jul 15 and Sep 5 1990. There they drilled out some samples, put them in insulated containers to keep them frozen, and returned the samples to their laboratories in the United States and in Moscow.

When thawed out, the samples revealed more than well-preserved organic remains. They produced living bacteria that the scientists believe have been dormant in the permafrost for the past 3 million years. Furthermore, the US team speculates that there is even older permafrost on Mars that might contain still living but dormant life forms from 3.5 *billion* years ago.

18 Million Years Ago

Magnolias, or more accurately the ancestors of magnolias, are believed to be among the first flowering plants. Michael T. Clegg, Edward M. Golenberg, and coworkers were able to study photosynthesis in one of these ancestors.

Golenberg began the explanation of their work in a talk in the spring of 1990 this way: "I have used the PCR technique to make copies of genes isolated from leaves that fell from some trees during an autumn windstorm. The windstorm happened 17 million years ago." (Later the date was placed at 18 mya.) Some of the leaves were so well preserved that they were still green or autumnal red when first found, although they quickly turned dark and shriveled, as if they had just been removed from Shangri-la.

DNA degrades over time and it is difficult to imagine how the researchers obtained the gene even from such well-preserved leaves. The secret was the PCR technique (see "Multicopying DNA" on p 218), which enabled them to produce enough of the DNA in the gene to study. Even so, not every expert on ancient genes thinks they succeeded. If they did, they hold the record for the oldest gene ever cloned. The scientists have also used the same technique on another plant preserved in the same formation and are studying leaves from another formation that is 100 million years old.

The ancient magnolia gene apparently came from a chloroplast, the part of a leaf in which photosynthesis takes place. The team compared the gene with the equivalent one from modern magnolias. They determined that the gene underwent an average of one mutation every million years, a mutation rate that is sometimes called the tick of the DNA clock. The DNA clock is a controversial way of determining how long ago various species separated from each other.

65-Plus Million Years Ago

George O. Poinar of the University of California at Berkeley has proposed a way to use PCR to obtain genes from dinosaurs, who have, of course,

---

MULTICOPYING DNA

PCR stands for polymerase chain reaction, a method for amplifying genes or complete stands of DNA (see "Who Owns PCR?" p 593). The method involves separating the paired strands of DNA and then using each strand as a template to create other strands. Appropriate chemicals are added to the mixture as a feedstock. With PCR, a single molecule of DNA can be reproduced in as great a quantity as one needs for analysis. Often, instead of using the entire DNA molecule, a single gene is isolated. Then the copies of the gene can be used in genetic engineering. With additional copies to insert into a bacterium or into an egg cell, the chances of success for such engineering projects are increased.

---

been extinct for 65 million years. So far, his proposal has not been acted on, partly for the reasons outlined below.

DNA can be preserved for longer periods of time when it is still in cells that are protected by some coating. The ancient leaves from 17 million years ago, for example, were preserved in airtight shale, which kept the cells intact. Many very old fossils that appear to be remarkably well preserved are small creatures that have been trapped in tree resins that, through geologic time, became the gemlike substance we know as amber. Like the original resins, amber is sufficiently transparent that one can see the creature trapped inside. Most often it is an insect or spider, but rarely one finds a frog or lizard. Such insects and other creatures appear through the amber to have been engulfed recently, even when they are millions of years old. They would seem to be good candidates for DNA analysis through PCR.

Poinar has an even more ingenious idea. Many insects obtain some of their energy by sucking the blood of other creatures; we are most familiar with various mosquito species in which the female obtains the blood from a mammal, often a human, to get the energy needed to produce and deposit eggs. Biologists think that mosquito species that use blood in this way first preyed on cold-blooded animals, reptiles, and amphibians, as some species of mosquitoes do today. Whether dinosaurs were cold-blooded or warm-blooded, they probably were preyed on by mosquitoes.

Suppose you find a female mosquito trapped in amber from a time before 65 mya. Such a mosquito is likely to have had a blood meal. A little of the DNA from the meal could be extracted from the mosquito's digestive system. With PCR, an unlimited supply of the DNA could be obtained, enabling one to study the genes of whatever creature the mosquito had bitten. Given the number of dinosaur fossils, it would seem entirely possible that the DNA might be from a dinosaur. Other biting insects or arachnids might also be used for such an experiment.

There are several hitches in this plan. For one thing, perfectly preserved specimens in amber are extremely valuable to collectors and even used as exotic jewels. Poinar once found a female lizard preserved in amber whose eggs he would have liked to test for well-preserved DNA, but a Japanese collector bought it for $20,000, far beyond what Poinar could pay. Secondly, it has not been proved that DNA will be preserved in amber for tens or hundreds of millions of years. Finally, there is the difficulty of locating the right biting insect or arachnid, one that has recently fed on a dinosaur.

Nevertheless, the idea captures the imagination. In 1990 a popular book, *Jurassic Park* by Michael Crichton, was based on carrying the idea a step or two further. In the book, complete dinosaur DNA is located and produced in quantity. Complete DNA contains the entire genetic makeup of the creature. The DNA is removed from crocodile eggs and dinosaur DNA is used to replace it. What hatches from the eggs would kill you! Soon to be a major motion picture.

150 Million Years Ago

W. Dale Spall of Los Alamos National Laboratory in New Mexico started with core samples from the vertebra of seismosaurus, a much studied fossil of what is believed to be the largest dinosaur ever. Spall's specialty at the lab is isolating proteins from petroleum or coal to learn the history of these fossil fuels, but his hobby is paleontology. He was delighted in 1988 to participate in the seismosaurus dig.

Spall was not looking for DNA to replicate (although Erika Hagelberg of John Radcliffe Hospital in Oxford, UK, has recently found a way to extract tiny amounts of DNA from bone). Spall thinks there is no chance that any DNA could survive the 150 million years since seismosaurus died. Instead, Spall was chasing proteins, which he thinks might survive in tiny quantities as a result of lucky chances. Furthermore, Spall thinks he has found some seismosaurus proteins, including one that controls the synthesis of hydroxyapatite, the hard stuff in bone that surrounds living cells.

As usual in this field, there is skepticism over whether protein could survive that long without degrading. A protein is a complex molecule that tends to fall apart because of its large size and in reaction to the increased activity of its parts caused by heat. Thus, some scientists suspect that much more recent proteins must have contaminated Spall's sample. He replies that he has performed the same tests on surrounding rock, where he finds some proteins, but not the one that he thinks comes from seismosaurus.

(Periodical Sources and Additional Reading: *Discover* 1-91, p 78; *Planetary Report* 3-4-91, p 8; *New York Times* 5-4-91; *Science News* 5-18-91, p 318; *Science* 6-1-91, p 1074; *New York Times* 6-25-91, p C1; *Discover* 8-91, p 8; *Science* 9-20-91, p 1354)

WEATHER AND CLIMATE, NOW AND THEN

EL NIÑO AND LA NIÑA

On Feb 16 1990 the US National Weather Service's Climate Advisory Center (CAC) in Washington DC released an El Niño/Southern Oscillation (ENSO) advisory—a prediction that an ENSO was about to take place. This was based on a rise in sea-surface temperatures, low air pressure, and a weakening of the trade winds in the central Pacific. The advisory contradicted the predictions of three well-known computer models in the United States that had successfully predicted the 1986-1987 ENSO. Another computer model, that of the Max Planck Institute for Meteorology in Hamburg, Germany, agreed with the CAC advisory. The CAC meteorologists, charged with making a 90-day prediction of weather for the United States, stuck by their forecast. For the actual outcome, see the final section of this article.

ENSO Thermostat

In the May 2 1990 *Nature,* V. Ramanathan and William Collins of the Scripps Institution of Oceanography in La Jolla CA proposed that an important role of the ENSO is to act as a feedback mechanism to keep the surface of the world's oceans from becoming too warm. The central Pacific warming that starts a typical ENSO increases the amount of water vapor in the air. This water vapor, combined with increased convection from the surface heat, produces giant cirrus clouds. The clouds, however, block enough sunlight to cool the surface temperature. This warming initiates cooling. The whole process is much like a reverse household thermostat, one set to 31°C (88°F).

La Niña

El Niño is not the only weather phenomenon that starts in the middle of Earth's largest ocean. Just as El Niño causes unusual weather when central Pacific surface temperatures rise, worldwide weather is affected when those temperatures are unusually cold. Several names have been proposed for this weather pattern, but the name La Niña seems to have stuck.

An Aug 31 1990 report in *Science* from Thomas W. Swetnam of the University of Arizona and Julio L. Betancourt of the US Geological Survey suggests that knowledge of one of the weather effects of La Niña could be

> *Background*
>
> El Niño was first recognized as a change in conditions off the coast of Peru and Ecuador in which at the same time as normally dry inland areas received torrential rains, fish and the birds that fed upon them died. The phenomenon was named El Niño after the Christ child because it usually happened around Christmastime. Scientists determined long ago that the immediate cause of the peril to fish and the local weather changes was a shift in the position of the cold Humboldt Current, which usually washes the coast.
>
> In 1982-1983 the strongest El Niño of the century occurred and spurred intensive study of the phenomenon. During the 1982-1983 event, weather seemed to be abnormal worldwide. India and Australia experienced droughts that were thought to be caused by the event. Animal and kelp die-offs around the world were also related to the unusual weather. During this period, meteorologists realized that a central Pacific phenomenon they had named the Southern Oscillation caused the changes in the Humboldt Current. They also suspected that it caused global changes. Among meteorologists, the event came to be called the ENSO (El Niño/Southern Oscillation), but most people continue to say "El Niño."
>
> Studies of historical records coupled with examinations of weather preserved as ice in South American glaciers have shown that an ENSO typically takes 12 to 15 months from start to finish and occurs every 4 to 7 years.

important to officials in the American Southwest. After comparing historical records of the Southern Oscillation since 1860 and Peruvian El Niños from periods before 1860 with fire scars on trees throughout the Southwest, they conclude that a La Niña precursor in the central Pacific in the winter is very often followed by abnormally large forest fires in the Southwest. In contrast, El Niños are associated with enhanced tree growth and fewer fires. The US Forest Service in Albuquerque NM is planning to use the theory as a guide in requesting increased funding and resources in appropriate years.

The Climate Mambo

Predictions for El Niño and La Niña could be made more than a year in advance if scientists were certain of the causes of these patterns. Various oceanographers, meteorologists, and climatologists put forward tentative theories in the last part of the 1980s and the early part of the 1990s. Most of the theories refer to an underlying 2-year cycle that was most strongly detected by Eugene M. Rasmusson and Xueliang Wang of the University of

Maryland along with Chester F. Ropelewski of CAC. Their findings were issued in a report in 1990. The scientists see this underlying 2-year beat combined with a weaker rhythm of 4 to 5 years. This combination causes El Niño to do the mambo instead of the fox trot.

If there is an underlying rhythm or two, what is the cause of the beat? A theory that has been around for the past 5 years says that the Pacific Ocean is just the right size for slow waves to take 2 years to cross it, after which the waves rebound and start heading the other way. Rasmusson doubts this, joining another camp that says that the annual seasonal cycle somehow produces a biennial beat; the problem with this explanation is that no one has offered a convincing theory of what makes the beat go on. A third school of thought, led by William Gray of Colorado State University and Tetsuzo Yasanari of Japan's University of Tsukuba, is that neither the ocean nor the seasonal cycle is involved. It is the air—specifically a cyclic wind shift in the lower stratosphere called the Quasi-Biennial Oscillation. The only point of agreement is the underlying 2-year cycle.

The Next El Niño

With regard to CAC's specific prediction for an ENSO warming in 1990, the US computer models were right and the CAC and the German model were wrong. The warming did not happen. In 1991, however, computer models at Scripps Institution of Oceanography and Lamont-Doherty Geological Observatory, both correct in 1990, predicted ENSO warming for 1991. The Lamont-Doherty model has steadfastly predicted since 1989 that 1991 would see an El Niño. Stephen Zebiak of that institution has proposed that El Niño would not only arrive in 1991, it would also be moderately strong. By the summer of 1991 the National Meteorological Center's computer model had joined in predicting an El Niño for the winter of 1991-1992.

By Aug it appeared that the predicted 1991 El Niño would appear, but it would be weak. Surface waters in the tropical Pacific were 1° to 1.5° C (2° to 3° F) higher than normal. Winds blowing from east to west had begun to slacken. By Dec the signs were stronger. The central Pacific had warmed enough to generate thunderstorms that redistribute heat from the ocean into the atmosphere. Although it was far from full-blown, it looked as if an El Niño was starting. In California, people were looking forward to an El Niño. After 5 years of drought, El Niño was expected by many to bring a wet winter in 1991-1992, although the US National Climatic Data Center (NCDC) in Asheville NC said that rain in California was not necessarily a consequence of El Niño. Ski-lift operators in the US Northeast could expect more snow, however, according to the NCDC.

(Periodical References and Additional Reading: *Science News* 3-3-90, p 135; *Science*, 8-31-90, pp 967 & 1017; *Science News* 9-1-90, p 132; *Science* 9-14-90, p 1246; *Science* 3-8-91, p 1182; *Science News* 5-11-91, p 303; *Science News* 8-10-91, p 87; *Science News* 12- 14-91, p 389; *New York Times* 12-24-91, p C4)

PREDICTING HURRICANES

The observed correlation between the weather off the coast of Peru, in Australia, and in South Africa suggests that other worldwide patterns must also exist. Understanding long-term weather patterns is important, but most of us are more concerned about predictions of short-range weather. For example, it has long been recognized that the typical pattern for hurricanes in the North Atlantic is for them to increase in late summer, to disappear virtually in winter, and to strike the Caribbean and Gulf of Mexico more often than New England. This knowledge is only slightly helpful to people who live along the western coast of the North Atlantic. What people really want to know is exactly when a hurricane is on its way, how big it will be, and where it will go. The past 30 years of advances by the US National Hurricane Center have improved predictions by 14 percent, not a very large amount.

A possible step ahead in the tricky art of hurricane prediction was observed in 1990 when weather forecasters reviewed the performance of the trial run of a new Yugoslavian-designed computer model called ETA (named after the Greek letter that was its model designation). ETA takes data on existing weather conditions and forecasts what the conditions will become after 24 and 48 hours. Although not specifically developed to predict hurricanes, ETA analyzed five that reached North American waters during its 9-month trial in 1989, and did a much better job of doing so than any of the other models in use by the National Hurricane Center.

Among the hurricanes that occurred during the trial period was Hugo, the most destructive one to strike the United States in many years. ETA was more accurate in general than previous models at predicting Hugo's path and was 320 km (200 mi) more accurate in describing the path after Hugo's landfall near Charleston SC.

A hurricane is powered by heat produced as clouds release rain. ETA apparently beats other models in hurricane prediction partly because it assumes more abundant rainfall under given conditions, more accurately reflecting what happens during a hurricane. Additionally, because it is devoted solely to predicting for North America, it has more computational power available for that task than programs that deal with the whole world. With more power, ETA can use a smaller 80-km (50-mi) grid of data points, improving the sharpness of its predictions. This precision will be increased in future versions.

Overall trends also help in making specific predictions. In 1990 William M. Gray of Colorado State University reported on a newly discovered weather pattern that has become another tool for hurricane predictors. He recognized that a well-known weather pattern in the western Sahel region of Africa, just south of the Sahara, is directly related to a previously unobserved pattern in hurricanes reaching the southeastern United States. The Sahel, for reasons that are not currently known, alternates between wet and dry periods on a cycle of about 20 years. The evidence for such a cycle stretches back to the seventeenth century. Although tropical storms that develop into hurricanes can form over any stretch of warm water, those that form off the coast of West Africa and remain coherent enough to reach the east coast of North America are generally stronger and more dangerous than those that form closer to North America. During rainy weather in the western Sahel, also known as a serious monsoon, conditions are excellent for forming strong, coherent tropical storms. The easterly jet stream is slower, giving storms a longer time to form; it propels the storms slowly but surely toward the Caribbean and the Gulf. Consequently, some weeks after a storm forms off Africa, a strong hurricane arrives in North America. The difference in frequency of hurricanes can be astonishing. During the wet Sahel cycle from 1947 to 1969, thirteen strong hurricanes struck the United States, compared with one during the dry cycle from 1970 through 1987.

When 1988 started the beginning of another wet cycle in the Sahel, it seemed to start a new strong hurricane cycle in North America, with three powerful hurricanes in the first 2 years of the cycle. But the weather dried up somewhat in Africa in 1990 and Gray's prediction of two strong hurricanes for 1990 was not realized: there was only one strong hurricane in the Atlantic and it did not make landfall. As the drier weather continued, Gray issued a prediction at the start of the 1991 hurricane season for no strong hurricanes. He continued to expect the wet weather to return to Africa for a long period and for the concomitant strong hurricanes in the Atlantic to return with it.

Gray has been issuing hurricane-season predictions since 1984. He revised his method considerably in 1990 to factor in sea-surface temperatures, atmospheric pressure over the ocean, shifts in major wind currents, and, of course, rainfall in Africa. With the new criteria, Gray predicted six Atlantic hurricanes (there were six), two of them strong (there was one), and a total of 25 days when hurricanes would be present in the Atlantic (there were 27.5 such days). For 1991, in addition to predicting no strong hurricanes, he predicted a total of three Atlantic hurricanes for the season and only 10 days with hurricanes present.

(Periodical References and Additional Reading: *Science* 2-23-90, p 917; *Science News* 5-5-90, p 281; *Science* 9-14-90, pp 1223 & 1251; *Science News* 9-15-90, p 164; *Science* 10-5-90, p 29; *New York Times* 8-6-91, p C5)

CLIMATE CLUES

Before the recognition in the nineteenth century of the recent ice age, scientists and people in general thought that Earth's climate stays fairly much the same—although older people have probably always claimed that it was colder or hotter or wetter when *they* were young. Once Louis Agassiz had popularized the notion that an ice age had produced a dramatic climatic change, other scientists began to devise ways to examine past climates. As more of the fossil record became known, it was easy to recognize that in some periods vegetation or animal life corresponded to what we now associate with the tropics or with cool climates. Geologists were also able to recognize characteristic scars and deposits left by glaciers from various periods much earlier than the recent ice age. These clues produced a general impression of long warm periods intermingled with a number of ice ages.

Recently, scientists have developed more subtle means for detecting specific climatic changes over shorter periods of time. Everything from the way the shells of tiny protists coil to the ratios of various isotopes has been enlisted in this endeavor. Many methods are quite clever. For example, R.C. Capo and D.J. DePaolo of the Lawrence Berkeley Laboratory reported in Jul 1991 that the ratio of two isotopes of strontium could be used to determine past climates. This ratio is different in continental rocks such as limestone and granite from what it is in general in Earth's crust. Therefore, when weathering is high, the ratio in ocean waters will be affected by strontium that has been released from continental rocks and carried into the ocean. Weathering is more powerful when precipitation and erosion are high, which happens during ice ages. Hence, the strontium ratio in deep-sea cores of calcium carbonate, which tend to pick up stray strontium due to the close chemical resemblance between strontium and calcium, records climate. Capo and DePaolo's study using this method tends to confirm other studies of the timing of the ice ages, but it also reveals a few climatic changes missed by others, probably because of gaps in the geological record.

R. Klein of Tel Aviv University and coworkers used the presence of humic acid in ancient coral reefs along the shores of the Red Sea to study climate in the Sinai Desert about 100,000 years ago. Humic acid is familiar to gardeners as the acid component of humus, which indicates decaying vegetable matter. Hardly any humic acid is formed in the Sinai today, since it is too dry for most plants. Klein and coworkers found high humic acid levels for interglacial periods, mainly in coral bands that indicate summer growth. Thus, the Sinai probably once had a summer rainy season during which humic acid washed into the Red Sea and was absorbed by growing corals.

Corals have also figured extensively in dating studies (see "Dating the Ice Ages," p 229).

B. Molfino and A. McIntyre of Lamont-Doherty Geological Observatory reported in Aug 1990 that the abundance of fossils of the golden-yellow alga *Florisphaera profunda* could be used to determine indirectly the relative temperature of upper layers of the ocean. Their analysis of fossils in a core representing the past 200,000 years shows that temperature varies regularly over a 23,000-year cycle. This is related to the length of the cycle known as the precession of the equinoxes, a gradual change in the direction of Earth's rotational axis. The actual precession takes about 26,000 years, but the time when the longest period of daylight in the northern hemisphere corresponds to the farthest distance between the Earth and the Sun varies with a slightly shorter period, calculated at 22,000 to 23,000 years. The highest implied temperatures from the *F. profunda* abundances are about 1500 years behind where one would expect them to be based on the relation of Earth to the Sun, a lag presumably caused by the length of time the effect takes to warm Earth. All Molfino and McIntyre's observations were taken near the equator, a location less likely than others to be influenced directly by glaciation.

In 1991 Paul Olsen and Dennis Kent of Lamont-Doherty Geological Observatory began an impressive study of a 7-km (4.3-mi) deposit of sediments from an ancient lake bed in what is now New Jersey. The Newark Rift Basin was laid down when the supercontinent Pangaea was splitting apart, some 200 to 230 mya. During that period, the rift lakes, similar to those in the Rift Basin of Africa today, filled and emptied and refilled over and over as the climate changed. Unlike many other methods for evaluating past climates, examining lake bottom sediments reveals rainfall, not temperature.

Olsen and Kent did not have to drill a 7-km hole to examine the 30-million-year record. Since the Newark Basin formed, it has been tilted by tectonic forces. As a result, six comparatively shallow boreholes drilled from Princeton to Martinsville NJ could be used to sample virtually the whole record.

Olsen and Kent found clear indications that rainfall some 200 mya varied according to several changes in the relationship between Earth and the

Sun. Both the climatic patterns most easily observed today, reflecting the 23,000-year precession of the equinoxes and the 100,000-year change in the shape of Earth's orbit, were present during the time of Pangaea. A more prominent cycle of 400,000 years, long predicted by astronomers but not observed during recent climatic patterns, also appeared. Furthermore, although analysis is not complete, there appears to be another astronomical cycle that shows up. This one is about 2.3 million years long.

What is peculiar is that the 400,000-year cycle has not been evident in the past 2 million years. Oceanographers who have done much of the climate dating think this may have something to do with the continuing presence of ice sheets in the Antarctic and Greenland during geologically recent time. Paul Olsen, however, suspects that something may be missing in the way ocean sediments record climate. Of course, the ocean record and the Newark Basin record are not only using different methods, they are also looking at different aspects of climate. The oceanographers are indirectly measuring how much water is locked up in ice, while the Newark Rift Basin study is concerned with how much rain is falling near the equator.

Another pair of drilling projects is using a more conventional method to study climate and a host of other factors. A US ice-core operation called Greenland Ice Sheet Project 2 (GISP 2), and a European one called the Greenland Ice Core Project (GRIP), are operating about 30 km (19 mi) apart at the summit of the Greenland ice cap. The operations have been running since 1988 and will complete the coring task in 1992. The ice sheet where they are drilling is about 3200 m (10,500 ft) thick and represents a record that goes back about 200,000 to 250,000 years. Greenland was chosen because its mean annual temperature is too low for summer melting while its annual accumulation of 0.25 m (10 in.) of the water equivalent of snow and ice is large enough to allow analysis of the record year by year. Not only can these cores be used to determine temperatures from ratios of oxygen and hydrogen, but also atmospheric gases trapped in the ice can be used to determine greenhouse effects in the past. The first such result from GISP 2, released in the Aug 1990 *Geophysical Research Letters,* shows that carbon dioxide levels during the colder climates of the sixteenth and eighteenth centuries were about the same as during the whole period from 1530 to 1810. Levels then began to rise, at first probably as a result of clearing of forests and more intensive agriculture. Carbon dioxide levels rose even faster as the Industrial Revolution picked up steam.

(Periodical References and Additional Reading: *Science* 4-6-90, p 31; *Science News* 5-19-90, p 318; *Science News* 6-9-90, p 356; *Science* 6-15-90, p 1314; *Science* 6-22-90, pp 1467 & 1529; *Science* 6-29-90, p 1607; *Science News* 12-22/29-90, p 388; *Science* 6-28-91, p 1834; *Science News* 9-14-91, p 168)

Climates of the Past

Era or Eon	Period	Epoch	Climate and Probable Cause	Time Before Present
Archean eon			Thought to be hot due to carbon dioxide in the atmosphere, causing a greenhouse effect	4600 mya
Proterozoic eon				2500 mya
	Sinan			800 mya
	Vendian			610 mya
Paleozoic era			Studies of atmospheric carbon dioxide suggest warm temperatures during the early Paleozoic and cooler temperatures in the late Paleozoic	
	Cambrian	Lower Middle Upper		570 mya
	Ordovician	Lower Upper	A small ice age occurs	500 mya
	Silurian	Lower Upper		425 mya
	Devonian	Lower Middle Upper		395 mya
	Carboniferous		The Permo-Carboniferous ice age occurs	
		Lower Upper		
	Permian	Lower Upper		290 mya
Mesozoic era			Studies of atmospheric carbon dioxide show warm temperatures throughout the Mesozoic; global temperatures are thought to have ranged from 17° to 25° C (63 to 77° F); the formation of Pangaea may have had something to do with climate during this era	
	Triassic	Lower Middle Upper		235 mya
	Jurassic	Lower Middle Upper		190 mya
	Cretaceous	Lower Upper		130 mya
Cenozoic era				
	Tertiary	Paleocene	Cool global temperatures of around 13° C (55° F); Antarctic ice sheet forms	66.5 mya

Era or Eon	Period	Epoch	Climate and Probable Cause	Time Before Present
		Eocene	Sudden warming of about 6° C (11° F) causes a major extinction of deep-sea organisms	57.8 mya
		Oligocene		36.6 mya 30.0 mya
		Miocene		23.5 mya
		Pliocene		5.2 mya
	Quaternary	Pleistocene	An ice age with glaciation in both the northern and southern hemispheres; global temperatures around 10° to 12° C (50 to 54° F)	1.6 mya
		Holocene	Because of temporal proximity, the details of 11,000 years ago the Holocene are better known than most; the start of the epoch is a cold period known as the Younger Dryas, thought to be caused by a sudden influx of glacial meltwater from Lake Agassiz into the North Atlantic; temperatures were definitely low in Europe, Greenland, and the northern east coast of North America; studies in 1990 and 1991 suggested that climate was also affected around the world	
			It is generally believed that medieval times were relatively warm, at least in Europe, with the period lasting from 1100 to 1300 CE, although a 1990 study of Scandinavian tree rings by K. R. Briffa from the University of East Anglia in Norwich, England, and co-workers reported that the early 1100s were cool in Scandinavia at least	
			Another cold period in Europe and possibly North America was the Little Ice Age, from 1550 to 1800 CE, although the Scandinavian tree-ring study does not confirm this	

DATING THE ICE AGES

Progress in earth science since the eighteenth century has often come from finding better dates for events in the past. For example, the realization that Earth is older than a few thousand years was essential for developing the uniformitarian concept of geology. One of the first applications of natural radioactivity after its discovery was finding the dates of rocks; this began in

1906 and the method has improved continually since then. Carbon-14 dating is the most famous dating technique, although it is not useful for dates from the truly distant past. Another famous dating technique for recent events has been the use of tree rings. What earth scientists need, however, is a dating technique that falls somewhere between the hundreds, thousands, and ten thousands of years of tree rings and carbon-14 and the millions and billions of years of many dating methods based on long-term radioactivity.

Learning the ultimate cause of ice ages is intimately associated with correct dating. As early as 1842, 5 years after the existence of an ice age first came to be accepted, Joseph-Alphonse Adhémar put forward the argument that ice ages were caused by changes in the alignment of Earth and the Sun that resulted in less heat reaching Earth. This is verifiable because astronomers are able to determine the dates of both past and future astronomical events with great precision. Based on his knowledge of astronomy, Adhémar claimed an ice age occurred in the northern hemisphere every 11,000 years, a period related to the precession of the equinoxes. Twenty-two years later, James Croll refined the theory to call for an ice age lasting 11,000 years every 100,000 years; his calculations were based upon combining precession of the equinoxes and periodic changes in the shape of Earth's orbit. Croll's theory was widely accepted until more exact dating of the most recent ice age showed that instead of being at its height 80,000 years ago, as Croll would have it, its strongest effects were much more recent. In 1920 Milutin Milankovitch put forth a new version of the idea that ice ages were caused by the relation of Earth and the Sun. He combined precession and changes in the shape of the orbit with a periodic shift in the inclination of Earth's axis. The interplay of the three accounted for both increased glaciation and periods of glacial retreat, known as interglacials. This theory was widely accepted until the advent of carbon-14 dating in 1946. When used to date the recent ice age stages more accurately, the method produced dates that failed to agree with the Milankovitch theory, causing it to be dismissed. In the 1970s the pendulum swung the other way. Oceanographers started using radioactive decay of uranium to thorium and another technique based on oxygen isotopes to date calcium carbonate, a common compound found in seashells of various types. By the early 1970s these dates were found to agree with Milankovitch's combination of Earth-Sun cycles, at least for the recent ice age.

For a time, the oceanographers' dates held sway. In 1988, however, Isaac Winograd and coworkers of the US Geological Survey used the uranium/thorium technique to date calcium carbonate deposits in Nevada at a place called Devil's Hole. They got dates for the recent ice age stages that were different from those of the oceanographers. In particular, Winograd's date for the previous interglacial differed by 20,000 years.

An interglacial is marked by higher ocean levels, since less of Earth's water is tied up in glaciers and ice caps. As a result, an interglacial can be identified by ancient coral reefs that now are above the waterline, since coral does not grow unless it is underwater. Improved forms of uranium/thorium dating were used by various oceanographers in 1990 to date the top layer of ancient coral reefs in Barbados. These dates confirmed

Background

All chemical elements occur in varying forms called isotopes. A given element, such as oxygen, is determined by the number of protons in its nucleus—eight in the case of oxygen. The number of neutrons in the nucleus can vary, however, producing different isotopes. An isotope is identified by a number that is the sum of the protons and neutrons in the nucleus. Thus, the most common form of oxygen, which has eight neutrons as well as the obligatory eight protons, is called oxygen-16 (in symbols $_8O^{16}$). Most elements found naturally are a mixture of different isotopes. Adding neutrons to a nucleus can make the atom unstable, or radioactive. Oxygen-16, oxygen-17, and oxygen-18 are stable, but oxygen-19 and oxygen-20 are radioactive. Nature finds it easier to form atoms with even numbers of protons and neutrons. In a natural sample of oxygen, about 99.759 percent is oxygen-16, about 0.204 percent is oxygen-18, and only about 0.037 percent is oxygen-17.

Radioactive isotopes change into other forms in various ways until they find a form that is stable. Radioactive carbon-14, with six protons, becomes stable nitrogen-14, with seven protons, when one of the neutrons in carbon-14 emits an electron and turns into a proton. Other radioactive isotopes may go through several different forms before becoming stable elements.

Most methods of dating based on decay of radioactive isotopes rely on ratios of the original amount of an isotope to the stable isotope that is the end product of the decay or to an isotope with a relatively long half-life. For example, the first such method used the ratio of radioactive uranium to its ultimate product, lead.

Carbon-14 dating is based on a different process. Cosmic rays are subatomic particles that bombard Earth from space. The energy from cosmic rays converts some nitrogen in air to carbon-14, producing throughout the atmosphere a definite ratio of radioactive carbon-14 to the much more common stable carbon-12. Atmospheric carbon interacts with oxygen to becomes carbon dioxide; plants incorporate carbon dioxide during photosynthesis; some animals acquire the carbon by eating plants, others by eating animals that eat plants. As long as all the organisms are living, they continue to incorporate carbon-14 into their tissues in a specific ratio to carbon-12. When an organism dies, however, this process stops. Since carbon-14 is radioactive, it decays at a steady rate, going back to nitrogen. Half returns to nitrogen after about 5730 years. Thus, the amount of carbon-14 that remains, as compared with the amount originally present, can be used to date anything that once was alive. The method can be used only for fairly recent dates (less than about 40,000 years), as the amount of carbon-14 eventually becomes too small to measure.

the earlier oceanographers' dates, restoring the 20,000 years Winograd removed. Specifically, the last interglacial was found to be between 122,000 and 130,000 years ago, as opposed to Winograd's date of 147,000 years ago.

Additional studies of coral reefs that were reported in 1990 confirm agreement between intensity of the recent ice age and the Milankovitch cycles. The studies also can be used to improve carbon-14 dating. Tree-ring studies had previously shown that older carbon-14 dates were systematically off—1000 years too young at 8000 years ago. But tree-ring dates go back only about 10,000 years. Coral can be dated with both the uranium/thorium method and with the carbon-14 method. Good uranium/thorium dates from coral reefs show that carbon-14 dates older than 10,000 years are farther off than anyone expected. By 20,000 years ago, carbon-14 dates are 3500 years too early. This implies that all older carbon-14 dates need to be moved back several thousand years.

The reason for the carbon-14 discrepancy may be a story in itself. Apparently, there was about 40 percent more carbon-14 during parts of the recent ice age. The most obvious cause would be that there was more carbon-14 formed 20,000 years ago than there is today. It seems unlikely that the universe produced more cosmic rays a few tens of thousands of years ago. But it is possible that more of the particles were reaching Earth. Today, many charged particles are deflected by Earth's magnetic field and do not reach the atmosphere. Perhaps there was less of a magnetic field 20,000 years ago, allowing more cosmic rays to interact with the atmosphere and form carbon-14.

In Papua New Guinea another sequence of the former tops of coral reefs (called raised marine terraces by oceanographers) was dated by a French, German, and Dutch team using various methods. In their 1991 report, the team theorized that the sequence of terraces reflects a half-dozen interglacials, including one that is about a million years old. Their dates also closely resemble those found from ocean sediments by oceanographers.

A group consisting of Fred M. Phillips, Marek G. Zreda, and Stewart S. Smith from New Mexico Tech, David Elmore from Purdue University, and Peter W. Kubik and Pankaj Sharma from the University of Rochester measured ice age dates by a different method. In the Jun 21 1991 *Science* they reported dates in general agreement with the uranium/thorium dates from corals. Specifically, they were dating several glacial moraines in the Sierra Nevada mountains of California, but it is assumed that glaciers descended from high mountains at the same time as, and rather more rapidly than, ice sheets invaded the North American continent.

As in carbon-14 dating, dating the moraines depends on cosmic rays. Cosmic rays convert chlorine-35 (about three-quarters of all chlorine) to chlorine-36, an isotope with a half-life somewhat more than 300,000 years. Although this half-life may seem long, it is short enough that one can be certain that most chlorine-36 around has been caused by geologically recent cosmic-ray bombardment. Any other chlorine-36 is the result of natural radioactivity in the substance. Furthermore, although chlorine-36 caused by natural radioactive elements pervades rock, cosmic-ray generated-chlorine-36 is only on exposed surfaces. Thus, one can determine how long the surface of a rock has been exposed by measuring the amount of chlorine-36 on the surface and subtracting the background chlorine-36 produced by natural radioactivity (and also correcting for the radioactive decay of chlorine-36). In dating a glacial moraine, the important thing is to use really big boulders near the top of the moraine. Their size prevents them from having rolled over during the last hundred thousand years or so. Their high location means that they are not likely to have been covered by snow most of the winter.

But the classic oceanographer's method for dating ice ages comes from studies of two isotopes of oxygen. There is a small amount of oxygen-18 mixed in with the oxygen-16 that makes up more than 99 percent of the oxygen on Earth. Because of the two extra neutrons in oxygen-18, that isotope is slightly heavier than the common one. The heavier oxygen-18 is also slower and, one might say, lazier than ordinary oxygen. In evaporation, for example, oxygen-16 is more likely to make it from the liquid to the gaseous state. This effect was used in the classic studies dating the ice ages that confirmed the Milankovitch cycle during the 1970s. That work used the oxygen-18/oxygen-16 ratio in the sediments formed by the shells of tiny sea organisms. The idea behind it was that during an ice age, oxygen-18 would be left behind in seawater as oxygen-16 evaporated, fell as freshwater rain, and was locked up in glaciers. The more oxygen-18 in the sediments, therefore, the more ice on the continents.

In 1990, *glasnost* brought US researchers a new tool to use in checking this idea. Soviet scientists made available a long core of Antarctic ice they had acquired at their Vostok ice station. It holds the record of the past 160,000 years. The ice itself is not especially useful for dating ice ages since it forms whether there is an ice age or not and it contains a higher amount of oxygen-16 as a regular course. But the ice has air bubbles trapped in it. Oxygen in the air comes from photosynthesis, most of which takes place in the ocean. Plants and protists in the ocean get the oxygen from seawater, which, as we have already seen, is enriched in oxygen-18 during an ice age. Thus, an increase in glaciation is marked by more oxygen-18 in the air bubbles trapped in ice. Happily, an analysis of the Vostok core by Todd Sowers and Michael Bender of the University of Rhode Island and coworkers from France and the USSR confirm dates from fossil sediments.

Analysis of the air in ice produces other results of interest. The air bubbles also reveal the amount of carbon dioxide and methane, two "greenhouse gases" that can produce global warming by trapping solar energy. Although it is known that there are more greenhouse gases in the atmosphere during an interglacial than during a glacial advance, the mechanism involved and the timing are still uncertain. If greenhouse gases in the atmosphere increase before an interglacial starts, then they could be the cause of the warming; but if they increase afterward, the interglacial may somehow produce more of the gases. The study of the Vostok records shows that the greenhouse gases increase before the ice melts. Now the problem is to determine what causes carbon dioxide and methane to increase when glaciation is intense.

An interesting footnote to the most recent ice age is the Younger Dryas mystery. In Europe, shortly after the glaciers had retreated to about their present locations about 11,000 years ago, they suddenly sprang forward and produced a 200-year-long cold snap known as the Younger Dryas. During this period the oak forest that had moved north into Germany moved back south, for example. The principal puzzle about the Younger Dryas is what caused it. This is still not clear, although a leading theory is that melting glaciers produced enough fresh water to affect the currents in the North Atlantic. One of the major causes of ocean currents is differences in density, and fresh water is less dense than salt water.

The other mystery has been what was happening in North America while the Younger Dryas was occurring in Europe. If the ocean-current theory is correct, North America might still have been warm even though northern Europe was cold. In 1986, however, pollen studies showed that plants that favor cool climates were dominant in North America. This climatic picture has been confirmed by a clever study announced in the Aug 30 1991 *Science* by Ian R. Walker and Jon P. Smol of Queen's University in Kingston, Ontario, and Robert J. Mott of the Geological Survey of Canada in Ottawa. Instead of plants, they studied fossil midges (two-winged flies related to gnats and familiar to anglers as a favorite food of trout). Different species of midges live in cold and in warm conditions, and accurately reflect surface temperatures of the lakes and ponds where their larvae grow. Furthermore, unlike oak trees, midges have short lives and can fly from place to place. Thus, fossil midges reflect current conditions better than pollen does. The Canadian scientists found that the fossil midges in sediments in Splan Pond in New Brunswick show that eastern Canada was cold during Europe's Younger Dryas. In the article, the scientists also proposed that techniques based on insect fossils can be used to determine climatic changes that would help understand global warming if it is occurring today.

(Periodical References and Additional Reading: *Science News* 1-13-90, p 20; *Science News* 5-19-90, p 318; *Science* 7-6-90, pp 7 & 51; *Science* 8-17-90, pp 719 & 766; *Science* 8-24-90, pp 835 & 863; *Science News* 8-25-90, p 126; *Science* 9-14-90, pp 1223 & 1251; *Science* 9-21-90 pp 1355 & 1382; *Science News* 9-22-90, p 184; *Science News* 10-6-90, p 217; *Science* 12-7-90, pp 1315 & 1383; *Science News* 12-22&29-90, p 388; *New York Times* 1-10-91; *Scientific American* 3-91, p 66; *Science* 5-31-91, p 1254; *Science,* 8-30-91, p 1010; *New York Times* 9-3-91, p C2)

IMPACTS AND THEIR IMPACT

IMPACT CRATERS LOCATED

The Moon and Mars feature signs of giant impacts made by large meteorites, asteroids, or comets. Such impacts are also known to have occurred on Earth, but there are few obvious signs to be seen of the about 110 craters that have been located. On Earth, erosion and movement of the active crust soon—in terms of geologic time—erase craters. Furthermore, even the best known crater, the Barringer Crater (or Meteor Crater) in Arizona, formed only 25,000 years ago, is small by comparison with the scars on other planets and satellites.

Aside from two possible impact sites from about 2 billion years ago, all of the known sites were formed during the past 600 million years. Consequently, it was somewhat of a surprise when Gary Byerly of Louisiana State University and Donald Lowe were able to show that four giant impacts occurred between 3.45 and 3.25 billion years ago (bya). They did not find the craters, but they located in a South African rock formation glass spherules with a high iridium content, considered by most earth scientists as near proof of formation by an impact (a minority thinks the iridium can be explained by volcanic activity). The spherules represented four different age groups, suggesting the four different impacts.

The K/T Boundary Crater

But the most interesting crater for many people, including much of the general public, is the one that is thought to have been made 65 mya when a large object struck Earth and caused the major mass extinction known as the K/T boundary extinctions. First evidence for such an impact was the 1980 discovery of a layer of clay enriched in iridium (see Background). Many scientists think that the 65-mya impact changed Earth's climate for a time and caused mass extinctions, including the extinction of the dinosaurs (see "New Views of the Dinosaurs," p 209, and later in this article for more on whether this actually happened).

The search for a suitable crater has engaged scientists all over the world. Candidates have included Pacific, Indian, and Arctic ocean sites, partly on the grounds that an impact at sea would leave a less detectable crater. An ocean impact also might be more destructive as a result of heating vast amounts of water to boiling. On land, the favorite site until recently has been the Manson Crater in Iowa, which seems to have been created about the right time, although with a 35-km (22-mi) diameter, it is too small to have caused the postulated worldwide damage. Manson has been known since the 1950s, but has been recognized as an impact crater only since the 1960s. It cannot be seen from the surface as it is covered with glacial debris. The iridium layer, which is worldwide, is supposed to be a mix of 1700 km^3 (400 cu mi) of extraterrestrial and earthly material that vaporized and entered the stratosphere, from which it rained fairly evenly on the crust thereafter. Minerals in this layer are partly characteristic of oceanic rock and partly of continental rock. A 1991 study of tektites— small glassy objects that apparently originated when melted rock flung into the atmosphere by the impact cooled on its way back to Earth—that were formed by a late Pliocene impact known to have been in the Pacific Ocean, an impact that has been studied intensively by Frank T. Kyte of the University of California, Los Angeles, and colleagues, suggests strongly that the K/T impact was at least partially in an ocean basin, but composition of the layer does not completely support either the sea or the land.

North America contains a layer of glassy material of debris just below the iridium layer. This 2-cm (¾ in.) layer of debris was first noted in 1985 by Jan Smit of the Free University in the Netherlands. It contains many tektites. The debris also includes shocked quartz that can be caused only by the high temperatures and pressures of a major bolide strike. The tektite layer is presumed to be mostly earthly rock directly ejected and transformed by the impact. Such material—the ejected layer—would fall to Earth nearer the impact site instead of being scattered worldwide. Thus, the impact crater should be in or near North America.

Planetary scientists and chemists Alan R. Hildebrand and William V. Boynton of the University of Arizona have been searching for the crater since 1984. Their studies have been instrumental in showing that the tektite layer is just under the iridium layer in much of North America and that the iridium layer's chemical composition suggests an impact site on the continental shelf. Several studies by other scientists in the late 1980s revealed large debris deposits in Texas and Arkansas that might have been caused by giant ocean waves 65 mya. Following these clues, Hildebrand and Boynton began examining in Apr 1988 core samples taken from the Caribbean Sea floor in the 1970s. The samples looked like pay dirt. Hildebrand took a field trip to the Caribbean, and in Feb 1990 found a ½-m- (20-in.-) thick layer of clay in Haiti filled with glassy debris. This is by far the thickest tektite deposit ever found, which would imply that the crater must be nearby. Furthermore, the tektites in Haiti, instead of being a few mm (hundredths of an in.) in size, included some as big as a centimeter (½ in.). Subsequently, in 1991 G.A. Izett, G.B. Dalrymple, and L.W. Snee of the US Geological Survey used a radioactive isotope of argon to date the tektites at 64.5 mya, right at the K/T boundary. This also makes the tektites by far the oldest ever accurately dated.

Background

In 1979 Walter Alvarez, Frank Asaro, and Helen Michel discovered a thin layer of clay marking the boundary between the Cretaceous and Tertiary geologic periods, known as the K/T boundary. Analysis of this clay by Luis Alvarez showed that it was enriched in iridium, an element uncommon in Earth's crust, but much more common in meteorites. Since then, other workers have discovered the iridium enrichment at sites around the world, adding up to about 200,000 tons of iridium in a layer 25 mm (0.1 in.) thick. To many, this suggests that a large meteorite (or asteroid or comet—the generic term for all such large objects that strike explosively is bolide) struck Earth and caused the change from the Cretaceous period to the Tertiary.

Geologic time periods are marked in two ways: by specific rock formations and by the fossils they contain. When there is a sudden shift between one set of fossils and another, a new geologic time period begins. But this implies that the boundary between two periods is generally marked by extinction of many previously common species. There are many such mass extinctions in the geologic record. Perhaps the most extensive occurred at the end of the Permian period and marked the dividing line between the Mesozoic and Paleozoic eras. But most popular interest focuses on the mass extinction at the K/T boundary, which is also the boundary between the Mesozoic and Cenozoic eras. The reason for this interest is that the K/T extinction is related to the still somewhat mysterious demise of the dinosaurs, everyone's favorite extinct life-form. Although the issue of the cause of the extinctions is still unresolved, many scientists during the 1980s came to believe that the evidence strongly favors a giant impact as the cause of the iridium. The main problem is locating the remains of a suitable crater—one of the right size and the right age.

Somewhat tentatively, Hildebrand and Boynton located the crater in the Caribbean off the coast of Colombia. Searching the literature, they found evidence for a semicircle that could be the edge of such a crater in the Colombian Basin—not conclusive evidence, but suggestive.

Soon after their announcement, and before publication of the Hildebrand-Boynton report in the May 18 1990 *Science,* a US Geological Survey geologist, Bruce F. Bohor, and Russel Seitz of MIT, proposed in the Apr 12 1990 *Nature* that Hildebrand and Boynton had the right idea, but that the actual crater, based on previously reported geology, was much farther north, on Cuba's Isle of Pines. Hildebrand and Boynton replied that the Cuban geologic features were *caused* by their bolide, and not by anything striking Cuba itself. Later, Robert S. Dietz of Arizona State University and John McHone visited Cuba and found no telltale glassy fragments there, tending to confirm Hildebrand and Boynton's interpretation. Still, evidence pointed toward an impact somewhere in the Caribbean.

In Mar 1990 Hildebrand learned from a reporter on the *Houston Chronicle* that two oil geologists had claimed in 1981 that they had discovered a giant impact crater in Yucatán. They had reported their find, based on gravitational and magnetic anomalies, to a geology meeting in Houston, where the reporter learned of the presumed crater. Hildebrand reached one of the geologists, Glen Penfield, and together they were able to find some oil company drilling samples from the region. Hildebrand, Boynton, Penfield and coworkers found glassy fragments in the samples, which were from a site called Chicxulub ("devil" in the local Maya language). At this point, Chicxulub emerged as a third Caribbean candidate. Study of the Chicxulub evidence by Hildebrand, Boynton, and David Kring, also of the University of Arizona, showed that the shocked quartz fragments were embedded in previously melted rock, further evidence that Chicxulub is the actual impact site.

In May 1991, Kevin O. Pope of Geo Eco Arc Research in La Canada CA, Charles E. Duller of the NASA Ames Research Center in Mountain View CA, and Adriana Ocampo of the Jet Propulsion Lab reported that surface markings can be used to identify the part of the Chicxulub Crater that is on land. Yucatán today is covered with a flat bed of limestone, somewhat like parts of Florida. In both regions, water charged with carbon dioxide tends to eat away the limestone, producing sinkholes. When these fill with water in Yucatán, they become the famous cenotes of the Maya. Pope's group noticed a large semicircle of cenotes that seemed to be in the right place to be the rim of the crater. Seen in satellite images in the mid-1980s, the cenotes looked like the edge of a giant paw print, but no one then had any idea what could have caused such a formation. Pope now surmises that the sinkholes were formed when water seeped through cracks at the edge of the crater.

Further work with the Haitian tektites from the K/T boundary reported in June 1991 by Florentin J-M. R. Maurrasse and Gautam Sen, both of Florida International University in Miami, is consistent with an impact at Chicxulub. An intriguing aspect of this report is that it seems to show that the tektite layer was mixed by giant tidal waves, both at the time of the impact and then subsequently, perhaps as a result of a second impact.

All proposed craters in the Caribbean region are about 200 km (120 mi) in diameter, three orders of magnitude larger (in volume) than the Manson Crater.

As Walter Alvarez noted at the Dec 1990 meeting of the American Physical Union: "A few years ago, our problem was that we didn't have any craters to point our fingers at, whereas now we've got several. So now, maybe instead of a smoking gun we've got a smoking firing squad." Alvarez went on to point out that the asteroid 1989PB, which approached Earth in 1989, showed up in radar images taken by Steve Ostro of the Jet

Propulsion Laboratory as a double object, suggesting that multiple objects may be common. Candidate craters at Manson and Chicxulub and at Popigay in Russia's high arctic are geographically close enough that a multiple-body impact in the northern hemisphere's summer could have caused them all within minutes of each other!

Causes of Extinction

If the impact of a giant bolide caused a mass extinction, how did it go about it?

The Alvarezes and their coworkers (see Background) attributed the dinosaur extinction to their purported bolide, which they calculated could have been as much as 10 km (6 mi) in diameter. They proposed that the impact disrupted the weather on the entire planet, causing green plants to die and leaving the dinosaurs with inadequate food supplies. (Pierre de Maupertuis, a French evolutionary theorist, had proposed similar ideas as early as 1750.) Not everyone thought this was reasonable. Traditional paleontologists cited evidence that the dinosaurs had gradually become extinct before the impact event, if there was one. Specialists in volcanoes claimed the iridium-enriched layer for their own field, and suggested that enormous volcano eruptions caused both the layer and the extinctions.

Jack A. Wolfe of the US Geological Service in Denver has proposed a scenario similar to the original concept of "impact winter," but with some added ideas. His analysis of fossilized water-lily and lotus leaves from the Teapot Dome site in eastern Wyoming has led him to conclude that a large impact in late Jun caused serious cooling as a result of impact debris in the stratosphere blocking sunlight. During this period, the lake, which contained lilies with fruit and lotuses without fruit (giving the time of year exactly), froze. Wolfe experimented with modern plants and claims that the buckling he observes in the fossils also occurs in his experiments.

As the floating debris cleared, the temperature of the atmosphere was raised sharply by the greenhouse effect of water vapor or other gases flung into the air by the impact. Wolfe supports this supposition with fossils of ferns found above the water-lily and lotus layer. The fern fossils are intermingled with another layer of debris, which Wolfe believes comes from a second impact—perhaps the Manson bolide, which struck near enough to Wyoming that it is conceivable that some of its immediate debris might have fallen on the lake. If so, the Manson impact must have been several months after the initial impact, to allow time for the warming and for the ferns to grow.

Wolfe's ideas, while consistent, are thought by many to involve more extrapolation than is reasonable.

Other recent ideas include a rain of extremely hot particles the size of sand grains over all Earth's surface (from a computer simulation from the

University of Arizona) and a rain of nitric acid (espoused by Ronald G. Prinn of MIT). If the impact was in the limestone of Yucatán, there might be yet another mechanism for extinction. A theory of John D. O'Keefe and Tom A. Ahrens of Caltech predicts that an impact in a limestone region would release vast amounts of carbon dioxide into the atmosphere, causing greenhouse warming that might have contributed to mass extinctions.

Other Impact Craters

Preliminary evidence gathered by David Bice of Carleton College while he was a graduate student of Walter Alvarez at the University of California, Berkeley, suggests that there is yet another ancient crater to be sought. Bice found shocked quartz in rock forming the boundary between the Triassic and Jurassic periods, some 190 mya. Meanwhile, Sarah Fowell of the Lamont-Doherty Geological Observatory has found evidence for an abrupt extinction of vertebrates and plants at that same boundary. Like all boundaries between geological time zones, the Triassic-Jurassic boundary is marked by great changes in flora and fauna; previously, paleontologists believed that extinctions at that time proceeded much more slowly.

In 1991 C. Wylie Poag and coworkers at the US Geological Survey in Woods Hole MA discovered along the Virginia coast a 60-m- (200-ft-) thick layer of boulders, tektites, and shocked crystals buried under 370 m (1200 ft) of other sediment. This matches a similar find in 1983 off Atlantic City NJ. They interpret this to mean that 40 mya a bolide hit the Atlantic Ocean that caused a tsunami that washed out to sea a region the size of Connecticut.

Planetary astronomers Clark R. Chapman of the Planetary Science Institute and David Morrison of the NASA Ames Research Center have calculated your chance of being killed by one of these giant impacts. It is 1 in 6000, which is about six times as great as your chance of being killed in an airline crash. This surprising result is based on the idea that infrequent (once in every 300,000 years) impacts by large bolides kill practically everyone that is around at the time, while airline disasters in the United States kill only about 130 people each year.

(Periodical References and Additional Reading: Science News 2-3-90, p 79; Discover 4-90, p 22; Scientific American 4-90, p 46; Science News 4-28-90, p 268; Science 5-18-90, p 815; Science News 5-19-90, p 311; Discover 9-90, p 32; Science News 9-1-90, p 133; Science News 11-17-90, p 319; Science 3-29-91, pp 1543 & 1594; Science 4-12-91, p 377; Discover 5-91, p 40; Astronomy 6-91, p 24; Natural History 6-91, p 47; Science News 6-1-91, p 351; EOS 6-4-91, p 249; Science 6-14-91, pp 1467 & 1539; Science 6-21-91, p 1690; Astronomy 7-91, p 30; Science 7-12-91, p 176; Science News 7-13-91, p 20; Planetary Report 7/8-91, p 29; Scientific American 8-91, p 22; Science News 8-3-91, p 71; New York Times 8-6-91, p C2; Science News 11-2-91, p 286)

REVERSALS OF MAGNETISM

The state of Earth's magnetic field is captured by any molten rock that solidifies or by grains of magnetite in sedimentary rock. The magnetic record in rocks reveals that from time to time, the magnetic field of Earth reverses. The north and south magnetic poles change place. This has happened about 300 times in the past 170 million years. Although this process was first noticed by Bernard Brunes in 1906, it was so surprising that most geologists did not accept the phenomenon until the 1960s. Even today no one knows for sure why the reversals occur.

Some periods have almost no reversals. The Cretaceous and Permian periods are among those in which the poles stayed the same for most of the time. Scientists who favor impact theories for mass extinctions and also impacts as affecting Earth's magnetic field may be cheered by that information, since both the Cretaceous and the Permian ended with mass extinctions, followed by long periods of geomagnetic reversals.

Magnetic reversals have become useful tools for dating past events. Knowledge of the fossil magnetic orientation of a rock along with that of rocks above and below it, for which the magnetic orientation is often different, can be combined with other information to determine a date within a range that averages about half a million years, long by some standards, but short by geologic standards. Since geomagnetic reversals are worldwide events, there is no problem correlating events on one continent with those on another.

For the past hundred years or so, Earth's magnetic field has been getting weaker, but it is not clear whether this portends another reversal.

What could cause Earth's magnetic field to reverse? A possible answer is the shock of being hit by a giant bolide. The magnetic field is thought to be caused by circulation of the iron and nickel that forms Earth's outer core. A big impact would be comparable to shaking canned soup to stir up the sediments. Ever since this idea was first proposed by Billy P. Glass of the University of Delaware in 1979, it has intrigued scientists. In the Feb 1990 *Geophysical Research Letters,* however, David Schneider and Dennis Kent of the Lamont-Doherty Geological Observatory refuted the theory. They demonstrated (to Glass's satisfaction) that an analysis of four key impact-reversal situations showed that the reversal preceded the impact for three of them. Although not everyone was convinced (Richard A. Muller and Donald Morris of the Lawrence Berkeley Laboratory, for example, still think that impacts cause cold weather that causes reversals), it looks as if some other explanation for geomagnetic reversals is needed.

(Periodical References and Additional Reading: *Science* 2-23-90, p 916; *Science News* 3-10-90, p 158; *Discover* 5-90, p 28; *Science* 6- 21-91, p 1617)

WHERE DOES THE WATER COME FROM?

In 1986 physicist Louis A. Frank of the University of Iowa ventured into the field of astronomy. He proposed that satellite data indicate that about twenty small comets, dirty snowballs the size of frame houses, strike Earth every minute. Astronomers have been extremely skeptical of Frank's ideas. Even when one group of astronomers thought they saw the comets in their optical telescope, others remained doubtful.

Earth scientists are interested in this issue because Frank's theory involves their field as well. The postulated number of comets of that size would provide a steady increase in the amount of water on the planet. Instead of water being produced from volcanic eruptions, currently the most popular origin theory among earth scientists, the small-comet hypothesis says water comes from outer space. While not as skeptical of Frank's idea as the astronomers, earth scientists too have viewed them with alarm.

Just as the small-comet theory originated from a satellite study that was looking for something else, a group studying the sky with another purpose in mind found evidence, reported in 1990, supporting the small-comet theory. They say that water appears in the upper atmosphere, seemingly out of nowhere. John J. Olivero and coworkers at Pennsylvania State University have been using a sensitive microwave radiometer to study gases in the upper atmosphere. Over a period of almost 2 years, they found 111 times when the region under observation suddenly blossomed with water vapor. Every 4 days or so, about 10^{29} to 10^{34} molecules of water vapor suddenly appeared and gradually faded over a period of about 20 minutes. One possible explanation is that a watery comet vaporized as it hit Earth's atmosphere. On the surface, this would seem to be dramatic confirmation of Frank's theory. Even the frequency of the bursts and the amount of water are right.

Nonetheless, there is a problem. Comets, like other bodies in the solar system, revolve in the same direction about the Sun as Earth does. A comet striking Earth must approach it from behind, so most comet impacts should occur during the a.m., not the p.m., hours. This pattern is not evident in the observed bursts of water vapor. Perhaps some other explanation needs to be sought.

(Periodical Reference and Additional Reading: *Science News* 6-9-90, p 365)

THE DYNAMIC EARTH

GEOTHERMAL VENTS IN LAKE BAIKAL

Since the discovery of vents of very hot water (variously called hydrothermal or geothermal vents) in the ocean bottom in the 1960s and 1970s, the importance of such vents to both ocean biology and geology has been increasingly recognized. Until 1990, however, all geothermal vents were found at ocean-spreading centers—places where new ocean floor was being created. In Jun and Jul 1990, Kathleen Crane from Hunter College in New York City and a US-USSR team discovered the first known freshwater geothermal vents in the floor of Lake Baikal, just north of Mongolia in Russia. Lake Baikal is the largest single pool of fresh water on Earth and Earth's oldest lake, containing 20 percent of the world's supply of liquid fresh water (Lake Superior has a greater area, but Baikal is much deeper). Baikal was formed about 25 mya, whereas most large lakes date from the most recent retreat of glaciers in the current ice age.

The US-USSR team used underwater cameras and various sensors to locate potential vents, and then sent down submersibles, with and without people, to check further. It took 2 weeks to find the vents, which were at a depth of 400 m (1350 ft). Fluids from the vents were at least 13° C (24° F) higher in temperature than the normally cold waters found at this depth. While this temperature change is significant, it is far lower than the hundreds of degrees difference between vent waters and surrounding ocean in the previously known vents.

As with geothermal vents in the ocean, the Baikal vents are surrounded by unusual forms of life—freshwater sponges, bacterial clusters, transparent shrimp, previously unknown fish, and snail-like gastropods. Further studies are needed to tell whether the newly found organisms draw their energy from chemicals released by the vents, as those at ocean vents do, or whether they rely on more common sources of energy and are just drawn to the vents by the higher temperatures. The bacteria and shrimp, however, seem to be similar to those found at ocean vents.

Background

The first geothermal vents were found in the Red Sea in the 1960s, but it was not until 1977, when John B. Corliss and Robert D. Ballard, using the research submersible Alvin, discovered the geothermal vents at a mid-ocean rift in the Pacific near the Galapagos Islands, that scientists began to realize how odd and important the vents are. Geothermal vents are found where two tectonic plates are splitting apart (although similar vents are caused by another mechanism in the Gulf of Mexico). Such places are

(continued)

called rifts. Most rifts are found under the oceans, although there are rifts on land in Africa and Iceland. The crust is thin and cracked under the rifts, and magma is not far from the surface. For underwater rifts, this means that there are places where water fills the cracks and is heated to high temperatures by the magma. In the oceanic rifts, these temperatures may reach 340° C (650° F), but the pressure is so great that the water does not boil. The superheated water does dissolve minerals from the crust, especially sulfur compounds. Eventually, the superheated water is transported back into the ocean. When the mineral-laden water emerges into the cold ocean, it precipitates the compounds and also feeds an animal community that uses bacterial metabolism of the sulfur compounds as its energy base. Similar vents in the Gulf of California use heated hydrocarbons as an energy base.

The vents confirm the accepted belief that Lake Baikal is a spreading area, a place where new crust is forming. If the spreading continues, Asia will gradually split apart and the lake will become an ocean between the two parts. Such a process would take several hundred million years.

(Periodical References and Additional Reading: *Science News* 8-18-90, p 103; *Discover* 12-90, p 12; *EOS* 12-24-91, p 585)

ADVANCES IN PLATE TECTONICS STUDIES

An Oceanic Plate Plunges

Today the largest of Earth's crustal plates is the Pacific plate, which covers about a quarter of the surface of the planet. This was not always so. Two hundred million years ago (mya), in the late Jurassic period, the ancestral Pacific plate was only about the size of the United States; a giant ocean plate larger than the present Pacific covered almost all of the planet, with the supercontinent of Pangaea occupying the rest. Except for the ancestral Pacific plate, whatever plates made up that ancient ocean have long since gone. Early in 1990 Roger Larson of the University of Rhode Island in Narragansett announced that he and coworkers on the Ocean Drilling Program's Leg 129 had located rock from the ancestral Pacific plate. As Larson had predicted almost 20 years earlier, the oldest Pacific plate section is near the Marianas trench, working its way toward reimmersion into the mantle. The rock obtained by ocean cores, found below a thin part of the basalt that now covers the ancestral plate, is being studied for clues to the climate and biology of the late Jurassic ocean.

Background

The original theory of plate tectonics as developed by a number of earth scientists in the 1960s described Earth's crust and upper mantle as composed of a half-dozen giant plates with a few smaller subplates filling in the larger cracks. An important part of the theory is that at one time in the past, the plates had moved to positions that created one giant continent, called Pangaea, which broke into two supercontinents, Gondwanaland and Laurasia, about 200 million years ago. Later, these supercontinents separated into the present continents.

Although overwhelming circumstantial evidence for this theory piled up in the next quarter century, it was not until 1987 that the theory was definitely established by actual measurements from space that showed that the plates are slowly moving with respect to one another. For example, the Pacific plate moves to the northwest about 48 mm (⅕ in.) a year. Although convection currents in the mantle are thought to provide the energy for plate movements, the movement is influenced by different conditions at plate boundaries. Some plates are diverging as new warm material is added at their ocean-ridge edges while cooler heavy material at the opposite edges plunges into the mantle in ocean trenches. Such plunging occurs when two plates meet under the oceans or where one plate carries a continent at its leading edge but the other one does not. If both plates are carrying continents where they meet, something different and less well understood occurs.

Plates Bend As They Scrape

Where, instead of one plate plunging beneath another, two plates rub against each other, they are partly locked together, but not tightly. One plate moves somewhat independently of the other and the difference in motion is absorbed by deformation of the plates' boundaries. Eventually, strain from the deformation overcomes resistance and the plate motions are equalized.

Where the Pacific plate slides past the North American plate in northern Mexico and along the US West Coast, the San Andreas Fault, right at the boundary, was thought to be the repository of the deformation, accounting for the great earthquakes that rock the region. In Feb 1990, however, Roy K. Dokka and Christopher J. Travis of Louisiana State University in Baton Rouge reported to the American Association for the Advancement of Science's New Orleans meeting that their studies show 9 to 20 percent of the deformation takes place away from the fault. The region they analyzed was the Mojave-Death Valley area of California. Using Landsat satellite images, they were able to identify the movement of the crust in that region.

Continental Plates Collide

One of the early successes of plate tectonics was the explanation of why the Tibetan Plateau and the Himalayan Mountains form the highest region on Earth, mostly higher than 5000 m (16,400 ft) above sea level: They have been lifted to these heights by a rapid shove from the plate that contains India as it piles into the Eurasian plate with great force, wrinkling the crust. Closer examination of this explanation occasions new questions even as it answers old ones.

The complex history of crustal movement in the region has produced a major controversy between two groups of scientists studying this collision. Since 1975, Paul Tappionier of the University of Paris has tried to show that the Indian plate, or the Indo-Australian plate (see the following paragraph), is basically plunging through Eurasia like a ship plowing through the water, building up a bulge in front of it, but sending most of the Eurasian plate off to the edges. In the Feb 1 1990 *Nature*, he and coworkers argued in particular that Indochina was muscled southeast about 500 km (300 mi) by the collision. He also thinks Tibet is dashing eastward at 3 to 5 cm (1 to 2 in.) a year, an extremely fast rate geologically. Others, notably Peter Molnar of MIT and Philip England of Oxford University, think any sidewise slippage is negligible. In their view, the collision is not like a ship plowing through water, but more like a piston pushing into some soft clay, thickening the clay immediately in front of the blow, but not deforming the clay much away from the immediate vicinity of the blow.

A related concern is the large number of earthquakes that take place in the middle of the Indian Ocean. Although standard plate tectonics holds that earthquakes occur at plate boundaries, some do not, such as the occasional large earthquake in the eastern United States or the very large 1811-1812 earthquake centered at New Madrid MO. While such earthquakes are difficult to fit into the theory, they are at least rare. The Indian Ocean quakes are plentiful. One possible explanation, put forward in *Tectonics*, in 1990 by Richard Gordon and coworkers from Northwestern University in Evanston IL, is that India is not on the same plate as Australia after all. But the proposed boundary between an Indian plate and an Australian plate does not behave like the one between any other pair of plates, so this idea is controversial. Specifically, earthquake data suggest that the plates are separating at one end of the putative boundary while pushing slowly together at the other end. Many earth scientists prefer an alternative explanation of the data: The collision with Eurasia, some 2500 km (1500 mi) away, is bending an Indo-Australian plate. The exact mechanism for such an effect remains mysterious, however. Any bending lies farther away from the plate boundaries than the effect of the Pacific plate on the deserts of California, for example. Analysis of studies of the floor of the Indian Ocean will be required to sort out the truth.

Although the collision at Tibet is the largest occurring today, the Alps have risen from a similar collision as the African plate moves north into the Eurasian plate. Some other mountain ranges, notably the Appalachians, were formed by this mechanism when there were different boundaries between the plates in the past. Even earlier ranges caused by plate impacts have been eroded away or wiped out by other geological processes. An unlikely place for the fossil of such a mountain range is the Bay of Bothnia, between Finland and Sweden, but an international group of earth scientists reported in the Nov 1 1990 *Nature* that it was the scene of such a mountain-building collision some 1.8 to 1.9 billion years ago. As in the discovery of the base of the Eurasian plate (see below), the scientists found the signs of the collision using seismic-reflection profiling.

Other Places Plates Diverge

When new crust is formed, plates on either side diverge; but direct evidence has been scarce. All the "new" crust found along the mid-ocean ridges, where most such divergence takes place, seems to be thousands of years old. Oceanographers finally caught the crust in the actual act of forming in Apr 1991.

First a chance discovery by Robert Embley, William Chadwick, and Chistopher Fox of Oregon State University's Hatfield Marine Science Center in Newport OR brought the achievement of this goal somewhat closer. The oceanographers noticed that the appearance of the part of the mid-ocean ridge system called the Juan de Fuca Ridge had changed between 1981 and 1989. Further comparison with another survey showed that the change occurred between 1981 and 1987. Sometime during that period a string of mounds 17 km (10 mi) long, with a total volume of 50 million m^3 (1765 million cu ft), had formed. There is good reason to think that the mounds formed in 1986 and 1987, when giant plumes of warm, mineral-laden water were observed in the same area. Edward Baker of the National Oceanic and Atmospheric Administration reported in Dec 1990 that his calculations show that the giant plumes formed when the sea floor split open for a few days, releasing 100 million m^3 (3531 million cu ft) of hot water along with the lava that formed the mounds.

In Mar and Apr 1991 the submersible *Alvin* engaged in an operation on the East Pacific Rise in the same region that had been imaged by the *ARGO* system in Dec 1989. The crew of *Alvin* observed many signs of recent volcanic activity that had not been present on the *ARGO* images. One site, called either the "tubeworm barbecue" or the BBQ site, was covered with what appeared to be recently killed tubeworms and mussels amid a glassy new flow of lava. Radioactive dating of rock samples from the BBQ site suggested that the eruption had taken place during the series of dives during the last week of Mar or the first of Apr. The *Alvin* dives took place in the first two weeks of Apr. A visit to the BBQ site on May 23 by Daniel J. Fornari of Lamont-Doherty Geological Observatory and Michael R. Perfit of the University of Florida showed that crabs had come to feast on the dead worms by that time, although none had been present in Apr. Thus, all indications are of an active eruption early in Apr 1991.

As land dwellers, we tend to focus more on continents than on plates, which always are at least partly beneath the oceans. Many people even refer to plate tectonics as the theory of "drifting continents." From the continental point of view, the plate movements change the map of the world by fracturing old continents or by shoving continents together to form new ones. These changes have had a profound effect on the evolution of life and on Earth's climate. But they are poorly understood and various new theories frequently are put forth to account for the evidence.

New light was shed on the breakup of the former supercontinent of Pangaea by a study conducted by Hans-Dieter Sues of the US National Museum of History and Paul E. Olsen of Lamont-Doherty Geological Observatory. Previously, it was believed that the animals that inhabited what was to become the northern hemisphere continent of Laurasia had become different from those in the southern hemisphere, which was to become Gondwanaland, well before the breakup of the two supercontinents about 200 million years ago. Studies of rock formations at the Tomahawk locality of Chesterfield County VA reveal that what was thought to be differences in the animals was caused by an artifact of incorrect dating and the incompleteness of the fossil record. New fossils found by Sues and Olsen help correct this situation and show that the types of animals were not different at that time, suggesting that the fauna of Pangaea was essentially continent-wide until the divergence.

Finding the Base of a Plate

J.E. Lie and coworkers on the research vessel *Mobil Search* reported in the Jul 12 1990 *Nature* that they had used seismic-reflection profiling to observe the base of the Eurasian plate in the North Atlantic Ocean. In seismic-reflection profiling, air guns are used to create pressure waves that travel through the crust. As in radar, the reflections of these waves can then be studied to reveal characteristics of unseen layers. In this case, the scientists were able to find interfaces deep in the mantle section of the plate.

(Periodical References and Additional Reading: *Discover* 2-90, p 26; *Science News* 2-3-90, p 69; *Science* 2-16-90, p 808; *Science News* 3-17-90, p 175; *Scientific American* 6-90, p 72; *Nature* 7-12-90; *Science News* 7-14-90, p 25; *Science News* 7-28-90, p 62; *Science* 8-31-90, pp 967 & 1020; *Science News* 12-1-90, p 347; *Science* 12-21-90, p 1661; *New York Times* 3-28-91; *Science News* 4-27-91, p 267; *Science* 6-7-91, p 1409; *EOS* 6-18-91, p 268; *EOS* 11-12-91, p 505)

PLUMES, HOT SPOTS, WET SPOTS, AND SUPERPLUMES

Plumes

The main question currently being debated about plumes is "Where do they start?" If the mantle is relatively homogeneous, plumes may start as low as the core and traverse the entire mantle. If the mantle is layered, then it is most likely that plumes originate in the top layer, thought by most believers in mantle layers to start around 660 or 670 km (410 or 415 mi) below Earth's surface.

A sizable majority of today's earth scientists now appear to believe plumes start at the core-mantle boundary, as revealed in the reports and reactions at a 1990 Plume Symposium run by Caltech that attracted 125 earth scientists. Key points in evidence were the amount of heat brought to the surface by plumes and the size of the largest plumes, the super-plumes (see below).

Geoffrey Davies of the Australian National University in Canberra theorizes that plumes confined to an upper layer would have to carry 70 percent of the heat from Earth's interior to the surface. But if that were happening, he claims, the bulges caused where plumes hit the surface would be at least two or three times as large as those observed. On the

---

Background

In 1963 J. T. Wilson proposed that chains of oceanic islands, such as the Hawaiian chain, were caused by stationary long-lasting features below the moving tectonic plates. In 1971 W. Jason Morgan of Princeton University extended this idea by suggesting that hot material rising from the core through the mantle at about twenty different locations accounts for several noticeable features on Earth's crust, including the Hawaiian Islands, Iceland, and the Yellowstone Park region. This concept has come to be accepted by most earth scientists. The columns of rising heat and rock are called plumes and the places where they appear on the surface are known as hot spots. Both 1990 and 1991 produced important refinements and extensions of Wilson's and Morgan's original concepts.

A hot spot is caused by a plume that is relatively stationary with respect to the moving crust above it. When magma from the hot spot breaks through the crust, it forms a volcano that grows into a seamount or island. As the plate moves away from the hot spot, the volcano becomes extinct and the island or seamount begins to erode and subside. The "big island" of Hawaii is currently over the largest hot spot known. This hot spot is now forming what will become a new island in the Hawaiian chain (already named Loihi), while still causing eruptions on the "big island." It has left behind a chain of extinct volcanoes that stretches from Hawaii to the northwest, reaching almost to the Kamchatka Peninsula, including the Emperor seamounts as well as the Hawaiian Islands themselves.

other hand, he thinks plumes that originate at the core would not need to be responsible for such a large percentage of the heat reaching the surface, so the bulges caused by such plumes would match the sizes observed. The second line of argument concerns the superplumes. These are simply too big for anyone to imagine appearing within a relatively shallow layer at the top of the mantle.

Not everyone is convinced by these arguments, however. Superplumes may originate at the core, for example, while more ordinary plumes, like the one under Hawaii, could be confined to a top layer. Donald Turcotte of Cornell University thinks that Davies's heat calculations are "nonsense."

Hot Spots

In 1991 a team consisting of M. Liu of Pennsylvania State University, D.A. Yuen and W. Zhao of the University of Minnesota, and S. Honda of Hiroshima University in Japan concerned itself with one of the puzzles of the hot spot theory. A plume is expected to be a long-lasting feature, while a tectonic plate moves in small jerks that are virtually continuous when viewed at scales comparable to the size of a volcano. Given those conditions, one would expect a single, long, continuous volcano to be formed over a hot spot instead of the chain of discontinuous volcanoes that actually occurs. Some mechanism breaks the plume up into individual pulses. But what mechanism?

Liu and coworkers constructed a computer model of the plumes. They assumed that phase transitions occur in the mantle depths of 670 km (415 mi) and 400 km (250 mi) (see "Understanding Earth's Mantle and Core," p 256). Phase transitions in rock include such familiar phenomena as melting as well as less familiar changes from one crystalline form to another. It is the latter type that was used in the computer model.

According to their model, if a plume is assumed to be sufficiently hot and a reasonable size, it behaves like the well-known mushroom cloud rising from a nuclear explosion. At a given height—in the plume's case, height above Earth's core—the column will form a mushroom shape. The top part corresponds to the cap of a common field mushroom. Geologists call a formation of fluid rock surrounded by less fluid rock and shaped like this a *diapir*. In the computer models, the diapir begins to form around 670 km (415 mi) below the surface, where major phase changes are assumed to take place. It forms more fully at about 400 km (250 mi) below the surface, at yet another phase change.

Here the analogy with the nuclear mushroom cloud falls apart. In the computer model, after a time (representing a few hundred million years), the cap of the mushroom breaks off. Under certain circumstances, it trails a very thin segment of plume behind it, making it look more like a parasol than a mushroom cap. Since the diapir remains hotter than surrounding material, it continues to rise, finally reaching the surface and producing a volcano. The volcano is active for a few hundreds of thousands of years, then becomes extinct. The tectonic plate moves along to a spot some distance from the extinct volcano. In the meantime, another cap is forming down around 670 to 400 km (415 to 250 mi) below the surface. Eventually, the new cap breaks free and forms another volcano. In between the two volcanoes is undisturbed crust.

Wet Spots

Earth is not a perfectly smooth sphere. In addition to the obvious mountains and river valleys, there are large regions that are farther from the center of Earth than average, even when the flattening at the poles and bulging at the equator predicted by gravitational theory is taken into account. Some regions, such as continents, are higher than others because they are made of lighter rock that floats in a plastic or partly fluid region called the asthenosphere. The asthenosphere includes the bottom of the crust and the top of the mantle. For raised regions smaller than continents, other explanations must be found. For example, the Tibetan Plateau and the Himalaya Mountains are both caused by one tectonic plate pushing under another. Other high regions are harder to explain. Since the 1960s many of these have been thought to be caused by plumes of material from the mantle pushing the crust up, sometimes erupting through it to form new layers of crust above the old.

There are two possible ways to produce a plume of material rising through the mantle to reach the surface, both requiring that some parts of the mantle be less dense than other parts. One way is for the less dense material to be hotter than the material around it. The other is for it to have a different composition, one with more light elements than are found in the surrounding material. Both ways are familiar from currents that rise and fall in the atmosphere and oceans. In all these cases, the heavier material actually pushes the lighter material up; gravity is the force supplying the energy.

In the atmosphere or in the ocean it is fairly easy to tell which of the two mechanisms—temperature or composition—is the cause of a rising current. You can measure the temperature of the air or the salinity of the water. The mantle is inaccessible directly except for a few rock samples believed to have been brought up from the mantle by deep-rooted volcanoes. Most of what we know of the mantle comes from studies of seismic waves. Unfortunately, seismic waves travel faster in both hotter and lighter rock. Thus, the presence of faster waves in a portion of the mantle tells that that portion is hotter or less dense, but not which of the two.

Seismic waves are observed in columns below regions that are higher than the average level of Earth's surface. For the past 20 years it has been generally assumed that such fast waves are caused by hot plumes of mantle material. As early as 1980, however, J.G. Schilling, M. B. Bergeron, and R. Evans suggested that the data might have another interpretation: lighter material, not hotter, might be rising to produce the higher elevations and the volcanoes that often accompany them.

In 1990 Enrico Bonatti of the Lamont-Doherty Geological Observatory studied this issue at the hot spot that is currently under the Azores. Like other hot spots, this one has left its track as the tectonic plates moved over it. At one time it was below Newfoundland. Today it is on the mid-Atlantic ridge. Bonatti's approach was to make a careful examination of the rock from the floor of the ocean near the Azores and from ocean cores to determine the rock's composition. His idea was that the composition could be used to determine the rate at which the rocks cooled. Really hot material would rapidly rise up through the mantle and cool quickly. Less dense material that was about the same temperature as the mantle around it would rise and cool more slowly. Comparison of peridotite rock that is presumed to be closely related to mantle material with peridotites from places away from any suspected hot spot region suggested to Bonatti that both had cooled at about the same rate, implying that the plume that caused the Azores is not especially hot. Similar comparisons of basalt from the same sources reinforce this idea.

Instead of being hot, Bonatti proposes, the plume is wet. The plume is lighter than the substance around it because the rock that formed the plume contains more water than surrounding material, which makes it lighter. Bonatti also thinks that the plume contains light compounds other than water, such as carbon dioxide, but his shorthand name for the phenomenon, a *wet spot,* caught on and helped give the theory greater currency. Bonatti thinks that further research will find more wet spots in the mid-Atlantic, but he does not believe that all plumes are wet. Many, such as the one under Hawaii, probably are hot.

Although his data for the Azores are not in dispute, his interpretation is. Henry J.B. Dick of Woods Hole Oceanographic Institution has made similar measurements of peridotite, but he thinks that a slower plume—the primary fact to emerge from evidence of slower cooling—does not necessarily mean a wetter one. It might just be slower because the crust is thicker above the plume, slowing it down.

Superplumes

A "permanent" hot spot, or even a wet spot, is caused by a plume that stays in one place in the mantle for a great length of time. A *superplume* is a giant blob of hot material that reaches the crust in a burst that is relatively short in terms of geologic time. The most recent such superplume started about 125 mya, reached its peak about 120 mya, was still active 115 mya, and was gone entirely by about 60 mya. According to Roger Larson of the University of Rhode Island, this superplume may have changed not only the shape of the crust of Earth, but also the climate, the map of Earth, the movement of tectonic plates, the present-day supply of energy, the processes of geomagnetism, and the course of evolution. A different superplume of about 250 mya probably had similar effects. Using the geologic time periods as labels, the more recent event can be called the Cretaceous superplume, while one of about 250 mya can be termed the Carboniferous superplume.

Evidence for the Cretaceous superplume comes from undersea cores, drilled in 1990 into a region more than twice Alaska's size, called the Ontong-Java Plateau. This plateau lies across the equator east of New Guinea and north of the Solomon Islands. Five cores extracted by the Ocean Drilling Program's *Joides Resolution* were supplemented by three cores drilled earlier by the *Glomar Challenger*. The Ontong-Java Plateau was formed by giant lava outpourings with a volume thought to be between 24 and 91 million km^3 (between 6 and 22 million cu mi), fifty times greater than that of the more famous Deccan Traps in India. Its production may have been the largest volcanic event of the past 200 million years. But the Ontong-Java Plateau is only one of four giant lava outpourings caused by the Cretaceous superplume. As tectonic plates moved above the superplume, it first formed the Parana-Etendeka Platform in the South Pacific and eventually formed the Kerguelen Plateau.

Enough lava was deposited on the floor of the ocean to raise sea levels significantly all over Earth. One result of this was the formation of shallow continental seas, an excellent place for life to prosper. Such a proliferation of organisms occurred under conditions where the remains could form great fossil beds. Calcium carbonate, the main constituent of many protist shells as well as mollusk shells, does not form great layers in the deep ocean because it dissolves in high-pressure water. Shallow seas produced the chalk beds of England, formed during this period, and the layer of chalk and ooze 900 m (3000 ft) thick that overlies the Ontong-Java Plateau.

The huge number of living creatures involved in these formations may also have been affected by a giant outgassing of carbon dioxide that accompanied the lava. Michael Arthur of Pennsylvania State University estimates that atmospheric carbon dioxide was as much as twelve times as high in the mid-Cretaceous as it is today. Ken Caldeira and Michael Rampino of New York University calculated in 1991 that the greenhouse effect produced by the carbon dioxide raised global temperatures between 2.8°and 7.7° C (5° and 15° F) at that time.

Not only does carbon dioxide cause global warming, but it also increases plant growth by speeding up photosynthesis. Among other side effects was formation of deposits of organic material that would later be turned to petroleum and natural gas through the action of heat and pressure (but see "Searching for Gas and Oil," p 602). A similar pattern during the Carboniferous superplume could have caused Earth's major deposits of coal.

During the Cretaceous superplume period, the pattern of Earth's geomagnetic reversals halted for about 41 million years. A superplume is thought to take a lot of heat away from Earth's core. This may have disrupted whatever cycle causes geomagnetic reversals. A similar halt occurred during the Carboniferous superplume. The time during magnetic reversals may possibly be a time of increased evolution, with mutations caused by radiation from space, radiation that is usually turned away from Earth by the magnetic field. Thus, if the field did not reverse for a long time, evolutionary change might be slowed.

Fossilization of immense layers of shells and other organic material removes much of the carbon dioxide from the air, building it into calcium carbonate or hydrocarbons. The climate becomes cooler as a result of lessening of the greenhouse effect. Glaciers form and remove some of the water from the sea, reducing sea levels. Tectonic movements, speeding at three times their usual rate as a result of the superplume, pull some of the ocean plates down into the mantle, lowering the levels of the ocean further. Eventually an ice age begins. The upheaval of another superplume might be needed to start the cycle anew.

Not everyone agrees with this scenario. John Mahoney of the University of Hawaii has pointed out that dating of the plateaus is not definite, and other geophysicists think Larson may be exaggerating the effects.

(Periodical References and Additional Reading: *Science* 10-5-90, pp 7 & 107; *Science News* 10-6-90, p 214; *Science* 2-15-91, p 746; *EOS* 5-21-91, p 236; *Science* 5-24-91, p 1068; *New York Times* 6-11-91, p C1; *Science* 6-28-91, p 1836; *Science* 7-12-91, p 176; *EOS* 7-16-91, p 307; *Science* 10-11-91, p 263; *Science* 10-18-91, p 399)

SUPERDEEP HOLES

A deep Siljan hole in Sweden reported on in the Technology Section (see "Searching for Gas and Oil," p 602) is far from the only deep well being drilled into continental crust or planned by geologists, although the others are drilled in search of knowledge, not energy.

Many oil wells are only 1000 m (3250 ft) deep or even shallower. The deepest water well is 2231 m (7320 ft) and oil wells of more than 3000 m (10,000 ft) are rare and considered deep. The superdeep German borehole project started drilling with a pilot hole of somewhat more than 4000 m (12,800 ft) that was completed in Apr 1989. On Sep 8 1990, the project began its main hole about 200 m (650 ft) east of the pilot, near the Bavarian village of Windischeschenbach. This borehole is expected to reach 10,000 m (33,000 ft) by 1994. Geologists call the drill holes the KTB holes after the jawbreaker name of the project—*Konitinentales Tiefbohrprogramm der Bundesrepublik Deutschland,* roughly "German Continental Drilling Project." US scientists are expected to participate in the project as well.

Geologists have developed a concept of what rock layers form the center of Europe from various indirect means of measurement. One of the goals of the KTB holes is to test whether this concept is accurate. Some of the results from the pilot borehole have been surprising already, notably the high temperatures encountered and the presence of cracks filled with very salty and gassy water, even near the greatest depth reached. There is also some indication that the gases in the water were caused by living organisms. The superdeep borehole is planned to provide the greatest percentage of its cores from below 9000 m (30,000 ft), a region that is still largely unexplored.

The record holder for superdeep holes remains the one drilled by the USSR in the Kola Peninsula, which is nearly 20,000 m (50,000 ft) deep.

Holes in Volcanoes

Other deep boreholes are planned with specific purposes in mind. One that has been pushed since 1986 by scientists concerned with the nature of hot spots (see "Plumes, Hot Spots, Wet Spots, and Superplumes," p 249) is a hole deep into the island of Hawaii to determine what is really going on in that amazing volcano. Evidence from nearby Oahu's extinct Koolau volcano, reported in 1990 by Gautam Sen of Florida International University and R.E. Jones of the University of California, Los Angeles, suggests that liquid magma collects about 90 km (56 mi) below the surface of Hawaiian volcanoes in ponds that are surrounded by still-ductile mantle. As the liquid magma is forced upward, it reaches less ductile regions around 60 km (37 mi) below the surface, producing earthquakes. Earthquakes from that depth are reported from Hawaii's Kilauea volcano. The suggestion of magma pools at these depths is surprising, since such pools were previously thought to form no lower than the base of the crust, roughly 10 km (6 mi) below oceans and 35 km (22 mi) below continents. The proposed Hawaii borehole will not reach any of these depths, but it might further clarify the processes that are occurring.

Scientists also want to investigate another volcano by drilling a pair of not very deep 1.2 km (0.75 mi) boreholes. The largest single volcanic explosion of the twentieth century removed most of Mount Katmai in

Alaska and left behind the Valley of Ten Thousand Smokes and a volcano cone named Novarupta. John Eichelberger of Sandia National Laboratories and coworkers conducted two preliminary expeditions in 1989 and 1990 to investigate the possibility of drilling at the site. Despite August temperatures in the single digits Celsius (forties, Fahrenheit), and winds of 160 km (100 mi) an hour, the US National Academy of Science has certified that the region, which is part of Katmai National Park's wilderness area, is uniquely suited to scientific drilling.

(Periodical References and Additional Reading: *Science* 8-3-90, p 475; *Science* 9-7-90, pp 1083 & 1152; *EOS*, 4-23-91; *EOS* 5-21-91, p 236)

UNDERSTANDING EARTH'S MANTLE AND CORE

The year 1991 saw great progress in understanding the interior mechanisms of Earth, along with the inevitable controversies that new ideas bring. The key issues revolve around how the mantle is structured, what dynamic events take place within it, how the mantle reacts with the core, and what formed the core.

The prevailing theory is that convection currents in the mantle move the plates that form Earth's crust (see "The State of Earth Science," p 191). But there are considerable differences about the nature of the currents. Some theories of mantle structure restrict convection currents to part of the mantle; others assume that some, but not all, convection currents traverse the whole mantle; yet others allow most currents to involve the whole mantle. Determining which view is correct is essential for progress in understanding Earth's dynamics.

Another unresolved issue in crust-mantle interactions is related to how the leading edge of a plate is destroyed as it plunges into the mantle. On the one hand, some earthquakes, called deep-focus earthquakes, occur farther into the mantle than expected. But evidence from seismic waves suggests that tectonic plates run into some obstacle about 660 or 670 km (410 or 415 mi) down. What is it about the mantle at that depth that causes this effect? Direct examination is impossible because the deepest borehole so far is only about 20 km (12 mi) deep. A correct understanding is related to the problem of the convection currents as well. This section will cover the problem and its implications from the outside in, as a giant borehole to Earth's center would do.

Mantle and Crust

The deep-focus earthquake evidence poses a problem for several reasons. Most rock becomes soft under great heat and pressure. For the rocks common in the crust, the temperatures and pressures thought to be found below 70 km (43 mi) should be sufficient to produce a ductile solid. Indeed, one of the assumptions of plate tectonics is that such a layer, called the asthenosphere, allows plates to move. The paradox is that many earthquakes originate at depths below 70 km, even though it is difficult to imagine how earthquakes are formed in a ductile or plastic layer of rock. Looking at "Earthquakes of the Year," pp 275 and 281, however, suggests a partial answer to the mystery. Most deep-focus earthquakes occur where one plate is plunging down into the mantle. If the plate, or "slab," as it is usually called in this context, takes a while to become ductile, it could maintain its brittleness for a time—and could break, causing earthquakes, so long as the plate remained rigid. In that simple view, the cessation of earthquakes below 670 km (415 mi) might be caused by the more complete melting of the slab at that depth. Furthermore, this would seem to explain the locations of earthquakes along the descending inclined slab.

Experimental work with rock samples as well as studies of rock extruded by volcanoes near plate boundaries suggests that this simple view does not describe what actually happens. The pressure transmitted by the overlying material is just too great. Even if the slab maintains considerable solidity, it becomes far too ductile to crack the way a brittle rock does. At best, it can bend, which does not cause earthquakes.

One theory suggests that, as with so many events on Earth, water is involved. Descending oceanic crust is heavily hydrated—that is, a lot of water has chemically combined with the minerals in the rock. As the slab descends into the mantle, high temperatures and pressure force the water out, or dehydrate it. This dehydration may be a more important process than partial melting.

A 1991 report by Charles Meade and Raymond Jeanloz of the University of California, Berkeley, attributes all the enigmatic deep-focus earthquakes to dehydration. Their diamond-anvil studies (see "The State of Earth Science," p 191) have been concerned especially with the phase transition that occurs with the dehydration of the hydrated mineral serpentine. Since 1989, Meade and Jeanloz have been listening to sounds made by small samples of minerals as the minerals are subjected to diamond anvil pressure and laser heating. The idea is that rapid fracturing of the sample causes the sounds and that such fracturing corresponds to the large-scale fractures that form earthquakes. Meade and Jeanloz heard sounds from serpentine samples that make them believe that dehydration, which can occur at various depths depending on circumstances, causes a phase change that produces rapid fracturing and accounts for deep-focus earthquakes. The disappearance of such earthquakes below 670 km (415 mi) is not caused by melting of the slab, but by the completion of dehydration at that depth. In their view, the slab maintains its integrity well below that depth.

Meade and Jeanloz also note that the water removed by dehydration must go somewhere. Initially, it would go into the upper part of the mantle, so it could be a factor in the question of how the mantle composition varies with depth.

A totally different point of view on the deep earthquake mystery is espoused by Stephen H. Kirby and Laura A. Stern of the US Geological Survey, along with William B. Durham of Lawrence Livermore National Laboratory. They do not believe that the sounds heard by Meade and Jeanloz necessarily indicate faulting. Kirby and colleagues have worked with ice to determine what they think really happens. As any reader of Kurt Vonnegut's *Cat's Cradle* knows, ice comes in more than one form. Kirby and colleagues examined the transition from ice I to ice II under a combination of rising pressure and rising temperature. Such a transition takes place at pressures two orders of magnitude below those used by Meade and Jeanloz (that is, roughly a hundredth of the pressure) and at temperatures only slightly colder than one might experience on Earth's surface. When the phase change takes place, it begins in small regions called anticracks. As long ago as 1945, Percy Bridgman [US: 1882-1961], a pioneer in high-pressure experiments, had suggested that anticracks from phase transitions similar to those observed in ice might account for deep-focus earthquakes. The experiments by Kirby and coworkers tend to confirm Bridgman's hypothesis, showing both anticracks and transform faults (perpendicular to the boundary between the plates) caused by them.

These two investigations were hampered by the tiny samples used by Meade and Jeanloz and materials not representative of those thought to constitute the bulk of the materials involved (used by Kirby and colleagues). Harry Green and Thomas Young of the University of California, Davis, working with David Walker and Christopher Scholz of the Lamont-Doherty Geological Observatory, developed a new device that got around some of these limitations. With their new press they were able to subject a fairly large sample of olivine, the main mineral of the upper mantle, to pressures corresponding to a depth of 400 km (248 mi) and a temperature of 1900° K (1600° C or 3400° F). When they opened the press, the olivine showed anticracks and a transformational fault. The time scale involved, however, is not clear. The anticracks and fault need to have happened quickly to produce the equivalent of an earthquake, and there is no evidence so far that this occurred. More experiments are being devised to settle this issue.

At this point, the cause of deep-focus earthquakes is still uncertain. The easy explanation, that such earthquakes are caused by ordinary strain as the slab descends into the mantle, is almost certainly not true; the phase-change alternative of Bridgman, Kirby, Green, and others is established mainly by analogy and may take place too slowly in any case; and the dehydration explanation of Meade and Jeanloz extrapolates from tiny sounds in tiny samples to giant earthquakes.

To complicate matters even further, two groups of seismic tomographers have found evidence that some slabs, but not all, can penetrate the 670-km (415-mi) "barrier" and enter the lower mantle. Yoshio Fukao and coworkers at Nagoya University found that the slabs they have studied beneath Japan and the regions immediately adjacent to Japan indeed do something at the 670-km level: They flatten out and parallel the 670-km depth for as much as a thousand kilometers (600 mi). But other slabs, those beneath Java, the Northern Kuriles, and the Marianas, ignore the 670-km threshold. They appear to keep on getting deeper. The Java slab shows an especially distinct picture of something going down as far as 1200 km (745 mi). Rob van der Hilst of Leeds University, Guust Nolet and Wim Spakman of the University of Utrecht, and Robert Engdahl of the US Geological Survey, agree with the findings, seeing an especially clear image of deep penetration by the slab beneath the Northern Kuriles.

Mantle Alone

Is the mantle homogeneous or formed in two or more layers? The implications of this question may not be immediately obvious, although it is clear that a two-layer mantle is one way to resolve part of the deep-focus earthquake problem.

If the mantle has layers, then convection currents are necessarily restricted within a layer. Otherwise, the currents would mix the mantle and destroy the layers. But convection is one of three ways to move heat (along with radiation and conduction); and it is by far the most efficient of the lot. Thus, convection is the main way that heat from Earth's interior reaches the surface, where we can measure it. A layered mantle that limits convection suggests that there is far more heat trapped deep in Earth than we know about. This, in turn, has various implications, most of them still to be worked out by geophysicists.

A layered mantle also suggests that the bottom part is denser than upper regions, not just because of compression from gravity, but also because the lower part is chemically denser—meaning that there are fewer light elements and more heavy ones. At surface conditions, many of these light elements are gases, so a lower mantle layer must be degassed. That would mean that the gas was eliminated when Earth formed. Thus, there are implications for the formation of Earth and the formation of the solar system.

Yet another subject that can be viewed differently depending on whether or not the mantle is layered is that of hot spots, plumes, and superplumes, discussed on page 249. Plumes are essentially rising parts of convection currents that are especially strong. Superplumes are, of course, even stronger.

In the late 1980s the weight of evidence on mantle layering seemed on the side of at least a two-layer structure. The principal indications came from diamond anvil studies by Raymond Jeanloz and his group at the University of California, Berkeley. They used the diamond anvil and high heat to produce a combination of pressure and temperature typical of that thought to exist about 800 km (500 mi) below the surface, well into the mantle and below the mysterious 670 km (415 mi) transition zone. Into this unpleasant environment of about 2000 K (1725° C, or 3140° F) and 3 million gigabars, the scientists placed a sample of ordinary silicate rock, which became transformed into something with a structure similar to that of the mineral perovskite, thought to have the structure of crystals in the lower mantle. By comparing the density of the artificial perovskite with the known density of the lower mantle, the group concluded that more iron was needed to make the two match. Their interpretation was that the upper mantle, which has the same chemical composition as the original silicate sample, and the lower mantle, which they thought would have more iron in it, were demonstrably different; hence, two layers of mantle.

In 1991 Yanbin Wang and coworkers at the State University of New York at Stony Brook repeated the experiment with a slight but very important difference. The original Jeanloz experiment had only been able to measure the density of the artificial perovskite at 840 K (570° C or 1050° F). The researchers had to take the perovskite out of the diamond anvil to measure the density; at ordinary pressure, it decomposes above 840 K. Thus, they did not have the true density at 2000 K, but extrapolated from the lower temperature. In Wang's 1991 version of the experiment, the scientists found a way to measure the density at 7.3 gigapascals, which held the sample together up to a temperature of 1250 K (980° C or 1790° F). They were also able to reduce the measurement error in the experiment.

As a result, Wang's group discovered a phase transition, a change from one crystalline form to another, that takes place at 600 K (339° C or 620° F). This resulted in a lower rate of expansion for the mineral, suggesting that it would be denser under conditions corresponding to the proposed lower mantle layer. If denser, then no iron is needed to increase its weight. If no extra iron is needed, then the top layer and the bottom layer could have the same chemical composition.

The Wang group has reported on a single experiment only. Repetition of their work and studies by other laboratories will be needed.

The mantle problem has been attacked in various other ways as well. One study used how the crust has rebounded after the last major glacial period paused some 12,000 years ago to infer the viscosity of the mantle. This investigation showed a regular increase in viscosity, agreeing with the Wang determination that there is no abrupt change from the top of the mantle to the bottom. Other studies, conducted principally by Bradford Hager of MIT and various collaborators, come to the opposite conclusion.

Based on measurements of heat flow, tectonic plate movement, and seismic evidence, they conclude that the mantle is so stiff in its lower levels that it must be fundamentally different from the top level.

Although there seems to be little prospect of drilling deep enough to settle this issue completely, we can still obtain samples from deeper than one might expect. In 1986 minerals in diamonds from South African mines were shown to have been forced up from a depth of about 180 km (110 mi) in the mantle by ancient volcanoes. This spurred a search for minerals from even deeper layers. In 1990 Stephen Haggerty of the University of Massachusetts and Violaine Sautter of the University of Paris-South in Orsay identified a rock known as eclogite that had been carried up from about 300 to 400 km (190 to 250 mi). Then, in 1991, Sautter, Haggerty, and Stephen Field of Stockton State College in Pomona NJ also found the rock lherzolite in their diamond samples from the 400 km region.

Now the problem is this: Eclogite above 300 to 400 km is uncommon, suggesting that a transition from common lherzolite to eclogite is occurring at around 400 km, which is about where the first signs of the major transition at 670 km begin to be seen. So, in 1990, Haggerty and Sautter endorsed the two-layer mantle. But their 1991 analysis caused them to reverse their verdict, for if there is a lot of lherzolite, a kind of background mantle rock, then the mantle is probably composed of the same stuff throughout. In any case, it is clear that more samples are needed.

Ken Collerson from the University of California at Santa Cruz may not have located the deepest rocks from the mantle, but he thinks he has found the oldest. He obtained samples of a rock called komatite from an outcrop in northern Labrador. Radioactive decay shows the komatite to be about 4 billion years old. He dated the outcrop from the ratio of a radioactive isotope of samarium to its decay product, a stable light isotope of neodymium.

Collerson has also compared the ratio of this light neodymium isotope to a heavier isotope that should have been present when the rock formed. There is less of the heavier isotope than expected, which suggests that it may have migrated to a lower mantle layer at a time when the mantle was molten. This is thought to be about 4.3 billion years ago, only 300 million years before the komatite was formed.

Another nearby outcrop is only about 100,000 years younger. Its presence suggests that yet other outcrops of mantle in the region might exist with some possibly even older than the komatite. This could help fix the date of mantle separation—if it actually occurred—more precisely.

Mantle and Core

Lower down in the mantle there are unresolved problems as well. Just as the transition between the possible layers of the mantle is of interest, so is the transition between the mantle and the core. In March 1991 Elise Knittle of the University of California at Santa Cruz and Raymond Jeanloz reported in *Science* that their research with the diamond anvil showed that conditions at the interface between Earth's core and mantle are extremely sensitive to variations in heat and pressure. They used a laser to heat pressurized samples to 3500° C (6300° F), the temperature thought to exist at a depth of 2900 km (1800 mi), where the boundary of the mantle and outer core is located by seismography. Knittle and Jeanloz think this boundary is very irregular and speculate that the reason for the irregularity is a chemical variation that occurs as a result of slight variations of heat and pressure. They also found that silicon and iron, which do not normally mix, alloy under high pressure and heat. They think this accounts for an observed density of the outer core that is about 10 percent lighter than it would be if the outer core were pure iron.

Studies based on seismic tomography (see "The State of Earth Science," p 191) reveal to some—but not to others—layers in the outer core and a 200- to 300-km- (120- to 190-mi-) thick layer in the mantle that surrounds the core. Jeanloz has speculated that the mixing he has observed not only accounts for the structure of the core, but also explains the wrap around the core in the mantle. He proposes that iron from the core is drawn up as much as a hundred meters into the mantle by capillary action, and then moved even farther by convection currents to form the mixed zone. Not everyone who detects such a region around the core thinks that this idea is proven or even likely. For one thing, the samples that are used in the experiment are tiny, so the observed mixed layer is barely discernible.

An irregular mantle-core boundary may account for an amazing phenomenon first detected in 1991. A geomagnetic reversal takes from 4000 to 8000 years to complete. During that transition, the north geomagnetic pole wanders south while, on the opposite side of the planet, the south geomagnetic pole moves north. Since there are hundreds of thousands of years between reversals, one would expect the paths taken from north to south and south to north would vary randomly, depending on whatever disturbance caused the poles to switch. As earth scientists accumulated records of such pole wanderings, however, they found that the paths were more like the regular migration routes of birds than like random moves between north and south. Considering the mantle as stationary, the most common migration route is where the Americas are currently located. Less traveled is a route through Asia and Australia. These are not the only routes for geomagnetic reversals, since a third of the reversals take some apparently random route, but they are far too frequent to be accounted for by chance. One possibility—especially if Knittle and Jeanloz are correct—is that the shape of the boundary layer influences the flow of liquid metal, considered the most likely source of the magnetic field. In that case, the effect would be similar to the way that climate is affected by high mountains.

Note that something in the mantle must produce these pathways. The core is too liquid for features to persist over millions of years. The more rigid mantle, however, is known to contain regions that stay the same for at least that long (see "Plumes, Hot Spots, Wet Spots, and Superplumes," p 249). The mantle does not, however, have to be the shape of the mantle-core boundary that sets the paths. Other possibilities include the more direct one of variations in the electromagnetic properties of the mantle.

Carlo Laj of the Center for Studies of Weak Radioactivity and his colleagues think that the geomagnetic routes are set by temperature variations in the mantle. If you assume homogeneity of the mantle, seismic waves can detect temperature. On that basis, there are two large regions of cooler mantle, each along one of the paths that the poles take most often. These lobes could affect the climate of the core the way large, cool lakes and oceans affect climate in Earth's atmosphere.

If Laj's idea is correct, there is an intriguing connection between the core and the crust. It is thought that the cooler regions of the mantle are the remains of tectonic plates that have maintained some integrity 2300 km (1400 mi) below Earth's surface. If that is the case, then the crust has some effect on the deepest regions of the planet.

Core Almost Alone

A 1991 review in *Science* of core-related issues by Horton E. Newsom of the University of New Mexico, Albuquerque, and Kenneth W.W. Sims of the University of California, Berkeley, concluded that there is not enough data to answer the three major outstanding questions about the core (the third question also involves the lower mantle):

1. Did the core form because Earth was once mostly liquid?
2. Why is the core less dense than iron?
3. Does the core interact with Earth's surface using the lower mantle as its primary medium?

The first question bears largely on why the core should exist. Textbooks often explain the core by saying that Earth's heavy elements migrated toward the center under the influence of gravity, thus forming the core. It cannot be this simple, however. For one thing, if Earth were not molten, rock layers would be in the way. Also, in a molten Earth, many elements would combine with each other; however, some elements, called the siderophile elements, do not combine readily with the dominant oxygen and sulfur of Earth. Thus, if most of Earth were liquid, the siderophile elements, which include iron and nickel, would tend to fall toward the middle and create the core. This simple point of view is complicated by direct measurements that show that some siderophile elements are still found largely in the mantle. Furthermore, many recent explanations of how Earth formed do not provide for sufficient heating for the whole Earth to melt. Some earth scientists get around this by proposing such mechanisms as melting caused by radioactive elements after a cold formation or melting (or partial melting) as the result of the impact of a giant bolide, such as an asteroid or a comet, early in Earth's history. If an asteroid, the bolide might also plunge into the core and alter the core's size and chemistry.

One idea, proposed by D.J. Stevenson in 1990, includes the possibility that the upper mantle melts to form a magma ocean at the same time that the lower mantle is a partially solid barrier to the still-forming core. Iron droplets about a centimeter (half an inch) in diameter form in the liquid and rain down. They create ponds on the surface of the partially solid layer. When the ponds become sufficiently heavy, the iron drains down through cracks to reach the core region. A problem with this mechanism is that liquid iron would tend to stick to the sides of the cracks unless the cracks were unacceptably wide. If, however, iron alloys with oxygen at great pressure, for which there is some evidence, it could become less viscous and manage its way down through the cracks.

This touches on the second question: Why is the core less dense than iron? (Nickel, the other presumed metal in the core, is heavier than iron.) Alloying with oxygen would explain that, but so would alloying with silicon (as Jeanloz proposes) or alloying with sulfur, also a possibility. Determination of the correct explanation severely limits possible explanations of how the core was formed and how it interacts with the mantle.

An intriguing answer to the density question comes from diamond-anvil studies conducted by John V. Badding, Russell J. Hemley, and H.K. Mao of the Carnegie Institution of Washington DC. They have found that a combination of iron and hydrogen subjected to a pressure of 3.5 gigapascals (35,000 atmospheres) forms an iron hydride that expands instead of condensing further. The iron hydride expands up to 17 percent and changes appearance from metallic to grainy as the layers of iron atoms that form a lattice are infiltrated by the hydrogen atoms, preventing the iron lattice layers from lining up with the hills of one layer in the hollows of the next. Instead, the hydrogen pushes the layers aside to make for a hill-to-hill alignment with hydrogen atoms between each pair of hills.

The Carnegie researchers think that this effect could account for the observed density of the core if water has infiltrated the core and disassociated to provide the hydrogen. The observed density of the iron hydride is approximately the same as the inferred density of Earth's core.

This investigation began when it was observed the hydrogen under pressure tends to destabilize iron or steel and make it brittle, affecting the steel parts of the diamond anvil. In addition to clues to the composition of Earth's core, the investigation established how hydrogen degrades iron and steel. Formation of iron hydride under pressure causes expansion and weakening, while release of the pressure causes the metal to fall apart as it contracts.

Finally, the third question: Does the core interact with Earth's surface using the lower mantle as its primary medium? This is a version of: Does the mantle have layers? When considered in connection with the core, it relates to how the core was formed. For example, formation by the two-stage process proposed by Stevenson would inevitably result in a two-layer mantle. The top layer, which formed from the molten part, would be depleted in iron and other siderophile elements as compared with the lower layer.

V. Rama Murthy of the Carnegie Institution of Washington and the University of Minnesota proposed in Jul 1991 that all of these problems can be resolved by a single assumption: Earth was once completely molten at a temperature of about 3000 to 3500 K (5400° to 6300° F; Celsius is essentially the same as Kelvin for estimates this imprecise). He pictures the liquid state as starting when Earth is about a tenth its present mass. At that point, the temperature might be only 2500 K (4500° F), but the core would begin to form. If Earth stayed largely liquid until it reached the present size, its temperature during the process might have to climb to 4500 K (8100° F). Along the way, Earth would be between 3000 and 3500 K for a considerable portion of the time.

According to Murthy, under these assumptions, the most troublesome siderophile elements—gold, rhenium, and iridium, for example—would closely match observed abundances in the mantle. The ratios of one siderophile element to another, however, would still not be right. For that to be the case, he has to assume that the mantle forms two layers at a temperature of about 3500 to 4000 K. He also claims that these mechanisms tend to incorporate magnesium, silicon, oxygen, and sulfur into the core, reducing its density. Thus, a fairly simple temperature assumption for Earth's history can resolve several of the main questions about the mantle and core.

Ongoing studies relating to the mantle and core should result in further breakthroughs in the 1990s.

(Periodical References and Additional Reading: *Science News* 5-26-90, p 324; *Science* 1-25-91, pp 382 & 410; *Science* 3-22-91, p 1438; *Science* 4-5-91, p 68; *Science* 4-12-91, p 216; *Science* 4-26-91, p 510; *Science* 5-10-91, pp 755, 783, & 827; *Science* 5-17-91, pp 895 & 926; *EOS* 5-21-91, p 236; *Science* 5-24-91, p 1068; *Scientific American* 6-91, p 101; *New York Times* 6-11-91, p C1; *Science* 6-21-91, p 1617; *Science* 7-19-91, pp 247 & 303; *Science News* 8-3-91, p 70; *New York Times*, 9-10-91, p C5; *Science* 11-29-91, p 1295)

VOLCANOES AND EARTHQUAKES

VOLCANO PREDICTION SHOWS CONSIDERABLE SUCCESS

Throughout the 1980s earth scientists were concerned about the possible eruption of the Long Valley caldera around the town of Mammoth Lakes CA, just east of Yosemite National Park. In late May of 1982 they announced their concern to the public, prompting some panic. One newspaper in nearby Nevada even reported that lava was flowing down Main Street in Mammoth Lakes. Although the incident hurt tourism and land values, the volcano failed to erupt. Since then, geologists have treated volcano eruption predictions more cautiously.

Also since then, earth scientists have gotten much better at predicting volcano eruptions. Some recent successes include the Japanese prediction of the eruption of an offshore volcano in 1989, a US prediction of the 1989 eruption of Redoubt in Alaska, and predictions of the destructive eruptions in 1991 of Mt Unzen in Japan and Mt Pinatubo in the Philippines. One spectacular string of successes concerns Mt St Helens in Oregon; the US Geological Survey has predicted nineteen out of twenty-two eruptions of that volcano since 1980. In all cases, a pattern of long-period earthquakes was the key to the successful prediction. Such earthquakes are caused by the movement of magma under the volcano.

Other tools used in prediction include laser-based devices that measure the bulge caused by increased pressure from the volcano's interior, satellite photographs of volcanoes too dangerous to approach even by air, and chemical sensors that detect levels of sulfur dioxide. Knowing where the volcano is distending helps predict which part of the cone will break—volcanoes seldom erupt through a symmetrical crater at the top, although an explosion can produce the effect of such an eruption. Then, as the magma begins to reach the surface, gases boil out of the magma, so a sudden rise in sulfur dioxide implies an imminent eruption.

A problem with volcano prediction has been that some people refuse to believe anything can happen, especially to them. The well-known case of Harry Truman, the 84-year-old man who refused to be evacuated from the vicinity of Mt St Helens and was subsequently buried in mudflows, is an example. Less well known is the case of Nevado del Ruiz in Colombia. Despite predictions of disaster from volcanologists, no one told the nearby population, of whom 22,000 were killed. At Mt Pinatubo (see following article), however, the message was received loud and clear by most Filipinos and by the Americans who ran the military bases near the volcano.

Even with better prediction and better communication, the problems caused by volcanoes are getting worse. Until the twentieth century, the regions around many volcanoes were sparsely populated and many eruptions did not affect humans directly. According to Robert Tilling of the US Geological Survey in Menlo Park CA, only 315 persons per year were killed as a direct result of volcanic activity from 1600 to 1900; but since 1900, that number has soared to 845 per year, mostly because of increased human population.

Volcanoes in heavily populated regions have posed major problems at least since the explosion of Thera in the Mediterranean in 1635 BC. Continuing geologic unrest in Long Valley still has volcano watchers nervous, but so far they have failed to see the long-period earthquakes that have been associated with the recent successful predictions of eruptions. Some other volcanoes that are cause for concern include Mt Fuji, 100 km (62 mi) from Tokyo, and Popocatépetl near Mexico City, the world's largest urban region.

(Periodical Reference and Additional Reading: *Science* 8-2-91, p 514)

MOUNT PINATUBO ERUPTS

The successful prediction of the eruption of the long-dormant Mt Pinatubo on Luzon Island in the Philippines undoubtedly saved thousands of lives, but no human power could prevent the worldwide effects of the largest eruptions of the twentieth century in terms of amount of material ejected. Within 10 days the Pinatubo cloud had spread 11,000 km (7000 mi), reaching from Indonesia all the way to central Africa. Early estimates based on satellite data showed more than twice the sulfur dioxide in the stratosphere as was found after the 1982 eruption of El Chichón. It is now fairly well established that El Chichón lowered global temperatures by a small amount, so it is expected that the Pinatubo eruption will have an even greater effect. Given the recent string of very warm years (see "Global Warming UPDATE," p 382), such cooling might be considered beneficial. On the other hand, the cooling may be masked at first by the 1991-1992 El Niño (see "El Niño and La Niña," p 220).

Another problem with the atmosphere, the erosion of the ozone layer, probably will be accelerated by Pinatubo. It is thought that El Chichón destroyed about 15 percent of stratospheric ozone through reactions among the sulfur dioxide droplets, artificial chlorine compounds, and the ozone.

Pinatubo also affected viewing of the great 1991 solar eclipse (see "The Sun Puts on a Show," p 41). The cloud of sulfur dioxide had spread to Hawaii by the time of the eclipse, interfering with observations from the array of telescopes on Mauna Kea on the "big island" of Hawaii. On the other hand, the cloud also produced unusually brilliant sunsets in Hawaii, and ultimately around the world.

Pinatubo, dormant since 1380, began showing renewed activity in Apr 1991. By Jun about 12,000 people had been evacuated from the immediate vicinity of the volcano and Pinatubo was emitting small amounts of steam and ash. On Jun 9 the US Air Force moved some aircraft from Clark Air Base, about 16 km (10 mi) from the volcano, when Pinatubo's activity turned into a definite, but small, eruption. Scientists were warning the Air Force, however, that a very large eruption was about to happen. The Air Force began evacuating personnel at 6 a.m. local time the following day, with about 14,000 moving out and a caretaker staff of 1500 staying behind.

The first big explosion came at 8:42 a.m. (local time) on Jun 12. More followed, with two much larger explosions later that same day. On Jun 15 scientists discovered from satellite photographs that a 2.9-km (1.8-mi) crack had opened up between two craters, suggesting that the whole mountain might explode. At that point, the remaining staff at Clark left. In all, about 80,000 people fled the neighborhood of the volcano. But one of the craters collapsed the next day, relieving much pressure and removing the likelihood of the whole mountain exploding. The result of the explosions and collapse was a caldera 2.5 km (1.5 mi) in diameter near the top of the mountain.

Despite the evacuation around the volcano, 745 were killed by the eruption. Most deaths or injuries occurred from disease in evacuation camps or when heavy ash from the volcano, made heavier still by soaking up water from a tropical storm that hit Luzon during the eruption, caused roofs to collapse.

On Jun 17 the United States formally abandoned Clark Air Base for good.

(Periodical References and Additional Reading: *Time* 6-24-91, p 42; *Science* 6-28-91, p 1780; *Science News* 7-6-91, p 7; *EOS* 7-16-91, p 305; *Science* 8-2-91, p 504; *Science News*, 8-31-91, p 132; *New York Times* 11-7-91, p A1)

EARTHQUAKE PREDICTION UPDATE

Although earth scientists have developed fairly reliable means of predicting the imminent eruption of a volcano, the problem of short-time-scale earthquake prediction, despite some reported successes in the past, remains unsettled. Medium-time-scale prediction—for example, "the part of the San Andreas Fault south of Palm Springs will have a magnitude 7.5 earthquake in the next 30 years with a probability greater than 40 percent"—has not really proved itself either, although geologists are somewhat confident that such predictions are valid. Most reliable of all are long-time-scale predictions, such as "the eastern United States will experience a serious earthquake on the average of once in every 500 years." Of course, the longer the time scale, the less the practical advantage of the prediction.

Short Time Scale

The earthquake prediction that received the most publicity in 1990 was for one that failed to materialize. The late meteorologist Iben Browning predicted that southeastern Missouri would be struck by a near repeat of what might be termed the "big big ones," the great New Madrid earthquake sequence of 1811-1812, three quakes so big that they were felt over a region of 10 million km^2 (4 million sq mi), virtually all of the United States. The nineteenth-century Missouri quakes were felt in Boston and collapsed the scaffolding around the Capitol in Washington DC.

Browning based his prediction of a repeat on stronger tidal forces than normal for that date. He claimed only a 50 percent probability for his prediction of an earthquake measuring between 7.0 and 7.6 on the New Madrid Fault during the 48 hours centered on Dec 3 1990. Thus, when nothing happened, Browning could legitimately say that he had been correct.

Much of what is known about short-time-scale earthquake prediction is based on casual observations and folk science. The most famous example occurred in 1975 when Chinese peasants living near Haicheng saw snakes crawling about in abundance and noted changing levels of water in their wells. Reports of these events were passed on to a geologist who successfully predicted a 7.3 earthquake. Haicheng was evacuated before the earthquake, saving hundreds of thousands of lives.

A recent example of casual observation occurred in connection with the "little big one," the 7.1-magnitude earthquake that struck the San Francisco Bay area at 5:04 p.m. Pacific time on Oct 18 1989, during the World Series. Geologists call this earthquake the Loma Prieta quake because its epicenter was Loma Prieta Mountain in the Santa Cruz range.

Here is how the observation occurred. Wife and husband earth scientists Nathalie Valette-Silver of the University of Maryland and Paul Silver of the Carnegie Institution in Washington DC took advantage of a side trip from a San Francisco geophysical meeting to visit a geyser in Calistoga CA, 180 km (110 mi) north of San Francisco. Like Yellowstone's famous Old Faithful, the Calistoga geyser normally goes off on a regular basis, approximately once every hour and a half. The Silvers talked to the owner about the geyser and were surprised to learn that she felt that "the geyser is affected by earthquakes." Since the owner had records of the time of every geyser eruption going back to 1973, the Silvers were able to check out her theory. They found that two and a half days before the "little big one" the geyser's schedule changed from once each 90 minutes to once each 100 minutes. Similar changes were recorded preceding earthquakes in 1975 and 1984. The Silvers plan further research to see if they can confirm their theory that increased strain before an earthquake closes the cracks in the rocks through which water reaches the geyser, thus changing the timing of eruptions.

Other changes before the Loma Prieta quake were recorded by Malcolm

J.S. Johnston of the US Geological Survey and Alan T. Linde of the Carnegie Institution in early 1989 on a strainmeter 35 km (22 mi) south of the epicenter. Another nearby strainmeter also indicated changes in Aug 1989, noted by Michael T. Gladwin of the University of Queensland in St. Lucia, Australia. Because of the strainmeter's distance from the epicenter—although considerably closer to it than the Calistoga geyser—it was not clear if these changes were actually precursors. Other scientists found additional precursors to the quake. Antony C. Fraser-Smith of Stanford University found that his instruments recorded unusual electromagnetic measurements before the quake, while G. Michael Reimer of the US Geological Survey noted a sharp increase in the amount of helium at a soil-monitoring station about 60 km (37 mi) from Loma Prieta. Previously, however, medium-sized earthquakes were preceded by drops in helium. Clearly, geologists need a lot more experience with such measurements before they can be used as infallible predictors.

Medium Time Scale

The first widely noted medium-time-scale earthquake prediction came in 1979 when B.A. Bold and R.H. Jahns proposed that the odds of California getting a major earthquake before 1990 were even. People were not impressed, since California is a big state and fifty-fifty odds do not make disaster seem inevitable.

Formal medium-time-scale earthquake prediction in the United States did not get started until 1985, when the US Geological Survey (USGS), sanctioned by the National Earthquake Prediction Evaluation Council, predicted a twenty to one chance that a magnitude 6 earthquake would strike the Parkfield segment of California's San Andreas Fault before 1993. Parkfield is a tiny town of thirty-four people in the hills that border California's central valley, roughly halfway between San Francisco and Los Angeles. Since the 1985 prediction, Parkfield has undergone two C-level alerts—both in the fall of 1990—the lowest level at which the USGS notifies the public of what they believe to be increased danger. Neither alert produced an earthquake, although—despite efforts to prevent panic—each managed to frighten a few folks. As the deadline approached, some USGS scientists found evidence in old documents that serious earthquakes on the Parkfield segment have diminished in number since 1930 and that different quakes broke different segments of the fault. Both pieces of evidence suggest that a Parkfield quake may be harder to predict than expected.

Also in 1985, scientists from the Soviet Union persuaded Mikhail Gorbachev to pass on to US President Ronald Reagan their own medium-time-scale prediction—a giant quake in southern California before 1988. It did not happen, but the Soviet scientists did not give up. Led by Volodya

Kelis-Borok of the Institute of Earthquake Prediction Theory and Mathematical Geophysics in Moscow, they came back with claims of past successes and new predictions in 1991. Among other claims, they said they had predicted the Loma Prieta quake of Oct 1989. Their claims are viewed as suspect by most US geologists. For one thing, the Soviet scientists also predicted a 7.5 magnitude earthquake for *northern* California—their methods do not localize earthquakes very well—before 1988. But in Jun 1988, after nothing had happened, they revised their prediction to before 1991. When the "little big one" of 7.1 magnitude occurred, they called it confirmation of the theory. US scientists are also skeptical on theoretical grounds of the USSR prediction method, which relies on precursors hundred of kilometers from the main fault. The US geologists just do not think that events that far away from the fault affect it in any way.

In Dec 1990 Lynn R. Sykes and Steven C. Jaumé claimed that they had determined how to predict large earthquakes in California. They noted that small earthquakes along a fault increase rapidly before a large earthquake. This pattern was evident before the 6.8 earthquake on the Hayward Fault that occurred in 1868, before the big San Francisco quake in 1906, and before the Loma Prieta quake in 1989. Not everyone is sure that their evidence is more than a statistical anomaly, however. The test will come in seeing whether such a pattern can be successfully used to predict an earthquake *before* it happens.

Paul Segall and Mike Lisowski of the USGS, on the other hand, claim that prediction of earthquakes caused by the San Andreas Fault using any recurrence patterns cannot be effective. Their study of the Loma Prieta slippage that caused the 1989 earthquake revealed that it took a totally unexpected direction. Since this earthquake was caused by a different mechanism from that of the big 1906 San Francisco earthquake, they think that prediction methods based on "it happened once, so it'll happen again" will not work in California.

Nevertheless, in Jul 1990 the USGS upgraded its previous ·earthquake prediction for the San Francisco Bay area. A twelve-person panel announced that there was a two in three chance for an earthquake with a magnitude greater than 7 before 2020. The previous prediction by the same group had been a one in two chance for the same general time frame. The panelists thought the chance of a major earthquake was more likely than they had previously calculated because of new information about the Rogers Creek Fault north of San Pablo Bay and because of higher estimates for the stress buildup along the Hayward and San Andreas faults.

Long Time Scale

The basis of most long-time-scale earthquake predictions is the record of past quakes. If a region has been hit by an earthquake every 30 years or so, then it is likely to continue that pattern. Using that method, Geologists Lisa Leffler of Memphis State University and Steven Wesnousky of the University of Nevada looked for geological evidence that could help make a long-time-scale prediction for major earthquakes centered around the New Madrid Fault. Geological evidence of past earthquakes consists mainly of deposits of sand that large earthquakes eject through soil under the right conditions. Leffler and Wesnousky's search showed plenty of such deposits for the 1811–1812 quake sequence, but none older. Other geologists who have also looked for such evidence in the past, including Stephen Obermeier of the USGS and Roger Saucier of the US Army Waterways Experiment Station, also failed to find any older deposits. This suggests that earthquakes in the region are not common at all. If major earthquakes recur, it may be on a scale of tens of thousands of years.

Obermeier and a large group of coworkers also looked for earthquake evidence in southern Illinois and Indiana. The lower Wabash River valley contains numerous faults and is frequently prone to small earthquakes. The researchers found that two large quakes have occurred in the region, one about 1500 years ago and another about 7500 years ago. There was no evidence that these were connected in any way with the New Madrid Fault, but at a maximum of 6.7 magnitude, recurrence could be serious today. The record, however, would suggest recurrence several thousand years in the future. By then, one hopes that short-term prediction will have been perfected.

Similar searches in the Pacific Northwest by Brian Atwater of the USGS in 1987 also located evidence of past earthquakes, suggesting that large quakes occur there more often than suspected. Although there have been no sizable historic quakes in the region, history in that region begins only about 200 years ago. In 1991 James Savage and Michael Lisowski of the USGS reported on their study of the earthquake potential for the region. They used lasers to measure the distances between mountaintops in Olympic National Park near Seattle WA during the period from 1982 to 1990. What they found was that the mountains were moving—only 16 mm (0.6 in.) along a 30-km (18.6-mi) line over 8 years, but that is a lot of movement for a mountain. Savage and Lisowski also used tidal gauges to check the height of the coastline; they found it to be rising 4 mm (0.14 in.) a year. A study reported by Brian Atwater and coworkers in the Sep 12 1991 *Nature* notes that trees along the coastline were flooded catastrophically by sudden coastal lowering about 300 years ago. This is an almost certain sign that a major earthquake occurred between 1680 and 1720, before Europeans settled in the region but recently enough for an oral tradition to exist among Native Americans.

Both moving mountains and rising coasts suggest that the Juan de Fuca plate may be setting up the region for a major earthquake. This is the

Background

Geologic evidence of past earthquakes of magnitudes greater than about 6.2 occurs because such earthquakes shake water-saturated sediment, changing it briefly from a damp solid to a fluid. The suddenly liquefied sediment, usually called sand, can now flow through any cracks that are available. The earthquake itself may open some new cracks, either permanently or temporarily. If a crack is completely in the crust, the result of its filling by the liquefied sand is what geologists call a dike—a dike is any crack in the crust that is filled with some material different from the crust around it. If a crack leads to the surface, the liquid sand flows out and forms a cone around the opening like the cone of a volcano. When the shaking stops, the water-drenched sediment becomes solid again and remains in its new location. The dike or cone of sand may later be exposed when excavations are made for highway cuts or irrigation ditches. The presence of such formations is taken as an indication of a past earthquake. Dating the rock or soil layer where the formation is found—usually by carbon-14 dating of trapped organic materials—is used as evidence of the date of the past earthquake.

plate whose movement under a 1290-km (800-mi) stretch of the Northwest coast between Cape Mendocino CA and Vancouver Island, Canada, is responsible for the line of volcanoes that includes Mt St Helens. If the entire stressed region shifts during such an earthquake, the shock could match those in Chile in 1960 or Alaska in 1964. Both of those disasters were caused by similar crustal plate movements and both exceeded a magnitude of 8. This is a serious concern for the 75 percent of the population of Washington and Oregon that lives in the potential earthquake zone.

A spate of midsummer 1991 earthquakes off the coast of northern California and southern Oregon was also adduced as a possible indicator of an increase of seismic activity in the US Northwest. The quakes, which occurred between Jul 12 and Aug 17 and ranged from 6.0 to 7.1 in magnitude, appear to be a result of movements of the Gorda plate, which is trapped between the much larger Pacific and North American plates.

Geologic evidence shows that earthquakes in the region typically occur every 500 to 600 years. It also shows the last one at 300 years ago. Thus, although Northwesterners should be wary, not much is expected until around the twenty-second century.

Geologists are predicting earthquakes in places that may surprise some. David P. Schwartz and Stuart P. Nishenko of USGS used the geologic record to calculate a 3 to 8 percent chance of a greater than magnitude 7 earthquake in Utah during the next 50 years and a 7 to 15 percent chance in the next 100.

Another place where earthquakes occur more often than previously suspected is Charleston SC, according to a Feb 8 1991 report in *Science* by David Amick and Robert Gelinas of Ebasco Services in Greensboro NC. Their analysis of the geologic record calls for large earthquakes in South Carolina every 500 to 600 years, as opposed to previous estimates of one every 1500 years or so. Previous earthquakes in the area were all limited to coastal South Carolina. The only really serious earthquake in the eastern United States occurred in Charleston in Aug 1886. Mysteriously, geologists have never been able to locate the fault that caused it or the other South Carolina quakes.

Earthquakes in eastern North America are scary, not only because of the high population density but also because the nearly fault-free bedrock transmits powerful shock waves much farther than in the fault-ridden West Coast. In addition to New Madrid and Charleston, historic large quakes have hit Newfoundland and Boston, while there is geologic evidence of a monster quake in Oklahoma about 1200 years ago. Based on this kind of evidence, plus a survey of smaller historic earthquakes, Klaus Jacob of the Lamont-Doherty Geological Observatory estimated in 1990 that during the next 20 years there is a 10 percent chance of an earthquake with a magnitude greater than 7 striking North America east of the Rockies and a 61 percent chance of a magnitude 6 quake. A magnitude 6 earthquake, like the one that struck Boston in 1755, would probably cause over 300 deaths; a magnitude 6 earthquake in New York, believed possible on the basis of geologic evidence, would be a lot worse. A 1990 study of earthquakes in stable continental crust around the world by Arch C. Johnston and Lisa R. Kanter of the Center for Earthquake Research and Information at Memphis State University suggested strongly, however, that large earthquakes in such regions are virtually unpredictable. If the three New Madrid quakes are considered a single event, they found no earthquakes above a magnitude of 7 that had repeated. Thus, predicting earthquakes at any time scale in eastern North America is extremely difficult.

Prediction at all time scales in the United States should be improved by a planned network of a hundred new seismographs. The first two were installed in North Carolina and West Virginia on Apr 3, 1991.

(Periodical References and Additional Reading: *Science* 1-19-90, pp 263 & 286; *Science News* 1-20-90, p 47; *Science News* 2-17-90, p 104; *Scientific American* 3-90, p 68; *Science News* 3-17-90, p 175; *Science News* 5-5-90, p 278; *Science News* 6-20-90, p 47; *Science* 6-22-90, p 1490; *Science* 8-14-90, p 860; *Science* 9-21-90, pp 1355 & 1412; *Science* 9-22-90, p 1476; *Science* 11-30-90, pp 1191 & 1248; *Science* 6-22-90, p 1490; *Science News* 7-28-90, p 52; *Science News* 8-25-90, p 126; *Discover* 9-90, p 78; *Science News* 9-22-90, p 183; *Science News* 10-6-90, p 217; *Science News* 10-13-90, p 231; *Science News* 12-1-90, p 342; *New York Times*, 12-13-90; *Science News* 12-15-90, pp 374 & 380; *Science* 12-21-90, p 1660; *Scienc-*

News 12-22&29-90, p 398; *Science* 1-11-91, pp 139 & 169; *Science* 2-8-91, pp 603 & 655; *New York Times* 2-12-91; *Science* 3-1-91, p 999 & 1061; *Science* 3-15-91, p 1314; *Discover* 4-91, p 13; *Science* 4-5-91, pp 7, 28, & 101; *Science* 6-14-91, p 1497; *Science News* 6-15-91, p 376; *Discover* 7-91, p 8; *Science News* 7-28-90, p 52; *Science* 8-9-91, p 622; *Science News,* 9-7-91, p 156; *New York Times* 10-1-91, p C2; *Science* 10-11-91, p 197; *Discover* 11-91, p 18)

MAJOR EARTHQUAKES

Earthquakes have been among the most destructive forces of nature throughout time. Most earthquakes are caused when rock on one side of a crack (a fault) in Earth's crust moves with respect to rock on the other side of the fault. The motion sets up vibrations in the crust that travel as waves through the rock. When the waves reach Earth's surface, they cause it to move in various ways. Small earthquakes that accompany volcanic eruptions are caused by the motion of liquid rock (magma).

Unlike volcanoes, which at present seem to be beyond control, the amount of damage an earthquake does to human life and property can be largely controlled by proper construction of buildings. Note that the energy of an earthquake is poorly related to the destruction it causes.

The damage caused by a common result of undersea earthquakes—the tsunami, or (incorrectly) tidal wave—is probably beyond human control. A tsunami is caused when an earthquake raises or lowers a section of the seabed. This produces a wave that, while not generally noticeable at sea, can reach great heights as it approaches land. Related to the tsunami is the harbor wave. Usually, harbor waves are caused when a landslide falls into a bay or strait. Because of the confines of the land around the sea, a harbor wave can be more destructive and even higher than a tsunami. High waves are also caused by volcanic explosions or collapses, such as the explosion and collapse of Krakatau.

In the following table, all of the magnitudes before 1880, when John Milne developed the first modern way of measuring the power of an earthquake, are estimated, based on various forms of evidence, such as contemporary accounts or geologic changes. Where available, moment magnitude, based on the actual slippage of rock along a fault, instead of surface magnitude, based on the size of earthquake waves at a seismograph station, are reported. For large earthquakes, the moment magnitude may be slightly greater than the more traditional surface-wave measurement. For example, the greatest earthquake ever recorded by modern seismographs, the Chilean earthquake of 1960, is 9.5 on a moment magnitude scale, 8.3 on the surface-wave scale. Moment magnitudes are marked with a * in the table.

Date	Location and Remarks	Estimated Deaths	Magnitude
May 20 526	Antioch, Syria (now Turkey)	250,000	N.A.
856	Corinth, Greece	45,000	"
1036	Shanshi, China	23,000	"
1057	Chihli (now Hopeh), China	25,000	"
1170	Sicily	15,000	"
1268	Cilicia (now Turkey)	60,000	"
Sep 27 1290	Chihli (now Hopeh), China	100,000	"
May 20 1293	Kamakura, Japan	30,000	"
1356	Basel (Switzerland)	N.A.	7.4*
Jan 26 1531	Lisbon, Portugal	30,000	N.A.
Jan 24 1556	Shansi, China	830,000	"
1604	Taiwan Straits	N.A.	7.7*
1605	Hainan Island, China	N.A.	7.3*
Feb 5 1663	St. Lawrence River	N.A.	N.A.
Nov 1667	Shemakha, Azerbaijan (USSR)	80,000	"
Jun 7 1692	Port Royal, Jamaica	30,000	"
1693	Naples, Italy	93,000	"
Jan 11 1693	Catania province, Sicily, Italy	60,000	"
1707	Tsunami hits Japan	30,000	"
1727	Boston MA	N.A.	
Dec 30 1730	Hokkaido Island, Japan	137,000	"
1731	Peking (Beijing), China	100,000	"
Oct 11 1737	Calcutta, India	300,000	"
Jun 7 1755	Northern Persia	40,000	"
Nov 1 1755	Earthquake with tsunami hits Lisbon, Portugal	60,000	8.7
Nov 18 1755	Boston MA (centered at Cape Ann)	0	6
Feb 4-5 and Mar 28 1783	Calabria, Italy	35,000	N.A.
Feb 4 1797	Quito, Ecuador, and Cuzco, Peru	41,000	"
Dec 16 1811	New Madrid MO	10-	8.1*
Jan 23 1812	New Madrid MO	0	8.2*
Feb 7 1812	New Madrid MO	0	8.3*
Dec 21 1812	Traverse Range CA	N.A.	N.A.
1819	Kutch, India	N.A.	7.8*
Sep 5 1822	Aleppo (now in Syria)	22,000	N.A.
Dec 28 1828	Echigo, Japan	30,000	"
Jun 1838	San Francisco CA	N.A.	"

Date	Location and Remarks	Estimated Deaths	Magnitude
Dec 24 1854	Tokai, Japan	3,000	8.4
Oct 1855	Edo (now Tokyo) Japan	2,000+	N.A.
1857	East of Naples, Italy	10,000+	N.A.
Jan 9 1857	Fort Tejon CA	2	8.3
1858	Portugal	N.A.	7.1*
Nov 18 1867	Virgin Islands	N.A.	N.A.
1868	Hayward Fault CA	N.A.	6.8
Apr 2 1868	Hawaiian Islands	N.A.	N.A.
Aug 13-15 1868	Peru and Ecuador	40,000	"
Mar 26 1872	Owens Valley CA	60	"
May 16 1875	Venezuela and Colombia	16,000	"
Aug 31 1886	Charleston SC	83	7.6
Oct 28 1891	Central Japan	7,300	N.A.
Jun 15 1896	Tsunami hits Sanriku and Kamaishi, Japan	26,000	"
Jun 12 1897	Assam, India	N.A.	"
Sep 3 1899	Yakatanga AK	0	8.3
Sep 10 1899	Yakatanga AK	0	8.6
Apr 4 1905	Kangra, India	20,000	N.A.
1906	Exmouth Plateau, Indian Ocean NW of Australia	0	7.2*
Apr 18 1906	San Francisco CA	667	8.3
Aug 16 1906	Valparaiso, Chile	20,000	8.6
1907	Tadzikistan, Russia	40,000	N.A.
Dec 28 1908	Messina, Italy	75,000	7.5
Jan 13 1915	Avezzano, Italy	30,000	N.A.
Oct 2 1915	Pleasant Valley NV	0	7.8
Jan 13 1916	Avezzano, Italy	29,980	7.5
Oct 11 1918	Mona Passage in Caribbean Sea (Nanai)	116	7.4*
Dec 16 1920	Kansu, China	180,000+	8.6
Sep 1 1923	Tokyo and Yokohama, Japan	143,000	8.3
Jun 27 1925	Helena MT	0	6.8
Mar 7 1927	Tango, Japan	1,120	8.0
May 22 1927	Nan-Shan, China	200,000	8.3
1929	Newfoundland (Grand Banks), Canada	27	7.4*
Aug 16 1931	Mt Livermore TX	0	6.4

(Continued)

Date	Location and Remarks	Estimated Deaths	Magnitude
Dec 20 1932	Cedar Mountain NV	0	7.3
Dec 26 1932	Kansu, China	70,000	7.6
1933	Baffin Bay, between Greenland and Baffin Island, Canada	N.A.	7.7*
Mar 2 1933	Tsunami hits Sanriku, Japan	2,990	8.9
Mar 10 1933	Long Beach CA	120	6.3
Jan 15 1934	India, Bihar, and Nepal	10,700	8.4
Mar 12 1934	Great Salt Lake UT	0	6.6
1935	Libya	N.A.	7.1
May 31 1935	Quetta, India (now Pakistan)	50,000	7.5
Oct-Nov 1936	Helena MT	2	N.A.
Jan 24 1939	Concepción, Chile	30,000	8.3
Dec 27 1939	Erzincan, Turkey	30,000	7.9
May 18 1940	Imperial Valley CA	9	7.1
Apr 1 1946	Earthquake at Unimak Island AK causes tsunami in Hilo HI	173	7.2
Dec 21 1946	Honshu Island, Japan	2,000	8.4
Jun 28 1948	Fukui, Japan	3,500+	7.3
Oct 1948	Ashkhabad, USSR	20,000-	N.A.
1949	Olympia WA	8	7.1
Jul 10 1949	Tadzhikistan, USSR	120,000	7.5
Aug 5 1949	Pelileo, Ecuador	6,000	6.8
Aug 15 1950	Assam State, India	1,500	8.7
1951	South Tasman Rise, Indian Ocean SW of Tasmania	0	7.0*
Jul 21 1952	Bakersfield CA	12	7.7
Mar 18 1953	NW Turkey	1,200	7.2
Sep 9-12 1954	Orléansville, Algeria	1,660	N.A.
Dec 16 1954	Frenchman's Station NV	0	7.1
Jun 10-17 1956	Northern Afghanistan	2,000	7.7
Jul 2 1957	Northern Iran	2,500	7.4
Dec 13 1957	Western Iran	2,000	7.1
Jul 9 1958	Lituya Bay AK	3	7.9
Aug 17 1959	Hebgen Lake MT	28	7.1
Feb 29 1960	Agadir, Morocco	12,000	5.8
May 21-30 1960	Southern Chile; on May 22, a tsunami strikes various Pacific islands, including Hawaii, where 61 are killed in Hilo	5,700	9.5

Date	Location and Remarks	Estimated Deaths	Magnitude
Sep 1 1962	NW Iran	12,403	7.1
Feb 21-22 1963	El Marj, Libya	260	5.0
Jul 26 1963	Skopje, Yugoslavia	1,011	5.5
Mar 27 1964	Southern Alaska	131	8.5
Mar 28 1965	Central Chile	420	N.A.
1965	Seattle WA	N.A.	6.5
Mar 1966	Hsingtai region, China	N.A.	N.A.
Aug 19 1966	Eastern Turkey	2,520	6.9
Jul 27 1967	Caracas, Venezuela	250+	6.5
Dec 11 1967	Konya, India (caused by water filling a reservoir; the extra pressure resulted in the earth shifting)	117	6.5
Aug 31 1968	Khurasan, Iran	12,000	7.8
Jul 25 1969	Eastern China	3,000	N.A.
Jan 5 1970	Yunnan Province, China	10,000	7.7
Mar 28 1970	Gediz, Turkey	1,086	7.4
May 31 1970	Yungay, Ranrahirca, Huarás, and other cities in Peru	66,794	7.7
Feb 9 1971	San Fernando Valley CA	64	6.6
Apr 10 1972	Ghir, Iran	5,057	6.9
Dec 23 1972	Managua, Nicaragua	10,000+	5.6
Apr 26 1973	Hawaii	0	6.2
Aug 28 1973	Puebla, Mexico	527	6.8
Dec 28 1974	North Pakistan	5,200+	6.3
Feb 2 1975	Near Islands AK	0	7.6
Feb 5 1975	Liaoning Province, China (successfully predicted)	300	N.A.
Sep 6 1975	Lice, Turkey	2,312	6.8
Nov 29 1975	Hawaii	0	7.2
Feb 4 1976	Guatemala City, Guatemala	22,778	7.5
May 6 1976	NE Italy	946	6.5
Jun 26 1976	New Guinea and Irian, Java	8,000+	7.1
Jul 28 1976	Tangshan, China	750,000	8.0
Aug 17 1976	Earthquake and tsunami hit Mindanao, Philippines	8,000	7.3
Nov 24 1976	Eastern Turkey	4,000	7.9
Mar 4 1977	Bucharest, Romania	1,541	7.5

(Continued)

Date	Location and Remarks	Estimated Deaths	Magnitude
Aug 19 1977	Indonesia	200	8.0
Nov 23 1977	NW Argentina	100	8.2
Dec 20 1977	Central Iran	500+	N.A.
Jan 14 1978	Tokai region, Japan	25	7.0
Jun 12 1978	Sendai, Japan	27+	7.5
Jun 20 1978	Salonika, Greece	51	N.A.
Aug 18 1978	Tsunami strikes Acajutla, El Salvador	100+	N.A.
Sep 16 1978	NE Iran	25,000	7.7
Nov 29 1978	Oaxaca, Mexico	0	7.8
Jan 17 1979	Eastern Iran	200	6.7-7.5
Apr 15 1979	Yugoslavia and Albania	129	7.2
Nov 14 1979	Meshed, Iran	500	6.7
Dec 12 1979	Colombia and Ecuador	800	7.9
Dec 18 1979	Bali, Indonesia	100	6.1
Aug 1 1980	West Irian, Indonesia	120	5.9
Oct 10 1980	NW Algeria	4,500	7.3
Nov 23 1980	Southern Italy	4,800	7.2
Jul 28 1981	Kerman Province, Iran	8,000	N.A.
Dec 13 1982	Yemen	2,800	6.0
Mar 31 1983	Popayán, Colombia	200,000	5.7
Oct 28 1983	Challis ID	2	6.9
Oct 30 1983	Turkey	2,000	7.2
Mar 3 1985	Algarrobo, Chile	177	7.8
Sep 19, 21 1985	Mexico City	4,200	8.1
Jan 31 1986	NE Ohio	0	4.9
Oct 10 1986	San Salvador, El Salvador	1,000+	7.5
Mar 5-6 1987	Ecuador	2,000	7.0
Oct 1 1987	Whittier Narrows CA	0	5.9
Aug 20 1988	Nepal and India	700+	6.5
Nov 6 1988	Yunnan Province, China, and Burma	938	7.6
Nov 25 1988	Chicoutimi, Quebec	0	6.0
Dec 3 1988	Pasadena CA	0	4.9
Dec 6 1988	Mamasani, Iran	7	5.6
Dec 7 1988	Armenia, USSR	28,854	6.9
Jan 22 1989	Tadzhikistan, USSR	274	5.5
Mar 10 1989	Malawi	9	6.2

Date	Location and Remarks	Estimated Deaths	Magnitude
May 23 1989	Macquarie Ridge, 300 mi SE of New Zealand	0	8.2-8.3
Oct 17 1989	Loma Prieta in Santa Cruz Mountains of CA, but destructive in San Francisco and Oakland as well; original death toll of 67 revised to 63 on Oct 17 1991	63	7.1
Oct 18 1989	Shanxi-Hebei border, China	29+	7.1
Oct 19 1989	Cherchell, Algeria	30	6.0
Dec 15 1989	Off coast of Mindanao, Philippines	2	7.3
Dec 27 1989	Newcastle, Australia	10	5.57
Dec 30 1989	Papua, New Guinea	0	6.7

EARTHQUAKES OF 1990

About 52,000 people were killed as a result of earthquakes in 1990. Here are details on the major seismic events.

Jan 14 A 6.1 magnitude earthquake with a focus 17 km (11 mi) below Xing-hai Province in northwest China caused little damage and no casualties.

Jan 16 An offshore earthquake of magnitude 5.5 off northern California at a depth of 2 km (1 ¼ mi) caused slight damage in the towns of Honey-dew, Weott, Redway, and Whitehorse, and was felt as level V on the modified Mercalli scale in Eureka.

Feb 8 Fifteen construction workers were injured as a result of a 6.6 magnitude earthquake with a focal depth of 31 km (20 mi) in the southern Philippines. The tremors were felt on Negros, Cebu, and northern Mindanao islands.

Feb 19 A 6.3 magnitude earthquake at a depth of 22 km (14 mi) in the North Island of New Zealand caused minor damage in Dannevirke and Waipukrau, but no casualties. Farther north, in the island nation of Vanuatu, a shallow 6.8 magnitude earthquake caused no damage.

Feb 20 A delay of the "bullet train" was the only observable result of the shallow 6.5 magnitude earthquake in Japan.

Feb 28 Southern California was rocked by a 5.2 magnitude earthquake located at a depth of 10 km (6 mi) near Upland. Rock slides blocked roads in the Mt Baldy region. There were thirty-eight injuries from the quake, which was felt from Escondido to Santa Barbara.

<u>Mar 3</u> A shallow 7.5 magnitude earthquake struck an uninhabited region south of the Fiji Islands, causing no damage.

<u>Mar 4</u> Eleven deaths and forty injuries were caused by a 28-km- (17-mi-) deep earthquake in west-central Pakistan. The 6.1 magnitude earthquake also caused significant damage to structures.

<u>Mar 5</u> A shallow earthquake in the New Hebrides Trench southwest of Vanuatu and about 1600 km (990 mi) north of Brisbane, Australia, registered 7.1 on the Richter scale. Although large, there was no damage on land.

<u>Mar 25</u> Ten injuries, but no fatalities, were reported from a 7.1 magnitude earthquake centered about 100 km (60 mi) west-southwest of San José, Costa Rica, at a depth of 20 km (12 mi). There was considerable damage to older structures in Puntarenas, Alajuela, Cobano, and San José. Landslides blocked the main road between San José and the Atlantic.

<u>Mar 26</u> The island of Mindanao in the southern Philippines was struck by a shallow 5.5 magnitude earthquake that killed one person in Santiago, injured two others, and damaged several structures.

<u>Apr 2</u> Buildings were damaged in Manchester, Liverpool, Wrexham, Welshpool, Shrewsbury, and vicinity as a result of a 4.5 magnitude earthquake with a focal depth of 18 km (11 mi) that was felt throughout Wales, eastern Ireland, and England from Newcastle-upon-Tyne to Kent and Cornwall; no casualties were reported.

<u>Apr 3</u> A 6.4 magnitude earthquake with a focal depth of 53 km (33 mi) interrupted a meeting of Central American presidents at Montelimar, Nicaragua, causing slight damage. The quake was felt throughout much of Nicaragua and northern Costa Rica.

<u>Apr 5-6</u> A strong 7.5 magnitude earthquake with a focal depth of 32 km (20 mi) in the northern Mariana Islands with a 6.2 magnitude aftershock the next day generated tsunamis that traveled as much as 6000 km (3700 mi) from the epicenter, measuring 24 cm (9 ½ in.) at Kailaukona, Hawaii. The shock registered IV on the modified Mercalli scale on Guam, and was also felt on Saipan.

<u>Apr 17</u> At least 1300 homes collapsed and two persons were injured as a result of a 6.1 magnitude, 20-km- (12-mi-) deep, earthquake in south Xinjiang, China, felt at Kashi, Shufu, and Wugia.

<u>Apr 18</u> Extensive damage was caused in the Gorontalo area of Sulawesi's Minnahassa Peninsula in Indonesia by a 7.4 magnitude earthquake with a focal depth of 28 km (17 mi). At least twenty-five persons were injured by the quake, felt throughout the peninsula and in central Sulawesi.

<u>Apr 26</u> Two destructive earthquakes, registering 5.6 and 6.8 in magnitude,

struck 5 seconds apart in Qinghai Province in northwestern China. The shallow quakes left nearly 5000 homeless, killed at least 126, and injured many, causing considerable damage throughout the region.

May 5 A 5.4 magnitude earthquake with a focal depth of 26 km (16 mi) in southern Italy seriously injured one person and caused structural damage in Potenza.

May 8 No damage was caused by a 6.5 magnitude earthquake with a focal depth of 6 km (4 mi) in southern Panama, although the shock was felt in Panama City. Effects on the modified Mercalli scale rated a V in Santiago and a IV in David.

May 13 A 6.5 magnitude earthquake with a focal depth of 30 km (19 mi) caused some damage and disrupted electric power in the Dannevirke region on the North Island of New Zealand.

May 20-24 A series of earthquakes ranging from 6.6 to 7.2 in magnitude with focal depths around 10 km (6 mi) destroyed many buildings in Juba, Sudan, but caused little serious damage in the sparsely populated region. This was the largest known earthquake in Sudan's history.

May 29 At least 137 people were killed and more than 800 injured by a 6.4 magnitude shallow earthquake that struck the region that includes the towns of Rioja, Moyobamba, and Cajamarco in northern Peru. The epicenter was near Moyobamba.

May 30 At least thirteen people were killed and hundreds injured when a 6.5 magnitude earthquake with a focal depth of 91 km (56 mi) struck Romania about 100 miles north of Bucharest. Structural damage also occurred in Moldavia, USSR, where four of the thirteen people were killed.

Jun 7 A shallow earthquake in the Pacific about 180 km (112 mi) northwest of Papua New Guinea caused no damage despite a magnitude of 6.5.

Jun 9 An earthquake struck the same region in northern Peru that was hit on May 29. One additional person was killed as the shallow 5.5 magnitude earthquake caused heavy damage to structures already weakened by the May 29 earthquake in Moyobamba and Rioja.

Jun 14 A 7.1 magnitude shallow earthquake killed four and injured fifteen in the town of Culasi, Antique Province, Philippines, on Panay Island.

Jun 14 A shallow 6.8 magnitude earthquake near the border between the Soviet Union and China in Kazakhstan, USSR, killed one person in Kazakhstan and caused considerable damage in China's Xinjiang Province.

Jun 16 Houses in northwest Greece were damaged and one person was killed in Riza as a result of a 5.5 magnitude shallow earthquake.

Jun 20-21, Jul 6 A massive earthquake measuring 7.6 on the Richter scale struck northern Iran on Jun 20. With its epicenter at the edge of the Caspian Sea and a focal depth of 10 km (6 mi), the earthquake devastated about a hundred cities, towns, and villages in Gilan and Zanjan provinces, killing about 50,000, injuring another 100,000 to 200,000, and leaving half a million homeless. In terms of its effect on human life, it was the most devastating quake since the one on Mar 31 1983 in Papayán, Colombia, that killed 200,000. The devastation of the Iranian quake was compounded by giant aftershocks that by themselves would have been considered major earthquakes. Buildings that barely survived the main earthquake collapsed as the aftershocks rolled through day after day, sometimes killing rescue workers. A 6.0 magnitude aftershock on Jun 21 killed twenty people and caused landslides. More landslides occured on Jul 6 from a 5.1 aftershock, injuring two more.

Jul 6 At least 103 people were injured and more than 10,000 structures damaged by a 5.8 magnitude shallow earthquake that struck western Java.

Jul 9 A 6.5 magnitude earthquake with a focal depth of 10 km (6 mi) occurred in Sudan, but caused no damage.

Jul 13 Forty mountain climbers on Lenin Peak in the Panir Mountains of the USSR were killed when a landslide triggered by a 5.5 magnitude earthquake with a focal depth of 219 km (136 mi) raced through their 4850-m-(15,900- ft-) high camp.

Jul 16-17 Almost 1700 people were killed, almost 3500 injured, and more than 100,000 left homeless by a 7.7 magnitude earthquake with a depth of 36 km (22 mi) that struck Luzon in the Philippines in the region around Baguio and Cabanatuan on Jul 16. It was followed the next day by a shallow 6.5 aftershock that caused little additional damage.

Jul 27 An earthquake in Vanuatu injured two people and disrupted telephone services and water supplies in Espírtu Santo.

Aug 3 A 6.1 magnitude earthquake with a focal depth of 19 km (12 mi) in northwestern Xinjiang Province, China, near the border with the USSR, injured eight and damaged more than 500 structures.

Aug 5 A deep earthquake with a focal depth of 519 km (322 miles) occurred in southern Honshu, Japan. Although its magnitude was variously calculated at 5.9 and 6.5, it was felt only slightly.

Aug 11 A 5.0 magnitude earthquake in Ecuador with a focal depth of 5 km (3 mi) killed four and injured ten. Most damage was in the Pomasui area.

Aug 17 A shallow 5.9 magnitude earthquake in the Solomon Islands caused no damage.

Sep 2 In western Ecuador, a 6.1 magnitude earthquake with a focal depth of 25 km (16 mi) damaged buildings that were in poor condition.

<u>Sep 23</u> In the Pacific, about 100 km (60 mi) west of Japan's Bonin-Izu island chain, a shallow 6.3 magnitude earthquake produced a small tsunami.

<u>Sep 26</u> A 5.2 magnitude earthquake with a focal depth of 5 km (3 mi) killed two and injured five in a mine near Welcom, South Africa, 230 km (143 mi) southwest of Johannesburg.

<u>Oct 10</u> A deep 5.9 magnitude Pacific earthquake with a focal depth of 594 km (368 mi) caused no damage.

<u>Oct 15</u> A shallow 6.6 magnitude Indian Ocean earthquake about 500 km (310 mi) west of Sumatra caused no damage.

<u>Oct 17</u> A deep 6.7 magnitude earthquake in Brazil near the Peruvian border caused minor damage in Rio Branco and registered IV on the modified Mercalli scale in Pucallpa, Peru, and III in La Paz, Bolivia.

<u>Oct 20</u> One person was killed and two injured by a shallow 5.8 magnitude earthquake centered about 150 km (93 mi) northwest of Lanzhou in central China.

<u>Oct 25</u> A 6.0 magnitude earthquake centered in the Hindu Kush region near the Pakistan-Afghanistan border, about 150 km (93 mi) northeast of Kabul, Afghanistan, with a focal depth of 118 km (73 mi), killed eleven and injured more than a hundred in the Mardan-Malakand region of Pakistan, 250 km (155 mi) east of Kabul.

<u>Nov 6</u> At least 22 people were killed, 100 injured, and 21,000 left homeless by a 6.8 magnitude earthquake with a focal depth of 25 km (16 mi) in southern Iran's Darab area, 800 km (500 mi) south-southeast of Teheran.

<u>Nov 6</u> A 7.0 magnitude earthquake with a focal depth of 32 km (20 mi) in the region of the Kamandorsky islands of the USSR was felt on the western Aleutian islands of Attu and Shemya, but caused no damage.

<u>Nov 15</u> A 6.8 magnitude earthquake with a focal depth of 56 km (35 mi) in northern Sumatra, Indonesia, caused injuries and structural damage in the Blang Kejeran-Sutacane-Medan region.

<u>Nov 27</u> A 5.7 magnitude earthquake with a focal depth of 10 km (6 mi) injured ten and caused serious damage in west-central Yugoslavia.

<u>Dec 1</u> About 3000 people were left homeless and 940 buildings destroyed in the Kirghiz republic of the USSR by a 5.0 magnitude earthquake centered about 350 km (217 mi) east of Tashkent, near the China border.

<u>Dec 11</u> A 6.0 magnitude earthquake with a focal depth of 10 km (6 mi) in the Tonga Islands caused no damage.

<u>Dec 13</u> A shallow 5.1 magnitude earthquake in Sicily, Italy, killed at least

15 and injured about 200 in the Carlentini area, 20 km (12 mi) south of Catania.

Dec 16 A shallow 5.6 magnitude earthquake in Iran injured sixteen and caused landslides and damage in the Borazjan area, 40 km (25 mi) from the northeast coast of the Persian Gulf.

Dec 21 A 6.0 magnitude earthquake in Greece with a focal depth of 10 km (6 mi) killed one person and injured more than 60 in Greece and Yugoslavia.

Dec 22 A 5.6 magnitude earthquake southwest of San José, Costa Rica, with a focal depth of 9 km (6 mi) killed one and injured about 350.

Dec 30 A 6.7 magnitude earthquake with a focal depth of 187 km (116 mi) off the coast of New Britain Island in Papua, New Guinea, was felt in many parts of the country, but caused only minor damage.

EARTHQUAKES OF 1991

Jan 5 A strong 7.1 magnitude earthquake with a focal depth of 192 km (119 mi) was felt over most of Myanmar (Burma) as well as in parts of India and Thailand; although some landslides were reported, there were no reports of casualties.

Jan 31 Some 200 to 400 people were killed in Afghanistan and at least 300 in Pakistan and considerable damage and injuries were caused in the USSR, by a 6.8 magnitude earthquake with a focal depth of 62 km (38 mi) centered in the Hindu Kush mountains. The earthquake was felt as far away as Delhi, India, about 1000 km (620 mi) from the epicenter.

Feb 21 A shallow earthquake with a magnitude of 5.8 that struck the Bering Sea west of Alaska generated a tsunami with a maximum height of 25 cm (10 in); the tsunami struck Adak in the Aleutian Islands, causing no damage.

Mar 8 A 6.7 magnitude earthquake with a focal depth of 270 km (167 mi) in eastern Siberia, USSR, caused no reported damage.

Mar 25 One hundred thirty-one people were injured and more than 1300 houses destroyed by a 5.1 magnitude earthquake with a focal depth of 56 km (35 mi) in the Datong area of northeastern China.

Apr 4-5 About 110 people were killed and about 650 injured by two shallow earthquakes (6.4 on Apr 4 and 6.7 on Apr 5) in the Moyabamba-Rioja-Tarapoto region of northern Peru, most by the second shock.

Apr 6 No reported damage was caused by 6.7 magnitude earthquake with a 14 km (9 mi) deep focal depth in northern Tonga.

Apr 18 A 5.1 magnitude shallow earthquake on the border between Tadzhikstan, USSR, and Afghanistan killed one and injured eleven.

Apr 22 A 7.4 magnitude earthquake in Costa Rica with a focal depth of 10 km (6 mi) killed seventy-six (including twenty-nine in Panama), injured hundreds, produced landslides that blocked major highways, damaged structures and roads, and produced small tsunamis.

Apr 24 One person was killed and three injured by a 4.5 magnitude earthquake with a focal depth of 10 km (6 mi) in Erzurum Province, Turkey.

Apr 29 A 7.0 magnitude earthquake with a focal depth of 14 km (9 mi) in the western Caucasus mountains of Georgia, USSR, 90 km (59 mi) northeast of Kutaisi, killed 144, injured 430, and produced damaging landslides.

May 4 Thirty-six people were injured and 400 left homeless by a 6.1 magnitude earthquake with a focal depth of 15 km (9 mi) near the Panama-Costa Rica border.

May 19 A shallow 6.9 magnitude earthquake in Indonesia on Sulawesi's Minahassa Peninsula caused no damage.

May 24 Adobe walls in Tacna, Peru, on the border with Chile, were destroyed by a 6.8 magnitude earthquake with a focal depth of 125 km (75 mi), but no other damage was reported.

May 30 A 6.8 magnitude earthquake with a focal depth of 47 km (29 mi) in southern Alaska caused no reported damage.

Jun 15 A delayed 6.5 magnitude aftershock from the Apr 29 earthquake struck near Bakuriana, Georgia, USSR, killing eight and nearly destroying eight villages.

Jun 15 Two earthquakes struck Japan, one at each end of the archipelago; a 5.2 quake about 480 km (300 mi) northeast of Tokyo and a 5.7 quake on Amami-Oshima Island near Okinawa; neither caused any damage.

Jun 15 A 6.3 magnitude earthquake struck the South Sandwich Islands in the South Atlantic about 3000 km (1860 mi) southwest of Buenos Aires, causing no reported damage.

Jun 17 An early morning 3.9 or 4.4 magnitude earthquake centered 64 km (40 mi) west of Albany, NY, shook parts of New York, Massachusetts, and Connecticut, wakening some people but causing no damage.

Jun 20 A shallow 7.2 magnitude earthquake in Sulawesi, Indonesia, damaged 1500 houses.

Jun 28 A 5.9 magnitude earthquake originating under the San Gabriel Mountains struck Los Angeles and vicinity, killing two, but causing relatively little damage, in part because a 1987 quake in the same region destroyed many buildings prone to earthquake damage and encouraged people to reinforce other buildings.

Jul 4 A shallow 6.2 magnitude earthquake with an epicenter in the Banda Sea near the islands of Alor and East Timor, Indonesia, killed 23 and injured 181.

Jul 12 A 5.0 magnitude earthquake with a focal depth of 10 km (6 mi) along the Romanian-Yugoslavian border killed one and injured thirteen in Romania.

Jul 18 Another Romanian earthquake, this time of 5.6 magnitude with a focal depth of 13 km (8 mi) below southwest Romania, destroyed at least 615 houses.

Jul 23 A shallow 5.6 magnitude earthquake in southern Peru killed twenty and injured eighty in the Maca region.

Jul 24 A shallow 5.5 magnitude earthquake in Iraq near the border with Iran and Turkey killed at least twenty in Iraq and damaged a hundred or more houses in Iraq and Iran.

Aug 14 A shallow 6.6 magnitude earthquake near the Banks Islands in Vanuatu caused no damage.

Aug 17 Three earthquakes off the coast of northern California, roughly 100 km (60 mi) west of Crescent City, were capped by a shallow 7.1 magnitude quake which, however, caused no damage.

Sep 18 A 6.0 magnitude earthquake centered about 40 km (25 mi) southwest of Guatemala City, Guatemala, at a depth of about 6 km (4 mi) killed at least twenty-five people; about 200 more were injured as houses collapsed on them.

Sep 28 A shallow 6.6 magnitude earthquake about 150 km (90 mi) southwest of Rabaul, New Britain, New Guinea, caused no casualties or damage.

Oct 14 A shallow 7.1 magnitude earthquake off the north coast of Santa Isabel Island in the Solomons caused no reported damage.

Oct 20 A 7.1 magnitude earthquake centered in the Uttar Kashi district of India, near the Nepal border, was felt over a wide region of India, Nepal, and Tibet, killing at least 2000 people and injuring nearly as many

Nov 1 A 6.5 magnitude earthquake struck the Kermadet Islands.

Nov 19 A 7.0 magnitude earthquake killed two in western Colombia.

Nov 22 A 4.6 magnitude earthquake killed ten in Yemen.

Nov 28 A 5.0 magnitude earthquake killed one or more in western Iran.

Dec 22 A 7.4 magnitude earthquake strikes Kuril Islands, AK.

MEASURING EARTHQUAKES

The size of an earthquake is generally reported in the United States using the Richter scale, a system developed by geologist Charles Richter [US: 1900-1985] in 1935. It is based on measuring the waves on a seismograph, which directly reflect the disturbance of ground motion. Each whole number on the scale, called a magnitude, represents a wave that is ten times as large. Thus, a magnitude 6 earthquake produces a ground wave ten times as large as a magnitude 5, so long as both earthquakes are the same distance away and at the same depth.

This does not mean that a magnitude 6 earthquake has ten times the energy as one of magnitude 5. Measuring the actual energy requires instruments placed at the site of the earthquake. Various methods have been developed for obtaining the energy from the magnitude. These show that a difference of one magnitude corresponds to a change of thirty to sixty times in the energy of the earthquake. The energy of a magnitude 8 earthquake, a very serious event, can be as much as a million to 10 million times as much as that of a magnitude 4 earthquake, one that can be felt but that causes almost no damage.

A very different scale, developed by Giuseppe Mercalli [Italian: 1850-1914] in 1902 and modified by Harry Wood and Frank Neumann in the 1930s, is used for some purposes in the United States and on a regular basis in certain other countries. This scale, the Modified Mercalli scale, is based on the effects of the earthquake at a point above the earthquake. Comparing the Modified Mercalli scale with the Richter scale helps one to understand the relative energy of earthquakes.

Richter Scale		*Modified Mercalli Earthquake Intensity Scale*	
2.5	Generally not felt, but recorded on seismometers	I	Not felt except by a very few under specially favorable circumstances
		II	Felt by only a few persons at rest, especially on upper floors of buildings
3.5	Felt by many people	III	Felt noticeably indoors, especially on upper floors of buildings, but not necessarily recognized as an earthquake
		IV	Felt during the day indoors by many, outdoors by few; sensation is like a heavy truck striking a building
		V	Felt by nearly everyone; many awakened; disturbances of trees, poles, and other tall objects sometimes noticed

(Continued)

Richter Scale	Modified Mercalli Earthquake Intensity Scale
4.5 Some local damage may occur	VI Felt by all; many frightened people run outdoors; some heavy furniture moved; few instances of fallen plaster or damaged chimneys; damage slight
	VII Most everybody runs outdoors; damage negligible in buildings of good design and construction; slight to moderate in well-built ordinary structures; considerable in poorly built or badly designed structures
6.0 A destructive earthquake	VIII Damage slight in specially designed structures; considerable in ordinary substantial buildings with partial collapse; great in poorly built structures; chimneys, factory stacks, columns, monuments, and walls fall
	IX Damage considerable in specially designed structures; buildings shifted off foundations; ground cracks conspicuously
7.0 A major earthquake; about ten occur each year	X Some well-built wooden structures destroyed; most masonry and frame structures with foundations destroyed; ground badly cracked
8.0 Great earthquake; these occur and once every five to ten years above	XI Few, if any, masonry structures remain standing; bridges destroyed; broad fissures in ground
	XII Damage total; waves seen on ground surfaces; objects thrown upward into air

ACTIVE VOLCANOES AND MAJOR VOLCANIC EVENTS

A volcano is an opening in Earth's crust that emits melted rock (lava), hot gases, and rocks of various sizes; it is also the mountain that forms as solidified lava and ejected rocks pile up around an opening—called a vent if it resembles a crack or a crater if it is fairly circular.

Volcanoes do not occur at random across the world. Almost all are found at plate boundaries, such as the famous Ring of Fire around the Pacific Ocean. A few, such as the volcanoes of Hawaii and the volcanic region (without a volcano) of Yellowstone Park, appear to be over a "hot spot," where liquid rock flows with sufficient force to burn through crust.

SOME VOCABULARY TERMS

active Describes volcano that has erupted in historic times. Thus Tambora, which has not erupted since 1815, is considered active.

ash Ejected rocks larger than dust but smaller than about half a centimeter (a quarter inch).

bombs Ejected rocks larger than several centimeters (inches) across. Some may be as heavy as 100 tons.

caldera a large depression formed by the collapse of a volcano when its magma reservoir flows back into the crust. A caldera may have several vents or craters on its floor.

cinders Ejected rocks larger than ash, but smaller than bombs; also called scoria.

dormant Not active, but showing signs, such as hot springs or mild earthquakes, that it may become active.

extinct Neither active nor dormant, but showing signs that is actually a volcano, not some other feature. It is usually not clear whether a volcano is dormant or extinct. Some volcanoes thought to be extinct become active.

hot spot A place where magma continually flows upward, sometimes causing a volcano by burning a hole in the crust. Yellowstone Park, Iceland, and Hawaii are currently over hot spots. Since tectonic plates move over the hot spot, it appears that the hot spot moved in the past, although the spot stayed still while the plates moved.

lahar A landslide or mudflow of volcanic debris, often caused when the dam to a lake is opened or overflows or when a glacier or snowcap melts.

lava Molten rock that has flowed from a volcano's vent or crater. Because it loses gases to the air and quickly changes composition as it nears the surface, lava is not the same substance as molten rock that remains deep below the surface.

magma Molten rock that remains beneath Earth's surface, where it powers all of the phenomena associated with a volcano.

nuée ardente A mixture of hot dust or ash and very hot gases that is ejected from a volcano and flows downhill at great speed.

phreatic explosions Violent steam blasts.

pyroclastic flow Same as a nuée ardente.

About 600 volcanoes are now active. The list that starts on p 292 contains more than a third of all the known active volcanoes, with emphasis on volcanoes that have been active in recent years and volcanoes that have had famous eruptions.

Volcano	Location	Height in m (ft) above sea level	Last reported eruption	Remarks
Africa and the Indian Ocean				
Cameroon Mt	Cameroon	4070 (13,354)	1982	
Erta-Ale	Ethiopia	503 (1650)	1973	
Fournaise, Piton de la	Réunion Island	2631 (8631)	1991	Brief eruption on Jul 19-20
Karthala	Grande Comore, Comoros Islands	2438 (8000)	1991	Jul 11
Kimanura	Zaire		1989	New cone in Nyamuragira-Nyirangongo complex
Lengai Ol Doinyo	Tanzania	2886 (9469)	1991	
Nyamuragira	Zaire	3053 (10,016)	1991	Starts Sep 20 and is a continuation of active period since 1976 that has seen 7 eruptions; previous eruption, Apr 24 1989, produces lava flow 8 km (5 mi) long
Nyirangongo	Zaire	3465 (10,056)	1977	Part of Virunga chain with Nyamuragira
Antarctica				
Big Ben	Heard Island	2745 (9007)	1986	
Deception Island	South Shetland Islands	576 (1890)	1970	
Mt Erebus	Ross Island	3795 (12,450)	1990	
Asia				
Agung	Bali, Indonesiah	3142 (10,308)	1964	May 17-21 1963 eruption kills 1584
Akan Caldera	Hokkaido, Japan	1503(4931)	1988	
Akita Komaga-take	Japan	1661(5449)	1970	
Alaid	Kuril Islands, Russia	2339 (7674)	1972	
Anak Krakatau	Indonesia	100 (330)	1972	New volcano at site of Krakatau (see below)

Volcano	Location	Height in m (ft) above sea level	Last reported eruption	Remarks
Anak Ranakah	Flores Island, Indonesia		1989	Dec 28 1987 eruption begins in area of no known volcanic activity
Amburombu	Indonesia	2149 (7051)	1969	
Apu Siau (Karangetang)	Sangihe Islands, Indonesia	1784 (5853)	1989	Jul; forty reported eruptions in past 300 years
Asama	Honshu, Japan	2530 (8300)	1983	Apr 8 eruption produces ash fall as far away as 200 km (120 mi)
Aso	Kyushu, Japan	1592 (5223)	1991	3 large vents open in 1990, ejecting ash, cinders, and bombs
Avachinsky	Kamchatka, Russia	2751 (9026)	1991	Jan 13 eruption not as large as Feb 1945 eruption or huge explosion about 3500 BC
Awu	Indonesia	1326 (4350)	1968	
Azuma	Honshu, Japan	2024 (6640)	1978	
Banda Api	Banda Sea, Indonesia	685 (2247)	1989	May 9-15 1988 eruption forces evacuation of 7000 people
Bandai	Japan	1891 (6204)	1888	Jul 15 eruption causes mudslides that kill 500
Barren Island	Andaman Islands, India		1991	Eruption that starts in later Mar or early Apr is first since 1803; continuing as of Oct 24, but appeared to moderate in Nov
Batur	Bali, Indonesia	1717 (5633)	1974	Frequent eruptions since first historic eruption in 1804
Bezymianny	Russia	2800 (9186)	1986	May 20 1956 eruption is most energetic of 20th century

(Continued)

Volcano	Location	Height in m (ft) above sea level	Last reported eruption	Remarks
Bulusan	Philippines	1560 (5115)	1988	
Canlaon	Negros Islands, Philippines		1989	Oct 25
Catarman (Hibokihibok)	Mindanao, Philippines	1330 (4364)	1951	Dec 4 eruption kills 500
Chokai	Honshu, Japan	2225 (7300)	1974	
Dukono	Halmahera, Indonesia	1087 (3566)	1991	Jun 8
Ebeko	Kuril Islands,Russia	1138 (3734)	1989	Starts Feb 3
Fukutoku-okanoba	Volcano Islands, Japan		1986	Jan eruption builds a temporary island
Galunggug	Java, Indonesia	2168 (7113)	1982	Eruptions Oct 8 and 12 1822 cause mudslides that kill 4000
Gamalama	Ternate Island, Indonesia	1715 (5625)	1990	No serious damage from Apr 25 eruption, despite 4.5 km (2.7 mi) pyroclastic flow
Gamkonora	Indonesia	1635 (5365)	1981	
Gede	Java, Indonesia		1957	Mar
Gerde	Indonesia	2958 (9705)	1949	
Ijen	Java, Indonesia	2386 (7828)	1952	1952 eruption is 3rd in 20th century
Iliboleng	Lesser Sunda Islands, Indonesia	1659 (5443)	1988	Oct 2; since 1885, has had at least 20 moderate eruptions
Ivan Grozny	Kuril Islands, Russia	1158 (3799)	1990	
Karymsky	Kamchatka, Russia	1484 (4869)	1982	
Kelimutu	Indonesia	1640 (5381)	1968	Jun 3 eruption features 100-m (300-ft) water fountain
Kelut	Java, Indonesia	1731 (5679)	1990	May 1919 activity causes crater lake to spill, killing 5000+; Feb 10 1990 eruption kills 32; 4 more killed by lahars in Nov
Kerinci	Sumatra, Indonesia	3800 (12,467)	1987	
Kirishima	Japan	1700 (5577)	1982	

Volcano	Location	Height in m (ft) above sea level	Last reported eruption	Remarks
Kliuchevskoi	Kamchatka, Russia	4850 (15,912)	1991	Large summit eruption Jan 29 1990 is preceded by 78+ eruptions since 1697; small eruption Apr 8 1991
Komaga-take	Hokkaido, Japan	1140 (3740)	1942	Jun 17 1929 eruption kills 2 and damages houses
Koraykskaya	Russia	3456 (11,339)	1957	
Krakatau	Indonesia	813 (2667)	1883	Aug 26-27 eruption blows up island, producing giant waves that kill 36,000; 1991 study by Haraldur Sigurdsson of University of Rhode Island and coworkers shows that waves, 30-40 m (100-130 ft) high, were caused by pyroclastic flows into the ocean
Kusatsu-Shirane	Honshu, Japan	2176 (7139)	1989	Jan 6 eruption is small
Lewotobi Lakilaki	Flores, Indonesia	1590 (5217)	1991	Eruptions start May 11 and continue into Jun; last previous eruption Jan 1971
Lokon-Empung	Sulawesi, Indonesia	1581 (5187)	1991	Main eruption starts May 17-18 and continues at least until Nov 20; volcanologist Vivianne Clavel reported missing on Oct 24 and not found by end of Nov; 11,000 villagers evacuated; last previous eruption 1988
Marapi	Sumatra, Indonesia	3045 (9485)	1981	Apr 30, 1979, landslide kills 82+

(Continued)

Volcano	Location	Height in m (ft) above sea level	Last reported eruption	Remarks
Mayon	Philippines	2462 (8077)	1978	
Me-akan	Japan	3001 (9846)	1966	
Merapi	Java, Indonesia	2911 (9550)	1988	Not to be confused with Marapi on Sumatra
Nasu	Japan	1893 6210	1977	
Niigata-yake-yama	Honshu, Japan	2458 (8064)	1989	
On-Take	Kyushu, Japan	3063 (10,049)	1979	Volcanic earthquake on Sep 14 1984 starts landslide that kills 29
Oshima	Izu Island, Japan	758(2487)	1990	Oct 4 eruption first since Jan 1988
Papandayan	Java, Indonesia	2665 (8743)	1772	Explodes on Aug 11-12, killing 3000+
Pinatubo	Luzon, Philippines	1600 (5250)	1991	Explosion Apr 2 signals renewal of activity; major eruption starts Jun 9 and peak lasts more than 15 hours Jun 15-16, with tephra injected 30 km (18.6 mi) into atmosphere; second largest volcanic eruption of century after 1902 eruption of Santa Maria in Guatemala; 745 killed or missing as a result of 1991 eruption, of these, 358 from disease in evacuation camps; previous major eruption c. 1380
Raung	Java, Indonesia	3332 (10,932)	1991	Sep-Oct eruption produces vigorous ash emissions and plumes; 6 eruptions from 1586 through 1817 result in small numbers of fatalities; more than fifty eruptions in nineteenth century

Volcano	Location	Height in m (ft) above sea level	Last reported eruption	Remarks
Rindjani	Indonesia	3726 (12,224)	1981	
Sakura-jima	Kyushu, Japan	1118 (3668)	1991	Eruption begins in 1955; falling cinders damage many car windshields; cone explodes 37 times in Mar 1991
Sangeang Api	Indonesia	1936 (6351)	1988	
Sarychev	Kuril Islands, Russia	1559 (5115)	1986	
Semeru	Java, Indonesia	3676 (12,060)	1991	More-or-less continuous eruption since 1967; lahars kill 252 in May 1981
Shiveluch	Kamchatka, Russia	3395 (11,138)	1991	Crater forms in 1964; lava dome extrudes 1980-81; brief explosive eruption on Apr 8 1991
Siau	Indonesia	1784 (5853)	1976	
Sinila	Indonesia	2134 (7000)	1979	Feb 21 eruption kills 175
Slamet	Java, Indonesia	3428 (11,247)	1919	Weak eruption in Jul 1989 preceded by tephra eruption to 500 m (1640 ft) in Jul 1989
Soputan	Sulawesi, Indonesia	1827 (5994)	1989	Apr 22-24 eruption damages 500 houses
Sorikmarapi	Sumatra, Indonesia	2145 (7037)	1986	Jul eruption emits ash cloud to 700 m (2300 ft) above summit
Suwanose-jima	Japan	799 (2621)	1990	Frequent explosive eruptions since 1957
Taal	Luzon, Philippines	400 (1312)	1970	1911 eruption kills 1300; has erupted 24 times since 1572; Sep 28 1956 eruption kills 200 and 1965 eruption kills at least 150

(Continued)

Volcano	Location	Height in m (ft) above sea level	Last reported eruption	Remarks
Tambora	Sumbawa, Indonesia	2851 (9354)	1815	Eruptions starting Apr 5 cause immediate death of 10,000; additional 80,000 die worldwide as a result of climatic changes
Tangkuban Prahu	Java, Indonesia	2023 (6637)	1967	
Tengger Caldera	Java, Indonesia	2329 (7641)	1980	Jun eruption emits ash to 900 m (3000 ft)
Tiatia	Kuril Islands, Russia	1833 (6013)	1973	
Tjarme	Indonesia	3078 (10,098)	1938	
Tokachi-dake	Hokkaido, Japan	2077 (6814)	1989	Eruption begins Dec 16 1988 after 26 years of dormancy; 1926 eruption kills 144
Unzen	Kyushu, Japan	1360 (4462)	1991	Last previous eruption, in 1792, causes harbor wave that kills 15,000; Jun 3 1991 eruption is largest in Japan in half a century, killing 42, including 3 volcanologists; 12,000 evacuated
Usu	Japan	728 (2390)	1978	
Yake Dake	Kyushu, Japan	2458 (8064)	1963	
Zhupanovsky	Kamchatka, Russia	2958 (9490)	1959	Preceded by 6 explosive eruptions since 1776
Central America and the Caribbean				
Acatenango	Guatemala	3560 (12,992)	1972	
Arenal	Costa Rica	1552 (5092)	1991	Current eruption begins in 1968
Concepción (Ometepe)	Nicaragua	1556 (5106)	1986	
Conchagua	El Salvador	1249 (4100)	1947	
El Viejo (San Cristóbal)	Nicaragua	1780 (5840)	1987	

Volcano	Location	Height in m (ft) above sea level	Last reported eruption	Remarks
Fuego	Guatemala	3835 (12,582)	1987	Jan eruption produces glowing ash
Ilopango	El Salvador	450 (1476)	260	Destroys early Maya civilization
Irazú}	Costa Rica	3432 (11,260)	1965	Several dozen people killed by lahars in 1963-65 eruptions
Izalco	El Salvador	2362 (7749)	1966	
Kick-'em-Jenny	Subocean, off Grenada	-49 (-160)	1990	Mar 26 eruption not as definite as preceding eruption, Dec 29-30 1988
Masaya	Nicaragua	635 (2083)	1989	Probably Jun 2
Momotombo	Nicaragua	1280 (4199)	1905	
Pacaya	Guatemala	2544 (8346)	1991	4 large lava flows form in 1991 eruption, but lava flows stop by Apr 10; new phase includes 16 eruptions from Jun 6 to Sep 20
Pelée	Martinique	1372 (4500)	1930	May 8 1902 eruption wipes out St. Pierre, killing 29,000
Póas	Costa Rica	2722 (8930)	1991	Acid rain from eruption damages vegetation and causes human health problems
Rincon de la Vieja	Costa Rica	1900 (6234)	1991	Eruption on May 8 is small; at least 16 eruptions noted since first historic activity in 1851
San Miguel	El Salvador	2132 (6994)	1986	
San Salvador	El Salvador	1886 (6187)	1923	
Santiaguito Dome (Santa Maria)	Guatemala	3768 (12,362)	1991	Continued strong explosions since 1922 eruption of Santa Maria;1902 eruption is the largest of century; four hikers killed Jul 19 1990

(Continued)

Volcano	Location	Height in m (ft) above sea level	Last reported eruption	Remarks
Soufrière	St Vincent and the Grenadines	1234 (4048)	1979	Eruption May 7 1902 kills 1500 to 2000
Tacaná	Guatemala	3780 (12,400)	1988	
Telica	Nicaragua	929 (3409)	1987	Dec
Turrialba	Costa Rica	3335 (10,941)	1866	
Europe and the Atlantic Ocean				
Askja	Iceland	1400 (4594)	1961	
Beerenberg	Jan Mayen Island, Norway	2277 (7470)	1985	
Eldfell	Iceland	99 (327)	1973	
Etna	Italy	3290 (10,794)	1991	Dec 14; eruption in 1683 kills more than 60,000
Fogo Islands	Cape Verde	2835 (9300)	1951	
Heimay	Iceland		1972	Fissure eruption causes city to be evacuated
Helka	Iceland	1491 (4892)	1991	Jan 17 eruption lasts until Mar 11
Ischia	Italy	789 (2589)	1883	Earthquakes from Jul 28 eruption kill 2000+
Krafla	Iceland	654 (2145)	1984	Sep
Leirhnukur	Iceland	654 (2145)	1975	
Skaptar (Lakagigar)	Iceland	500 (1640)	1783	Laki fissure emits gases that kill crops and livestock and interfere with fishing, killing 9800
Stromboli	Italy	926 (3038)	1991	In almost continuous eruption for over 2000 years
Surtsey	Iceland	173 (568)	1967	Creates a new island during eruptions starting in 1963
Thera	Mediterranean near Crete	131 (430)	1628 or 1645 BC	Explosion destroys island and maybe affects Minoan civilization on Crete

Volcano	Location	Height in m (ft) above sea level	Last reported eruption	Remarks
Tristan de Cunha	St Helena	2060 (6760)	1961	
Vesuvius	Italy	1281 (4203)	1944	Eruption in 79 CE destroys Pompeii and Herculaneum, killing 2000+
Vulcano	Aeolian Islands, Italy	500 (1640)	1890	Frequently active in 19th century; breaks submarine cable in 1892

North America

Akutan	AK, US	1300 (4265)	1991	Sep 15-Oct 18
Amukta	AK, US	1064 (3490)	1963	
Aniakchak	AK, US	1356 (4450)	1931	
Augustine	AK, US	1218 (3995)	1986	Mar 27-31 eruption interrupts international air traffic
Bogoslof	AK, US	45 (150)	1931	
Carlisle	AK, US	1620 (5315)	1838	
Cerberus	AK, US	780 (2560)	1873	
Chiginagak	AK, US	2434 (7985)	1929	
Cinder Cone	CA, US	2105 (6907)	1851	
Cleveland	AK, US	1740 (5710)	1987	
Colima	Mexico	4268 (14,003)	1991	Lava extrusion starts Mar 1 and increases on Apr 16-17, along with collapse of dome, which produces dangerous avalanches; lava still advancing in Aug
El Chichón	Mexico	2225 (7300)	1983	Mar 28 and Apr 3-4 1982 eruptions kill 2000 and send cloud of ash around globe
Fisher	AK, US	1081 (3545)	1826	
Gareloi	AK, US	1637 (5370)	1982	
Great Sitkin	AK, US	1760 (5775)	1974	
Iliamna	AK, US	3091 (10,140)	1978	
Isanotski	AK, US	2495 (8185)	1845	

(Continued)

Volcano	Location	Height in m (ft) above sea level	Last reported eruption	Remarks
Kagamil	AK, US	898 (2945)	1929	
Kanaga	AK, US	1396 (4450)	1933	
Katmai	AK, US	2298 (7540)	1974	1912 eruption one of century's largest
Keniuji	AK, US	270 (885)	1828	
Kiska	AK, US	1227 (4025)	1990	4th small eruption since 1st eruption, seen in 1962
Korovin	AK, US	1489 (4885)	1987	
Lassen Peak	CA, US	3186 (10,453)	1914-21	Signs of restless-ness noted in 1982
Little Sitkin	AK, US	1202 (3945)	1828	
Mageik	AK, US	2224 (7295)	1946	
Makushin	AK, US	2048 (6720)	1987	
Martin	AK, US	1844 (6050)	1960	
Mt Baker	WA, US	3427 (10,778)	1870	Begins steaming in 1975, but activity soon ceases
Mt Hood	OR, US	4392 (11,245)	1801	Shaken by earth-quakes in 1980
Mt Rainier	WA, US	4316 (14,160)	1882	Eruption could loose floods on Tacoma and Seattle
Mt St Helens	WA, US	3285 (9671)	1991	May 18 1980 erup-tion kills 61; on and off activity continues
Mt Shasta	CA, US	4316 (14,160)	1855	Earthquake swarms recorded in 1981 and 1982
Novarupta	AK, US	841 (2759)	1912	Eruption of Katmai, which gives birth to Novarupta, one of largest in century
Okmok	AK, US	1079 (3540)	1988	
Parícutin	Mexico	457 (1500)	1952	Volcano grows in cornfield, starting in 1943; lava over-runs 2 villages
Pavlof	AK, US	2731 (8960)	1988	
Pavlof Sister	AK, US	2149 (7050)	1786	
Peulik	AK, US	1533 (5030)	1852	

Volcano	Location	Height in m (ft) above sea level	Last reported eruption	Remarks
Pogromni	AK, US	2300 (7545)	1964	
Popocatépetl	Mexico	5452 (17,887)	1943	40 million people live in valleys on each side of volcano
Redoubt	AK, US	3129 (10,265)	1990	Eruption starts Dec 14 1989 but largely subsides in late spring 1990
Sarichef	AK, US	614 (2015)	1812	
Seguam	AK, US	1056 (3465)	1977	
Shishaldin	AK, US	2874 (9430)	1987	
Spurr	AK, US	3374 (11,070)	1953	
Tanaga	AK, US	2138 (7015)	1914	
Tobert	AK, US	3478 (11,413)	1953	
Trident	AK, US	2082 (6830)	1974	
Veniaminof	AK, US	2576 (8450)	1987	
Vsevidof	AK, US	2123 (6965)	1880	
Westdahl	AK, US	1541 (5055)	1991	Ash plume more than 6 km (4 mi) long spotted on Nov 29
Yunaska	AK, US	603 (1980)	1937	

Oceania—Australia, New Zealand, and the Pacific Islands

Volcano	Location	Height in m (ft) above sea level	Last reported eruption	Remarks
Agrigan	Mariana Islands	965 (3166)	1990	Residents evacuate in Aug, but volcanic activity remains low
Ambrym	Ambrym Island, Vanuatu	1334 (4376)	1990	Almost continuously active since 1774 discovery by James Cook
Anatahan	Mariana Islands	790 (2592)		
Aoba	Aoba Island, Vanuatu	2496 (8189)	1670(?)	Evidence of a return to activity in 1991
Bagana	Bougainville Island	1702 (5584)	1990	
Gaua	Santa Maria Island, Vanuatu	797 (2615)	1977	
Haleakala	HI, US	3056 (10,025)	1790	
Hualalai	HI, US	2515 (8251)	1801	
Karkar	Papua New Guinea	1500 (4920)	1979	Summit denuded of vegetation by 1979 eruption

(Continued)

Volcano	Location	Height in m (ft) above sea level	Last reported eruption	Remarks
Kavachi	Solomon Islands	25 (82)	1991	Temporary island built by eruption first observed May 4; 8 previous eruptions have formed such islands since 1950
Kilauea	HI, US	1242 (4077)	1991	Continuous eruption begins Jan 2 1983 and localizes in the Pu'u `O'o vent in Jun, fountaining, destroying houses, and cutting off roads with lava
Lamington	New Guinea	1780 (5840)	1951	Jan 15 eruption kills 3000 to 5000
Langila	New Britain, Papua New Guinea	1189 (3901)	1991	Small explosions and light ash falls observed almost daily since May 16, with signs of slackening in Oct
Lopevi	Lopevi Islands, Vanuatu	1413 (4636)	1982	
Macdonald Seamount	Central Pacific		1989	
Manam	Off coast of Papua New Guinea	1725 (5659)	1991	Continuous eruption since 1974
Mauna Loa	HI, US	4170 (13,678)	1984	Lava flows present danger to city of Hilo
Motmot	Long Island, New Guinea	1304 (4278)	1974	
Ngauruhoe	North Island, New Zealand	2091 (7515)	1975	
Pagan	Mariana Islands	570 (1870)	1988	Islanders evacuated during May 1981 eruption still unable to return
Rabaul Caldera	New Britain, Papua New Guinea	229 (751)	1990	6 people suffocate from carbon dioxide on Jun 24-25

Volcano	Location	Height in m (ft) above sea level	Last reported eruption	Remarks
Ruapehu	North Island, New Zealand	2796 (9173)	1990	1945 eruption creates dam, which gives way Dec 25 1953, killing 150
Rumble III	Off Kermadeo Island, New Zealand	-200 (-660)	1986	
Soretimeat	Vanua Lava Island, Vanuatu	931 (3054)	1966	
Tarawera	North Island, New Zealand	1111 (3645)	1886	
Ulawun	New Britain, Papua New Guinea	2300 (7546)	1990	Low level of activity since Jan 1990
White Island	New Zealand	321 (1053)	1991	Sep 16-18 eruption produces 4-5-km (3 mi-) column of ash; new crater named Wade forms in Oct; eruption started Dec 1976
Witori	New Britain, Papua New Guinea	724 (2375)	1933	
Yasur	Tanna Island, Vanuatu	350 (1148)	1991	Continuous activity since observed by James Cook in 1774
South America				
Alcedo	Galapagos Islands, Ecuador	1097 (3599)	1960?	From photographs of a lava flow, an eruption occurs sometime between May 30 1946 and Oct 30 1960
Cotacachi	Ecuador	4939 (16,204)	1955	
Cotopaxi	Ecuador	5897 (19,347)	1975	
Cumbal	Colombia		1926	
Fernandina	Galapagos Islands, Ecuador	1495 (4905)	1991	Eruption that starts on Apr 19 1st since Sep 1988
Galeras	Colombia	4482 (14,705)	1991	Eruption starts Feb 1989 and grows more intense throughout 1990 and 1991; more than 10 eruptions in past 100 years

(Continued)

Volcano	Location	Height in m (ft) above sea level	Last reported eruption	Remarks
Guagua Pichincha	Ecuador	4784 (15,696)	1990	Apr 16 eruption most recent in decade-long series of occasional phreatic explosions
Guallatiri	Chile	6060 (19,882)	1960	Produces strong plumes in 1985 and 1987
Hudson	Chile	2615 (8580)	1991	Aug 8 eruption melts glacier; larger eruption on Aug 12 produces plume of gas and ash that travels completely around world in 7 days; ash blankets about 65,000 sq km (25,000 sq mi) of Argentina; main eruption ends Aug 15, but there is another eruption in Oct; most victims are sheep and cattle; previous eruption in 1971 leads to 11 deaths in Jan 1973 when ice melt suddenly flows down valley
Lascar	Chile	5600 (18,372)	1991	Oct 21; Lascar has had more than 15 historic eruptions; this eruptive phase started Sep 1986
Llaima	Chile	3121 (10,239)	1984	Apr 20 eruption, 30th since 1640, produces dense ash column
Lonquimay	Chile	2865 (9400)	1990	13 months of eruptive activity ends between Jan 22 and 25
Marchena	Galápagos Islands, Ecuador		1991	Sep 25; uninhabited 12-by-16 km (7.5-by-10 mi) low island

Volcano	Location	Height in m (ft) above sea level	Last reported eruption	Remarks
Planchón-Peteroa	Chile	3977 (13,048) Pln 4101 (13,455) Pet	1991	Feb 9 eruption kills large numbers of fish in Teno and Claro rivers
Puracé	Colombia	4757 (15,604)	1977	Hot springs bring many visitors
Reventador	Ecuador	3485 (11,434)	1976	Jan eruption produces 300 m- (9800-ft) ash column and 2.5+- km (4+-mi) lava flows
Ruiz	Colombia	5401 (17,720)	1991	Nov 13 1985 eruption causes mudslides that kill 25,000
Sabancaya	Peru	5976 (19,606)	1991	Eruption starts May 28 1990 and kills more than 20 when earthquakes cause houses to collapse on Jul 23-24 1991; church records show last previous eruption in Jul 1784
Sangay	Ecuador	5230 (17,159)	1976	
San José	Chile	5830 (19,127)	1895	Erupts 4 times in 19th century
Shoshuenco	Chile	2360 (7743)	1960	
Tupungatito	Chile	5640 (18,504)	1986	
Villarrica	Chile	2840 (9318)	1985	Oct 1984 to Nov 1985; has erupted more than 50 times since 1558, with four eruptions resulting in fatalities from mudflows

(Periodical Reference and Additional Reading: *Bulletin of the Global Volcanism Network* 1-90 to 10-91)

P L A T E   T E C T O N I C S

The Ring of Fire, and much else about Earth, can be explained by a theory known as plate tectonics (the word *tectonics* in this context refers to anything connected to movements of Earth's crust). This theory has its basis in two earlier ideas, continental drift and seafloor spreading.

Scientists often noticed that the shorelines of Africa and South America look as if the two continents had once been joined. Gradually, other evidence emerged to support this idea as well as the idea that other continents had also once been one. Similar rocks and fossils were found where the continents appeared to have met. Despite this evidence for joined continents, most geologists before 1960 rejected this theory because no one could propose a mechanism that would allow continents to "drift" through solid rock.

A series of discoveries in the 1950s and 1960s changed most minds. Magnetic patterns in the seafloor and volcanic activity at mid-ocean ridges showed that the floor of the Atlantic Ocean was spreading. Since Earth is not growing larger, the new ocean floor must be plunging back through the crust at other sites. Studies of earthquake waves have established that the crust is disappearing in deep trenches in the ocean. Mapping the ridges where new floor is appearing and the trenches where old floor is plunging showed that Earth's crust is made up of a number of plates that move with respect to each other. Recently, the slow movement— 5 to 12 cm (1 to 5 in.) a year—has been established by measurements from satellites.

These observations of plate movement explain many features of Earth's crust. Most high mountain ranges, such as the Himalayas and Andes, are caused by one plate plunging under another. The deep ocean trenches are the result of plates plunging back below the crust. Movement of one plate with respect to another is a common cause of earthquakes (the San Andreas Fault in California, Mexico, and the Pacific is the boundary between two plates), although not the cause of all quakes. Plate movements cause new seas to form, such as the Red Sea and the Gulf of California. The Ring of Fire, with its volcanoes and earthquakes, is a direct result of plates in the Pacific Ocean that are plunging under the Americas, Asia, New Zealand, and Antarctica.

Unexpected results also emerged to confirm plate tectonic theory. Where plates are moving apart, there are vents of superheated water rising under the ocean, a result discovered only after the theory had been developed.

Early in 1991 Ian Dalziel from the University of Texas at Austin and Eldridge Moores of the University of California at Davis presented evidence that showed that an ancient rock formation called the Grenville Belt, long known from the east coast of North America, was once continued in Antarctica. When combined with data from the magnetic orientation of continental rocks of 570 million years ago (mya), this indicates that at that time Antarctica was joined to North America. The configuration indicated is shown on the map below. Note that the future site of Las Vegas was at that time just a thousand miles from what would become the south pole.

Shortly after 570 mya, the continents broke up. As the gap between Antarctica, Australia, and India on the one hand and North America on the other became larger, the Pacific Ocean was formed. Today Australia and India are still moving farther from Antarctica, while North America is heading northwest, into the Pacific.

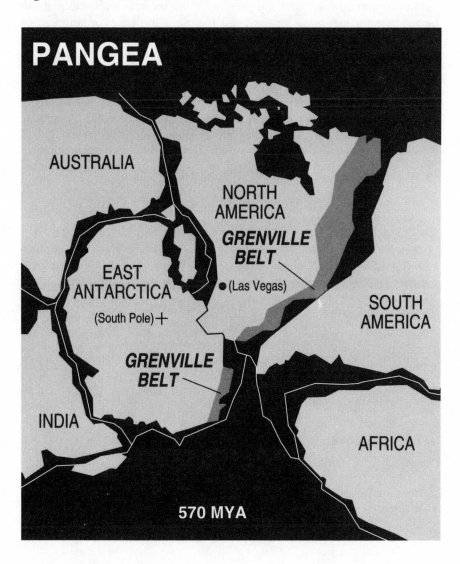

PANGEA

AUSTRALIA

NORTH
AMERICA

*GRENVILLE
BELT*

EAST
ANTARCTICA

• (Las Vegas)

(South Pole) +

SOUTH
AMERICA

*GRENVILLE
BELT*

INDIA

AFRICA

570 MYA

C O M P O S I T I O N   O F   E A R T H

Earth is built of layers. From the inside out:

Core A mixture of iron and nickel in two layers, a liquid outer core and a solid inner core. The inner core is more than 5000 km (3100 mi) below the surface and the outer core is about 2900 km (1800 mi) below the surface.

Mantle The bulk of the planet. Nearly half of it is thought to be silicon dioxide, which is quartz when found in the crust, but it may have very different properties under heat and pressure. Another large part is thought to be magnesium oxide. After that, the mantle is probably composed of oxides of various metals in fairly small amounts, although iron oxides may account for about 8 percent of the total. The mantle extends to about 90 km (55 mi) below the higher mountains and to as little as 5 km (3 mi) below parts of the ocean. Part of the upper mantle is somewhat plastic or ductile and is known as the asthenosphere.

Crust or lithosphere The mostly solid part above the mantle. It consists of light rocks in the continents, which are 25 to 90 km (15 to 55 mi) thick, and heavier rocks in the thin oceanic crust, which is 5 to 8 km (3 to 5 mi) thick.

PRINCIPAL ELEMENTS IN EARTH'S CRUST

Continental crust is very different from oceanic crust. Not only is the crust of the continent much thicker than the oceanic crust, but also the composition is different. Continental crust is mainly granite and other relatively light rocks. The crust under the oceans is mainly basalt, a relatively heavy rock. In addition, there is considerable variation from place to place.

Element	Percent by Weight*
Oxygen	45.6
Silicon	27.3
Aluminum	8.4
Iron	6.2
Calcium	4.7
Magnesium	2.8
Sodium	2.3
Potassium	1.8
Hydrogen	1.5
Titanium	0.6

*More than 100 percent due to rounding.

Hydrosphere A layer of water covering about 70 percent of the crust. By far, most of the water is a salty mix that fills the oceans. Most of the pure water is locked in ice caps near the poles and glaciers at the tops of high mountains.

COMPOSITION OF EARTH'S OCEANS

By weight, pure water is 11 percent hydrogen and 89 percent oxygen. Ocean water varies in composition from place to place; on average it is 3.5 percent salts and other compounds by weight. The largest amount of salt in the ocean is sodium chloride, ordinary table salt, which is 2.7 percent of the total amount of water by weight. The following is the typical composition by weight of all the solids dissolved in ocean water.

Compound	Percent by Weight
Sodium chloride	77.8
Magnesium chloride	10.9
Magnesium sulfide	4.7
Calcium sulfate	3.6
Potassium sulfate	2.5
Calcium carbonate	0.3
Magnesium bromide	0.2
Other compounds	trace

Atmosphere Also separated into layers. The lowest, extending upward 10 to 18 km (5 to 11 mi) above the surface, is the *troposphere*. In this region, temperature falls with increasing height. Up to an altitude of about 50 km (30 mi), the next layer is the *stratosphere,* where temperature rises with increasing height. This pattern alternates; between 50 km (30 mi) and 80 km (50 mi), in the *mesosphere,* temperatures fall again, while above 80 km (50 mi), in the *thermosphere*, temperatures rise. The upper part of the mesosphere and all of the thermosphere contain many charged particles, or *ions,* so this region is also called the *ionosphere.* Above the mesosphere, the gases that make up air begin to separate, with mostly oxygen out to 1000 km (600 mi), followed by a layer of mostly helium that extends to about 2500 km (1500 mi) above the surface. Above the helium, a layer of hydrogen gradually thins out into the vacuum of space.

COMPOSITION OF EARTH'S AIR

Air is nearly the same all over Earth at the same altitude, but its composition changes at higher altitudes, where there is more of the lighter gases. There is, however, a variable amount of water vapor from place to place at the same altitude, which is generally reported in terms of relative humidity. The composition of air has changed slowly over time, although human activities have accelerated some of these changes. Carbon dioxide, for example, has gone from 0.0316 percent to 0.0340 percent in the last 25 years. Methane is also increasing at a rate that could cause it to double in the next hundred years. Both of these gases trap heat the way a greenhouse does, leading to fears that the greenhouse effect will change the climate. Ozone is formed from oxygen by sunlight, so there is more ozone at ground level in the summer than in the winter. There is also more ozone in cities than in rural regions because its formation is facilitated by auto exhaust gases. Ozone high in the atmosphere is being depleted at a rate of about 5 percent a year over northern mid-latitudes by interactions with chlorine from chlorofluorocarbons.

Constituent	Percent by Volume*	Parts per million
Nitrogen	78.084	780,840
Oxygen	20.946	209,460
Argon	0.934	934
Carbon dioxide	0.034	34
Neon	0.00182	1.82
Helium	0.00052	0.52
Krypton	0.00011	0.11
Hydrogen	0.00005	0.05
Nitrous oxide	0.00005	0.05
Xenon	0.0000087	0.0087
Ozone (summer)	0.000007	0.007
Ozone (winter)	0.000002	0.002
Methane	0.0000017	0.00017

*Based on dry air at sea level.

M O H S ' S C A L E O F H A R D N E S S

Rocks and minerals are identified by color, luster, crystallization, and other means. One of the most useful tools is relative hardness. In 1822 Friedrich Mohs [German: 1773-1839] devised a ten-level scale for judging hardness in terms of common minerals of varying hardness. Since then, geologists have added some common objects that are not minerals, but which may be at hand, to judge hardness when a standard mineral is not available.

The basic idea of Mohs' scale is that each mineral will scratch any mineral that has a lower number in the scale. The ease and amount of scratching is also relative to the distance apart in the scale—that is, diamond (10) just barely scratches corundum (9), for instance, but it scratches apatite (5) easily.

Hardness	Mineral	Tests using common objects
1	Talc	Easily scratched with fingernail
2	Gypsum	Scratched with fingernail or copper wire
3	Calcite	Scratched with copper penny
4	Fluorite	Easily scratched with steel knife
5	Apatite	Scratched with steel knife or window glass
6	Orthoclase (feldspar)	Scratches window glass but not steel; steel file will scratch it
7	Quartz	Scratches both window glass and steel
8	Topaz	Hardest common mineral
9.	Corundum	Minerals above 8 are rare, and there are no common tests
10	Diamond	Hardest mineral known

GEOLOGIC TIME SCALE

Geologists and other earth scientists break the history of the planet into periods of varying length based on the fossils in rock strata.

Not shown on the chart is one very large division. Geologists often speak of the period before 570 million years ago (mya) as Precambrian Time.

The closer to the present, the more disagreement there is about appropriate divisions. Some do not accept such terms as Quaternary, for example (preferring to use Paleogene for the subdivisions of the Cenozoic era, and grouping the Miocene, Pliocene, and Pleistocene into the Neogene subera), or they fold the Holocene into the Pleistocene.

For recent times, the two systems of North American and European ice ages are also used. They are presented below as a separate chart.

Era or eon	Period	Epoch	Stage	Organisms	Time before present
Archean Eon				Monerans: bacteria and blue-green algae	4600 mya
Proterozoic Eon					2500 mya
	Sinan				800 mya
	Vendian			Protists, algae, and soft-bodied creatures similar to jellyfish and worms	610 mya
Phanerozoic Eon					570 mya
Paleozoic Era					
	Cambrian	Lower	Tommotian Atdabanian Lenian	Tiny fossils with skeletons followed by animals with shells, notably the trilobites	
		Middle	Solvan Menevian		
		Upper	Maentwrogian Dolgellian		
	Ordovician	Lower	Tremadocian Arenigian Llanvirnian Llandeilian	Brachiopods (shellfish similar to clams), corals, starfish, and some organisms that have no modern counterparts, called sea scorpions, conodonts, and graptolites	500 mya
		Upper	Caradocian Ashgillian		
	Silurian	Lower	Llandoverian Wenlockian	Snails, clams and mussels, ammonoids (similar to the nautilis), jawless fish, sea scorpions, first land plants and animals (club mosses and land scorpions)	425 mya
		Upper	Ludlovian		

Era or eon	Period	Epoch	Stage	Organisms	Time before present
	Devonian	Lower	Gedinnian Siegenian Emsian	Spiders, amphibians, jawed fish, lobe-fined fish, lungfish, and ferns	395 mya
		Middle	Eifelian Givetian		
		Upper	Frasnian Famennian		
	Carboniferous	MISSISSIPPIAN		Insects, land snails, amphibians, early reptiles, sea lilies, giant club mosses, and seed ferns	350 mya
		Lower Visean Namurian	Tournaisian		
		PENNSYLVANIAN			
		Upper	Westphalian Stephanian		
	Permian	Lower	Asselian Sakmarian Artinskian Kungurian	Mammal-like reptiles, fin-backed reptiles, and conifers	290 mya
		Upper	Ufimian Kazanian Tatarian		
Mesozoic Era					249 mya
	Triassic	Lower Middle	Scythian Anisian Ladinian	Sharks, marine reptiles (plesiosaurs and ichthyo- saurs), crocodiles, frogs,	
		Upper	Carnian Norian Rhaetian	turtles, cycads and ginkgos, early mammals and early dinosaurs	
	Jurassic	Lower	Hettangian Sinemurian Pliensbachian Toarcian	Dinosaurs (such as stego- saurus), pterosaurs (such as pterodactyl), early birds, dinoflagellates,	190 mya
		Middle	Aalenian Bajocian Bathonian Callovian	diatoms, and early flower- ing plants	
		Upper	Oxfordian Kimmeridgian Titonian		
	Cretaceous	Lower	Berriasian Valanginian Hauterivian Barremian Aptian Albian	Dinosaurs (such as tyranno- saurus, triceratops, and bron- tosaurus), salamanders, modern bony fish, moasaurs (marine lizards), flowering plants, placental and marsupial	130 mya
		Upper	Cenomanian Turonian Coniacian Santonian Campanian Maastrichtian	mammals	

(Continued)

Era or eon	Period	Epoch	Stage	Organisms	Time before present
Cenozoic Era					
	Tertiary				
		Paleocene	Danian	Early primates and horses,	66.5 mya
			Selandian	rodents, and sycamores	62.4 mya
		Eocene	Ypresian	Whales, penguins, roses,	57.8 mya
			Lutetian	bats, camels, early	52.0 mya
			Bartonian	elephants, dogs, cats,	43.6 mya
			Priabonian	and weasels	40.0 mya
		Oligocene	Rupelian	Deer, pigs, saber-toothed	36.6 mya
			Chattian	cats, and monkeys	30.0 mya
		Miocene	Aquitanian	Seals, dolphins, grasses,	23.5 mya
			Burdigalian	daisies, asters, sunflowers,	21.0 mya
			Langhian	lettuce, giraffes, bears,	16.5.mya
			Serravallian	hyenas, and early apes	15.0 mya
			Tortonian		10.5 mya
			Messinian		6.5 mya
		Pliocene	Zanclean	Apes, australopithecines	5.2 mya
			Piacenzian	(early hominids), *Homo habilis* (the first human species), mammoths, giant sloths, and armadillos	3.4 mya
	Quaternary				
		Pleistocene	Calabrian	*Homo erectus* (ancestor of	1.6 mya
			Emilian-Sicilian	modern humans), modern humans, and Neanderthal	1.0 mya
			Cassian*	humans	0.9 mya
			Flaminian*		0.45 mya
			Tyrrhenian*		0.3 mya
			Nomentanan*		0.2 mya
			Tyrrhenian II*		120,000 years ago
			Tyrrhenian III*		80-60,000 years ago
			Pontian*		20,000 years ago
		Holocene		Modern humans and the flora and fauna of today	11,000 years ago

*These stages are European sea-level changes; the others are based primarily on changes in organisms.

ICE AGES

While there have been several periods of global glaciation, the most recent one has been the most severe. In common parlance, the phrase ice age can refer to any one of the following: (1) any past period when temperatures were significantly lower than normal, of which there have been about six, each lasting 2 to 3 million years, since the start of the Cenozoic Era, 66.5 mya; (2) the most recent such period, which is thought to have started the Pleistocene Epoch, as long ago as 1.6 mya; or (3) the most recent, and probably the coldest, of the stages of the Pleistocene ice age.

This last, which is often called the Ice Age, started about 122,000 to 130,000 years ago and lasted until about 11,000 years ago. To an American geologist, the Ice Age is known as the Wisconsin glacial stage, while its manifestation in Europe is called the Riss-Würm glacial stage. Each such glacial stage has been followed by an interglacial stage, during which the ice retreated as temperatures became warmer.

Although exact dates are disputed, and the correlation between North American and European stages is not clear, the stages (from the earliest to the present) are as follows:

North America	Europe
Nebraskan glacial	Danube glacial
Aftonian interglacial	Unnamed interglacial
Kansan glacial	Günz glacial
Yarmouth interglacial	Unnamed interglacial
Illinoisan glacial (190,000 to 130,000 years ago)	Mindel glacial
Sangomon interglacial (130,000 to 74,000 years ago)	Unnamed interglacial
Wisconsin glacial (74,000 to 12,000 years ago)	Riss-Würm glacial
Recent stage (Recent interglacial?) (12,000 years ago to the present)	Either end of ice age or unnamed interglacial

THE BEAUFORT SCALE OF WIND STRENGTH

In 1806 British Royal Navy Commander Francis Beaufort developed a method of defining wind speeds at sea based upon how much sail a square-rigged ship would need. A simple number, called a Beaufort number or "force," was used to designate winds ranging from not enough to make the ship move to those that required removing all the sails to save them (and, one hoped, the ship). Over the nineteenth century, the scale was modified to apply to observations of sea surface and conditions on land. As wind-measuring instruments improved, it was correlated with wind speeds in knots, miles per hour, and kilometers per hour. Therefore, one can use the Beaufort Scale to estimate wind speed.

The correlations to speed do not all agree in details. The miles-per-hour figures given below are those used by the US Weather Service, and the kilometer figures also reflect those speeds. The Weather Service reserves the term *gale* for what are called *fresh* or *strong gales* (force 8 and 9 winds) in the Beaufort system, and the Weather Service refers to both a Beaufort force 10 and force 11 as a *storm*. A *tropical storm*, however, can be what the weather service would call either a gale or a storm if it were on land or not in the tropics.

Beaufort number	Type of Wind	Wind speed in knots	Wind speed in km (mph)	Description and effect
0	Calm	Less than 1 (1)	Less than 1.6	Still; smoke rises vertically; sea like mirror
1	Light Air	1-3	1.6-4.8 (1-3)	Wind direction shown by smoke drift; weather vanes inactive; sea shows ripples with appearance of scales; no foam crests
2	Light Breeze	4-6	6.4-11.3 (4-7)	Wind felt on face; leaves rustle; weather vanes begin to move; sea shows small wavelets and crests of glassy appearance, not breaking
3	Gentle Breeze	7-10	12.9-19.3 (8-12)	Leaves and small twigs move constantly; wind extends light flags; sea shows large wavelets; crests begin to break

Beaufort number	Type of Wind	Wind speed in knots	Wind speed in km (mph)	Description and effect
4	Moderate Breeze	11-16	20.9-29.0 (13-18)	Raises dust and loose paper; moves small branches; sea shows small waves and whitecaps
5	Fresh Breeze	17-21	30.6-38.6 (19-24)	Small trees in leaf begin to sway; at sea, moderate waves, many whitecaps, and some spray
6	Strong Breeze	22-27	40.2-49.9 (25-31)	Large branches move; overhead wires whistle; umbrella difficult to control; at sea, whitecaps every-where; more spray
7	Moderate Gale	28-33	51.5-61.1 (32-38)	Whole trees sway; some what difficult to walk against wind; sea heaps up; foam from breaking waves begins to form streaks
8	Fresh Gale	34-40	62.8-74.0 (39-46)	Twigs broken off trees; walk-ing against wind very diffi-cult; waves are longer and moderately high; foam is blown in well-marked streaks
9	Strong Gale	41-47	51.2-86.9 (47-54)	Slight damage to buildings; shingles blown off roof; high waves at sea; sea begins to roll; dense streaks of foam form; spray may reduce visibility
10	Whole Gale	48-55	88.5-101.4 (55-63)	Trees uprooted; consider-able damage to buildings; occurs more often at sea, where there are very high waves, dense streaks of white foam, rolling is heavy and visibility poor
11	Storm	56-63	103.0-115.8 (64-72)	Widespread damage; rarely occurs inland; exceptionally high waves at sea; sea cov-ered with white foam patches; very poor visibility
12	Hurricane	64 or greater	117.5 or greater (73 or greater)	Extreme destruction near coasts; at sea, air filled with foam; sea completely white with driving spray; visibility greatly reduced

RECORD BREAKERS FROM THE EARTH SCIENCES

The Earth is thought to be about 4.6 billion years old.

BY AGE OF ENTITY

Record	Holder	Age in Years
Oldest minerals	Zircons from Australia	4.3 billion
Oldest rocks	Granite from northwest Canada found in 1989 by Samuel Bowring of Washington University in St. Louis MO	3.96 billion
Oldest fossils	Single-celled algae or bacteria from Australia	3.5 billion
Oldest slime molds (bacteria)	Slime molds, also called slime bacteria, gather together to form multicellular bodies for reproduction	2.5 billion
Oldest petroleum	Oil from northern Australia	1.4 billion
Oldest fish	*Sacabambasis,* found in Bolivia	470 million
Oldest land plant	Moss or algae	425 million
Oldest land animals	Trigonotarbids and centipedes found in the Ludlow Bone Bed at Ludford Lane, Ludlow, Shropshire, England	414 million
Oldest insect	Bristletail (a relative of the modern silverfish)	390 million
Oldest spider	Known only from a spinneret used to make silk, found in Gilboa NY	380 million
Oldest tetrapod (ancestor of amphibians that walked on land)	Known only from scattered fossils collected in Scotland before 1950 and thought until 1991 to be fish	370 million
Oldest reptile	"Lizzie the Lizard," found in Scotland in 1988 by Stan Wood (probably to be named *Westlothiana curryi*)	340 million

Record	Holder	Age in Years
Oldest dinosaur	*Herrerasaurus,* found in Argentina by Osvaldo Reig in 1959	230 million
Oldest bird	Either *Protoavis,* fossils found near Post TX by Sankar Chatterjee, or *Archaeopteryx lithographica* (see "Early Birds," p 206)	225 million 175 million
Oldest flower	An unnamed tiny dicot similar to a pepper plant found in Koonwarra, Australia	110 million
Oldest mushroom	*Coprinites dominicana,* found in Haitian amber from the La Toca mine	35 to 40 million

BY SIZE OF ENTITY

Record	Holder	Size
Longest dinosaur	*Seismosaurus,* who lived about 150 million years ago in what is now New Mexico	About 50 m (160 ft) long
Tallest and heaviest dinosaur	*Ultrasaurus,* who lived in Utah	More than 70 metric tons (80 short tons)
Largest land animal skull	The skull of *Chasmosaurus,* found by University of Chicago senior Tom Evans in Big Bend National Park TX in Mar 1991	1.5 to 1.8 m (5 to 6 ft) long
Largest land mammal	*Indricotherium,* also known as *Baluchitherium,* an ancestor of the rhinoceros	About 5.5 m (18 ft) high at shoulder; weighed 5 times as much as largest elephant
Largest exposed rock	Mt Augustus in western Australia	At 377 m (1237 ft) high, 8 km (5 mi) long, and 3 km (2 mi) wide, it is twice the size of the more famous Ayers Rock
Largest island	Greenland (also known as Kalaallit Nunnaat)	About 2,175,000 sq km (840,000 sq mi)
Largest ocean	Pacific	166,236,100 sq m (64,186,300 sq mi)
Deepest part of ocean	The *Challenger* Deep in the Marianas trench in the Pacific Ocean	11,000 m (35,640 ft), or 11 km (6.85 mi)

(Continued)

Record	Holder	Size
Greatest tide	The Bay of Fundy between Maine and New Brunswick	15 m (47.5 ft) between high and low tides
Largest geyser	Steamboat Geyser in Yellowstone Park	Shoots mud and rocks about 300 m (1000 ft) in the air
Longest glacier	Lambert Glacier in Antarctica (upper section known as the Mellor Glacier)	At least 400 km (250 mi) long
Longest known cave	Mammoth Cave KY, which is connected to the Flint Ridge Cave system	Total mapped passageway of over 550 km (340 mi)
Largest single cave chamber	Sarawak Chaper, Lobang Nasip Bagus, on Sarawak, Indonesia	700 m (2300 ft) long, 300 m (900 ft) wide on average, 70 m (230 ft) or more high
Largest canyon on land	Grand Canyon of the Colorado, in northern Arizona	Over 350 km (217 mi) long, from 6.5 to 31 km (4 to 13 mi) wide, as much as 1615 m (5300 ft) deep
Highest volcano	Cerro Aconagua in the Argentine Andes (extinct)	6960 m (22,834 ft) high
Most abundant mineral	Magnesium silicate perovskite	About 2/3 of the planet
Highest artificial pressure by a diamond anvil	Arthur L. Ruoff and coworkers at Cornell University	4.16 megabars
Deepest hole in oceanic crust	Hole 504B in Pacific, 800 km (500 mi) west of Panama	2 km (1.2 mi)

Environment and Issues

THE STATE OF THE ENVIRONMENT

The environment is not a branch of science, but public policy issues that relate to environmental problems often can be solved only by the application of science and technology. Thus, it is useful to have a section devoted to issues concerned with the environment in *Current Science*. These issues include what to do about various environmental hazards, how to supply safely the energy needed to maintain civilization, how to dispose of the various kinds of wastes we produce, and, in general, how best to deal with the many changes taking place in the environment as a result of human activity.

Concern about the environment is not new, but it became what many saw as a fad in the 1960s and early 1970s. The "fad" refused to go away, however. Instead, environmental priorities shifted from local ones, such as preservation of a lake or pollution of a specific site, to such problems as global warming and loss of the ozone layer. Of course, many local problems remain, but they seem less serious vis-á-vis a new perception of planet-wide change as a result of human activity. Among evidence of renewal of interest in the environment is the hoopla surrounding the Biosphere 2 project, a privately financed scheme to demonstrate, among other things, that clean living can sustain the environment. The idea was to lock four men and four women into a closed shell containing five sample Earth environments for 2 years. They were to be cut off from everything outside except sunlight and externally generated electric power. The experiment began when 3800 species of life (including the humans) were locked in on Sep 26 1991, but it was marred somewhat when an accident resulted in one of the women leaving Biosphere 2 for treatment soon afterward. Order restored, the experiment continued. In its next incident, however, a Biosphere spokesperson announced that about 600,000 cu ft of air had been pumped into the "closed" environment on Dec 9 1991. The spokesperson denied that the air was intended to control a buildup of carbon dioxide; it was to replace air that had leaked from the building. For $9.50 a head it is possible to visit the site in Oracle AZ and to observe activities through the glass walls.

Ecological and Environmental Priorities

During the time that early stirrings of the environmental movement began with publication of *Silent Spring* in 1962, and swelled to the first Earth Day in 1970, many people encountered the word *ecology* for the first time; to this day, the science of ecology remains closely linked in people's minds with preservation of the natural environment. Strictly speaking, ecology is the study of the interactions of living things with their environments and is not involved with issues of preservation.

The identification of conservation with ecology may become even stronger in the future, however, if the 6200-member Ecological Society of America follows its own announced priorities in the 1990s. After a 2-year study, a sixteen-member committee reported in Apr 1991 that the number one goal for society members should be research on changes that affect the entire Earth, especially the way in which global climatic changes interact with other aspects of the environment. A second level of priority would be assigned to studies relating to the importance of biological diversity in interactions of populations with their environments. A third-level priority is to understand how existing ecological systems are affected by specific stresses. In other words, the priorities suggested are

1. Global warming
2. Saving complex ecosystems such as rain forests
3. Handling pollution and similar disturbances.

In this instance, the ecologists' priorities are being set more by environmental concerns than by scientific puzzles. Nevertheless, research along the lines indicated also involves many poorly understood processes that ecologists would want to study even if there were no possible practical payoff.

The ecologists are not the only ones setting priorities. The following environmental hazards were ranked by the Science Advisory Board of the US Environmental Protection Agency (EPA) late in 1990:

Relatively High-Risk Ecological Problems

1. Habitat alteration and destruction (from soil erosion, deforestation, and similar incursions)

2. Loss of biological diversity from species extinction

3. Depletion of the ozone layer of the stratosphere

4. Global warming as a result of the increasing amount of greenhouse gases in the atmosphere

Relatively High Risks to Human Health

1. General outdoor air pollution

2. Worker exposure to chemicals used in industry and agriculture

3. Indoor air pollution

4. Pollution in drinking water

Relatively Medium-Risk Ecological Problems

1. Herbicides and pesticides

2. Pollution of surface water

3. Acid rain and other forms of acid precipitation

4. Airborne toxic substances

Relatively Low-Risk Ecological Problems

1. Oil spills

2. Groundwater pollution from hazardous wastes, from underground tanks, or from landfills

3. Escape of radioactive materials

4. Changes in lakes caused by acid precipitation

5. Heat pollution of water or the atmosphere

In Jan 1991 William K. Reilly, administrator of the EPA, told the US Congress that the EPA will revise its priorities to follow the order set forth by its advisory board. Since the United States will spend, according to an EPA estimate, nearly 3 percent of its gross national product on environmental protection by the end of the 1990s, such priorities are vital in making sure that all that money is well spent.

Individuals also set their own priorities. Recycling continues to make significant gains around the world in the early 1990s. In the United States, a survey in Jun 1990 asked 313 women nationwide if they would pay more for products that had environmental benefits. Of the respondents, 78 percent would pay 5 percent more, while 47 percent would pay as much as 15 percent more to protect the environment. Manufacturers of all sorts tried to capitalize on this impulse by labeling products "recyclable," "printed on recycled paper," "biodegradable," "ozone friendly," and even "dolphin safe." In the early 1990s soap powder manufacturers stopped adding fillers previously used to make their products look like "more for the money," in part because they thought consumers now want products that have less packaging.

Not all the commercial claims are justified. Some "ozone safe" sprays and "biodegradable" diapers and trash bags were found to be fraudulently advertised by the US Federal Trade Commission (FTC), by the states, or by environmental organizations. Because US states often are tougher on environmental claims than the federal government is, the consumer products industry in 1991 actually asked the FTC to provide national guidelines for acceptable claims.

All such priority setting is becoming easier as more and more supercomputers become available to predict the consequences of environmental change. In the early 1990s, environmental projects had become the largest and fastest-growing business for some supercomputer manufacturers.

Renewed Awareness of the Population Bomb

In the late 1960s and early 1970s many people became aware for the first time that Earth's growing population is a problem, although Thomas Malthus had proclaimed the evil consequences of growth as early as 1798. Best-sellers in the United States during the first awakening of environmental concern included Paul Ehrlich's *The Population Bomb* (1968) and the Club of Rome's *The Limits of Growth* (1973), while many other prophets of gloom were heard around the world as well. The message was that there would not be enough food for everyone, nor enough energy in any form; such shortages would have unpleasant results for many people. It would be just as Thomas Malthus had predicted around the turn of the nineteenth century, only worse.

The message of the gloom-and-doom books went largely unheeded. Population continued to increase in most of the world at a rapid rate (although much more slowly in developed nations than in the countries that were already poor and overcrowded). Over the course of the 1970s and 1980s, about 1.3 billion people were added to the total world population.

But the Malthusian famines did not come. It appeared that, as had been the case in the nineteenth and twentieth centuries, technological advances would continue to stay ahead of population growth. Throughout the 1980s this was explained by the concept that more people would produce more ideas that would solve any problems caused by there being more people. Furthermore, it seemed that all the additional consumers in a higher population group would generate prosperity. And everyone knows that prices go down with mass production; similar economies of scale could apply to any situation caused by high population. In this view, the larger the population, the better.

So why did people start fretting about population growth again in the 1990s?

Global warming was one reason. People realized that reduction of carbon dioxide from smokestacks might not make much difference if all the people in China and India got the cars, refrigerators, and perhaps even air conditioners that people in the developed world enjoy. On the other hand, since more industrialized nations produce more carbon dioxide per person than the less developed nations that have the population problems, reductions in population might result in more global warming, not less.

Another factor causing alarm is that some economists think that many regions have already reached a point of complete population saturation. For those regions there is already no more room, energy, or food for further growth, even though population growth continues.

Meanwhile, the United Nations has estimated that the population of less developed countries will rise by almost 3 billion by 2020.

Parks, Population, and the Environment

Large national parks were originally instituted primarily to protect unusual parts of the environment. Later they were used as reserves for animals or plants. Around the world in the early 1990s, the mission of such parks has become more difficult, in part because of the growing world population. Since parks are set aside as protected places, their failure to protect can be taken as an early warning that places not so set aside may be in trouble. Despite the problems that parks are facing, new ones continue to be created to meet specific needs, such as a giant protected area in Brazil intended to preserve a native population and their culture

Typical of the parks' problems are those of Amboseii National Park in Kenya, intended to preserve a savanna ecosystem of lions and elephants. In the early 1990s about 200,000 tourists came to the park each year, cameras in hand. If they wanted to photograph an elephant, they had no problem. Unlike other African parks, Amboseii has been well protected by park rangers against poachers who want ivory. As a result, there are too many elephants and they are tearing the park apart. In other African parks, where the rangers are less successful, poachers have decimated the elephants. This contrast highlights the difficulty of getting all components into balance. And in all the large African game parks, the pressure from Africa's fast-growing population, who move into border regions of the parks, is a factor along with the increasing number of people who want to visit the parks.

In the United States, population pressures mainly affect parks in terms of the number of visitors. Yosemite, fairly close to the San Francisco Bay area and to southern California, has such severe visitor overload that from time to time it must restrict the number of people entering. Even so, about 250,000 people congregate on the floor of Yosemite Valley, just 18 km^2 (7 sq mi) in area, each day in Jul and Aug. Other parks in the US system have also had to close from time to time to avoid being swamped by visitors.

Visitors are not the only problem in US national parks. Several parks are suffering from air pollution of one kind or another. This is not surprising for the virtually urban Gateway National Park, just outside New York City. But the worst problem is in Shenandoah National Park, which extends 130 km (80 mi) along the Blue Ridge Mountains of northern Virginia. There the haze has reduced what used to be a 100-km (65-mi) view to just 25 km (15 mi). Shenandoah ranks first in the amount of sulfur compounds in its air, resulting in serious acid rain; but the haze is photochemical smog that starts with ozone, for which Shenandoah ranks second. Twice rangers have had to issue ozone alerts to hikers, climbers, and joggers in the park. Both the acid rain and the ozone are damaging plants and animals in the park. The acid rain comes mainly from power plants in the Midwest, but the National Park Service has opposed building new coal-fired power plants in Virginia on the grounds that the proposed plants will make the situation worse. The local office of the EPA agrees with Virginia that new plants will not aggravate conditions in the park.

Great Smoky Mountains National Park in Tennessee, Acadia National Park in Maine, and Grand Canyon National Park in Arizona are also experiencing serious haze or other forms of air pollution. The haze that is obscuring the view of the Grand Canyon has been attributed by many to the Navajo Generating Station in Page AZ, 25 km (15 mi) north of the entrance to the canyon. The station began operations in 1976 and in 1987 chemical tracers showed that sulfur dioxide in the air at Grand Canyon originated there. Despite conflicting evidence that showed the haze may have other causes—some say photographs show that it started 20 years before the power station was built—the EPA pressured the plant into agreeing to cut emissions of sulfur dioxide by 90 percent by 1999, assuming that the $430 million worth of air scrubbers are as effective as predicted. According to a 1991 study by the Environmental Defense Fund, the group that first sued the Navajo in 1984, the plan will also reduce pollution at seven other national parks and monuments—Arches, Bryce, Canyonlands, Capitol Reef, Mesa Verde, Petrified Forest, and Zion.

Pollution may or may not be the cause of the cacti dying at Saguaro National Monument outside Tucson. There is some evidence that suggests that increased ultraviolet radiation caused by thinning of the ozone layer may be killing off the giant cacti for which the park is named, but no one is certain.

In addition to the problems of national parks, the 472 US national wildlife refuges are suffering environmental degradation from mining, oil drilling, military use, ranching, and visitors. There are 91 million acres in the refuge system, which was instituted in 1903. Some examples include Sheldon in Nevada, established in 1931 to protect pronghorns, but until the early 1990s also used for cattle grazing; D'Arbonne in Louisiana, established in 1975 to help protect about 145 species of birds but now contaminated by saltwater brine from natural gas wells; Chincoteague, a 1943 refuge for snow geese and other migratory birds that now gets fifteen times as many tourists as it did in the early 1970s; and Prieta, established to protect bighorn sheep in 1939, but routinely in the path of military exercises, including live fire exercises near its border.

The Collapse of Eastern Europe and the End of the Soviet Union

It is not surprising that the collapse of most communist governments in the face of serious economic problems dominated international politics in the early 1990s, but the ecological effects were unexpected. Because the communist governments had paid almost no attention to the environmental problems their economic activity caused, that same activity was a major contributor to large-scale acid rain and global warming. As the activity wound down when the economies began to fall apart, so did the emissions to the atmosphere. This produced the first global reduction in carbon dioxide emissions in years. The United States still leads in total production of carbon dioxide and acids that result in acid rain. On a per person basis, however, in 1988 E Germany produced nearly four times as much sulfur dioxide as did the United States.

What is good for the whole globe or large sections of it, however, is not necessarily good for localized problems. The end of communism revealed that many of the plants and sites in Eastern Europe and the former Soviet Union were in far worse condition than corresponding plants and sites in the West. Local residents often have to put up with incredible pollution.

On the other hand, the collapse of communism might result in less arctic haze. The haze is mostly caused by sulfur particles produced by industrial regions in the northern part of Russia.

Environmental Concerns in the United States

Although originally a leader in the environmental movement, especially in terms of level of concern, the United States has fallen behind Europe in its actions. Many current US policies, especially as they affect the environment beyond the borders of the United States, are viewed as excessively conservative by organizations in the environmental movement. This has been particularly true with regard to global warming and acid rain, although less so for problems affecting the stratospheric ozone layer. For example, the

United States has refused to sign proposals favored by the European Community and by Japan for reducing carbon dioxide emissions to the 1990 level in the year 2000. Consequently, many have questioned President Bush's claim to be "the environmental president."

A typical list of complaints from environmental liberals about US policy at the end of 1991—adapted from columnist Tom Wicker writing in the *New York Times*—includes support by President Bush for drilling for oil in the Arctic National Wildlife Region (see "Oil as a Hazard," p 352); resistance to sufficient controls on automobile emissions and higher taxes on fuels (see "Alternatives to Gasoline," p 353); lack of support for birth control nationally and, perhaps more important, internationally; refusal to attend the UN-sponsored "Earth Summit" on environmental problems that is scheduled for Rio de Janeiro in Jun 1992; and insufficient efforts to cut carbon dioxide production (see "Global Warming UPDATE," p 382).

A major controversy during the early 1990s centered around the definition of a wetland. President Bush had promised that wetlands would not be decreased in size in the United States. His administration, however, proposed a new definition on Aug 9 1991 of a wetland; that definition would drastically reduce the area that is legally protected. In the face of widespread criticism, the change in definition was abandoned on Nov 22 1991 and the issue left unresolved. The main factor was the government's own study that showed that the new definition would have eliminated about a third of the previously recognized wetlands, including all wetlands in twenty-nine states.

Another policy issue was smog. Although much of the visible smog caused by burning fossil fuels in the United States has been controlled since the end of World War II, photochemical smog has gotten worse during that period. In Los Angeles CA smog was the butt of hundreds of jokes on nationwide radio programs of the 1940s, but most of the listeners at that time had never seen photochemical smog. Today episodes of photochemical smog are familiar in many major US cities, although the region around Los Angeles continues to lead the United States (international honors go to Mexico City). Efforts to control photochemical smog have been based on the knowledge that its main cause is the action of sunlight on automobile tail pipe emissions of unburned hydrocarbons and nitrogen oxides.

When the Clean Air Act of 1990 was passed, better control of photochemical smog was among the main goals, so the act required an in-depth study of the problem. The US Environmental Protection Agency (EPA) sponsored the review at a cost of $430,000, jointly funded by the Department of Energy, the American Petroleum Institute, and the Motor Vehicle Manufacturers Association. Results, leaked by someone from the business community on Dec 14 1991, can be presumed to have pleased the sponsors, since they support the contention of the oil, automobile, and trucking industries that not enough is known about the cause of photochemical smog to change the way fuels are formulated or engines are built. The report also suggested that efforts to control hydrocarbons have been misplaced because the main contributor to smog is nitrogen oxides. No one really has a good way to control nitrogen oxides at present. Nitrogen oxides are present whenever combustion occurs in air (which is about 80 percent nitrogen and 20 percent oxygen) at a high enough temperature, a condition that seems necessary for the operation of an internal combustion engine. Furthermore, the report states that mismeasurements have led to an overly optimistic estimate of success in reducing volatile hydrocarbons. (For further information on changes in automobiles and automobile fuels as a result of the Clean Air Act, see "Alternatives to Gasoline," p 353.)

Landfills, sludge, and water pollution were also in US news in the early 1990s. New US rules for landfills were issued on Sep 11 1991 (originally scheduled in a 1984 law for Mar 1988, but successfully evaded by the EPA until it was forced into compliance by a lawsuit from environmental protection organizations). Most rules would still not go into effect until 1993. The most likely result of the rules, which will require advanced technology to prevent leakage of pollutants, will be to close most small local landfills and put landfill operations into the hands of big business.

In Nov 1991 scientists from Woods Hole proposed that sewage sludge be carefully placed at the bottom of the sea in regions of the open ocean. Sludge would be carried to a depth of 5000 m (16,000 ft) in leak-proof containers and then left there. The plans were later detailed at a Jan 1992 conference.

Although states were supposed to develop rules for controlling toxic water pollution by 1990, nineteen of them failed to do so by the end of 1991. The EPA said on Nov 6 that it would impose federal rules on the states early in 1992.

Nuclear Waste

The end of the cold war brought plans to dispose of large stockpiles of nuclear, chemical, and biological weapons. The prospects for disposal of the active ingredients from these weapons do not look good, based on difficulties the United States is having in getting rid of the waste produced during the manufacture of some of them.

Low-Level Waste If radiation from waste is low enough, the US Energy Department plans to use this waste as an ingredient in concrete, which probably would become free of noticeable radiation sometime in the twenty-sixth century. Waste with somewhat higher levels of radiation will be used as an ingredient in glass logs, then encased in steel cylinders about 3 m (10 ft) high and 0.6 m (2 ft) across. A $1.3 billion plant for this purpose was dedicated on Nov 27 1990 in Aiken SC. It was not decided at that time, however, where the wastes would be ultimately stored, since locating permanent storage sites has proved to be elusive.

Plutonium The only proposed permanent site for nuclear waste, intended primarily for about a million barrels of plutonium waste from the-manufacture of nuclear bombs, was kept from opening by pressure from the state of New Mexico, where the Carlsbad Waste Isolation Pilot Plant is to be located. As of the end of 1991, the site had not opened.

Other High-Level Nuclear Waste The latest move in a long-debated plan for disposal of high-level nuclear waste came in Dec 1991 when a panel of five geologists split three to two against the possibility that an earthquake might result in water pollution if such waste was stored at the Yucca Mountain site in Nevada, as planned. In 1992 a National Academy of Sciences panel will issue a separate report on the matter. The Yucca Mountain site is not scheduled to open until sometime in the twenty-first century.

Turn Nuclear Waste into Different Elements One can use a particle accelerator to change one element or isotope into another. For example, the isotope iodine-129 has a half-life of 16 million years. Bombard it with low-energy neutrons and it becomes iodine-130, which decays in a day to nonradioactive xenon. A report on whether such transmutations are feasible on a large scale is due from a panel of the US National Academy of Sciences and Engineering around 1993.

(Periodical References and Additional Reading: *Science* 2-23-90, p 911; *National Wildlife* 4/5-90, p 28; *New York Times* 11-28-90; *New York Times* 12-9-90, p E3; *International Wildlife* 1/2-91, p 27; *New York Times* 1-31-91; *New York Times* 1-26-91, p 11; *New York Times* 1-29-91, p C4; *New York Times* 1-31-91, p D1; *New York Times* 2-1-91, p D1; *New York Times* 2-2-91, p 10; *Science* 2-15-91, p 721; *New York Times* 4-1-91, p D1; *Science* 4-5-91, p 26; *New York Times* 4-18-91; *New York Times* 4-19-91, p D1; *Science* 4-26-91, p 5-4; *New York Times* 5-2-91; *New York Times* 5-4-91; *New York Times* 5-26-91; *New York Times* 7-14-91; *New York Times* 8-9-91, p A1; *New York Times* 8-10-91; *New York Times* 8-11-91; *New York Times* 8-25-91, p E4; *New York Times* 9-12-91, p A18; *New York Times* 9-22-91, p L19; *New York Times* 9-27-91, p A1; *Science News* 10-5-91, p 215; *New York Times* 10-10-91, p A25; *New York Times* 10-29-91, p C1; *New York Times* 11-23-91, p 1; *New York Times* 12-2-91, p A1; *New York Times* 12-3-91, p C2; *New York Times* 12-14-91, p 1; *New York Times* 12-21-91, p 46; *New York Times* 12-29-91, p E5)

Timetable of the History of the Environment to 1990

c 9000 BC	Shortly after the Clovis hunters enter the lower part of North America, all large species of mammals become extinct
1627 CE	The auroch, a wild relative of cattle, becomes extinct
1681	The dodo becomes extinct
1775	Sir Percival Potts [English] observes that chimney sweeps develop cancer as a result of their contact with soot, the first recognition that environmental factors can cause cancer
1872	Robert Angus Smith [Scottish: 1817-1884] describes acid rain
	Yellowstone becomes the first US national park
1885	Canada establishes its first national park
1874	Othman Zeidler [German] prepares DDT, but does not know that it is a potent insecticide
1892	John Muir [Scottish-US: 1838-1914] founds the Sierra Club
1900	Asbestos is linked to a type of type of lung disease called asbestosis
1905	The US national forest system is begun
1906	Upton Sinclair [US: 1878-1968] writes the novel *The Jungle,* which alerts Americans to contamination by the meat-packing industry
	Congress passes the US Pure Food and Drugs Act
1914	The passenger pigeon and the Carolina parakeet, both of which had been abundant, become extinct
1916	A systematic approach to US national parks is started with the founding of the National Park Service
1939	Paul Müller [Swiss: 1899-1965] discovers insecticidal properties of DDT
1948	Smog lasting 5 days kills 20 people in Donora, PA
1952	The great London smog is believed to kill 4000 people
1953	Smog in New York City is believed responsible for 200 deaths
1955	The link between exposure to asbestos and lung cancer, suspected since 1935, is definitely established
1957	Nuclear waste stored by the Soviet Union in a remote mountain region of the Urals explodes, contaminating thousands of square miles with radioactivity and causing a number of villages to be permanently evacuated
1961	Acid rain and its consequences are rediscovered in Sweden by Svante Odén; subsequent investigations in Scandinavia and the US Adirondacks confirm that it is causing some lakes to lose all species living in them
1962	*Silent Spring* by Rachel Carson [US: 1907-1964] attacks pesticide use and stimulates a US wave of environmentalism
1964	Congress passes the US Wilderness Act, setting up the National Wilderness Preservation System

1965	Congress passes the US Highway Beautification Act
1966	Congress passes the US Rare and Endangered Species Act
1967	S. Manabe and R.T. Wetherald predict that rising amounts of carbon dioxide in the air will cause temperatures to rise as a result of the "greenhouse effect"
	Thirty million gallons of oil are spilled off the southern coast of England from the wreck of the *Torrey Canyon*
1968	Congress passes the US Wild and Scenic Rivers Act
1970	The first Earth Day is celebrated on April 22
1972	Congress passes the US Clean Air Act
1973	The first US Endangered and Threatened Species List is issued
1974	F. Sherwood Rowland and Mario Molina warn that chlorofluorocarbons are destroying the upper atmosphere ozone layer
1976	Congress passes the US Toxic Substances Control Act
1977	The United States signs the Convention of International Trade in Endangered Species
1978	The community of Love Canal NY is evacuated because of hazardous wastes dumped in the local landfills in the 1940s and covered over in 1953
1979	A nuclear reactor at Three Mile Island, near Harrisburg PA suffers a partial meltdown, but radiation is confined
1980	Congress passes the US Comprehensive Environmental Response, Compensation, and Liability Act (the "Superfund") to clean up hazardous waste sites
1983	Scientists predict that a large-scale nuclear war could produce a "nuclear winter," cold global temperatures that might destroy most living creatures
1984	More than 2000 die and thousands others are injured by toxic gas from an industrial accident in Bhopal, India
1985	British scientists discover that a "hole" in the ozone layer develops over Antarctica each year
1986	A worldwide ban on whaling (with limited exceptions) begins
	The Chernobyl nuclear reactor number 4 explodes and burns, causing 31 deaths in the short run
1987	An international treaty, including the United States, agrees to halve production of chlorofluorocarbons
	The last wild California condor is trapped and moved to a zoo in an effort to keep the species from becoming extinct
1989	The *Exxon Valdez* leaks 35,000 tons of oil into Prince William Sound AK
	Thirteen industrial nations, including the United States, agree to halt chlorofluorocarbon production by the year 2000

HAZARDS TO YOUR HEALTH

ENVIRONMENTAL ILLNESS

For years, people who have been concerned about the safety and nutritional value of what they eat have sworn by the products of Deaf Smith County TX, a purer-than-most region of the Texas Panhandle. But recently, deep in the heart of Texas, another community has attained fame as the place to go when the world is not only too much with you, but literally makes you sick. Wimberley TX, between Austin and San Antonio, is the place to be if you have become—or think you have become—allergic to the twentieth century.

Since 1978 a dozen families have moved to Wimberley to escape not only pesticides, but also tap water, treated wood, deodorants, electromagnetic fields, concrete, some dyes, and other hazards too numerous to list. These people are convinced that they suffer from a problem known as multiple-chemical sensitivity, or as environmental illness. The main cause for this condition is believed to be artificial chemicals—substances produced by human activity that otherwise do not exist in nature. The symptoms of multiple-chemical sensitivity range from rashes and headaches to simple fatigue and confusion. The theoretical basis of the syndrome is the sure knowledge that in some circumstances chemicals can enter the brain or other parts of the body and cause toxic effects. Furthermore, it is well known that different people respond in different ways to the same amounts of the same chemicals. The practical problem is establishing whether or not there are some people who are responding to the very small amounts of artificial chemicals in our environment that so far produce no noticeable effects in other people. Victims of environmental illness may be diagnosed and treated by regular physicians, but more often their cases are handled by specialists called chemical ecologists.

Donald W. Black and coworkers from the University of Iowa College of Medicine investigated multiple-chemical sensitivity and reported their findings in the Dec 26 1990 issue of the *Journal of the American Medical Association*. They interviewed sixty-nine people using standard psychiatric techniques and concluded that 65 percent of the people diagnosed by chemical ecologists as suffering from environmental illness showed symptoms of mental disorder, as opposed to only 28 percent of the members of the same community who were thought not to have multiple-chemical sensitivity. In many cases Dr. Black felt that the mental illness was simple clinical depression; however, most of the people claiming environmental illness had been unsuccessfully treated for depression before turning to chemical ecologists for help. Also, the cause of clinical depression is not known; it could be, speculates Max Costa of New York University Medical School, environmental chemicals.

The people who come to Wimberley to avoid artificial chemicals often take drastic measures to change their life-styles, staying in specially lined rooms to avoid chemicals from the walls and wearing masks on the infrequent occasions when they leave their havens. Most other people do not feel that modern living conditions have become serious enough for such steps. Many, however, are concerned about chemicals from the environment, radiation, electromagnetic fields, or other artifacts of modern life. The following articles address several of these perceived environmental hazards, as well as some natural chemicals in the environment that may be a more serious cause of illness. The articles focus on what we actually know about these chemicals, how dangerous they are, and what can be done to minimize adverse impacts.

(Periodical References and Additional Reading: *New York Times* 12-2-90, p A1; *New York Times* 12-26-90)

DANGER FROM MERCURY

In 1990 and 1991 problems with the toxic element mercury once again became apparent.

Mercury in Paint

The first clue in the recent spate of problems with mercury poisoning occurred in Aug 1989 when a four-year-old boy in Michigan developed mercury poisoning from fumes given off by a latex-based interior paint. The paint was loaded with phenylmercuric acetate, which was used as a preservative. The amount detected was two and a half times the amount permitted by the 1989 standards of the US Environmental Protection Agency (EPA). Almost immediately, the EPA completely banned the use of mercury-based preservatives in interior paints. No recall was initiated, however, so mercury-laden interior paints continued to be sold even though no more were being manufactured. Mercury compounds are still used in exterior paints.

The US Centers for Disease Control (CDC) in Atlanta GA also became involved. Its investigators, Mary M. Agnocs and Ruth A. Etzel, along with John L. Hesse of the Michigan Department of Public Health, studied the effects of using high-mercury paint in interior rooms, a practice common in the Detroit MI region. Air samples from the nineteen homes the investigators monitored revealed high concentrations of mercury vapor in the air at an average level about 600 times greater than that of outdoor air. Most of the samples were taken about a month after paint had been applied.

Urine samples from people living in the newly painted homes also revealed in three individuals mercury concentrations high enough to cause illness, although health problems were not investigated in the study. This study was reported in the Oct 18 1990 *New England Journal of Medicine.*

Etzel suggested that anyone concerned about the possibility of mercury in interior paints should keep windows of painted rooms open as much as possible after painting.

Voodoo Mercury

Mercury, the only common metal to be liquid at room temperature, has always seemed somewhat mysterious. Droplets of mercury seem to move around as if they have a life of their own, which gives mercury its other name, "quicksilver." *Quick* in this context means "living," not "speedy." Tiny drops of mercury that seem almost alive could appear to be infused with spirits to people who do not know that this "behavior" is caused by surface tension.

On the same day as the report on mercury hazards in Michigan from paint was issued, *Nature* published a study by Arnold P. Wendroff of Columbia University that showed that 86 percent of 115 *botanicas* (shops selling potions and ritual objects, mostly to Spanish-speaking immigrants to the United States from the Caribbean or from Latin America) sold capsules containing elemental mercury. They were to be used by breaking the capsules open and sprinkling the liquid mercury around the floor of a room to disperse evil spirits. Repeat as necessary.

Because such a practice would release toxic mercury vapor into the air, it is dangerous. Also, small children might ingest globules of liquid mercury. So far, no action has been taken by the EPA or other agencies to halt the practice.

Mercury in Lakes

In Sep 1990 Greg Mierle of the Dorset Research Center in Ontario, Canada, reported that his study of a lake in Ontario showed that rain was depositing much more mercury in the lake than previously expected. He believes that most of the mercury in rainwater comes from air pollution caused by combustion of materials containing mercury. Much of this is released by cement and phosphate plants, while most of the rest comes from incinerators.

Because it falls from the sky, mercury can pollute even remote lakes, far from industrial sources. One example is Crane Lake in Minnesota, where almost every fish tested comes close to or exceeds the FDA standard of one thousandth of a gram of mercury per kilogram (one part in a million).

Background

Depending on how hydrogen and arsenic are classified, 87 to 89 of the 109 known elements are metals. Several metals in fairly large amounts are essential for life—notably sodium, potassium, and calcium. Others, such as iron, zinc, and cobalt, are needed in small amounts, but become toxic in large quantities. A third group, which includes lead, cadmium, antimony, tin, and mercury, is toxic even in small quantities. Since the metals in the last group all have atomic numbers greater than 47, they are often grouped as the "heavy metals." For example, it is frequently noted that incinerator ash may be contaminated with heavy metals (principally cadmium and lead). In terms of actual density, not all of the heavy metals are especially heavy, with tin and antimony being especially light, but all are far heavier than sodium, potassium, and calcium.

None of the heavy metals is particularly common. Mercury is only about a sixth as common as lead or cadmium, the two that are most abundant in Earth's crust. But mercury is easily concentrated into rich deposits by geologic processes. As a result mercury has been known since ancient times and was the first of the heavy metals to be recognized as toxic; its dangers are mentioned by Pliny the Elder in the first century ce. (Lead was not known to be poisonous until 200 to 300 years ago, while the dangers of cadmium have been apparent only in the past 50 years or so.)

Mercury is dangerous in all forms, as an element and in all compounds, but it is also extremely useful in industrial processes, so it continues to be used. The exact mode of mercury poisoning is not clear, but it seems likely that mercury removes sulfur from key enzymes, making them useless. It also destroys kidney cells.

Mercury can cross the blood-brain barrier; once in the brain, it causes dementia. Since the sixteenth century, fur used to make felt for hats has been treated with nitrate of mercury. Before the dangers of this procedure became known, hatters often succumbed to mercury dementia; thus, the phrase "as mad as a hatter" entered the language and became the basis of Lewis Carroll's Mad Hatter in Alice in Wonderland. Even earlier, mercury was used by alchemists because gold and silver dissolve in it. Mercury fumes are thought to have caused dementia in alchemists who used it, perhaps even in Isaac Newton, who experimented extensively with alchemy and who became noticeably deranged in his later years. The ability to act as a solvent for other metals has led to the extensive use of mercury in mining, as well as to the development of an alloy of mercury used in dental work, amalgam.

Mercury in water is often converted by bacteria to the compound methylmercury, in which a carbon atom and three hydrogen atoms are chemically bound to each mercury atom. Methylmercury is much more easily absorbed by living organisms than elemental mercury. In water, it moves up the food chain, where it is concentrated in predator fish. Older, and therefore larger, predator fish are especially prized by humans for food. Also, methylmercury tends to accumulate in the parts of fish that people eat, the muscles. Fish are not especially affected by methylmercury poisoning, but humans, especially young ones, can be harmed. At least twenty-one US states, two Canadian provinces, and Sweden have had to blacklist lakes in which fish have high levels of methylmercury. Typically, pregnant or nursing women and young children are advised not to eat fish from such lakes, while others are told to limit their consumption. The worst lakes for methylmercury contamination are small, shallow ones, or lakes newly created by impounding water. In the smaller lakes it is warmer water and in the younger lakes organic debris that produces conditions suitable for bacterial action. Another factor is acidity; acid lakes develop more methylmercury.

Mercury gets into air not only from industrial processes, but also from incineration of batteries, paints, and switches containing mercury.

Perhaps the most recent case of widespread mercury poisoning occurred in and around the town of Minimata, Japan, on Kyushu Island. In the early 1950s fishers and other villagers in the region began to go mad and develop many neurological problems. There were many stillbirths and children born with serious defects. About a thousand people in all are thought to have died from what came to be known as Minimata disease. The "disease" was recognized in 1959 as mercury poisoning caused by industrial wastes that the Chisso Corporation dumped in Minimata Bay, polluting the fishing grounds of a people whose main source of protein is fish. Fish caught in Minimata Bay had mercury levels eleven times as great as would be permitted by the US Food and Drug Administration (FDA). As recently as the early 1990s about 2000 people living around Minimata Bay continued to suffer from the effects of mercury poisoning.

Because mercury is toxic to all forms of life, it has frequently been used as an additive to prevent growth of fungi or other organisms, especially in paper manufacture and in paint.

The Minnesota standard is the strictest in the United States, about a sixth of the FDA standard. Consequently, restrictions on fish consumption from the lake have been advised by the state for Crane Lake fish, although these are merely warnings, not prohibitions.

Echoes of Minimata

In 1990 the then Environment Minister of Japan renewed investigations into the Minimata disaster. His interest was fueled by a series of court decisions that ordered the Japanese government to aid about 2000 people who claim to be victims of mercury poisoning. Another 8000 or so who have so far not pressed their claims in court also say that they have grounds for compensation. Despite the court orders, the Japanese government has not taken action on the problem. The Environment Minister has since been dismissed, while the regional government of Kyushu has negotiated with the victims for compensation.

Although the mercury poisoning was identified as being caused by ocean dumping of wastes by the Chisso Corporation in 1959, fishing in the bay was not banned until 1968, when there was a second wave of poisonings. Chisso settled with several hundred survivors in 1973, a settlement that left the plant nearly bankrupt. Chisso today employs about 900 people in producing much of the liquid crystal that is used in various displays, including monitors for laptop computers. Despite the company's bad environmental record, local residents do not want Chisso to leave because people in the economically depressed region need the jobs.

Mercury in Your Mouth

In 1990 the Swedish Health Administration recommended that as soon as possible dentists cease using fillings made from amalgam, which is half mercury and half other metals. The Health Administration called amalgam "unsuitable and toxic." Furthermore, the Swedish government banned outright the use of amalgam fillings in pregnant women, starting in 1991. Researchers in Sweden and at the University of California Dental School had tested cadavers that had several amalgam fillings for mercury. The tests showed the cadavers to contain as much as three times the mercury in the brain and nine times the amount in the kidneys as controls with no fillings.

This finding prompted the first review of amalgam by the US Food and Drug Administration (FDA) since the 1970s, when the FDA approved its use on the basis that amalgam is widely employed and is not known to have caused problems other than rare allergic reactions. Amalgam has been used in dental fillings for at least 150 years; more than one American in ten has amalgam fillings.

Concern about amalgam was further increased by animal studies conducted by Fritz J. Lorscheider and Murray J. Vimy of the University of Calgary Medical School in Alberta. Their results, published in Aug 1990 and in Nov 1990, showed that monkeys and sheep with amalgam dental fillings accumulated enough mercury in body tissues to cause an observable decline in kidney function.

Despite these concerns, the FDA reported in 1991 that its review showed that amalgam is safe and could continue to be used in dental fillings. Alternative materials—gold, ceramics, and plastics—are generally more expensive and harder for dentists to install, so it is expected that in the United States amalgam fillings will continue to be the popular choice.

On a cheerier note, either the widespread use of fluorides in various forms or other health practices has resulted in less need for dental fillings of any substance. In the United States, children now develop fewer than half as many cavities as they did in the 1970s. Opponents to fluorides think, however, that fluorides were shown to cause cancer in several studies in the late 1980s. The US Environmental Protection Agency announced on Jan 3 1991 that it would take a renewed look at the fluoride issue.

(Periodical References and Additional Reading: *Science* 1-19-90, p 276; *Science News* 10-20-90, p 244; *New York Times* 12-13-90; *New York Times* 1-16-91, p A3; *Science News* 3-9-91, p 152; *New York Times* 8-26-91, p A1)

DANGER FROM RADIOACTIVITY

There are several well-known ways that a person can be exposed to hazards from radiation, ranging from damage caused by the relatively low-energy photons of ultraviolet radiation to damage caused by various kinds of subatomic particles. Perhaps because radiation damage was unrecognized until early in the twentieth century, and perhaps because most forms of radiation are undetectable by human senses, many view radiation as scarier than other forms of pollution. People given to black humor make jokes about those exposed to radiation as "glowing in the dark." Since radiation can result in damage to DNA or other genetic mechanisms, another association in people's minds connects radiation to monstrous births. Much of this concern is exaggerated, but the threat of damage from radiation of one kind or another is real.

Types and sources of radiation vary widely. These include cosmic radiation, which probably originates in violent processes in the Milky Way galaxy; solar radiation, which originates with fusion of hydrogen to helium in the Sun; natural radioactivity that is caused by the transmutation of elements in Earth's crust; radiation produced by people, including that from X rays and from radioactive elements, both used primarily to observe the interior of material opaque to light or to damage tissue; incidental artificial radioactivity produced as a result of energy production by nuclear fission or weapons research and development; and other forms of radiation incidental to modern technology. While the incidental forms of artificial radioactivity cause the most concern—largely because of the potential for devastating damage in nuclear accidents—for most people the natural forms of radiation are more damaging to their health.

In the early 1990s reports reminded people of the danger from cosmic and solar radiation to air travelers and of the danger of excessive use of diagnostic X rays. But the main problems that were extensively discussed involved natural radiation from Earth's crust, incidental radiation from computers, and nuclear reactors, working or broken, used to produce electricity.

Radioactivity from Earth

In Oct 1990 a conference of more than 230 scientists convened in Richland WA to grapple with problems connected with radon. Radon is an elemental gas that seeps into people's homes as a by-product of natural radioactivity of minerals in Earth's crust. Most commonly, radon is released along the way as uranium decays into stable lead. Because radon is a gas at temperatures and pressures that prevail in the upper crust, unlike other common radioactive elements, it leaks out of mineral deposits. If it reaches the outside air, radon disperses and causes little or no damage. The problem occurs in places, such as mines or home basements, where radon can collect. In such situations, breathing radon-enriched air leaves behind particles of its radioactive daughter products in the lungs, which are thought to cause cancer.

It has long been clear that miners who work in hard-rock tunnels have an increased risk of lung cancer as a result of radon exposure; it is less clear as to whether the much lower levels of the gas found in people's homes could also increase risk. A study by Janet B. Schoenberg of the New Jersey Department of Health, reported at the Richland conference and in the Oct 15 1990 *Cancer Research,* did establish a link, albeit a weak one because of the small size of the sample, between household radon exposure and lung cancer.

The US Environmental Protection Agency (EPA) has mounted a campaign against household radon throughout the latter part of the 1980s and into the 1990s. However, a study by the National Research Council of the US National Academy of Sciences that was released on Feb 1 1991 suggested that the EPA overestimated the risk of radon by 20 to 30 percent, although that lower level would still be considered serious. Another study, conducted by Naomi H. Harley of New York University Medical Center, demonstrated that people in homes contaminated by radon are less exposed to the gas than previously believed. Despite these results, it appeared that the EPA campaign to eliminated radon from homes would continue to be pressed.

Radioactive minerals are common in Earth's crust by some standards and uncommon by others. For example, thorium, uranium, and potassium-40 (the most common radioactive elements) are each a thousand to four thousand times as common as gold; but they are each about one ten thousandth or less as common as iron. The radioactive elements are common enough that any exposure to large amounts of material from beneath the topmost layer of the crust is potentially as dangerous as working in a

---

RADON

Radon is the heaviest of the noble gases (the others are helium, neon, argon, krypton, and xenon). All of these gases are exceptionally inert, meaning they combine with other materials only under unusual circumstances. Thus, radon does not present a chemical hazard in the way that sulfuric acid, carbon monoxide, and ozone do. Although it is highly radioactive, radon decays quickly, with a half-life of 3.82 days for the most common isotope and less for other natural isotopes. Thus, radon per se is not much of a hazard, but its immediate decay products are several radioactive elements in sequence: polonium-218, lead-214, bismuth-214, polonium-214, lead-210, and so forth (these are also sometimes known as radium A, B, C, C', and D). Collectively, these have a half-life measured in years, not days, and they cause most of the damage associated with radon.

---

nuclear power plant. In 1989, for example, Kerry M. St. Pé, a Louisiana state biologist, found that waste water from an oil field was twenty times as radioactive as the maximum allowed for waste water from a nearby nuclear power plant. Material that sticks to oil drilling equipment has also been found to be radioactive, as have tailings from fertilizer production. Natural gas and ash from coal are other common sources of radiation. All such radiation sources can be grouped as naturally occurring radioactive material, or NORM.

NORM, known since at least 1948, was propelled into the news in 1990 and 1991, partly by lawsuits by people or groups who found themselves exposed to radioactivity and partly as a result of publicity by the nuclear power industry fighting what it saw as overregulation. Another major factor was the increase in the regular use of radiation detectors. For example, detectors at landfills and scrap yards went off in response to loads of brine sludge from oil and gas drilling (Apr 1991) and paper mill wastes (Jul 1991), as well as hundreds of other triggers. In May 1991 the EPA began circulating a draft risk assessment for NORM, reporting, for example, that oil and gas drilling produces about 360 m^3 (12,713 cu ft) of NORM-laden waste each year. Despite this, no one is sure what to do about it, although both Louisiana and Mississippi have instituted regulation of oil-drilling related NORM.

Deliberately Produced and Accidental Radioactivity

Not all the material that trips the radiation detectors is NORM, however. Scrap metal dealers also encounter problems with metal containers containing radioactive isotopes developed for medical uses. On Mar 5 1990 the

Nucor Steel Company detector registered a problem that led to more than $2 million in costs to clean up the result of melting a container of radioactive cesium. A similar event that went undetected at a plant in Calcutta, India, resulted in fencing contaminated with radioactive cobalt-60. The contaminated fencing was first noticed in Aug 1990 in California and Oregon.

In 1991 the EPA devoted considerable effort to tracking down houses in and around Landsowne PA that had been built with sand contaminated with waste from radium production in the 1920s. The sand, now found in stucco, concrete, and plaster, is still emitting radon at levels that caused fourteen people to be evacuated from ten buildings in the area. Some houses are so badly contaminated that they may have to be torn down.

The US Department of Energy (DOE) has a noticeably bad record in handling radioactive waste. A typical example is the handling of tritium, the radioactive isotope of hydrogen. An internal review released on Nov 8 1991 showed that tritium losses had contaminated areas around plants and other DOE installations in at least seven states, resulting in some California wine and Georgia milk being contaminated, although the levels were low. Tritium is often taken lightly by workers because it is not as dangerous as some other radioactive isotopes, but it does add to the cumulative burden of radioactive exposure.

Regulation of radioactivity produced in nuclear plants in the United States will change significantly in 1993. Late in 1990, the US Nuclear Regulatory Commission (NRC) revised exposure limits for humans for the first time since 1957, with the new rules to start in Jan 1993. Instead of as many as 20 rems a year allowed for some workers in nuclear power plants, all workers will be limited to fewer than 5 rems a year, and pregnant workers will be exposed to only 0.5 rem. Also, nuclear power plants will for the first time be required to keep radiation doses to their workers "as low as reasonably achievable." Exposure of the general public to radiation caused by power plants is also to be reduced from 0.5 rem a year to 0.1 rem.

Although the new rules do not go into effect until 1993, they were first proposed by scientists in the field as early as 1977. The NRC, however, maintains that the new rules are only needed to provide an extra margin of safety, and that there was no danger to workers or to the general population from the previous rules. Nevertheless, a study by Steve Wing of the University of North Carolina School of Public Health, published in *The Journal of the American Medical Association* on Mar 20 1991, showed that workers at the Oak Ridge National Laboratory who received more than 4 rems had an increase in deaths from cancer over workers with less exposure. The rate among Oak Ridge workers studied for leukemia was 63 percent higher than for a similar population that was not exposed to known sources of radioactivity. On the other hand, a study released in Sep 1991 of 70,000 workers who repaired and refueled the US Navy's nuclear reac-

Background

Radiation that causes damage to living organisms is often called ionizing radiation, because knocking electrons out of atoms to form ions or unpaired electrons is frequently the first step in tissue damage. A compound that has an unpaired electron, which can occur as a result of chemical processes as well as from radiation, is called a free radical. Free radicals have been implicated in tissue damage, in degenerative disease, in cancer, and in aging. Certain isotopes of elements release various forms of radiation as they change from one isotope or element to another. This kind of radiation is commonly called "radioactivity," and it consists of various types of radiation in different proportions for each isotope.

Common name	Identity	Description	Damage to tissues
Alpha rays	Helium nuclei	Positively charged particles formed from two protons and two neutrons bound together	Because they are slow and massive, they cause the most damage, but do not penetrate very far into tissues
Beta rays	Electrons	Negatively charged particles	High-energy beta rays pass through tissues with little damage; lower energy ones cause much ionization and form many free radicals
Gamma rays	Photons, or electromagnetic radiation	Neutral particles of high energy that can often be more easily thought of as waves	Penetrate deeply, but have too much energy to produce more than widely separated interactions, so cause minimal damage
Positrons	Antiparticles of electrons	Positively charged particles	Similar to electrons
Neutrons	Neutrons	Neutral particles easily captured by atomic nuclei	Can change elements into radioactive isotopes
Neutrinos	Neutrinos	Neutral particles of little or no mass	No damage

Radiation is measured in rads and radiation-damage-potential in rems. A rad is based on the amount of energy absorbed per gram of exposed material, while a rem (an acronym for "roentgen equivalent man") is based on the different health effects of the different types of radiation. For example, 1 rad of slow alpha particles is equivalent to 1 rem, but so is 10 rads of fast electrons. A medical X ray generally produces exposure to about 6 to 7 millirems (a millirem is a thousandth of a rem). A lethal dose of radiation is about 500 rems, but substantial damage occurs at lesser amounts. Repeated exposure to radiation results in additional rems; the average person receives about 360 millirems each year or about 20 rems of radiation over a lifetime.

tors showed that the workers who had been exposed to low levels of radiation actually had fewer cancer deaths than the ones who had not been exposed.

The US rules apply to nuclear power plants in regular operation. But how do they apply to a nuclear power plant that has exploded and burned with the products going into the atmosphere? How do they apply to the Chernobyl disaster of Apr 26 1986? In 1990 and 1991 the changing political situation in the Soviet Union and in the Ukraine resulted in a steady flow of new information about Chernobyl, nearly all of it revealing that conditions there were and are much worse than anyone knew during the 1980s. Late in 1990, for example, Dmitri M. Grodsinsky of the Ukrainian Academy of Sciences in Kiev said that data indicated that about 150,000 people suffered some sort of thyroid illness as a result of the explosion and fire. Despite an official death toll that still stands at 31, Pripyat, the government agency charged with cleanup of the site, estimated in Oct 1990 that the true death toll is probably more than 300.

The official safe dosage of radiation for the general population in the Soviet Union has been 5 rems per year, the same as the official safe dosage for workers in nuclear power plants in the United States after 1993. According to an Oct 18 1991 letter to *The New York Times*, published Oct 28, by Alexander Shlyakhter and Richard Wilson of Harvard University's physics department, the average dosage received by cleanup workers at Chernobyl was 25 rems during the first year and 12 rems in subsequent years. Perhaps 1 percent of the workers received as many as 100 rems during the first year. The leader of the fire fighters, Leonid Teliatnikov, received more than 200 rems, but survived. A dose of 400 rems causes destruction of bone marrow and other organs, usually resulting in death.

Meanwhile, problems at the Chernobyl complex of reactors continued in 1991. While efforts to provide a permanent tomb for the reactor that exploded are still incomplete, the other two reactors at the site have continued to operate. On Aug 10 the working reactors suffered a major leak of water; On Oct 11 a fire destroyed the roof over one of the reactors; on Nov 1 another fire broke out in a generator room. Meanwhile, on Oct 29 the Ukrainian parliament voted to close the Chernobyl complex completely, although it was far from clear whether this would result in shutting the reactors down.

(Periodical References and Additional Reading: *Science News* 10-27-90, p 260; *New York Times* 12-24-90; *New York Times* 12-26-90; *New York Times* 12-30-90; *New York Times* 1-5-91, p 8; *New York Times* 2-21-91; *New York Times* 3-20-91; *New York Times* 3-29-91; *Science News* 10-26-91, p 264; *New York Times* 10-28-91, p A16; *New York Times* 11-3-91, pp L1 & L29; *New York Times* 11-9-91, p 10)

Timetable of Lead Poisoning

c 8500 BC	Lead beads are produced at Catal Huyuk in what is now Turkey
c 400 BC	Hippocrates of Cos [Greek: c 460 BC-c 370 BC] documents illness caused by lead mining
1621 CE	Lead mining in the United States starts in Virginia
c 1760	George Baker establishes that lead-lined cider presses cause a stomach illness
c 1890	Australian physicians discover that children are being poisoned by ingesting old paint that has turned to powder
c 1917	Physicians in the United States find that children develop lead poisoning by eating paint chips that have peeled off interior walls
c 1930	Leaded gasoline is introduced
1955	US paint manufacturers limit lead in paints to 1 percent or less
1971	US government orders stripping buildings of paint applied before 1955 (Lead Paint Poisoning Prevention Act)
1977-80	US government gradually imposes a ban on use of lead shot for hunting migratory fowl because of poisoning of wounded birds
1978	US government institutes first steps to ban lead from gasoline, mainly to protect catalytic converters in automobile exhausts
1979	In Mar, Herbert Needleman releases an influential study that shows that low levels of lead in children's blood and teeth correlate with lower IQ test scores
c 1980	US government bans paints containing lead
1983	A panel of experts from the EPA at first objects to the way the 1979 Needleman study was conducted; however, its final report states that the study's methodology was essentially correct
1984	EPA moves to ban leaded gasoline to reduce lead concentrations in the lower atmosphere
1986	EPA lowers its standards for the permissible amount of lead in air and bans the use of solder containing lead
1989	US government mandates tests for lead in children from low-income families
1990	A report from the US Department of Housing and Urban Development states that 75 percent of all US homes built before 1980 contain lead paint, amounting to 3 million tons of lead
	For the first time since the mid-1970s, a child who has eaten paint chips dies from lead poisoning
1991	On May 7 US mandates tests for lead in tap water; it will take about 20 years for tests to be completed and problems corrected
	In Jun Clifford Weisel and coworkers at the University of Medicine and Dentistry in New Jersey-Robert Wood Johnson Medical School report that lead ink used on bread wrappers could be a hazard; lead ink also contributes about 0.8 metric ton (0.9 short ton) of lead each day to the US waste stream

1991 (Cont.)	Donald Smith of the University of California, Santa Cruz, finds that California sea otters have up to 40 times the lead in their teeth compared with fossil sea otters of 1000 to 2000 years ago

On Oct 7 the United States lowers the definition of lead poisoning in children from 25 micrograms of lead per deciliter of blood to 10 micrograms

On Oct 11 the California government settles a lawsuit out of court by agreeing to provide tests federally mandated in 1989 for lead poisoning in poor children

In Oct Claude F. Boutron of the *Laboratoire de Glaciologie et Geophysique de l'Environemènt* in St. Martin-d'Hères, France, and coworkers report that lead concentrations in Greenland ice have dropped to levels not seen since the early 1900s; between 1967 and 1985 lead levels in the ice dropped by more than 85 percent; this is largely attributed to a decline in the use of leaded gasoline (the study also reports declines of 60 percent since the 1960s in cadmium and zinc)

In Nov the US government recommends new guidelines limiting leaching of lead from dishes to 3.0 micrograms of lead per milliliter of a 4 percent solution of vinegar over a 24-hour period and from cups and mugs to 0.5 microgram per milliliter; although not binding, the guidelines are set at about half the previously recommended limits

On Nov 12 California sues ten manufacturers of tableware for violating its limits for lead in dishes and mugs, which call for about one-fiftieth the amount of leaching permitted under federal guidelines; the California law does not ban tableware violating its permissible levels, but calls for clear labeling

On Dec 7 California obtains an agreement from major wine makers in the state to stop using lead caps over the corks on wine bottles

The US Office of Scientific Integrity asks the University of Pittsburgh to investigate charges that Herbert Needleman's 1979 study violates rules against scientific misconduct

1995	Lead is scheduled to be eliminated from all US gasoline

(Periodical References and Additional Reading: *Science News* 4-28-90, p 261; *New York Times* 12-20-90, p A1; *New York Times* 12-22-90; *New York Times* 2-21-91, p B8; *New York Times* 3-29-91; *New York Times* 5-8-91, p A1; *Science News* 5-18-91, p 308; *International Wildlife* 5/6-91, p 28; *Science News* 6-8-91, p 367; *Science* 8-23-91, p 842; *New York Times* 9-3-91, p C10; *New York Times* 8-24-91; *Science News* 9-14-91, p 166; *New York Times* 10-7-91, p C3; *New York Times* 10-12-91, p 7; *New York Times* 10-15-91, p C4; *Science* 10-25-91, p 500; *New York Times* 11-13-91, p A18; *New York Times* 12-8-91, p L39; *Science* 12-13-91, p 1575)

SYNTHETIC CHEMICALS UPDATE

For many people, the perceived main environmental hazard in the early 1990s was pesticide residues in their food, although there was little evidence that those constitute a major health hazard. Still, no one would suggest that pesticides are good for humans. Pesticides are not the only synthetic chemicals with a bad reputation. Classed along with pesticides in many minds are the dioxins, which are indeed contaminants in herbicides, and alar, a ripening agent that, like pesticides, is sprayed on food.

The Changing View of Dioxin

Dioxin is the general name for a group of related chemicals. The most common of the dioxins is 2,3,7,8-TCDD, short for 2,3,7,8 tetra-chlorodibenzo dioxin, a contaminant in the herbicide Agent Orange, or 2,4,5-T, which was used as a defoliant in Vietnam. Dioxins are poisonous for many animals, but in humans they have been proven to cause only the skin disease chloracne. There is, however, a great deal of circumstantial evidence and some scientific studies linking dioxins with cancer.

Timetable of Dioxin to 1990

1949	Monsanto's manufacture of pesticides in its plant at Nitro WV exposes its workers to dioxin
1951	Diamond Shamrock Corporation starts manufacturing Agent Orange
1971	Used oil contaminated with dioxins is sprayed on the roads in and around the small town of Times Beach MO; birds and animals die and at least one child becomes ill
1974	The US Centers for Disease Control in Atlanta GA (CDC) investigates the health problems in Times Beach
1976	Many around the globe learn of dioxin for the first time when an explosion at a chemical plant in Seveso, Italy, contaminates a large region with dioxin, resulting in the death of birds, farm animals, and pets, along with cases of chloracne in humans
1978	Dow Chemical releases a study that shows that dioxins are produced when garbage is burned
1980	A recycling plant in Hempstead NY is shut down because of high dioxin levels produced in incineration
1982	Soil samples collected by the EPA in Times Beach in Dec reveal high levels of dioxins
1983	The EPA agrees to buy Times Beach and move the residents out of it; eventually, it spends $37 million on the resettlement through 1991, with another $150 million needed to clean up the site

1984 US veterans of the Vietnam War reach in May an out-of-court settle-
     ment with chemical companies that once manufactured Agent Orange;
     the companies agree to set up a $250-million fund for veterans with
     health problems from exposure to the chemical

The years 1990 and 1991 brought a renewal of controversy about dioxins.
In Nov 1990 a conference at the Banbury Center in Cold Spring Harbor
Laboratory, later revealed to have been sponsored by the chemical industry, concluded that the dangers of dioxins to humans have been exaggerated; by Feb 1991 it appeared that many of the conferees had failed to
agree with that reported conclusion and that some felt that their names
had been used without permission or agreement. However, the main conclusion of the conference—that a low-enough level of dioxins might be
harmless—was not disputed at that time.

Despite the defections from the Banbury conference report, the EPA
seemed to take the report seriously and in May called for further study of
the significance of dioxin. In Aug 1991 the EPA touted its incomplete
year-long study of the health hazards of dioxins with widely publicized
statements that suggested that the hazards of dioxins would turn out to be
seriously exaggerated. The main goal of the study was to determine the
background level of dioxins to which the public is exposed, and the EPA
comments apparently were still based on the Banbury conference.

However, other signs pointed toward a change in attitudes towards
dioxins in 1991. On May 28 Vernon N. Houk, director of Environmental
Health and Injury Controls for the CDC, told the twenty-fifth annual Conference on Trace Substances in the Environment that he had been mistaken in recommending the evacuation of Times Beach MO in 1982. This
recommendation had resulted in the first large-scale public knowledge of a
dioxin problem in the United States. Shortly before the admission by
Hook, a federal district judge approved on Jan 1 1991 plans to tear down
buildings in Times Beach and to build an incinerator for burning the soil.

In the face of all this optimism, research showed for the first time in
1991 that dioxin does influence cancer in humans. On Jan 24 1991, Marilyn A. Fingerhut reported that the largest study ever of the effects of
dioxin in the workplace showed a small, but significant, increase in cancer among workers exposed to high levels of dioxin; the study was interpreted by many, however, as showing that the danger of dioxin to
humans was less than previously had been believed, since only high levels of the chemical were involved.

Another study, by Alfred Manz of the Center for Chemical Workers'
Health in Hamburg, Germany, released on Oct 19 1991, also concluded that
exposure to dioxins produced during the manufacture of chemicals leads to
an increase in the cancer rate in humans while not causing any specific
cancer. Also, in late Sep Georg Lucier and Chris Portier of the National

Institute of Environmental Health Sciences in North Carolina announced a new study of the effects of dioxin at low doses. The results suggested that low doses also might cause cancer, although further research was needed.

Alar UPDATE

Alar was the bad-news synthetic chemical of 1989. National television programs and famous movie stars denounced it. The EPA agreed that alar could cause cancer; by the end of 1989 the manufacturer, Uniroyal Company, had removed it from the market. Apple growers—alar was sprayed on apples—and the general public were distraught.

In Oct 1991 the EPA finally completed an extensive analysis of the toxicity of alar. The study showed that alar is about half as harmful as was thought, but that in high doses it causes cancer in rodents. At that point, however, the issue was largely moot—no one had used alar in almost 2 years and the apple industry was prospering as never before.

Alar may return to the news if not to the orchard in 1992. Families claiming to represent all apple growers in Washington State are suing the Natural Resources Defense Council, the instigator of the 1989 publicity, for misrepresentation. The US Department of Agriculture thinks that apple growers lost about $120 million in 1989, although more was lost in New York State than in Washington. The growers think the loss was more like $200 million in Washington.

Pesticides in the News

Just as the risk from dioxin remains controversial, so also does the situation with regard to pesticide residues in food. Although a 1954 US law still bans sale of any food with any level of pesticides that could cause cancer, on Feb 15 1991 the EPA announced that it did not have to follow that law for pesticides that pose no more than a one in a million chance of causing cancer. As a result, the EPA permits residues of benomyl on raisins and tomato products and trifluralin in peppermint and spearmint oil.

Fish kills probably caused by a pesticide in southern Louisiana in Jul and Aug 1991 were reminiscent of fish kills that were among the first major news stories of the environmental movement of the 1970s. By Aug 20 there had been nineteen major kills in the state according to the EPA. The suspected source was a pesticide used to combat the sugarcane borer, aziniphos-methyl.

(Periodical References and Additional Reading: *New York Times* 1-2-91; *New York Times* 1-24-91; *Science* 2-8-91, p 624; *New York Times* 2-17-91; *Science* 2-22-91, p 866; *Science* 5-17-91, pp 895, 911, 924, & 954; *Science News* 5-18-91, p 308; *New York Times* 5-26-91, p 20; *New York Times* 8-15-91, p A1; *New York Times* 8-21-91, p A20; *Science* 10-4-91, pp 7 & 20; *New York Times* 10-18-91, p A7; *Science* 10-18-91, pp 377 & 415; *New York Times* 12-3-91, p B6)

ENERGY PROBLEMS
AND SOLUTIONS

OIL AS A HAZARD

The early 1990s saw the setting of the world's record oil spill into the ocean—mostly as the result of a deliberate act. On Jan 19 1991 Iraqi soldiers released millions of gallons of oil into the Persian Gulf as an act of war. Some oil may also have been accidentally released by bombing by the US-led coalition against Iraq. Still more was spilled during a battle near a refinery in Saudi Arabia. The total amount that found its way into the Persian Gulf ultimately is believed to be 950 million L (250 million gal), twice the volume of the previous world record holder, oil from the blowout in 1980 of a Mexican well known as Ixtoc I. Previously, as much as 300 million L (80 million gal) had been spilled into the Gulf during 1983 as a result of the Iran-Iraq War.

Due to vagaries in wind and coastline topography, nearly all the spilled petroleum in 1991 wound up on the shores of Saudi Arabia, where it coated some 560 km (350 mi) of coastland. Cleanup operations have proceeded slowly. The Saudi Arabian Meteorology and Environmental Protection Administration estimates that the oil has killed about 30,000 birds and damaged many forms of marine life in uncountable numbers.

According to the US Coast Guard, about 8.7 million L (2.3 million gal) of oil is spilled into the oceans each year from all sources. Tanker accidents, such as the infamous *Exxon Valdez* rupture in Mar 1989 (48 million L or 13 million gal) account for only 5 percent of the total, although the concentration of oil from such spills makes them more dangerous than the more diffuse releases from urban runoff, industrial waste, tanker cleaning, and various small accidental spills. The year before the *Exxon Valdez* accident, four times as much oil was released into the ocean from tanker spills as from the *Exxon Valdez*, but it came from some 16,000 separate spills. A 1985 report from the National Research Council of the US National Academy of Sciences estimated that 33 percent of oil entering the ocean comes from tanker operations as release of oily water used as ballast, while 36 percent comes from runoff from cities and factories.

In 1990 the United States moved against oil spills with the Oil Pollution Act, but it failed to provide funding to enforce it. The act called for double-hulled oil tankers that are equipped with their own cleanup equipment. Such tankers are gradually being phased in, and it is expected that all tankers in US waters will conform to such standards by 2015. In the meantime, the International Maritime Organization is working on its own set of standards to prevent, or at least minimize, damage from spills.

The location of a large accidental release is important with regard to its consequences as well as the concentration of oil. The *Exxon Valdez* spill, about one-eightieth the size of the big Persian Gulf release, may have killed eight times as many birds—mainly because Prince William Sound, where the *Exxon Valdez* accident occurred, contains exceptionally abundant numbers of organisms and has a richer ecosystem than that of the Persian Gulf. Also, the *Exxon Valdez* spill actually coated a longer stretch of shoreline—650 km (400 mi)—and affected 1900 km (1200 mi). According to a government summary of fifty-eight scientific studies of the spill that was released Apr 9 1991, 350,000 to 390,000 seabirds were killed by the oil from the *Exxon Valdez*.

In the United States in 1991 the president, formerly a successful Texas oil producer, proposed an energy program based on increasing petroleum production and on additional capacity for energy from nuclear fission. The program's most controversial feature was opening the 77,000 km^2 (30,000 sq mi) Arctic National Wildlife Refuge in Alaska, created in 1980, to oil exploration and possible subsequent production.

One place, however, that will not suffer from oil spills for the next 50 years is Antarctica: on Oct 4 1991 a group of twenty-four countries with interests in that continent agreed to ban oil exploration until then. While the treaty had not been completely ratified by the end of 1991, it appeared that it would become law.

(Periodical References and Additional Reading: *New York Times* 4-10-91, p A1; *Science* 4-19-91, p 371; *Scientific American* 10-91, p 102; *New York Times* 10-5-91, p 3; *New York Times* 10-31-91, p D8)

ALTERNATIVES TO GASOLINE

In the United States motorized vehicles account for more than half the nation's oil consumption, so efforts to reduce national dependence on petroleum often focus on the fuel that cars use, mostly in the form of gasoline. Following are some of the ideas that were prominently suggested or even enacted into law in the early 1990s.

Electric Cars

Instead of gasoline in each auto's tank, oil can be burned in powerhouses (along with coal and natural gas) or, alternatively, nonpolluting sources of energy such as hydropower or nuclear energy can be used to produce motive energy. The resulting energy can be transported to autos and stored in the car batteries instead of fuel tanks.

The pace car for the 1991 New York Marathon was an experimental electric-powered Mercedes-Benz with a sodium-nickel-chloride battery, but that was one of the few electric cars on New York (or any other) streets in the early 1990s, approximately a hundred years after the first electric-powered vehicles took to the road. Plans persist for introduction of high-volume consumer electric vehicles in the 1990s, however. Ford Motor Company has announced that it will sell its Ecostar van to utility companies in early 1992; General Motors (GM) announced in Jan 1991 that its two-seater electric Impact will go on sale to consumers some time in the mid-1990s; Chrysler is planning an electric van with a nickel-iron battery for some time in the 1990s. State legislatures, led by California, are trying to accelerate the process with regulations and incentives, most of which will go into effect in 1998. The California Air Resources Board has mandated that 2 percent of all new cars sold in the state be electric by 1998 and 10 percent by 2003. An Anglo-Swedish company, Clean Air Transport of Goteborg, Sweden, plans to beat the deadline by introducing an all-electric car in Los Angeles in 1993, based on the car's having won a Los Angeles City Council contest for best design.

The key problem facing electric-powered automobiles is the battery. Existing batteries take up sixty times more space and weigh eighty times as much as the amount of gasoline needed to produce the same amount of power. Most batteries must be charged overnight, although Nissan announced in the summer of 1991 that it had developed a battery that could be charged in 15 minutes. Furthermore, in 1991 it cost about $0.60 a km ($1 a mi) to operate a battery-powered vehicle, but only $0.24 a km ($0.38 a mi) to run a gasoline burner. In Oct 1991 a consortium of US automobile makers (Ford, GM, and Chrysler), partly supported by the US Department of Energy, agreed to work together on solving the battery problem and pledged $35 million for the effort.

Liquid Fuels Other Than Gasoline

The very first internal-combustion automobiles were not propelled with gasoline, but with natural gasoline or even alcohol. Those options and others still exist today.

Compressed Natural Gas Around the world, more than half a million vehicles in the early 1990s were powered by compressed natural gas (CNG). These vehicles produce from 40 to 90 percent less of the unburned hydrocarbons, which contribute to smog and low-level ozone, than gasoline does, depending on driving conditions. They also produce about half as much carbon monoxide, a reactive gas that reduces the oxygen-carrying capacity of the blood. They even produce less of the well-known greenhouse gas carbon dioxide, which contributes to global warming.

On the other hand, they produce more nitrogen oxides, which contribute to acid rain and to smog. CNG-driven vehicles would cost about $1000 more than gasoline-powered vehicles if they were mass-produced, which they are not. But the higher initial cost would be paid back after 2 or 3 years of driving due to lower fuel costs. Finally, a tank of CNG runs a vehicle about half the distance of a tank of gasoline, an important consideration in some parts of the world, although only an annoyance in most of the United States.

Methanol Anything containing carbon can be converted chemically or biologically to methanol, also known as wood alcohol because it results when wood is subjected to destructive distillation. Like CNG, methanol is already in use in vehicles worldwide and produces about 10 percent of the amount of unburned hydrocarbons as gasoline does, although it does not have the other pollution advantages of CNG. Indeed, combining methanol burning with its production, methanol can result in substantially more carbon dioxide entering the atmosphere than gasoline burning and production does. Also, like CNG-powered vehicles, methanol-powered vehicles cost more initially and travel fewer miles on each fill-up.

As in experimental solutions to so many environmental problems, California is leading the way in practical methanol-power studies. Starting in 1992, the California Energy Commission will use a fleet of 2000 General Motors cars equipped for both methanol and gasoline fuel. Ford has announced mass production of such vehicles starting in 1993.

Synthetic Diesel Fuel Diesel fuel burned commonly in trucks and buses in the United States is known as No. 2, although a better grade—less polluting, less smelly, and easier to burn—called No. 1 is available at a higher cost. The Rentech Corporation of Denver CO has developed a process to make the equivalent to No. 1 diesel fuel synthetically. While the process itself is just an improvement on a method developed in the mid-1920s (and used by the Nazi government of Germany during World War II to make fuel from coal), the raw material is something that is almost free and definitely unwanted. The fuel is made from the gas that arises from decaying organic matter in landfills, mainly the methane that is about half the gas. The first plant producing the fuel uses gas from a landfill in Pueblo CO and produces about 240 barrels of fuel a day. Rentech thinks that it can make a profit, since it pays Pueblo only $0.60 for the gas needed to produce a barrel of fuel, but sells the product for $60 a barrel. The plant, however, cost $16 million to build.

If it can find a market, Rentech plans to build more such plants. This would suit the US Environmental Protection Agency, which proposed in May 1991 that methane gas emitted by about 600 landfills should be controlled.

Propane Although not highly touted, propane, which is produced from wells that also produce natural gas and which can also be produced by refining petroleum, is an alternate fuel that is even now less expensive as well as less polluting than gasoline. Propane is familiar to many as the fuel for gas barbecue grills or as the fuel for some camping equipment. Houses in remote locations often use propane as a cooking fuel as well.

The main advantage of propane for automobiles and trucks is that it produces 70 to 80 percent of the carbon monoxide that gasoline does when it burns. Some locations, such as Denver, have more carbon monoxide than others, but carbon monoxide is something of a problem anywhere that there is sufficient traffic and stagnant air. Carbon monoxide causes adverse health effects by binding to sites on red blood vessels that normally carry oxygen through the body.

The main disadvantage of propane is that not enough of it is produced for it to be used nationwide. Also, it does not offer all the advantages for pollution abatement that some other alternative fuels do.

Nevertheless, Conoco, a gasoline-selling subsidiary of E.I. du Pont de Nemours & Company, began selling propane for use in automobiles and trucks in Denver and in Colorado Springs CO on Aug 29 1991. It also plans to open additional stations in Colorado, Missouri, Oklahoma, and Texas. Conoco sells propane for about 85 percent of the cost of gasoline, but the same amount of propane yields only about 85 percent of the distance that gasoline does, so there is no real cost advantage. Conversion of existing gasoline engines to propane or to mixed use of propane and gasoline costs from $1800 to $3000, so the only real advantage is in reduction of carbon monoxide.

Hydrogen Hydrogen would seem to be the perfect fuel, since it can be made from water and returns completely to water vapor when it is burned. Its main pollution drawback is that caused by the production of electricity needed to free the hydrogen in water, although there are proposals to use nonpolluting solar power for that purpose.

Hydrogen, however, is hard to store and automobiles using it would need to be redesigned from the ground up. To store hydrogen as a liquid, it must be kept at 20 K (-253° C or -423° F). The alternative is to store it in some form, such as a metal powder, from which it can be chemically released as a gas.

Gasoline as an Alternative

Some state and local governments, and even some petroleum refiners, have suggested that newer "good" gasolines are the best alternative to the old "bad" gasolines. This idea is somewhat controversial, in part because dramatic reductions in tailpipe emissions so far have come from engine changes, not from new fuel formulas. The best estimate is that in the 1970s the average US car emitted about 14.5 g per km (0.3 oz per mi) of unburned hydrocarbons. It is predicted that this average will reach 2.5 g per km (0.05 oz per mi) by 1995, starting with a Nov 1992 deadline called for by the Clean Air Act of 1990, which mandated the addition of oxygen-bearing molecules (oxygenates) in gasoline to reduce pollution from unburned hydrocarbons. But that is not enough for the nine cities and adjoining metropolitan regions in the United States with the worst ozone and smog levels, so the federal government has mandated reformulation of gasoline for those markets to lower emission levels further. The federally imposed gasoline standards for those regions after 1995 call for reducing all chemicals that contribute to smog, including the harder-to-control nitrogen oxides.

The new gasolines are expected to cost about $0.016 per L ($0.06 per gal) more, mostly as a result of reformulation, although a 1991 study by the Cambridge Energy Research Associates of Cambridge MA puts the projected costs at a little more than half that much. By the year 2000 another, even cleaner, reformulation is supposed to kick in, one that will raise the cost of gasoline by another $0.024 per L ($0.09 per gal). On Nov 22 1991 California adopted a statewide standard for reformulated gasoline to take effect in 1996; the standard is almost exactly the same as that for the nine metropolitan regions in the year 2000. A coalition of eleven northeastern states has vowed to follow California standards on auto pollution equipment and they are expected to follow California's lead on fuel as well, tending to cause such standards to become nationwide.

Oil companies are divided on whether the new gasolines are a good idea. ARCO and Ultramar, both centered in California, are for the new standards, partly because they already have equipment in place that can easily meet them. ARCO announced its new gasoline before the new standards were adopted, claiming it would cost $0.04 more per L ($0.16 more per gal). Texaco has led the fight against reformulated gasolines with claims that reduction of pollution from a ton of unburned hydrocarbons will cost $10,000 for the federal 1995 standards for heavily polluted urban regions, more than $200,000 for the 2000 federal standards for the urban regions, and $500,000 for the 1996 California standards. California regulators prefer to express the cost of their new standards as $0.003 per km ($0.005 per mi). In that case, a motorist driving 15,000 km (about 10,000 mi) a year would pay about an additional $50 for fuel.

In Oct 1993 diesel fuel gets some new restrictions as well. Its sulfur content must be cut to 0.5 percent, which will probably add $0.04 to $0.07 a gallon to its cost.

Just Use Less

Since 1975 the US federal government has been imposing increasingly stronger rules concerning how far automobiles must travel on a given amount of gasoline. From the beginning US automakers have claimed that such requirements can be enforced only by making automobiles unaccept-ably tiny, weak, and dangerous. Although these predictions have not been borne out by events, US automakers continue to make them. In the mean-time, fuel efficiency of cars has doubled since 1976, saving 2.5 million bar-rels of gasoline daily (according to the American Council for an Energy Efficient Economy in 1991). In 1991 US requirements for its own automak-ers were for a corporate fleet average of 11.7 km per L (27.5 mi per gal).

US automakers were taken aback when Honda Motor Company announced the EP-X on Oct 16 1991, a two-seater aluminum prototype car that Honda claimed could travel more than 40 km per L (100 mi per gal) of gasoline. The EP-X achieves a high distance/fuel ratio by mixing much more air into its gasoline. In Dec 1991 General Motors (GM) announced that it also had developed an experimental automobile with the same high distance/fuel ratio, at least at highway speeds. The GM Ultralite uses a car-bon-fiber body to reduce its overall mass to 635 kg (about 1400 lb). The main problem is that carbon-fiber material currently costs more than a hundred times as much as a given amount as steel does, although GM has patented a process that it believes will reduce the cost to only about twenty times that of steel.

The advent of such innovative cars is still some way off. The Mas-sachusetts consulting firm of DMI/McGraw-Hill proposed in Oct 1991 that an immediate improvement in air quality can be made by buying up and scrapping the 9 million US vehicles made from 1967 to 1978 that are still on the road. This would have the immediate effect of raising the average distance traveled per amount of gasoline by about 50 percent.

There are still other ways to reduce pollution from gasoline and oil. For example, in Jan 1991 all vehicles were banned in a fifty-block area of downtown Mexico City, believed to have the worst air pollution of any urban region. Furthermore, the Mexican federal government announced on Mar 18 1991 the closing of a fifty-seven-year-old oil refinery in the city rather than permit it to continue to add to the air pollution. The closing throws 3206 workers out of a job. Mexico also indicated that it was taking other steps to ease air pollution from gasoline and oil, including passage of new laws, better enforcement of existing ones, and planting of trees in Mexico City.

(Periodical References and Additional Reading: *Science* 2-1-91, p 515; *Science* 3-15-91, pp 1287 & 1318; *National Wildlife* 8/9-91, pp 10 & 13; *New York Times* 8-20-91, p D2; *New York Times* 8-21-91, p D7; *New York Times* 9-11-91, p A16; *New York Times* 10-17-91, p D1; *New York Times* 10-24-91, p A20; *New York Times* 10-31-91, p A20; *International Wildlife* 11/12-91, p 25; *New York Times* 11-23-91, p 1; *New York Times* 11-26-91, p D1; *New York Times* 11-27-91, p D2; *New York Times* 12-25-91, p 47; *New York Times* 12-30-91, p D2)

ALTERNATIVE SOURCES OF ENERGY

Virtually all of the energy we use to run our machines or heat our dwellings comes originally from the Sun. One way to measure energy from the sun is by the *solar constant,* which is 429.2 Btu's per square foot per hour (1.94 calories per square centimeter per hour) at the outside of Earth's atmosphere. Clouds and the atmosphere itself diminish this energy by different amounts before it reaches Earth's surface. The amount of energy that reaches the surface also depends on the angle the Sun's rays make with a given place on the surface. In the United States, for example, the Sun is never directly overhead except in Hawaii. The angle is progressively less at higher latitudes, so less energy reaches northern regions than ones closer to the equator.

Fossil Fuels

In developed countries today, most of the solar energy used is in a form that was originally captured by plants millions of years ago and then changed by geologic processes into coal, petroleum, or natural gas. While the same processes continue today, it is clear that we are using these "fossil fuels" much faster than nature makes them. This situation is worse than previously believed. In Jan 1988 the International Energy Agency announced that the world was using a million more barrels of oil each day than any previous estimates had shown. Later that year the US Geological Survey cut their estimates of undiscovered oil and gas in the United States in half. In addition to the environmental problems, there is a growing need foreseen for additional electric power. Although total electric power consumed in the United States has remained stable for the last 15 years—largely due to conservation inspired by the oil crises of the 1970s—the North American Electric Reliability Council predicts that an additional 73 thousand-megawatt power plants will be needed in the United States by 1997. (One megawatt of power production can conveniently be thought of as the power production needed to supply a town with a population of 1000.) Similarly, the US Department of Energy's Energy Information Administration predicts an annual growth rate in power consumption of 2.4 percent each year in the 1990s. As a result, we know that at some time in the fairly near future it will be necessary to change to some other form of energy. Furthermore, use of fossil fuels causes several environmental problems, including acid rain, ground-level ozone, and other forms of air pollution; the greenhouse effect; and oil spills.

One fossil fuel, however, is a fairly attractive energy source. It is natural gas, which supplies 27 percent of US energy, but causes only 13 percent of the air polluting nitrogen oxides and virtually no sulfur dioxide. (These chemicals are the main causes of acid rain.) Natural gas produces less carbon dioxide for the same amount of energy as burning other fossil fuels, so it is less of a contributor to the greenhouse effect. Gasoline produces 40 percent more carbon dioxide and coal 85 percent more than natural gas. Natural gas also is innocent of causing ground-level ozone, a principal cause of lung damage from air pollution. Finally, natural gas does not produce any soot, a product of both coal- and oil-fired sources. Natural gas is also more abundant than petroleum.

One result of the many benefits of natural gas is that new uses are being found. A number of successful programs use natural gas to power automobiles. Although the natural gas refrigerator has been known for years, only recently has the same technology been adapted for air-conditioning and dehumidification. Natural gas is also one of the gases that can be used in fuel cells (see below).

Here are the available alternatives to the use of fossil fuels. All are renewable in one way or another except for nuclear fission and nuclear fusion.

Direct Solar Power

Photovoltaic cells These use the interaction of light (photons) and electrons to generate electric current. This effect was first discovered by Edmund Becquerel [French: 1820-1891] in 1840. The amount of current from a single cell is not very great and the efficiency of cells commonly in use is only about 10 percent, which means that a 100-square-centimeter cell produces about 1 watt of electricity on a clear day. The best silicon cells have an efficiency of 12 percent. More complex double-cell methods that are largely experimental can convert 34 percent of sunlight to electricity.

Currently, panels of such cells are used as power sources in various US spacecraft and isolated places on Earth where electric lines do not reach. More than 15,000 homes worldwide depend on photovoltaic cells as a power source, while the US Coast Guard uses more than 10,000 such systems to power lighthouses and buoys. Small cells or panels of cells are also used to power calculators, watches and clocks, and other small appliances or toys. In Phoenix AZ there is a small solar-powered community in which each of twenty-four homes gets some of its power from photovoltaic panels.

Since their introduction in 1954, the cost of electricity from such cells has dropped from $15 a kilowatt-hour to $0.30 a kilowatt-hour. It is predicted that a new thin-film cell can generate electricity at $0.13 an hour when the technology is mastered. Still, photovoltaics need to get below $0.12 a kilowatt-hour to be competitive with conventional power sources. The world's largest installation of cells, completed in 1985, is in Carissa Plains CA; it can generate 6.5 megawatts of electricity.

A promising new type of cell was introduced in Oct 1991 by Michael Grätzel and Brian O'Regan at the Swiss Federal Institute of Technology in Lausanne. It uses a liquid electrolyte to transfer electrons from a positive to a negative electrode; sunlight becomes involved when a dye that absorbs the Sun's energy uses that energy to lift an electron from the dye into the electrolyte, in a process analogous to photosynthesis. Although the cell's efficiency would be only about 12 percent, such a cell would be much cheaper to manufacture than those based on silicon or on gallium arsenide, the two main types of existing solar cells. Long-term tests to make sure there are no corrosion problems and to firm up manufacturing techniques are needed before it is clear whether the new type of cell is the breakthrough people have been seeking.

Solar thermal generators These devices use mirrors to focus the light and heat from the Sun on fluids that boil to produce power from otherwise conventional generators.

Large Solar Thermal Electric Systems

Project	Location	Capacity (megawatts)	Completion Date
Luz, SEGS 1	California	14	1984
Luz, SEGS 2	California	30	1985
Luz, SEGS 3	California	30	1986
Luz, SEGS 4	California	30	1986
Luz, SEGS 5	California	30	1987
Luz, SEGS 6	California	30	1988
Luz, SEGS 7-19	California	450	1989-92
Luz, Eliat	Israel	25	1990
Solar One	California	10	1982
Mysovoye	Soviet Union	5	1986
Solarplant 1	California	4	1984
Themia	France	2	1983
CESA-1	Spain	1	1983
Sunshine 1	Japan	1	1981
Sunshine 2	Japan	1	1981
Eurelios	Italy	1	1981
Solntsye	Soviet Union	1	1983

Solar heat panels Flat plates that usually contain water (although sometimes air is used) collect the heat of the Sun in the fluid and move it, often by convection, to a storage area or through pipes and radiators in a house. Water heated this way can be used either for hot water in bathing, cleaning, or swimming pools, or it can be used to heat a building. The temperature of the water produced in this way is not high enough to use for power generation.

Passive solar heating Buildings can be constructed to take advantage of solar energy, even fairly far north. A building in Wallasey, England, for example, that was designed in this way was found not to need a conventional furnace that had been installed as a backup; the furnace was eventually removed. This form of heating, called *passive solar heating,* is suitable only for the heating of buildings. The basic principle of passive solar heating is to orient the building and its windows so that the Sun's energy is let into the building in the winter and usually so that less energy enters in summer. Once inside the building, some of the energy is stored in concrete, rock, or water so that the building will stay warm at night and on cloudy days. Furthermore, heavy insulation keeps the interior heat from leaking out. Often a greenhouse is attached to the building to collect energy beyond the amount that could be collected with conventional windows. Sometimes a shallow pond is built on the roof to collect and store heat.

Indirect Solar Power

Moving water Perhaps the oldest way of turning solar energy into motion is through waterpower. The energy of the Sun converts liquid water to water vapor that rises into the air; as it rises, the water vapor is cooled by lowering air pressure, starting a process that leads to precipitation. Some of the liquid or solid water falls on high ground. Eventually, liquid water moves toward lower ground. If it is intercepted by a water wheel or a turbine, this moving water—an indirect form of solar energy—can be used to turn mills or generators. This idea was implemented somewhat in antiquity, but was not really a major source of power until the Middle Ages. By 1086 CE there were about 5000 waterwheels in England. People also learned early on to build dams to supply both a more consistent power source and a greater one, since the amount of power generated depends in part on the distance that water falls in a short time.

After networks of electric power began to be introduced, very large dams were built in part to use waterpower to produce electricity. Plants that transform waterpower to electricity are often called *hydropower* generators. In 1986 Venezuela built the largest dam in the world, complete with generators that produce 100,000 megawatts of power. Other countries, notably Brazil and China, are planning even larger dams.

In another approach to the generation of electricity, many very small plants are being built or planned. In some cases, these are in old mills and produce electricity from waterwheels and millponds. Others are on swift-moving rivers in Third World countries. All these plants generate less than 25 megawatts each.

Although using moving water does not cause air pollution, it often has a serious environmental impact. Dams can prevent fish from traveling upstream to spawn; can flood productive farmland, forests, or villages; and can result in endangering species, or even in making species extinct. Often the environmental consequences are unpredictable. The Aswan Dam across the Nile both ruined the sardine fishery in the eastern Mediterranean and caused a rise in schistosomiasis, a parasitic disease that requires snails that live in still waters for its spread.

In the United States there are about 2000 hydroelectric dams operating in the early 1990s—1400 of which are privately owned. The others were built by and are owned by the federal government. About 10 percent of the electric power in the United States is produced by these dams.

Ten Largest US Hydropower Projects

Name	Location	Power Capacity in Megawatts
Grand Coulee	Columbia River WA	6180
Chief Joseph	Columbia River WA	2457
John Day	Columbia River OR	2160
Bath County P/S	Little Black Creek VA	2100
Robert Moses-Niagara	Niagara Falls NY	1950
The Dalles	Columbia River OR	1805
Luddington	Lake Michigan MI	1657
Raccoon Mountain	Tennessee River TN	1530
Hoover	Colorado River NV	1434
Pyramid	California Aqueduct CA	1250

Wind Wind occurs as a result of unequal heating of air masses by the Sun in different parts of the world. It has been used as a source of power since about 600 CE, when horizontal-bladed windmills were introduced in Persia (now Iran) to grind grain. Vertical windmills for pumping water were important in the Netherlands and in the US Midwest until the advent of electrical pumps. Today the chief interest in wind power is in the generation of electrical current. A General Electric study in 1977 claimed that wind power could produce as much as 13.6 percent of US electric power by the year 2000 if available sites were used. Since then, windmills have gotten better and power consumption has not increased as fast as the study projected, suggesting that wind power could contribute even more.

Countries around the world use both small and large windmills to generate electricity. In the United States, most large-scale wind plants are in California, in "wind farms" that contain hundreds or thousands of windmills.
The world's largest single windmills, however, are in Oahu HI and Cap
Chat, Quebec. The Hawaiian windmill is vertical-bladed, like the propeller
of an airplane, but this single prop is 97.5 m (320 ft) long from tip to tip.
Installed in 1988, it produces 3.2 megawatts.

The Quebec windmill is vertical and looks like part of an eggbeater,
except that it is 94 m (308 ft) high, with two arc-shaped blades that are
76.2 m (205 ft) apart at the widest distance between them. It generates 3.6
megawatts of power.

The California wind farms, which produce 1 percent of the state's electricity, use three-bladed propellers with an average diameter of the circle
made by their tips of 17 m (56 ft). The average power-production capacity
of the individual windmill has gradually risen to 120 kilowatts.

Wind Farms in California Since 1981

Year	Number of Machines Installed	Total Capacity in Megawatts	Total Power Generated in Millions of Kilowatt-Hours
1981	144	7	1
1982	1145	64	6
1983	2493	172	49
1984	4687	366	195
1985	3922	398	670
1986	2878	276	1218
1987	1500	180	1600
TOTALS	16,769	1493	3739

Biomass The earliest known use of indirect solar power was the use of
biomass to produce light and heat around 1 million BC. The evidence suggests that our ancestor *Homo erectus* was using fire in the Swartkens caves
of South Africa at that time. Today, over half the wood cut each year
around the world is still being burned, primarily for heat.

Wood and other present-day plant products (as opposed to fossilized
plant products) are referred to in the energy field as *biomass*. Biomass is
an indirect form of solar energy, since energy from the Sun is stored in
green plants. Fire and other means can be used to liberate some of this
stored energy.

The wood-fueled stove or furnace has been seen in recent years as a
sensible way to use renewable biomass for heating instead of nonrenew-
able fossil fuels. In 1988 wood provided 10 percent of the residential heat

Selected Biomass-Fueled Electricity Generating Sources in the United States

Project	Capacity in Megawatts	Fuel Sources	Start-Up Year
Union Camp Franklin VA	96	Pulping waste, peanut shells	1937
Champion International Cantonment FL	78	Pulping waste, bark	1961
Manville Forest Products W. Monroe LA	72	Wood and pulping wastes	1961
Northern States Power Ashland WI	72	Forest residues (sometimes supplemented with conventional fuel) from logging	1983
McNeil Generating, Burlington Electric Burlington VT	50	Forest residues from logging	1984
Eugene Water & Electric Board Weyco Center OR	46	Sawmill residue	1983
Washington Water & Power Kettle Falls WA	46	Sawmill residue	1983
Ultrasystems Fresno CA	27	Forest residues from logging and agricultural residues	1986
Ultrasystems West Enfield ME	27	Forest residues from logging	1986
Lihue Plantation Kauai HI	26	Bagasse (pulp from refining sugar from cane)	1980
Louisiana Pacific Antioch CA	26	Wood waste	1983
Wheelabrater Energy Delano CA	25	Orchard prunings	1989
Dow Corning Midland MI	22	Wood chips	1982
Alternative Energy Decisions Bangor ME	17	Forest and industrial residues	1986
Farmers Rice Milling Lake Charles LA	11	Rice husks	1984
Procter & Gamble Staten Island NY	10	Industrial residues, wood chips	1983

in the United States, used exclusively in 5.6 million homes and in conjunction with other heat sources in another 21 million homes. Unfortunately, in some places, notably Aspen CO, the air pollution from wood stoves has been so bad that their use has had to be curtailed.

Biomass can also be converted to a liquid fuel, either ethanol (grain alcohol) or methanol (wood alcohol). Ethanol is generally used as an additive in gasoline, where it can replace lead to increase octane. It can also be used as a fuel by itself in cars that have been converted to use it. In Brazil, the nation most committed to ethanol fuel, ethanol provides about half the fuel used. Furthermore, since ethanol can be produced from sugarcane, the fuel can be produced in plants that are largely fueled by burning the sugarcane residues (bagasse).

While wood-fueled steam engines were used in the nineteenth century to convert biomass to power, the idea lost popularity as coal and later oil became the principal fuels. In the United States, there has been a recent trend toward using biomass that would otherwise be discarded to generate power in the form of electricity, which can be used either by the generating plant or sold to a utility company. Often both uses can be employed, by selling excess power when available and buying power from the company when there is not enough for the plant's needs.

It is estimated that there are also about a thousand wood-fired electric generating plants in the United States, most in the 10 to 25 megawatt range, and most producing electricity for the forest-products industry. A 20-megawatt plant burns 600 tons of wood a day. In 1990 coal cost 70 percent of what wood cost as a fuel, but geography can make wood economical for some plants that are located in forests with poor access to coal.

In Third World countries, biomass is often used to generate heat or light by another method. Manure is biomass that has passed through an animal. Under suitable conditions (proper amount of manure, proper confinement, and so forth), manure produces methane, a gas with properties similar to those of natural gas (which is principally methane). A manure-powered methane generator is cheap and effective when there is enough manure to keep it working.

Garbage also produces methane in landfills. Some landfills in the United States use garbage-generated methane to produce electric power, which is then sold to a utility.

Ocean thermal energy conversion (OTEC) Most of the Sun's energy that reaches Earth lands in the ocean, since about 70 percent of Earth's surface is covered with ocean. Water is a good medium for heat storage, so some of that energy remains in the ocean as heat. This is especially true in the tropics. But even in the tropics, water a half mile below the surface may be as cold as 6° C (43° F). It is the difference between a 27° C (80° F) or warmer surface temperature and a 6° C (43° F) or colder temperature farther down that is exploited in Ocean Thermal Energy Conversion, or OTEC. The minimum temperature difference to make OTEC practical is a difference of 20° C (37°F).

Pilot plants have demonstrated that the concept of OTEC can work (the first French plant was built off Cuba some 60 years ago). A volatile liquid, such as ammonia or propylene, is used to turn a turbine. At low temperatures the substance is a liquid, but at higher surface temperatures it becomes a gas. Since the chemical expands when it becomes a gas (just as water expands when heated to form steam, but at a lower temperature), the volatile liquid can be used to turn a turbine and generate electric power.

At present the economics of OTEC have not made it a good alternative to other energy sources in most places. An exception may be in Third World countries where electricity can cost $0.15 a kilowatt-hour, since projections show that OTEC in warm-enough waters can produce electricity at $0.06.5 per kilowatt-hour.

Experimental plants have shown that the main problem with OTEC is fouling of the machinery by the ocean and by creatures living in it. On the other hand, OTEC causes no pollution itself as long as the volatile liquid is confined.

The other limitation is that OTEC is effective only in the tropics. Hawaii is the only really suitable site in the United States, although Florida and the Gulf of Mexico are marginally suitable. The best sites are in the South Pacific. If this technology were to become a serious competitor, manufacturers might find it economical to move plants to places like Tahiti.

Ocean wave energy conversion Since at least 1911, when an ocean wave converter was installed at Atlantic City NJ, it has been known that the energy of ocean waves can be converted to power a generator. This energy is derived from the Sun by way of the wind. Various devices that can bob up and down or be turned by the movement of ocean waves are used to move camshafts that power a generator. Those that work completely submerged are environmentally less harmful and less prone to damage by storms. Although a number of devices have been patented and tested on a small scale, there have been no large-scale applications.

Nonsolar Power

Fuel cells In a sense, the process in a fuel cell is just the reverse of the well-known method of breaking water into hydrogen and oxygen by applying an electric current. Reversing the reaction means that you start with hydrogen and oxygen and wind up with electricity and water. To accomplish this, you need to combine hydrogen and oxygen in the presence of a catalyst, which also serves as an electrode.

The most successful fuel cells have been used in the US space program, starting with Gemini in 1965 and continuing with the Apollo program and the space shuttle. Fuel cells are the energy choice of the space program because they can produce the highest energy for the weight and volume.

Also, the space program normally uses liquid hydrogen and liquid oxygen as rocket fuel, so it is easy to use the same combination to produce electric power.

The environmental advantages of the fuel cell are intriguing. Both hydrogen and oxygen are extremely plentiful on Earth. The "ash," water, does not contribute to air or water pollution and is nontoxic. In space, where liquid hydrogen and liquid oxygen are used, there is some danger of a fuel cell exploding. In the ill-fated Apollo 13 mission of 1970, the oxygen tank of a fuel cell exploded, leaving the ship short of electric power and causing the crew to return to Earth without accomplishing a planned Moon landing. Such an explosion is less likely on Earth, where the most common fuel cells would run on either natural gas or methanol with gaseous oxygen from the air. Although such mixtures can explode, an explosion would be no more likely than that of a gas furnace in a home or a fuel tank in a car.

In the early 1980s there was considerable enthusiasm for using arrays of fuel cells either to produce electric power for distribution through the grid or to produce power for vehicles, such as tractors. Pilot power-generating stations were started in Tokyo, Japan, and in New York City. The US Army let out contracts for the development of transportable-fuel-cell systems. By the end of the decade, however, various technical problems that arose led to stagnation in the field. A 1989 survey of energy technology in *Science* magazine, for example, mentions fuel cells only twice in passing, and both times suggests that their routine use is far in the future.

Geothermal energy The interior of Earth is hot. For the first 18 to 21 m (60 to 70 ft) below the surface, the climate of the region influences the temperature of soil and rock; below that distance, however, the average temperature rises about 1° C (2° F) for every 45 m (150 ft) deeper you go. That is the average, but in many places the temperature rise is much steeper. In general, these are places associated with some form of present or past volcanic activity, including geysers and hot springs. Such hot spots are the most attractive places for using heat to produce power. At places other than hot spots, one would have to drill about 3 km (2 mi) down to achieve a temperature that would boil water. At some hot spots, boiling water is at the surface.

The source of this heat has been disputed, but most scientists today believe that nearly all of it comes from radioactivity. The small amount of naturally occurring radioactive elements that are part of the crust and lower layers of the planet give off heat as they decay. Because that heat is trapped under tons of rock, it gradually accumulates. Other sources of heat include heat left over from the formation of Earth by the collisions of many smaller bodies, heat caused by gravitational contraction of Earth, and heat caused by friction between moving tectonic plates.

Conversion of Earth's heat to electrical energy was first demonstrated in a 1904 experiment in Lardello, Italy, in which geothermal energy was used to power five light bulbs. Geothermal energy has long been used as a source of hot water and electric power in Iceland, which is on the hot boundary between the North American and Eurasian plates. Iceland is sometimes called the land of fire and ice because of its active volcanoes and cold climate. In California, geothermal power is produced from hot springs called The Geysers and sold to the utilities. An experimental program in Cornwall, England, has extracted geothermal power from a site that is not a hot spot. Two wells, each 6000-m (20,000-ft, or nearly 4-mi) deep, are connected at the bottom by cracks produced by explosions at the bases of the wells. Water put into one of the wells emerges from the other as steam at a temperature of 260° C (500° F). The steam can then be used to generate power in a conventional power plant. A similar experiment conducted in 1986 by the Los Alamos National Laboratory produced enough hot water at 190° C (375° F) to run a 4-megawatt power plant. The advantage of this method is that it can be used almost anywhere. The Los Alamos experiment required a well only 4-km (2.5-mi) deep. Los Alamos followed this experiment up with two deeper 3700-m (12,000-ft) wells at Fenton Hill NM in Nov 1991. Results from that test are yet to come.

One problem with geothermal energy is that there may not be as much of it as previously believed. A 1991 inventory of heat trapped in rock in California found only half as much heat as was believed to be there after a survey in 1978.

So far, no one has found any environmental damage from use of geothermal power. As costs of fossil fuels inevitably rise, this source of power seems more and more attractive. Although US funding for geothermal power experiments has fallen in recent years, a 1988 report by the National Research Council strongly recommended increased funding for geothermal projects.

Tidal energy Unlike most energy on Earth, which is powered mainly by the Sun, tidal energy is powered mainly by the Moon, although the Sun contributes to changing the heights of the tides. As Earth turns, water on the side of the planet nearest the Moon is lifted by the Moon's gravitational attraction. Although we think of the Moon revolving about Earth, both bodies move about their common center of gravity. This motion causes the ocean on the opposite side of Earth from the Moon to experience high tide at the same time as the high tide caused directly by gravity. Thus, there are two tides a day, with high tides about 12 hours and 50 minutes apart (they would be exactly 12 hours apart if the Moon and Earth were not moving with respect to each other). When it is high tide, the water has to come from somewhere, so it is low tide around the part of Earth halfway between the two high tides. The gravitational pull of the Sun also produces high and low tides. When the Sun and Moon are directly lined up, they work together to produce unusually high tides, called *spring tides*. These occur during the new Moon and the full Moon. At the quarter Moon (when the Moon looks like half a circle), the Sun and Moon pull in opposing directions, causing unusually low high tides known as *neap tides*.

The difference in shape of the ocean bottom and shoreline at different places can amplify the difference between high and low tides and affect the tides in many other ways as well. The difference between high and low tide, or *tidal range,* can be less than 60 cm (2 ft) in a large, broad basin, such as the Gulf of Mexico, but can reach about 15 m (50 ft) in the narrow, funnel-shaped Bay of Fundy in Nova Scotia. This means that the depth of the water changes in the Bay by as much as 15 m (50 ft) about every 6 hours and 25 minutes. Such tidal flows have been used to power mills in France and England since the twelfth century.

Installation of a turbine in the constantly changing flow can be used to produce electric power. However, calculations show that one has to be clever to obtain useful amounts of energy. Although the total energy of the ocean tides all over the world is equivalent to about a billion kilowatts, it is highly concentrated in narrow bays and straits, with perhaps as much as 70 percent of all the energy in the Bering Strait.

The world's largest tidal generating station is on the Rance Estuary in France. At spring tide, the tidal range on the estuary is almost 14 m (45 ft). The power plant, completed in 1967, has a capacity of 240 megawatts. A dam is used to control the tidal flow. Turbines that run in either direction produce electricity both while the water is flowing into the dammed region and while it is being let out.

In the United States suitable sites for tidal power occur only in Alaska (Cook Inlet) and Passamaquoddy and Cobscook bays in Maine.

Although tidal power initially seems harmless to the environment, a closer look reveals that modifying the tides in one site can change the tidal range of sites hundreds of kilometers away. A tidal plant on the Bay of Fundy, if not carefully sited to avoid these effects, might influence the tides in the Gulf of Maine.

If the tidal range increased, it might destroy some shore facilities, for example. Also, changes in tidal range could affect the ecosystem of the tidal zone, the part of the ocean shore that is alternatively covered and uncovered by the tides.

Nuclear fission This is the process of using the energy that is released when a large atom breaks into two or more smaller ones, a process that occurs naturally sometimes but that can be greatly accelerated under the proper conditions. One set of conditions is to get a sufficient mass of an unstable type of uranium into a small space.

This is the basic method used in a nuclear-fission reactor, the energy source for a nuclear-fission power plant. Some of the energy is released as heat, which is then used to power conventional steam-turbine generators.

Nuclear fission uses uranium as its primary fuel, so it is not completely renewable; the amount of uranium used and the known supplies suggest that nuclear fission has a virtually inexhaustible supply of fuel. Also, the process does not contribute to the greenhouse effect or to other forms of air pollution.

There are problems, however. The radioactivity released in nuclear fission can escape in a plant accident and raise cancer rates or, in high enough levels, cause death from radiation sickness. In a very serious accident, it is possible that a nuclear-fission plant could turn itself into a nuclear bomb (an accident that has not so far happened). Finally, although properly operating nuclear-fission plants do not release any radiation or toxic substances to the environment, they produce large amounts of nuclear waste that must eventually be disposed of. In 1957 improperly stored waste at a site in the Soviet Union exploded, permanently contaminating a large region around the explosion. At present, there is no satisfactory way to accomplish final disposal of such waste, although plans continue to be made for storage at underground sites in rock strata that are thought to be geologically stable and impermeable to water.

Properly speaking, nuclear fission is not an alternative energy source, since the 110 operating plants in the United States already supply about 18 percent of the nation's electric power. Worldwide there are 414 plants, accounting for 16 percent of the world's electric power. In the past decade, however, no new plants have been ordered in the United States and 108 orders have been cancelled. Since 1987, moreover, it has become cheaper in the United States to build and operate coal-fired electric plants. In other countries, however, especially those without the impressive coal production and reserves of the United States, nuclear plants continue to be built.

The outlook for nuclear fission may become brighter, however. New designs feature "passive stability"; this means that if something goes wrong, the plant will shut itself down. Modular designs and standardization are reducing plant costs.

Nuclear fusion Nuclear fusion is the source of most of our current power, since it provides the energy we get from the Sun. On Earth, however, the only successful attempts at nuclear fusion that everyone acknowledges are thermonuclear, or hydrogen, bombs (but see "Cold Fusion?," p 162).

Fusion occurs when two atomic nuclei merge to form a new nucleus. The new nucleus has less mass than the two nuclei had before fusion. The excess mass is released as energy, according to Einstein's famous equation $E = mc^2$. Although the mass (m) changed to energy is small, it is multiplied by the square of c, the speed of light in a vacuum, which is a very large number. Thus, a large amount of energy is released by each fusion, although not as much as in typical nuclear fission reactions.

Calculations show that the easiest fusions to achieve involve two heavy forms of hydrogen, deuterium and tritium. Deuterium differs from ordinary hydrogen in having a nucleus that contains a neutron as well as the single proton of hydrogen's nucleus. Tritium has two neutrons. (See "Cold Fusion?" p 162.)

The standard approach to trying to control fusion on Earth has been to create a high temperature; this causes nuclei to move so fast that they bump into each other at great speeds and fuse. In the thermonuclear bomb, the high temperature is supplied by a nuclear fission bomb. The problem with this approach for controlled fusion is that a temperature of at least 100 million degrees Celsius (180 million degrees Fahrenheit) is needed, a temperature that would melt any container. Two approaches have been taken to solving this problem. In one, a magnetic field is used to confine deuterium and tritium nuclei. In the other, high-powered lasers are used to heat small pellets of fuel without heating the container.

Neither of these approaches has produced even a break-even point, where the amount of energy produced is equal to the amount of energy used to start the reaction. Break-even is expected between 1992 and 1995 by various groups, most of whom are working with magnetic confinement. The next stage, ignition, would be the point at which the reaction is self-sustaining (just as the chemical reaction we call fire can be ignited and become self-sustaining at a high enough temperature). If the projected time schedule is maintained, it is anticipated that the first usable nuclear fusion reactor will start producing power sometime in the twenty-first century, possibly as late as 2050.

Fusion power is sought for several reasons. Deuterium, contained in ordinary water, is in great supply. Tritium has to be manufactured in special nuclear fission reactors or with particle accelerators, but such manufacture has been done for years for military uses and is well understood. The by-products and wastes of nuclear fusion are much less harmful and much less radioactive than the by-products of nuclear fission. It is important to keep in mind, however, that when nuclear fission was first introduced as a power source, scientists predicted that electricity produced by nuclear plants would be too cheap to even bother to meter. Similar claims for nuclear fusion should be greeted with skepticism.

Wild Ideas

All of the possible alternatives mentioned so far are plausible today if the price is right, except for controlled nuclear fusion. There are also a few wild ideas floating around that someday someone may work out.

Artificial photosynthesis Most of the energy we use today came originally from photosynthesis, the process green plants use to turn solar energy into carbohydrates. An artificial analog of that process, to use solar energy to produce hydrogen gas, has been tried experimentally. Hydrogen gas can be used to produce electricity in fuel cells and can be liquefied to fuel vehicles; the only result of burning it is the production of water, one of the few substances that we know does not cause harmful air pollution (although water vapor is a greenhouse gas). So far, however, only tiny amounts of hydrogen have been produced this way.

Solar power from space On Earth, solar power faces cloudy days, the darkness of night, and a general loss as the Sun's energy passes through the atmosphere. All this is routinely avoided by spacecraft that use solar power. Why not collect solar power in space and send it back to Earth in the form of microwaves? The microwaves could be reconverted to electricity on the ground.

A cluster of satellites in geosynchronous orbit, with each 5-million-kg (10-million-lb) satellite carrying several km^2 of solar cells, could beam their electricity to ground antennas of roughly the same size. A receiving antenna about 8 km (5 mi) in diameter could collect enough microwaves to amount to 5 million kilowatts of electricity, about the same as from five nuclear power plants.

This idea has been championed by Peter E. Glaser since the 1960s. The basic technology was proven in tests in the California desert in the 1970s. A government study in 1981, however, estimated that it would cost $3 trillion over 50 years to make it work, which dampened enthusiasm considerably.

Some also worry that the energy from space would turn Earth into a giant microwave oven, but careful aiming should prevent that. Even off beam, the low intensity of the microwaves should not cause short-run damage. Microwaves at low intensities have been claimed to cause various health hazards after prolonged exposure.

A laser beam—instead of microwaves—would reach a smaller region on Earth's surface, but would be subject to disruption by clouds and be harder to produce in space or to convert to usable electricity on Earth.

(Periodical References and Additional Reading: *New York Times* 12-5-90, p D9; *Discover* 8-91, p 10; *New York Times* 11-3-91, p F16; *Science* 11-22-91, p 1113; *New York Times* 12-25-91, p 49; *New York Times* 1-1-92)

S O M E T H I N G   I N   T H E   A I R

SAVING THE OZONE LAYER

Spring and fall of 1991 were times of bad news about the ozone layer that protects life of Earth from too much ultraviolet radiation.

In the spring the US Environmental Protection Agency reported that nonsummer declines of ozone over the United States during the 1980s were much worse than previously believed. The fall, winter, and spring ozone layer had declined between 4.5 and 5 percent during that decade according to the Total Ozone Mapping Spectrometer (TOMS) instrument aboard the Nimbus-7 weather satellite. This was about twice what ground-based measurements suggested.

The fall string of bad news started on Oct 9 1991. The US National Aeronautics and Space Administration (NASA) reported that TOMS detected on Oct 6 the lowest level of ozone in the Antarctic hole ever recorded, 110 Dobson units. A Dobson unit measures the ability of the atmosphere to absorb certain wavelengths of light, including ultraviolet light. Normally, the air above Antarctica measures about 500 Dobson units. TOMS has been measuring Dobson units in the air above Antarctica since 1978.

On Oct 17 1991 a study sponsored by the American Institute of Aeronautics and Astronautics claimed that chlorides from solid-fuel rockets were apparently depleting the ozone layer by a small but unnecessary amount. NASA estimates that each space shuttle flight deposits about 70 metric tons (75 short tons) of chlorine directly into the ozone layer. Reformulation of fuels could eliminate this completely.

On Oct 22 a United Nations (UN) panel reported for the first time that scientists had found that ozone levels in the northern hemisphere are weaker in the summer. This is especially unwelcome because more people tend to be out in the Sun.

The UN panel reported in Nov 1991 that thinning of the ozone layer by 10 percent, a degree of change expected by the year 2000, will cause 300,000 additional cases of skin cancer and 1.6 million extra eye cataracts annually around the world. It also projected an unknown amount of damage to plants. This latter prediction was reinforced by another Nov 1991 report, that of Susan Weller, head of the American Society of Limnology and Oceanography. She told a Senate hearing into ozone thinning that growth of small plant life in the Antarctic Ocean slowed by 6 to 12 percent when ozone in the Antarctic "hole" thinned by 40 percent during the Antarctic winter.

Background

Ozone is a gas like oxygen. In fact, ozone is oxygen—but oxygen with a difference. Ordinary oxygen always contains two atoms combined into a single molecule. This is the oxygen we breathe. With a slight energy boost, however, three atoms of oxygen can combine to form an ozone molecule.

Ozone has different properties from oxygen. For example, oxygen is odorless, but ozone has a distinct odor that can often be noticed near electric sparks or powerful ultraviolet lights. Oxygen is transparent, but ozone is blue.

Ozone is to oxygen rather like hydrogen peroxide is to water. Ozone is much more reactive than oxygen. Also, just as hydrogen peroxide gradually turns into water, ozone gradually turns into ordinary oxygen. The action of light speeds up these processes.

Although ozone is damaging when it interacts with life directly, ozone high in the atmosphere is important in protecting life. In the upper atmosphere, ozone is both formed by ultraviolet light and broken down by ultraviolet light. When ozone is broken down by light, one atom of oxygen quickly replaces the ozone molecule that was broken. In the process, the energy of the ultraviolet light is trapped by the electrons in the ozone. As a result, the ozone keeps some ultraviolet light from reaching Earth's surface, especially the light with higher energy levels. The amount of ultraviolet that does reach the surface is blamed for most skin cancers. Furthermore, high-energy ultraviolet light kills microorganisms.

Scientists believe that if more ultraviolet light reached the surface, the number of skin cancers, some of them fatal, would drastically increase. It is also thought that small ocean algae, which produce much of the oxygen in the air and break down much of the carbon dioxide, and bacteria important to crop production would be greatly reduced.

Chlorofluorocarbons (Freon is the most familiar type) are gases that have been used as spray propellants, in refrigeration, as cleaning agents, and in plastic foams, such as Styrofoam. Since 1974 it has been known that chlorine can be produced when chlorofluorocarbons break down in the upper atmosphere. The chlorine can then destroy ozone, turning it into ordinary oxygen. Each chlorine molecule destroys only one ozone molecule, but it does so in a process that leaves the original chlorine molecule intact, so it can then proceed to destroy another ozone molecule. Consequently, each chlorine molecule destroys many, many ozone molecules.

Ozone levels in the atmosphere are difficult to measure. The current best guess is that in the temperate zone of the northern hemisphere, summer ozone declined by about 1 percent in the 1970s and by about 3 percent in the 1980s. These changes are about double in winter, however, so the winter decline might be as high as 6.3 percent.

Since the discovery that chlorofluorocarbons can destroy high-atmosphere ozone, evidence has gradually accumulated that the amount of ozone in the upper atmosphere is decreasing. This process is especially noticeable in the Antarctic, where tiny ice particles increase the rate of breakdown, producing an ozone "hole" during the Antarctic summer, when there is nearly continuous sunlight. Increasing evidence shows that the same forces are also at work in the Arctic, where such a "hole" would be more dangerous because of the greater amount of life in that region— including human—than in the Antarctic.

Furthermore, chlorofluorocarbons are contributing to another problem. Although about half the greenhouse effect—the warming of the atmosphere by gases that trap the Sun's heat—is caused by carbon dioxide, the other half is caused by other gases. Carbon dioxide affects the temperature because there is so much of it, but chlorofluorocarbons are a thousand times more effective as "greenhouse gases" than carbon dioxide is. Consequently, even though there is not much chlorofluorocarbon in the atmosphere, the contribution of chlorofluorocarbons to the greenhouse effect is great.

For both ozone destruction and the greenhouse effect, it does not take many chlorofluorocarbon molecules to make a difference.

Steps are being taken to protect the ozone in the upper atmosphere. Chlorofluorocarbons have not been used as spray propellants in the United States for years, although they continue to be used in other ways. An international treaty in 1987, signed by the United States, calls for limiting production of chlorofluorocarbons and related gases, but in 1989 the European Economic Community agreed to eliminate chlorofluorocarbon production in its twelve nations completely by the turn of the century. It called on other nations to join in sharper reductions more immediately than the 1987 treaty requires and to work toward a complete ban, a call that the United States echoed the following day. Steps to find acceptable substitutes for chlorofluorocarbons have been undertaken in the United States and the United Kingdom. Substitutes found so far are less effective and more expensive to manufacture than chlorofluorocarbons, but nearly everyone agrees that anything is better than permanently damaging the worldwide environment.

(Periodical References and Additional Reading: Science News 3-24-90, p 183; Science News 4-7-90, p 215; Science 7-6-90, p 30; Science News 7-7-90, p 6; Science News 8-11-90, p 87; Science News 9-29-90, p 198; Science News 10-13-90, p 228; New York Times 4-5-91, p A1; New York Times 4-9-91, p C4; Science 4-12-91, p 204; Science 5-3-91, pp 623 & 693; Science 5-31-91, p 1260; Scientific American 6-91, p 68; National Wildlife 6/7-91, p 25; New York Times 7-23-91, p C4; Science News 9-28-91, p 199; Science News 10-5-91, p 214; New York Times 10-10-91, p A23; Science News 10-12-91, p 237; New York Times 10-18-91, p A18; Science 10-18-91, p 373; Science News 10-19-91, p 244; New York Times 10-23-91, p A1; Science News 10-26-91, p 270; EOS 10-29-91, p 474; Discover 11-91, p 13; Science News 11-2-91, p 278; Physics Today 12-91, p 34; Science News 12-7-91, p 380)

ACID RAIN UPDATE

On Oct 29 1991 the US Environmental Protection Agency (EPA) proposed regulations under the Clean Air Act of 1990, to go into effect in 1995, that were supposed to cut acid rain in the United States in half (Canada would also benefit). The final rules would be issued in May 1992 after public comment on the proposal. If approved, the new rules will cut pollution at 110 electric utility plants in the US East and Midwest in 1995. In later years, the coverage will expand until almost all plants that burn fossil fuels will be affected. The goal is to cut emissions of sulfur dioxide by 9.1 million metric tons (10 million short tons).

An unusual feature of the new rules is that one company would be able to sell its right to pollute to another company; that is, a company that exceeds the standards can sell that "excess" to another company that would otherwise not be in compliance. The main reason for this provision is to reduce the adverse economic impact of the regulations. The EPA believes that buying and selling pollution rights will reduce the cost of compliance from $3.8 billion to $1 billion. The futures market of the Chicago Board of Trade plans to trade in pollution rights just as it now trades in such commodities as grain or pork bellies.

Because acid rain was perceived as a problem in the United States in the late 1970s, but one that the federal government was reluctant to move against for economic reasons, a 10-year study of the problem, the $600-million National Acid Precipitation Assessment Program (NAPAP), was started. As NAPAP began to wind down in 1989, the US Environmental Protection Agency (EPA) set up an oversight panel to review the assessment. This meta-assessment was inspired by criticism of an interim NAPAP report released in the mid-1980s. The oversight panel released its own report in Apr 1991.

In short, the oversight report said that NAPAP was good science, but not very useful in terms of its original goal of clarifying what steps, if any, the US government should take to contain acid rain. Although the original NAPAP expired as scheduled in Dec 1990, a slimmed-down version was created to observe the effects on acid rain of the US Clean Air Act Amendments of 1990. There was some indication that the criticisms of past reports would be taken into account as the new agency began work.

The review of NAPAP was not the only EPA action on acid rain in the early 1990s. On May 24 1991 the major results of its National Surface Water Survey showed that acid rain is the main source of acidity in 75 percent of the acidified lakes and in 47 percent of acidified streams. The other causes of acidity are tailings from mines (mainly in streams) and natural acidity from decaying plants (mainly in lakes). The survey also showed that while organic sources dominated lakes in Florida and the upper Midwest and coal mines parts of the Mid-Atlantic Highlands, acid rain accounted for the

Background

Acid rain caused by industrial pollution has been known since 1872, when Robert Agnus Smith discussed its appearance in England in the wake of the Industrial Revolution. It was not until 1961, however, that acid rain reached public consciousness. The Swedish scientist Svante Odén rediscovered the phenomenon in Scandinavia and took his findings to the press instead of to obscure scientific journals. In 1976 Odén showed that acid rain is a regional phenomenon. By 1980 acid rain in the United States, Canada, and Western Europe was understood as a major environmental issue, and the US Acid Precipitation Act of that year initiated a 10-year study program. That year also marked the start of negotiations between the United States and Canada on halting acid precipitation that crossed over from one country into the other. More recently still, scientists have recognized that acid rain is found in parts of the world that are not industrialized. Reports in Jun 1989 showed that acid rain falls almost continuously on the African rain forest and seasonally on the South American rain forest.

"Acid rain" is the commonly used term to denote acidic precipitation of all kinds, as well as acidic dust particles, which may contribute as much as actual wet precipitation in the form of rain, snow, and fog. Although rainwater is normally slightly acid, precipitation is noticeably higher in acidity in certain regions. One result of the higher acidity is that small lakes become more acid than they were in the past—technically, they lose the ability to buffer the acidity with alkaline chemicals from the rocks and soil. As these lakes become more acid, they progressively lose populations of various types of organisms. Many small invertebrates are the first to go. This reduces the food supply for fish, frogs, and other vertebrates. Different species of fish stop breeding at different levels of acidity. Soon there are only a few adult fish left, and little for them to eat. Eventually, all forms of animal life are lost. This particular effect was the first to call widespread attention to acid rain. It has affected lakes in Scandinavia, the US Appalachian Mountains, and southeastern Canada.

Another clear effect of acid rain has been increased weathering of marble, limestone, and sandstone. Bronze is also attacked. Statues have lost their features and tombstones have become unreadable. This is particularly evident in the US Gettysburg National Military Park, which contains 1600 monuments and is in the highest acid rain state (Pennsylvania) in the United States.

More controversial is the effect of acid rain on forests, crops, and human beings. Forests at high altitudes in the United States and at both low and high altitudes in Europe are severely stressed and many trees are dying. Although this would seem to be evidence of acid rain, the situation is more complex. Acid rain may be one of the factors involved, but even that is not clear. Some tests have shown that acid rain injures leaves on some food crops, such as beans, broccoli, and spinach. However, this damage does not seem to be severe; furthermore, few major food crops are grown in regions of highly acid precipitation. As for human beings, it is clear that breathing sulfuric acid, the main component of acid rain, is not a good idea—but it is much less clear how much sulfuric acid from acid rain actually reaches the lungs. There is no definite indication that humans are directly injured by acid rain.

Sulfuric acid is a product of reactions of sulfur dioxide. Sulfur dioxide is released primarily by industrial plants that use coal or oil for fuel. High grades—the expensive grades—contain much less sulfur than lower grades.

Nitric acid, produced from nitrogen oxides, also contributes to the acidity. Nitrogen oxides are produced largely by reactions at high temperatures of the nitrogen in air with oxygen in air. Thus, nitrogen oxides are found in automobile exhausts as well as in emissions from plants that burn almost any kind of fuel at sufficiently high temperatures. Burning vegetation causes acid rain by a different mechanism, producing formic acid and acetic acid, as well as nitric acid.

A surprising effect of nitric-acid rain is that it can promote plant growth, since availability of nitrogen compounds is one of the factors that limit plant growth. A study released by the Environmental Defense Fund in 1988 revealed that about 25 percent of the excess nitrogen in Chesapeake Bay comes from acid rain. (The remaining 75 percent comes from crop fertilizer runoff and sewage.) This nitrogen is resulting in excessive growth of algae, which is choking out fish and shellfish production in the bay.

Acid rain can travel great distances from its source, with as much as 10 to 80 percent increases in acidity noted as far as 4000 km (2500 mi) from the source. In North America, the eastern Middle West, especially the Ohio valley and Great Lakes region, produces the emissions that cause the most damage. Acid rain from this region is most likely to fall on lakes with little buffering capacity (ability to reduce acidity). Most acid rain in this region is thought to be caused by emissions from electrical power plants, especially those that burn high-sulfur coal or oil.

In Europe, acid rain is produced in various industrial regions, including West Germany, northern England, and parts of the former Soviet Union. In China, some of the worst acid rain falls on the Xishuangbanna National Nature Reserve in the southwest. The reserve is home to several rare mammals, 35 percent of all of China's bird species, half its butterfly species, and 4000 types of flowering plants. In sub-Saharan Africa, acid rain falls on the tropical rain forest as a result of year-round burning of the savanna to make land suitable for agriculture. The Amazonian rain forest also receives acid rain from land clearing, although burning the rain forest in the Amazon region is seasonal, not year-round.

Reduction in sulfur dioxide can be accomplished in many ways. Among these are switching to low-sulfur coal or oil as a fuel. Switching to natural gas as fuel is even more effective, since natural gas contains almost no sulfur, but it requires new furnaces. Devices called scrubbers can be added to smokestacks to remove sulfur dioxide, but these are expensive to install and maintain. Encouraging conservation of electric power is one of the least expensive ways to reduce the need for fuel, and therefore reduce emissions. Various new technologies, based mainly on getting sulfur out of coal before it is burned, can also be used. Also, alternative energy sources can replace part of the generating capacity.

Reduction in nitrogen oxides is more difficult. Automobile emissions can be partly controlled by various means, including catalytic converters and use of alternate fuels. Converters can also be added to smokestacks.

acidity in the Adirondacks and in non-coal-mining parts of the Mid-Atlantic Highlands. Because of prevailing winds, these data tend to support what most scientists strongly believed before the survey: Most of the acid rain in the eastern United States is caused by air pollution from the Midwest. Further analysis of the National Surface Water Survey included a look at evidence for the presence of acidity at various times in the past. It had been suggested that regrowth of forests after clear-cutting could produce observed acidic conditions, but the timing of the onset of acidity did not coincide with clear-cutting or regrowth, and did coincide with industrialization. The proponents of organic causes for acidification of lakes and streams continued, however, to push their case in the face of overwhelming evidence that acid rain causes many acid lakes and streams.

(Periodical References and Additional Reading: *Science News* 3-3-90, p 143; *Science News* 9-15-90, p 165; *Science* 3-15-91, p 1302; *Science* 4-19-91, p 370; *Science* 5-24-91, pp 1043 & 1151; *Science* 9-20-91, p 1334)

THE KUWAIT OIL FIRES

Before the Iraqi invasion of Kuwait, its oil wells pumped about 1.5 million barrels of petroleum each day, oil that went into tankers and pipelines to be burned under controlled conditions at various sites around the world. In Feb 1991 retreating Iraqis set fire to about 650 of the Kuwaiti oil wells and blew up another hundred or so. As a result, the damaged wells released about four times as much oil as before, most of which burned freely within a small region, producing immense air pollution.

An early concern was that the smoke from burning fires would rise into the stratosphere where it might lower global temperature, the way that dust from a volcano sometimes does. On Jun 24 1991 a team of investigators from the US National Science Foundation reported that the gases and soot did not rise that high. The highest smoke found in thirty-five flights in May was 6700 m (22,000 ft) high, well below the 11,000-m-high (35,000-ft-high) stratosphere above the Persian Gulf. This result had been correctly predicted before the war in a study conducted by the Pacific Sierra Corporation using a supercomputer model.

The plume of smoke stretched 1300 km (800 mi) from its origin in Kuwait. Although global temperatures were unaffected, temperatures were lower locally as a result of blocked sunlight. Another early prediction was that the lower temperatures would weaken or halt the important monsoon winds that bring rain to India, but this did not occur. Something like the opposite may have happened, however. According to Thomas J. Sullivan of the US Lawrence Livermore National Laboratory, a cyclone in Bangladesh that killed more than 100,000 and flood-causing rains in China were both spawned by smoke from Kuwait.

In Aug 1991 George D. Thurston of the Institute of Environmental Medicine at New York University reported that ground-level air pollution in Kuwait would result in a 10 percent higher mortality rate than before the Gulf War, causing about 1000 additional deaths over the course of a year among the million or so inhabitants. Air samples taken in May were analyzed by the US National Toxic Campaign and found to contain higher levels of pollutants than would be allowed in Massachusetts, which is considered to have the most complete official list of permissible levels for airborne chemicals. By Oct 1991 the ground-level pollution in Kuwait was still judged to be about half the level considered very hazardous to humans, even though fewer than a hundred wells were still burning.

The oil that did not burn flowed across the desert, where the heat of the sun turned it into tarry asphalt. The asphalt paving is expected to remain in place for years to come. There were about 200 lakes of oil, some of which were more than 2-km (1-mi) wide and several feet deep before they evaporated or seeped into the soil. The part that seeps into the ground poisons it. The soil is also affected by the rain of black soot that has fallen on it from nearby oil fires. One estimate is that the fires produced about 5000 tons of carbon soot each day.

Efforts to cap damaged wells and to put out fires began immediately after the war. Eventually, about 10,000 workers from thirty-four countries were employed, bringing with them about 125,000 tons of heavy equipment. Predictions at first were highly pessimistic, however, with many authorities saying that the immense effort might take 2 years before all fires were put out. The pessimists were wrong. The last fire was out and the last well capped on Nov 7 1991, just 9 months after the end of the war. The first half of the wells were in good shape by Sep 8, so it took only 2 months for the remainder of the task. As a result of these efforts, the total amount of oil lost was only about 600 million barrels, or about 3 percent of Kuwait's oil reserves. Even so, this is about as much oil as the whole world uses every 3 months. Furthermore, Kuwait's oil production in Nov 1991 was about a fifth of what it had been before the war, although production was expected to surpass prewar levels by the end of 1993.

These predictions may be far too optimistic, however, if, as some observers believe, the fires depleted the oil reserves. Most evidence for this comes from suggestions that brine is beginning to replace oil at some wells and that oil pressure became lower as the fires burned.

(Periodical References and Additional Reading: *New York Times* 1-31-91, p D1; *Science* 6-14-91, pp 1467 & 1536; *Science* 6-22-91, p 1609; *New York Times* 6-25-91, p C4; *Scientific American* 7-91, p 17; *EOS* 7-2-91, p 289; *Science News* 7-13-91, p 24; *New York Times* 7-16-91, p A3; *Science* 8-30-91, p 971; *Scientific American* 10-91, pp 12 & 30; *New York Times* 10-19-91, p 1; *New York Times* 11-7-91, p A3; *Scientific American* 1-92, p 20)

GLOBAL WARMING UPDATE

As early as 1967 earth scientists S. Manabe and R.T. Wetherald warned that human activities are causing a global buildup of carbon dioxide in the atmosphere that could lead to global warming as a result of the green-house effect (see Background below). Many believed them, but others were skeptical, including the US government. When global temperatures in the 1980s reached record highs, many more became believers (although not the US government). As the world entered the 1990s, the potential for global warming and some of its possible causes—industrialization, defor-estation, other greenhouse gases such as chlorofluorocarbons and methane—had passed from the esoteric idea of earth scientists to the intel-lectual currency of the common man. But in the 1990s it also became apparent that global warming was, like many worldwide phenomena, much more complicated and less straightforward than had been suspected. Worldwide there were six operating computer models for global climatic change, and they frequently contradicted each other.

1990

The year 1990 continued a trend that was apparent throughout the 1980s. It was hailed as the warmest year on record, although the heat was unevenly spread around the world. The surface temperature was about 15° C (60° F) when averaged over the whole globe and the whole year. The four warmest years during the twentieth century are now thought to be 1990, 1981, 1987, and 1988 in that order.

In 1990, according to figures released at the end of 1991 by the World-watch Institute, global emissions of carbon dioxide fell by about 0.017 per-cent as a result of the collapse of the economies of Eastern Europe and the USSR. Using data collected by the US Oak Ridge National Laboratory and by British Petroleum, total carbon dioxide emissions fell from 5.813 billion tons in 1989 to 5.803 billion tons in 1990, a drop of 10 million tons. With the exception of the oil-shortage years of 1973 and 1979, emissions had previously increased steadily since 1950.

Although long-term reductions are probably still over the horizon, the continued collapse of the economies of the former communist nations should contribute to further drops in 1991 and 1992, despite the additional carbon dioxide burden caused by the Kuwait oil fires in 1991.

1991

Although 1991 started out with the promise of being even warmer than 1990, global temperatures had dropped sufficiently by the end of the year to make the average for all of 1991 lower than that for all of 1990. Accord-

ing to the climate model developed by James E. Hansen of the US NASA
Goddard Institute of Space Studies in New York City, 1991 would rank
somewhere from second to fourth worldwide in warmth when all the fig-
ures were in. This result was confirmed by the model used by a team
headed by Phil Jones of the University of East Anglia, UK, and David
Parker of the British meteorological office. A different model, used by the
US National Climatic Data Center at Asheville NC, showed that in the
United States 1991 had dropped to at least the eleventh warmest year by
Nov (Dec figures were not in when this account was written). The principal
reason for the drop was thought to be the Jun eruption of Mt Pinatubo in
the Philippines, which lowered temperatures by depositing a haze of sulfuric acid and water vapor in the stratosphere, a haze that blocked some
incoming energy from the sun (see "Mount Pinatubo Erupts," p 267). The
cooling was detected both by balloon observations and satellite data, which
showed that until Aug, 1991 was as warm as 1990, but rapid cooling started
in Sep and continued. The National Climatic Data Center study of temperatures in the United States obtained similar results. The changes in the economy of Eastern Europe were probably a factor throughout the year.

In Feb the first of five scheduled international conferences on global
warming opened in Chantilly VA, attracting delegates from 130 countries.
The United States, the world's largest producer of carbon dioxide, declared
at the start of the conference that no new steps were needed for it to
reduce its own production of greenhouse gases, but that other countries
might need to take measures similar to those already on the US books.
The United States produces a third of the approximately 3.5 million metric
tons (3.75 short tons) of atmospheric carbon dioxide that industrial nations
release each year, about the same amount as produced by all the undevel-
oped nations together. The last such conference is scheduled for Rio de
Janeiro in Jun 1992.

1992

Predictions for 1992 began to make news as 1991 came to a close. The
two principal influences foreseen for global temperatures in 1992 were the
continued effects of the Mt Pinatubo eruption and the effect on global climate of a developing El Niño (see "El Niño and La Niña," p 220). The
combination will produce changes that are hard to predict because the
phenomena tend to cancel each other out: Mt Pinatubo's haze cools, but
the overall effect of the global changes in weather patterns that are called
El Niño is to warm. The best guess at the end of 1991 was that the cooling
from Mt Pinatubo would prevail over the warming from El Niño, resulting
in global temperatures for 1992 that would be cooler than for 1990. As a
result of such cooling, James Hansen predicts that cherry trees will blos-
som about a week later than usual in both Tokyo and Washington DC.

Beyond 1992

The effects of Mt Pinatubo will gradually fade during the 1990s and cease to be an important factor by the middle of the decade. Its eruption, however, should serve as a caution for people who indulge in global weather prediction for medium-range time scales (such as decades). All bets are off in the case of other large eruptions, major impacts from asteroids, nuclear war, or even the outbreak of peace. Barring such happenings, however, there is already enough of various greenhouse gases in the atmosphere to promote the resumption of global warming on a scale similar to that found in the 1980s. Even if all the political fixes for the production of greenhouse gases work, warming should continue well into the twenty-second century. Then the next worry is when other forces will start the resumption of the ice ages.

Could It Be the Sun and Not the Greenhouse Effect?

Virtually all predictions of global warming or explanations of recent warm years have been based on the effects of greenhouse gases in the atmosphere, although some scientists have said all along that warm weather in the 1980s could have been caused by some other factor. Eigil Friis-Christensen and Knud Lassen of the Danish Meteorological Association reported in the Nov 1 1991 *Science* that they have identified the true cause of recent warming. Although they think it is the Sun, they do not think that the growth in greenhouse gases is involved. Instead, they think that solar energy, as measured by the length of the sunspot cycle, has increased.

They plotted the length of the solar cycle, which varies from 10 to 12 years (this is actually a half cycle), against land temperatures in the northern hemisphere during the late nineteenth and twentieth centuries; they also plotted the solar cycle lengths against sea ice records since 1740. The curves showed that warmer temperatures correspond to short sunspot cycles, implying that the Sun is more active and produces more radiant energy during short cycles.

Although the Danish scientists think that solar cycles can account for all the global warming so far, they expect that higher greenhouse gases will produce additional warming in the future, no matter what the length of the solar cycle.

> *Background*
>
> Recently, whenever there is a hot summer or a warmer-than-usual winter, people have tended to blame the greenhouse effect. The greenhouse effect is a real phenomenon, but its effect on current weather is not clear to most scientists. Here are the causes of the effect and what is known about how it is changing climate now, as well as expectations for future change.

Causes Gases tend to be transparent to electromagnetic radiation at visible wavelengths because light does not react easily with electrons in isolated molecules. At different wavelengths, however, gases are generally not transparent. One example is ozone, which is less transparent to ultraviolet radiation than other gases in the atmosphere. Another example is carbon dioxide, which is less transparent to infrared radiation, or heat, than it is to light. Both are interactions between a gas and electromagnetic radiation that have proved to be involved in worldwide environmental problems.

Carbon dioxide in the atmosphere permits solar radiation in the form of light to reach the surface of Earth, as do several other gases. There the radiation is absorbed by solids or liquids, although some is reflected in all directions. The light that is absorbed heats the solids or liquids. This heat is then emitted from the surface as infrared radiation. Gases that are not transparent to infrared radiation, such as carbon dioxide, collect this heat and keep it in the atmosphere. If all gases in the atmosphere were transparent to infrared radiation, the heat would escape. Mars, although its atmosphere is 95 percent carbon dioxide, has insufficient carbon dioxide to trap much heat (its air pressure is about seven thousandths that of Earth's air pressure). As a result, its surface temperature is -55° C (-67° F). Venus also has a 96 percent carbon dioxide atmosphere, but its air pressure at the surface is ninety times that of Earth. Consequently, it retains much more heat, resulting in a surface temperature of about 457° C (854° F). Earth has been comfortably in between these extremes. Our atmosphere is only 0.035 percent carbon dioxide, but that, combined with other gases, traps 88 percent of the Sun's energy, some of which is reradiated toward Earth's surface, while the rest is radiated into space. This process then repeats, with some reradiated heat collected by the gases and some escaping into space. Eventually, 70 percent of the infrared radiation is emitted toward space, while the remaining 30 percent stays in Earth's surface and atmosphere.

Because a similar process traps infrared radiation in a greenhouse, this process is known as the greenhouse effect, and gases that are less transparent to infrared radiation are called greenhouse gases. Besides carbon dioxide, the principal greenhouse gases in the atmosphere are methane, chlorofluorocarbons, nitrogen oxides, and low-level ozone. Methane is produced primarily by natural sources, such as digestive processes of cattle and termites; chlorofluorocarbons are synthetics; and the other two gases are forms of air pollution caused mainly by automobile exhausts and burning of wood or fossil fuels (coal, oil, and natural gas).

The greenhouse effect is an environmental problem because greenhouse gases are increasing in the atmosphere. Carbon dioxide has increased by about 25 percent since about 1850. Much of this increase is thought to be due to burning of fossil fuels, but it is also clear from the geologic record that carbon dioxide levels have varied from time to time in the past, resulting in periods when most of Earth was tropical and in periods where there were ice ages. A major source of carbon dioxide is deforestation, which releases the gas during burning and/or rotting of cut wood. Dr. Paul Crutzen, an atmospheric scientist at the Max Planck Institute for Chemistry in Mainz, Germany, estimates that two-thirds of the carbon dioxide being added to the atmosphere comes from burning fossil fuels and the remaining third comes from deforestation. Other greenhouse gases are also increasing, in most cases clearly as a result of human activity, although it is less clear what is causing the rise in methane levels. It is thought that cud-chewing animals such as cattle produce about 15 percent of the atmospheric methane by belching, while natural wetlands produce the largest share, about 20 percent. With more cattle and fewer wetlands, however, the amount of methane is rising about 1 percent (of the total methane) each year.

The Greenhouse Effect and Climate   Climate does change in response to the greenhouse effect. Recent evidence suggests that such dramatic changes at the end of the most recent ice age some 11,000 years ago was precipitated by a rise in carbon dioxide and methane, which resulted in a global warming of 5° C (9° F) since the retreat of the ice. This was enough to change drastically the climate and environment of North America.

Scientists have fairly good data for global temperatures in the past hundred years, during which carbon dioxide increased from about 0.028 percent of the air to 0.035 percent and other greenhouse gases also increased. Although the trend has not been constantly upward, overall the increase in temperature during this period has been about 0.5° C (0.9° F).

If trends in the use of fossil fuels are not changed, scientists expect the amount of carbon dioxide in the air to reach 0.06 percent in the twenty-first century. This would produce a global warming variously estimated from 1° C to 5° C (1.8° F to 9° F). Even the lowest projection of 1° C would result in considerable climatic change, while anything over a 2° C increase would drastically change the climate. This is because a global temperature rise is not evenly distributed and because changes in temperature affect weather patterns around the world.

At this time, different projections of regional effects arise from different scientific studies. Most predict that more of the global warming will occur in the temperate and arctic zones than in the tropics. They also predict changes in rainfall; for example, in some projections, the US Midwest becomes a semidesert.

Other Complications   Carbon dioxide is not a bystander in the ecology of Earth; it is an active player in many ways. Since it is used by green plants in photosynthesis, the size of the world's population of green plants affects the amount of carbon dioxide in the air. Trees, especially, tie up large amounts of carbon in their woody parts for years—as long as thousands of years in the case of a few species. They take the carbon dioxide from the air, use it to collect energy from the sun, release much of the oxygen, and keep the carbon. The carbon returns to circulation when the wood rots or is burned.

Sometimes carbon locked up by green plants does not return to the atmosphere for a longer period of time. If it is buried or formed into peat in bogs, it can become fossilized as coal, oil, or natural gas. Certainly, this process removed vast amounts of carbon from the atmosphere in the past.

The ocean also collects carbon dioxide. Some is merely dissolved in ocean water (where it ceases to contribute to the greenhouse effect) and some is incorporated into calcium carbonate, the principal component of most seashells. After the organism that made the shell dies, the shell either is buried, where it may later be converted to limestone or chalk, or it dissolves in ocean water. In that case, the calcium carbonate becomes available for other organisms to use in their cells.

It is believed that changes in oceans and ocean currents caused by movements of tectonic plates caused variations in carbon dioxide that resulted in vast climatic changes. Also, the tying up of carbon in vast forests or as fossil fuels created changes.

These factors make it difficult to predict the amount of carbon dioxide over long periods of time. In addition, cloud cover and the presence of snow change the climate in ways that are hard to predict. Thick clouds screen out solar radiation, but if the radiation is coming from Earth, they trap it. Snow is cold, but it also is white, which means that it reflects most light back into space.

Possible Consequences of the Greenhouse Effect With the difficulties of making accurate projections in mind, it is still useful to list some of the possible consequences of the greenhouse effect.

Climates will change, although it is not absolutely clear in what way. In the United States one concern is how much water will fall in various regions. One study suggests that a local temperature increase in the neighborhood of 3° C (5° F) might reduce runoff in the Colorado River basin by as much as 10 percent. This would affect water use over much of the US West. Reduction of rainfall or increased heat could change crop patterns all over the United States. Some fruits, such as apples, need a certain amount of winter cooling to flower and fruit, for example. Corn needs a lot of rain at the right time. Winter wheat needs the groundwater that comes with snow melt. Furthermore, rising temperatures would permit insect and fungal pests from the South to migrate into northern farming regions.

Ice caps will melt faster than at present. This, combined with water's increase in volume with increasing temperature, will cause sea levels to rise around the world. Predictions are that this rise will be from 0.5 m to 1.5 m (1.5 ft to 5 ft) over the next 50 to 100 years. Such a rise would affect coastal regions, wetlands, and fishing. The US Environmental Protection Agency (EPA) estimates that if sea level rises 1 m (3 ft), the United States might have to spend as much as $111 billion to protect critical shorelines, but it would still lose an area the size of Massachusetts. Worldwide, most coastal cities would have to build dikes, and low-lying countries, such as Bangladesh, might lose a much larger portion of their land.

The EPA also is concerned about the effects on crucial coastal wetlands. Its projections by region are on p 388. The projection assumes a 1.5-m (5-ft) rise in sea level and reveals that the nation would lose 30 percent of its coastal wetlands.

The National Academy of Sciences was not the only one to propose a fix for the rise in greenhouse gases. John Martin of Moss Landing Marine Laboratories in California got a lot of coverage for his idea of enriching the Antarctic Ocean with iron to promote growth of phytoplankton, small floating organisms that reduce carbon dioxide in the process of photosynthesis. Studies by other scientists suggested that the plan, even if it worked, would produce only a slight decrease in the greenhouse gas buildup.

Changing Coastal Wetlands with Sea-Level Rise

Region	Total wetlands in hectares (acres)		Net loss/gain by CE 2100 in hectares (acres)	
Northeast	48,500	(120,900)	-1,600	-4,000
Mid-Atlantic	296,800	(733,300)	-37,300	-92,200
South Atlantic	557,100	(1,376,600)	25,000	+61,800
Florida	298,000	(736,300)	85,700	+211,700
Alabama and Mississippi	162,400	(401,400)	14,600	+36,000
Louisiana	1,116,300	(2,874,600)	-933,600	-2,306,900
Texas	246,600	(609,400)	-34,600	-85,500
Pacific Coast	36,100	(89,100)	-14,700	-36,300
TOTAL	2,809,200	(6,941,600)	-895,700	-2,213,400

Patterns of Global Warming

Temperature Patterns

1. Continents should warm more than oceans. 2. Subarctic regions should warm more than tropics in the northern hemisphere. 3. The troposphere (lower atmosphere) should become warmer and the stratosphere (layer above the troposphere) should cool.

Humidity Patterns

1. The amount of water vapor in the air should increase. 2. Humidity would rise higher in the tropics than in temperate or arctic regions.

Sea-surface Patterns

1. There should be a uniform increase in sea-surface temperatures, even if land temperatures change variably.

Climate Patterns

1. Winters should be warmer; summers cooler. 2. Winters should be noticeably warmer in higher latitudes.

Possible Responses   The EPA has proposed the following ways to mit-
igate the greenhouse effect: raise prices on fossil fuels; increase use of
alternative energy sources, especially those such as solar and nuclear
power that do not produce greenhouse gases; grow new forests around
the planet; stop use of chlorofluorocarbons (an action already agreed to
because of the effect of these chemicals on the ozone layer); capture gases
now released by landfills (primarily methane); and change ways of raising
rice and cattle to reduce the production of methane.

The bad news is that the EPA says that if all this were done world-wide,
starting in 1990, it would result in the rate of gas buildup leveling off only
sometime in the twenty-second century. Still, by the year 2100, greenhouse
warming might be reduced to at little as 0.5° C to 1.5° C (1° F to 2.5° F).

On Apr 10 1991 a fourteen-member panel of the US National Academy
of Sciences issued recommendations for what it termed free to moderate-
cost steps that could be taken to reduce carbon dioxide emissions.

Recommendation	Potential reduction in metric Net tons (short tons) of CO_2	Cost per metric ton
Improve efficiency of buildings	900 million (990 million)	Free
Improve vehicle efficiency	300 million (330 million)	Free
Manage industrial energy better	500 million (550 million)	Less than $9
Manage transportation system better	50 million (55 million)	Less than $9
Improve power plant heat rates	50 million (55 million)	Less than $9
Collect gases produced by landfills	200 million (220 million)	Less than $9
Reduce halocarbons and chlorofluorocarbon use	1400 million (1540 million)	Less than $9
Change agricultural practices	200 million (220 million)	Less than $9
Reforest	200 million (220 million)	Less than $99

(Periodical References and Additional Reading: *Science* 2-20-90, p 521; *Science* 3-23-
90, pp 1379 & 1431; *Science* 3-30-90, pp 1527 & 1529; *Science News* 3-31-90, p
195; *Science* 4-6-90, pp 7, 33, & 57; *Science News* 4-28-90, p 263; *Science News* 5-
19-90, p 308; *Science* 6-8-90, p 1217; *Science* 8-3-90, p 481; *Science* 8-10-90, p
607; *Scientific American* 10-90, p 94; *New York Times* 1-15-91; *Science* 1-18-91,
pp 91 & 274; *New York Times* 2-5-91; *Science* 2-8-91, pp 615 & 621; *Science* 2-22-
91, pp 851, 868, & 932; *Science* 3-1-91, pp 999 & 1058; *Science* 4-12-91, p 204;
American Scientist 5-6-91, p 210; *Science* 5-17-91, p 912; *Science News* 5-18-91, p
310; *EOS* 5-21-91, p 234; *Science News* 5-25-91, p 327; *EOS* 6-4-91, p 249; *Science*
6-14-91, p 1496; *Science* 6-21-91, p 1608; *EOS* 7-2-91, p 290; *Science* 7-5-91, pp 7
& 64; *Science News* 7-13-91, p 27; *Science News* 7-21-90, p 46; *EOS* 7-22-91, p
313; *Science News* 8-10-91, p 96; *Science News* 8-24-91, p 119; *New York Times* 8-
27-91, p C4; *Science* 9-13-91, pp 1187, 1206, & 1266; *Science News* 9-28-91, p
207; *New York Times* 10-1-91, p C1; *New York Times* 11-5-91, p C4; *Science News*
12-7-91, p 380; *New York Times* 12-8-91, p L17; *New York Times* 12-24-91, C4)

ENDANGERED SPECIES AND ECOSYSTEMS

IS THIS A MASS EXTINCTION?

Many people believe that we live in a unique time, when the animals and plants that we know and love are fast vanishing, often as part of the loss of complex communities, or ecosystems. Ask any environmentally aware person and you will learn of the vanishing

whales	songbirds	sea turtles	lobsters	rain forests
elephants	American eagles	amphibians	wildflowers	prairies
rhinos	ducks and geese	snail darters	cacti	wetlands

or other species, group, or ecosystem; or all of the above. This is an age of mass extinction.

Or is it? It is a common phenomenon to think that one's own time is special in some way and that the world has previously never seen such dramatic events—but such is seldom the case. For example, species of large mammals disappeared from North America rapidly about 11,000 years ago, either from overhunting by the newly arrived hunters we call the Clovis people or from some other cause. Among the North American species to be lost at that time were the giant bison and beaver, mammoths, and horses the size of Clydesdales—all part of the group known as the late Pleistocene giant fauna, or megafauna. The Pleistocene megafauna seem to have disappeared faster than similar large species are currently disappearing from Africa. Furthermore, the geologic record tells us that species in general have finite lifetimes. Most species that ever lived are now, by a wide margin, extinct.

Yet, there are times when many species become extinct all at once. Such a phenomenon is called a mass extinction. The extinction of the late Pleistocene megafauna in North America, while dramatic, fails to qualify as a mass extinction. A mass extinction happened 65 million years ago, the K/T event that ended the reign of the dinosaurs and also resulted in the loss of species from every habitat, a loss amounting to about 76 percent of the species alive at that time (see "New Views of Dinosaurs," p 209). The more alarmed scientists compare the situation today with the K/T mass extinction. Even greater mass extinctions existed in the past, notably the one that ended the Permian Period about 249 million years ago. That extinction eliminated about 96 percent of the species then on Earth. Some think we could be facing a Permian-type mass extinction today.

The Case for Mass Extinction Today

Alarms are being sounded by organizations determined to preserve the existing environment or to return the world to an earlier state of ecological balance. For example, the World Wildlife Fund in 1991 estimated that the rate of extinction has increased from about one species a day around 1970 to about one an hour in the early 1990s. It compares this with one estimate of the "natural" rate of species extinction—"natural" because it assumes no humans on Earth—of one species becoming extinct every 27 years. But concern about speedy extinctions is not limited to environmental organizations. Many relatively conservative scientists are also alarmed.

Among the prominent scientists who believe that we live in a period of mass extinction is Edward O. Wilson of Harvard University, well known as an expert on ants and as the inventor of sociobiology. Wilson has created a mathematical model that uses as data the rate of reduction in different types of ecosystems, although he mainly uses the reduction of the tropical rain forest ecosystem (see "Forest and Desert Problems" below). Based on that model, Wilson estimated for *The New York Times* that 50,000 species a year, or about six per hour, are becoming extinct. He also told an interviewer for *Science* that "We're easily eliminating a hundred thousand [species] a year." In another estimate, Wilson and Paul R. Ehrlich of Stanford University predicted that a quarter of all species now on Earth will disappear in the next 50 years. While these estimates are not immediately comparable, all are interpreted by many to mean that we are living in an age of mass extinction.

In part Wilson bases his claims on the disappearance of tropical rain forest, which he thinks is taking place at the equivalent of the land area of Florida each year. This would result in a reduction of remaining rain forests to half their extent over the next 30 years. In turn, this would cause a loss of 10 to 22 percent of all rain forest species during that time. Since the rain forests are more diverse than other ecosystems, loss of that percentage of rain forest species would greatly reduce the total of all species. It is widely believed that more than half of all species live in rain forests.

Another prominent scientist, familiar to the general reader from his popular writings of considerable charm and thoughtfulness, is Jared Diamond of the University of California at Los Angeles. He thinks that if current trends continue, about half of all existing species will become extinct in the next century. In part he bases his observations on long-term research on birds in New Guinea.

In the following articles, *Current Science* examines numerous groups of species. Many species face massive problems and some are unlikely to survive.

The Case Against Mass Extinction Today

Suppose Wilson's *New York Times* statement is correct and that 50,000 species disappear each year, a supposition not universally accepted. Although only 1.4 million species have been identified by scientists, estimates are that the total of all existing species is more on the order of 100 million. At a rate of 50,000 a year, it would take 15,000 years to lose three-quarters of all species and match the record of the K/T extinction. Although 15,000 years is not long in terms of geologic time, measured by human standards it is about the time since our ancestors were making cave paintings in what are now France and Spain and hunting the European mammoth and cave bear into extinction. It should be noted, however, that the extinction of the late Pleistocene megafauna probably took 9000 years from beginning to end, and Wilson's higher estimate in *Science* would imply loss of three-quarters of all species in only 7500 years. The total number of species is only a wild guess in any case.

There is wide enough divergence in estimates by those supporting the idea that this is a time of mass extinction to call the whole concept into question. For example, Norman Myers of Oxford University predicted in 1979 the loss of a quarter of all species over the 20-year period from 1980 to 2000, but Ehrlich and Wilson predict the same amount of loss over the period from about 1990 through 2040, with only 2 to 3 percent gone by 2000. In short, there is too little known about extinctions to make definite statements.

As a result, most of the scientists claiming that we are in a time of mass extinction base their case on casual observation, not on scientific studies. Analyses of such claims by Ariel Lugo of the US Forest Service's Institute of Tropical Forestry in Puerto Rico and Richard Tobin of the State University of New York at Buffalo shows that eighteen out of twenty-two predictions of mass extinction have no clear-cut scientific basis.

Another problem with extinction estimates is that they are sometimes based on incorrect data. For example, Ehrlich and Wilson use a figure for the rate of loss of tropical rain forests that comes from a book published by the environmental organization Friends of the Earth. This rate of about 1.8 percent disappearing each year is central to much of their reasoning, but Cleber Alho, director of the Brazilian branch of the World Wildlife Fund, calculates actual forest clearing in Brazil, where much of the forest loss is supposed to take place, at 0.5 percent a year, less than a third the rate used by Ehrlich and Wilson.

There are other technical objections to the argument that we are currently experiencing—and causing—a mass extinction. For example, extinction models are based on data from islands and extrapolated to continents, but conditions for survival are better on continents than on islands. Also, people predicting extinctions carry extinction curves into regions where many think the curve is no longer a valid model.

A less technical argument against current mass extinction comes from Michael A. Mares, an expert on New World tropical rain forests from the University of Oklahoma. He compares the loss of tropical rain forests to the loss of the virgin forests in the eastern United States since the seventeenth century. Although the US forests were completely cut more than once (almost all wooded areas in the eastern United States were once farmers' fields), Mores notes that "we haven't had massive die-offs." Mares does not remark on the loss of a few species, such as the passenger pigeon and Carolina parakeet, that were once abundant, but these losses did not result in excessive degradation of the ecosystem. What we see today is that the secondary woods are being rapidly recolonized by white-tailed deer, black bear, raccoons, wild turkeys, and other species.

The Bottom Line

In historic times there have been few extinctions of species of major importance either economically or in other ways, including species important to humans or to other species. In 1969 the International Union for Conservation of Nature and Natural Resources, which has paid particular attention to species preservation, counted 36 full species of mammals out of 4226 as having become extinct since 1600 and 94 out of 8684 species of birds. At that time the organization singled out 88 species of mammals and 120 of birds that seemed to be endangered. One of the mammal species may have already become extinct in 1969 (the Tasmanian wolf, whose exact status is unclear to this day). One bird, the dusky seaside sparrow, has become extinct since 1969, and the California condor came very close. In that perspective, the people predicting mass extinction sometimes seem like alarmists.

But the populations of many species, including all of those thought to be endangered in 1969, remain very small, and for most species, populations seem to be shrinking. This change is not resulting in fewer individual organisms, for weedy species, such as cockroaches, white-tailed deer, starlings, Russian thistle, rats, poison ivy, house flies, kudzu, and so forth—not to mention humans—flourish in ever-increasing populations. Ehrlich has asked what the world would be like if every species were reduced to a single population of about 200 members, which has already happened to some species that are technically not extinct. The result would be not so much a mass extinction as a general loss of biodiversity in any specified region. In other words, even though the same number of species may continue to exist, the vast majority of individual organisms would be of just a few species, a decidedly undesirable situation.

(Periodical References and Additional Reading: *International Wildlife* 3/4-91, p 29; *Science* 8-16-91, pp 736, 754, & 758; *New York Times* 8-20-91, p C1; *Science* 10-11-91, p 175)

FISH AND OTHER WATER CREATURES IN TROUBLE

When a species on land is facing extinction, the most likely cause is a change that removes all or part of the ecosystem to which it is adapted. Thus, species are lost when rain forests are turned into cattle ranches. In the vast environment of the ocean, that is less likely to happen. Nevertheless, species from the ocean suffer from and even become extinct from other factors.

Corals Under Pressure

Corals are tiny creatures related to jellyfish and sea anemones that have a symbiotic relationship with algae. Despite their size, corals are credited with building the largest single feature on Earth ever constructed by living organisms (humans included), the Great Barrier Reef of Australia.

In the 1980s scientists became concerned that corals around the world were facing environmental challenges they could not overcome. The immediate cause of concern was a phenomenon known as bleaching (see Background) that often precedes the death of a coral colony.

In Oct 1990 some scientists, notably Thomas Goreau of the Discovery Bay Laboratory in Jamaica and Raymond Hayes of Howard University, put forth the idea that widespread bleaching in the Caribbean Sea in the 1980s was caused by global warming (see "Global Warming UPDATE," p 382). A panel from the National Science Foundation looked into the hypothesis and reported in Jul 1991 that the bleaching, while perhaps caused at least in part by higher temperatures, was not the result of global warming—yet. If global warming continues, as most think it will, corals will either have to adapt or die. Death of the reef-building corals would mean the loss of one of Earth's major ecosystems, essential to many fish and marine invertebrates as well as to humans living near coral reefs.

Corals are already in trouble without global warming. Water pollution caused by increasing human population is a major factor in many parts of the world, especially Southeast Asia.

A Species Becomes Extinct

For once it was not our fault. Humans had not even got around to naming a species of reef-building fire coral found only in the Gulf of Chiriqui, part of the Pacific Ocean south of western Panama, when warm waters resulting from the great 1982-1983 El Niño caused the fire coral to die out completely. The fire coral was of the genus *Millepora*, several other species of which were also badly affected by the warming of the Pacific. The unnamed species from the Gulf of Chiriqui, first described in 1970, has not been observed alive since 1983 despite extensive searches.

> ## Background
>
> Most true, or stony, reef-building corals and their somewhat distant relatives the fire corals (hereafter collectively referred to as "corals") depend on symbiosis with photosynthetic brown algae called <u>zooxanthellae</u> for part of their energy, capturing small plankton for their other food source. One consequence of corals' dependence on the zooxanthellae is that corals are restricted to the upper parts of the sea, where light needed by the zooxanthellae can penetrate. The pigments in the zooxanthellae give living corals their typical colors. The temperature range comfortable for the corals is small; above or below that range they expel their symbiotes. When the zooxanthellae are no longer present on the corals, which then become white, the corals are <u>bleached.</u> Corals also expel their zooxanthellae as a result of stresses other than temperature changes, including changes in salinity and water pollution.

Fire corals are close relatives of hydra. Fire corals live symbiotically with algae and build reefs just like the more familiar true, or stony, corals, which are more closely related to sea anemones.

Many species of fire coral and coral in the eastern Pacific Ocean were bleached when their symbiotic partners, the zooxanthellae, were expelled by the corals as water temperatures rose by about 2° to 3° C (4° to 5° F) over a period of nearly half a year. Other corals from the eastern Pacific, in addition to the unnamed fire coral, died out regionally during the warming, but survived in different parts of the Pacific Ocean.

Sea Turtles and Tortoiseshell Combs

Anyone who follows environmental news knows that most sea turtles are endangered, although strenuous efforts to save them may finally be paying off. More and more breeding grounds are being protected, for example.

Japanese manufacturers have continued to produce jewelry and useful objects from the shell, called tortoiseshell, of hawksbill turtles gathered in the wild. For this purpose, Japan in 1990 imported the shells of about 18,000 hawksbill turtles—an endangered species as all large sea turtles are.

The hawksbill was on the first US endangered list in 1973; it was also listed as endangered by the Convention on International Trade in Endangered Species of Wild Fauna and Flora (CITES) when the convention was signed in 1975. Although Japan joined more than a hundred nations already in CITES in 1981, it reserved the right to continue importing hawksbills. Under pressure from various environmental groups, the United States began active consideration in Mar 1991 of trade sanctions against Japan. The United States threatened to ban purchase of such Japanese items as pearls and goldfish if Japan did not change its ways (importation of hawksbill products into the United States had been banned since the turtle was first declared endangered in 1973). Finally, after weeks of negotiation, Japan on May 17 1991 agreed to phase out the tortoiseshell industry. The details of the phaseout were not officially made public, but it was understood that imports of the shells would be reduced from 20 metric tons (22 short tons) a year to just 5 metric tons (11,000 pounds) each year until 1994, and then stopped completely.

Clear-cutting the Sea

More progress was made in another area of concern to conservationists, the use of very large drift nets for squid and salmon. There are several problems with the drift nets, which can be 50 km (30 mi) long. The major fishing nations using drift nets were Japan (about 450 boats), Taiwan (about 150 boats), and South Korea (about 150 boats). According to one report, the Japanese drift nets in 1990, in addition to the squid that were their targets, killed 1758 whales and dolphins, 253,288 tuna, 81,956 blue sharks, 30,464 seabirds, and more than 3 million other fish, including salmon released by Columbia River hatcheries. A different report put the 1990 Japanese total at 106 million squid, 39 million unwanted fish, 700,000 sharks, 270,000 seabirds, 26,000 marine mammals, and 406 sea turtles. Conservation organizations compare drift net fishing to clear-cutting of forests in overall effect.

A major new development of 1991 was the promise by Taiwan to stop its fishers from using the drift nets as of Jul 1 1992. This would put the Taiwanese in compliance with a United Nations ban on drift nets scheduled for the summer of 1992. The United States has indicated that it will impose sanctions on nations that do not comply with the ban.

A Late Pleistocene Rerun in the Sea

One theory of the extinction of the late Pleistocene megafauna (see "Is This a Mass Extinction?" p 390) is that the early Clovis hunters of North America started with the largest animal they could kill and continued to eliminate animal populations in size order, stopping somewhere around the moose-bison-grizzly level, the level of the Native American enterprise when the Europeans joined the hunt. Whether or not this theory is true, something similar seems to be happening in the oceans. Technology has permitted humans to move into a new environment rich in fish and other marine animals, just as the Bering land bridge brought the Clovis hunters into the realm of the megafauna. Large species, such as whales, sharks, and giant bluefin tuna, are the first to feel the effects of the arrival of the rapacious fishers. For other species, ranging from lobsters to cod, removal of the larger individuals by fishing has resulted in a rapid genetic change toward smaller ones (a similar effect is thought to be involved in producing the bison and beaver of today from giant Pleistocene versions of the same animals).

But we are better equipped to recognize today what is happening, so fishing and marine hunting regulations are being changed to stop this process. The banning of whaling is a prime example. Conservationists think similar regulations are needed for the bluefin tuna, hunted to near extinction for sushi and sashimi, but in 1991 the US National Marine Fisheries Service recommended a 2-year delay in declaring the tuna endangered under the CITES treaty. Another country could still come to the tuna's rescue at the Mar 1992 meeting of the CITES signatories, ironically scheduled to take place in Japan, the prime consumer of tuna.

Tuna are not only disappearing, they are also getting smaller. One estimate is that there are now only 10 percent as many giant bluefin tuna in the western Atlantic as there were in 1970; giants are defined as tuna larger than 135 kg (310 lb), but some of the bluefins can reach 680 kg (1500 lb), making it the largest bony fish in the world. There is some current protection for bluefin tuna in terms of an overall limit by weight, but that does not discriminate between giants and year-old baby tuna that weigh only 10 kg (20 lb).

Salmon Spawn in Connecticut But Not in Idaho

The abundance of exploitable wildlife in pre-Columbian North America is legendary. People wrote of rivers so thick with fish that a person could walk across them without getting wet feet, for example. While these stories may often have been exaggerated, there is no doubt that Western civilization has resulted in a tremendous reduction in most native organisms used for food. Rivers, as small and fragile ecosystems that can be exploited for energy and water as well as for fish and other seafood, have been especially hard hit. Old place-names that refer to salmon in the Connecticut River or oysters in the Hudson reflect a reality that is long past and can never be recaptured.

Or can it? On Oct 30 1991 Atlantic salmon were observed breeding in Connecticut's Salmon River for the first time since Colonial days. If all goes well, the eggs will hatch in the spring of 1992 and the fry will gradually work their way down the Salmon to the Connecticut River, growing into parr along the way. After a couple of years they will mature into smolt to prepare to enter Long Island Sound and then pass from there to the Atlantic Ocean. After about a year or two they may return to the river system, either as nonspawning grilse or even as adult spawners. If all goes well, sometime around 1997 the second modern generation of native salmon on the

Salmon River will be started. Fish of the first modern native generation should return from three to six more times to spawn again. With a similar pattern from their parents, a continuous native spawning run might exist on the Salmon River by the start of the twenty-first century. Shortly thereafter, salmon fishing might once again be permitted in the river system.

Salmon were eradicated from the Connecticut system by a combination of pollution and dams that prevented spawning. At present there are 11 large hydropower dams on the main stream and perhaps 700 smaller dams in various parts of the system. Although these are now provided with fish ladders to permit spawning salmon upstream access, it will not be until 1994 that the hydropower dams will also have fish passageways for downstream travel, passageways officials hope will prevent parr from destruction by the turbines used to produce power.

In another move to help the Connecticut River salmon, the US Congress approved on Nov 25 1991 establishment of the Silvio Conte National Fish and Wildlife Refuge on the Connecticut, the second of about 400 planned national refuges for fish and other water life.

In the meantime, a different Salmon River, this one in Idaho, reports the loss of its sockeye salmon for the same general reason as the earlier loss of the Atlantic salmon on Connecticut's Salmon River—too many dams making it difficult for adult fish to swim upstream to spawn and almost impossible for fry to return to the Pacific Ocean. In 1990 only one Snake River sockeye returned to the spawning ground near the Continental Divide in Idaho, and the subspecies may now be extinct. Somewhat belatedly, the National Marine Fisheries Service proposed on Apr 2 1991 that the Snake River sockeye, which spawns in the Salmon River, be declared endangered; from proposal to formal declaration is likely to take a year, but few are thought likely to oppose the designation. Three kinds of chinook and one kind of coho are also on the verge of becoming extinct, but they have not been officially listed in any way.

Endangered Fish of the 1990s

From time to time, tiny fish that go unnoticed by any except biologists and larger fish intent on eating them, make the news because the small fish face extinction. While the snail darter, whose predicted extinction failed to stop construction of the Tellico dam in Tennessee in the mid-1970s and was later found to be illusory in any case, is the most famous, the delta smelt (*Hypomesus transpacificus*) is the tiny fish of the 1990s. When the US Fish and Wildlife Service proposed on Sep 27 1991 that the delta smelt, a 7.6 cm-(3-in.) denizen of the Upper Sacramento River and San Joaquin Delta region, be given threatened status, the water supply for millions of Californians was thrown into peril. The reason is that the main cause of a

Background

The Salmon River is part of the large Connecticut River system, which extends through Massachusetts to northern New Hampshire and Vermont. The Connecticut system was probably the southernmost home of the Atlantic salmon in North America at the time of European settlement, although some claim that the salmon ranged as far south as the Hudson or even the Delaware.

The fight to restore salmon to the Connecticut and to New England river systems in general began with the 1965 US Anadromous Fish Conservation Act. For the Connecticut specifically, the Connecticut River Atlantic Salmon Commission was formed in 1984; that group tied together federal agencies and the four states of the Connecticut watershed. The commission coordinates expenditures of $260 million a year in the effort to restore the salmon, including $600 million in pollution abatement efforts as well as more specific programs, such as the release of more than a million salmon fry into the river system each spring.

90 percent decline in the delta smelt population over the past 20 years appears to have been salination caused by pumping fresh water from the San Joaquin Delta. If in 1992 or 1993 the smelt is finally declared threatened, the pumping may have to be limited. As a result, about a third of the water that now is removed from the delta for use by California homes and ranches would pass into San Francisco Bay instead.

(Periodical References and Additional Reading: *International Wildlife* 1/2-90, pp 27-28; *New York Times* 2-19-91; *International Wildlife* 3/4-91, pp 40 & 42; *New York Times* 3-21-91, p A12; *New York Times* 4-1-91, p A1; *New York Times* 4-3-91; *New York Times* 5-17-91, p A1; *New York Times* 5-18-91; *International Wildlife* 7/8-91, p 25; *Science* 7-5-91, pp 7 & 69; *New York Times* 7-9-91, p C2; *Science* 7-19-91, p 259; *National Wildlife* 8/9-91, p 25; *New York Times* 8-25-91, p 18; *New York Times* 8-28-91, p A12; *New York Times* 9-17-91, p C1; *New York Times* 9-29-91, p 22; *New York Times* 10-27-91, p L16; *Science News* 12-8-90, p 364)

IT'S NOT EASY BEING GREEN

In 1989 herpetologists comparing notes at various scientific conferences around the world began to perceive a pattern that dismayed them. Many amphibian species had rapidly diminishing populations. A conference on the topic was organized for Feb 1990 at the University of California in Irvine. Predictably, a bunch of scientists gathered to proclaim not just an endangered species but an endangered class do so. The media paid attention, and soon the vanishing amphibians had joined the endangered whales and sea turtles in the consciousness of those who care. Kermit was in trouble.

The initial evidence seemed overwhelming. Golden toads, abundant in the Monteverde Cloud Forest Preserve in Costa Rica in 1983, were no longer much in evidence after 1987. Leopard frogs, familiar across the United States, were also scarce, and one species in Nevada seemed to have become extinct. Another frog, *Rana muscosa*, had disappeared from 98 percent of the ponds where it had been studied in Sequoia-Kings Canyon National Park in California. Tiger salamanders in the Colorado Rockies were being wiped out by acid snow. European forest amphibians were being lost as air pollution destroyed the forests. Gastric-brooding frogs (*Rheobatrachus silus*) in Australia, discovered in 1973, have not been sighted since 1979. During the 1980s about 10 percent of the 194 frog species in Australia became endangered and two of those species probably became extinct. Since 1981 eight of thirteen species of frogs found in the Reserva Atlantica, Brazil, have vanished.

Although many herpetologists claim that the decline in amphibians is an early warning of what will happen to other species—amphibians are thought to be more vulnerable to pollution and environmental stress than other animals—a few species are thriving. Along with the general devastation in Australia, for example, a marsh frog (*Limnodynastes tasmaniensis*) prospers, possibly as a result of a new habitat opened up by human activity.

By the summer of 1991 there was an amphibian-extinction backlash. A half-dozen biologists studying amphibian population changes in Rainbow Bay at the Savannah River federal nuclear plant site in South Carolina claimed that natural fluctuations accounted for the observable phenomena, not pollution or even, as some claimed, global warming. Although Rainbow Bay had been in a protected area since 1951, this evidence was taken by many to mean that the previous alarm about amphibian extinction was unnecessary. According to this line of thought, populations of amphibians vary so much from year to year that no overall trends can be seen.

(Periodical References and Additional Reading: *Science News* 2-24-90, p 116; *Science News* 3-3-90, p 143; *Science News* 3-10-90, p 158; *Discover* 5-90, p 36; *Science* 8-2-91, p 509; *Science* 8-23-91, pp 831 & 892; *Science* 9-27-91, p 1467; *Scientific American* 11-91, p 29; *International Wildlife* 11/12-91, p 4)

BIRDS IN TROUBLE

Unlike amphibians and desert fish, birds are warmly regarded by many and efforts to protect them have led the way for other species. Virtually everyone is aware of the passing of the dodo and the passenger pigeon. Among noted examples of species that have been saved from extinction by heroic efforts are the trumpeter swan, the Japanese crane, the whooping crane (currently at 140 in the wild after the mysterious disappearance of six wild cranes during the winter of 1990-1991), the osprey, and the bald eagle. Even an ugly bird with disgusting (to humans) feeding habits, such as the California condor, merits strong human intervention to prevent extinction.

Many in the United States decry the loss of native songbird populations. Until recently it was thought that most of the loss might be due to changes in the winter feeding grounds in Central America and the Caribbean, but a 1991 study by Richard T. Holmes of Dartmouth College and Thomas W. Sherry of Tulane University showed that suburbanization of the spring breeding grounds in the United States is the real culprit. It is thought that suburbanization results in greater predation by cats and by wild animals, such as raccoons and squirrels, that associate easily with humans, as well as increased brood parasitism by cowbirds.

California Condors; Taken in and Now Being Returned

The California condor (*Gymnogyps californianus*) is viewed by many as the last remaining member of the late Pleistocene megafauna (its corresponding normal-sized species, found throughout the Americas, is the turkey vulture). On Apr 19 1987, when it appeared that the California condor was about to become extinct, the US Fish and Wildlife Service removed the last known wild condor after a 4-year effort; this resulted in a population in California zoos of twenty-eight that contained all known California condors. The captive breeding program since then has cost more than $10 million and resulted in twenty-five chicks hatching and only one dying (the average life span of the condor is believed to be quite long, with breeding starting at about age six). Thus the total population reached fifty-two by 1991. This is better than originally expected.

In 1991 the breeding program entered its second phase, as two of the condor chicks were slowly returned to the wild in the Sespe Condor Sanctuary, owned by the US Forest Service, in Ventura County CA. Over the winter the condors were allowed more and more freedom, with all restraints removed on Jan 14 1992.

Northern Spotted Owl Officially Threatened

After several years of lawsuits by environmental groups, suits that resulted in judicial decisions to protect forests in the Pacific Northwest from logging, the US Fish and Wildlife Service sided with environmentalists on Jun 22 1990. At issue was the status of a population of spotted owl (*Strix occidentalis*) that reproduces only in forests that have been relatively undisturbed for more than 100 years (old-growth forests) or in certain limited

TAXONOMIC NOTES

The Distribution and Taxonomy of Birds of the World by Charles G. Sibley and Burt L. Monroe, Jr., lists one species of spotted owl along with two subspecies: Strix occidentalis occidentalis, which it calls the California spotted-owl (with the added hyphen); and Strix occidentalis lucida, variously known as the mountain spotted-owl or the Mexican spotted-owl. The book also notes that the spotted owl may be a subspecies of the barred owl, an owl that exists throughout most of North America. The spotted owl ranges along the west coast and adjacent mountains of North America from southwest British Columbia to California (previously to Baja California). From the book's distribution discussion, it is clear that the northern spotted owl is considered simply to be the northern population of the main spotted owl species and not a separate subspecies.

stands of somewhat younger redwoods. Various accounts place the number of remaining northern spotted owls most often at 3000 nesting pairs or sometimes as 500 individuals, numbers that are hard to reconcile with each other. When the Fish and Wildlife Service assigned a threatened status to the northern spotted owl, to take effect on Jul 23 1990, US environmental laws automatically swung into effect to protect the owl from habitat destruction as well as from direct hunting. It appeared that the arguments of the environmentalists had succeeded in stopping logging of the old-growth forest. Cynics suggest that this is what the environmentalists had in mind all along, with owl protection a ploy in the battle to save the forest.

The US Forest Service chose Christmas Eve 1991 to announce that reports from the timber industry that the spotted owl had been seen living outside the old-growth forest and limited regions of redwood forest were unwarranted. The unusual date for this announcement was not caused by government machinations, however. Instead, it was a response to an order by a federal district court issued the previous day. The judge denied a request by the Forest Service to postpone release of habitat maps that would show where loggers cannot operate; instead, the Forest Service maps had to be available by Jan 8 1992. The maps show that logging is restricted for about 8.3 million acres of forest in the Pacific Northwest.

However, the forest workers' jobs might still be saved by other branches of the US government. An executive committee known as the "God squad" because it can decide to allow a species to become extinct was to hold hearings in Jan on the spotted owl. The seven-member Cabinet committee headed by Interior Secretary Manuel Lujan, Jr., has the power to put economic interests ahead of the provisions of the Endangered Species Act of 1973.

In a related development, the US Fish and Wildlife Service proposed on Jun 17 1991 that the marbeled murrelet (*Brachyramphus marmoratus*) be added to the threatened list. These marbeled murrelets nest in the same old-growth forests of the US Northwest as the northern spotted owl and in California. Other populations can be found breeding in eastern Asia and Alaska. When they are not breeding, marbeled murrelets take to the open ocean or to seacoasts.

(Periodical References and Additional Reading: *New York Times* 2-5-91; *New York Times* 6-18-91; *International Wildlife* 7/8-91, p 28; *New York Times* 8-2-91, p A1; *Science* 8-16-91, p 740; *New York Times* 11-5-91, p C4; *New York Times* 12-26-91, p A19; *New York Times* 1-9-92, p A14)

MAMMALS IN TROUBLE

There are many mammals large and small in trouble around the world. The lemurs of Madagascar, some populations of African elephants, big cats in Asia and the Americas, the great apes everywhere, the giant panda, and whales are only some of the mammals that are endangered and threatened. Don't forget bats: According to Merlin Tuttle of Bat Conservation International in Austin TX, "bats are disappearing at a faster rate than any other group of vertebrates." Here are a few other mammals that were in the public consciousness in the early 1990s.

The Rhinoceros Story

Various species of rhinoceros, supposedly the original model for the unicorn, have suffered at the hands of humans throughout recorded history and probably before it. There is some reason to believe that *Homo erectus* hunters were a factor in the extinction of the forest rhinoceros in Europe nearly 100,000 years ago, while ice age humans helped eliminate the woolly rhinoceros population about 90,000 years later. After that there were five known species of rhino left, three in Asia and two in Africa. Until the nineteenth century, those five—the white and black rhinos of Africa and the Indian, Javan, and Sumatran rhinos of Asia (whose names are not accurately descriptive of their ranges)—were thriving. In the nineteenth century, however, colonialism brought European sport hunters into Asia and Africa, where they proceeded to reduce sharply populations of the three larger rhinoceroses, the Indian, white, and black, especially the Indian. Farming and population expansion affected the Javan and Sumatran rhinos as well as both African species.

But through it all, the rhinoceros was not much worse off than other large mammals of Africa and Asia. In the twentieth century, however, populations of three of the rhinos declined precipitously. Continued expansion of the human population in southeast Asia, combined with other pressures, reduced the Sumatran rhinoceros to about 700 individuals. The Javan rhinoceros, with fewer than 60 left, has become the rarest large mammal on Earth. The black rhino population, in better shape to begin with, declined to about 3000 from a population in the 1970s numbering as much as 70,000. Meantime, conservation efforts improved the numbers of the Indian and the white rhinoceroses.

Two factors affect the rhinoceros that do not affect other large mammals. One is the greatly increased prosperity of ethnic Chinese throughout much of Asia and the other is the oil wealth of the Persian Gulf region. Traditional Chinese medicine has always ascribed great powers to powdered rhinoceros horn; with prosperity, the Chinese market for the horns greatly increased. Similarly, a tradition of using rhinoceros horn for dagger handles in Yemen and neighboring states has also raised prices. Consequently, the black market in rhino horn is one of the main forces behind vastly increased poaching—so much so that some conservation agencies are dehorning rhinos and returning them to the wild to protect the animals from poaching.

It now appears that the only hope for the continued existence of any of the rhinoceros species will be in protected sanctuaries. It is not clear that efforts to protect the Sumatran and Javan rhinos will be in time to prevent extinction.

Optimistic News and Some Dissent

Some mammals are returning to regions where they have not been for years or even generations. Among them are the moose in New England, the gray wolf in the northern Rocky Mountains, and the bowhead whale in the waters of Alaska.

Of these returns, the only one that has everyone happy is that of the bowhead whale. A census taken during the whale's spring 1991 migration revealed a population of somewhere between 6200 and 9400, up from as few as about 1000 during the 1970s. Although bowheads, like all whales, are officially protected, Alaskan Eskimos are allowed to take forty-one each year. It seems likely that the population will continue to increase until it reaches its environmental limit.

On the other hand, the moose, which has rebounded in New England from near extinction to a population of about 26,000 in northern Maine, New Hampshire, and Vermont, is a mixed blessing. Like the California condor, the moose is sometimes considered a remnant of the late Pleistocene megafauna, since it is the largest representative of the deer family. The main problem with moose is that they cannot read traffic signs, or perhaps the human New Englanders are failing to note marked moose crossings. As a result, there were 711 motor vehicle collisions with moose in the three states in 1990, dramatically more than in previous years. Often people or moose are killed in such collisions.

There are about forty to fifty wolves in Montana, fewer in Idaho, and maybe some in Wyoming. These wolves have not been put there by human conservationists; they just moved on down from Canada after Canadian laws were changed to offer the wolf population more protection. Since the gray wolf is on the endangered list in the lower United States, the new arrivals cannot legally be killed. Ranch owners are not, however, happy to see the immigrants. To protect the wolves, a group called Defenders of Wildlife has been compensating ranchers for apparent wolf kills. In 1988-1990, the Defenders paid for eight cows, fourteen calves, and eleven sheep. Wolves have long ranged through northern Minnesota where statistics reveal that they kill about five cows out of 10,000 and twelve sheep out of 10,000.

The Squirrel and the Telescope

Rising more than 3000 m (10,000 ft) above the desert Coronado National Forest of southeast Arizona, Mt Graham has become known as the home of an endangered subspecies of red squirrel (*Tamiasciurus hudsonicus grahamensis*) in an important relic ecosystem. About 300 Mt Graham red squirrels are thought to be present today. At one time, it was believed that the entire population had become extinct. Mt Graham is also known as the site for construction by the University of Arizona of one of the world's most advanced astronomy complexes, with seven telescopes planned.

The two different uses for the same space have caused conflict since the late 1980s. In 1988 the US Fish and Wildlife Service declared that the observatory construction would harm the red squirrel and might lead to its extinction. This halted the project until the US Congress passed legislation exempting the project from the Endangered Species Act so long as careful monitoring demonstrated that the environment was not being seriously damaged. In particular, the legislation permitted construction of the first three of the planned telescopes. By the end of 1991 an access road had been built and foundations poured for the first two telescopes.

On Dec 11 1991 the US federal appeals court for the Ninth Circuit went further than the initial legislation. The court ruled unanimously that construction of the first three telescopes on Mt Graham could continue and that the remaining four could be built if environmental monitoring showed that their completion and operation would not harm the Mt Graham red squirrel. A coalition of organizations intended to protect the environment, led by the Sierra Club, was expected to appeal some parts of the decision, especially on the grounds that environmental monitoring practices and plans were not adequate.

Dolphin UPDATE

Although many aspects of drift net fishing practices are more environmentally destructive, another type of netting that affects mammals led to major public objections that resulted in changes in fishing practices. Schools of yellowfin tuna, a type commonly canned (as opposed to the bluefins that become tuna steaks, sushi, and sashimi), are often captured in large nets. For unknown reasons, schools of yellowfins often swim under dolphins. When the tuna are netted, the dolphins also are captured. Dolphins are mammals and need air to live, so they drown as the nets are being hauled in. When people learned of this as a result of publicity by environmental groups, led by the Earth Island Institute headed by David Brower, they began to support a boycott of canned tuna. Videotapes of dolphins dying, made by undercover activist Samuel LeBudde, were prominently featured on television news in the late 1980s.

On Apr 12 1990 the tuna industry in the United States caved in under the boycott. The three major canners, led by H.J. Heinz, announced that they would buy tuna only from boats carrying observers who could testify that the tuna was "dolphin-safe." Environmental groups supplied the observers, mainly in boats out of Thailand, the hub of the canned tuna industry.

There was another flurry of activity on the issue starting on Dec 4 1990 when Earth Island Institute accused one US tuna canner of buying fish that were not dolphin-safe. This died down when it became apparent that violations of the agreement had not been involved.

(Periodical References and Additional Reading: *Science* 6-22-90, p 1479; *Astronomy* 7-90, p 22; *Science* 7-6-90, p 26; *Science News* 7-7-90, p 7; *Science News* 8-11-90, p 84; *Science* 8-31-90, p 988; *New York Times* 12-7-90, p 50; *New York Times* 12-8-90, p 50; *International Wildlife* 1/2-91, p 26; *Science* 3-8-91, p 1178 & 1187; *International Wildlife* 3/4-91, p 39; *Science* 5-3-91, p 643; *New York Times* 5-7-91, p C1; *Science* 5-31-91, p 1257; *Astronomy* 6-91, p 26; *New York Times* 6-3-91, p A13; *Science News* 6-15-91, p 374; *New York Times* 6-25-91; *Science* 7-19-91, p 250; *Science* 8-16-91, pp 744 & 750; *Scientific American* 10-91, p 18; *New York Times* 10-29-91, p C1; *Science News* 11-2-91, p 279; *International Wildlife* 11/12-91, p 28; *New York Times* 12-13-91, p A28)

FOREST AND DESERT PROBLEMS

Not only are many species being lost, entire ecosystems are in trouble as well. In their place, of course, other ecosystems arise. If a forest disap-

DOLPHINS AND DOLPHINS AND PORPOISES

Some confusion exists in English between various species that are called <u>dolphin.</u> Small toothed whales, relatives of the sperm whale, orca, and narwhal, are generally known as dolphins, although members of one family are correctly called porpoises. The name <u>porpoise</u> is also used as a generic term for dolphins. All of these small whales are, like all whales, mammals.

Confusion between true dolphins and true porpoises is not the end of the story; two fish, <u>Coryphaena hippurus</u> and <u>C. equistis</u>, are also known as dolphins. Although both mammal and fish dolphins are edible, many people have an affection for the mammal that is like that for the horse, one that precludes eating its flesh. No such compunction exists for the fish. As a result, the fish today is usually sold under either the Spanish name <u>dorado</u> or the Hawaiian name <u>mahimahi.</u> As <u>mahimahi,</u> the fish has in recent years passed from being a regional specialty of Hawaii to a major menu item in US seafood restaurants.

pears, the land where the forest stood must be covered with something. Nearly always different living organisms colonize any land or water region left behind after an existing ecosystem has been removed. As human beings, we are often concerned largely because an ecosystem that we view as pleasant or productive is replaced by one that is unpleasant or economically useless. As citizens of the planet we are also concerned about extinctions engendered by changing ecosystems, even when there is no recognizable use for a given species. We often later learn that a species thought to be useless from a human perspective is of importance.

Rain Forest UPDATE

Rain forests are in trouble, a well-known fact among the environmentally informed. There is even an ice cream called "rain forest crunch," part of whose proceeds benefit tropical rain forests. Despite this attention, it is often not clear to the general public what a rain forest is. Frequently, news stories with "rain forest" in the headlines are about some other kind of tropical ecosystem. Temperate rain forests, on the other hand, are seldom known by that name.

Any action that might reduce rain forests is widely condemned, including a 1991 proposal by a Canadian corporation to log the temperate rain forest on the west side of Chile's Tierra del Fuego. Although the planners promised to cut only mature trees, freeing younger trees to grow more vigorously, conservationists were alarmed, since the lenga trees involved are slow growing and take 70 to 100 years to mature.

Background

A rain forest is a natural ecosystem in which rainfall is more than 200 cm (80 in.) evenly distributed over the year and temperatures are high enough to permit tree growth. Thus, there are both tropical and temperate rain forests. Tropical rain forests are sometimes called jungles, although strictly speaking a jungle is a disturbed forest that is then overgrown with shrubs and vines. One characteristic that all forms of rain forest share is extremely tall trees, two or three times as tall as those in other types of forests.

Tropical rain forests exist on every continent except Europe and Antarctica, but not every forest in warm regions is a rain forest. The largest existing tropical rain forests are in Brazil, South America; Zaire, Africa; and Indonesia, Asia. Tropical rain forests in Central America are perhaps better studied than any others, and conclusions about rain forests in general are often based on Central American forests.

Although opinions differ on how many different species exist on Earth, everyone seems to agree that half those species live in tropical rain forests. In part this is because a tropical rain forest has more absolute mass of living organisms (biomass) for a given unit of area than any other ecosystem; but there is more diversity in the tropical rain forest for each unit of biomass as well.

Temperate rain forests are a different kind of ecosystem entirely, although also important and endangered. They are found mainly along the northwest coast of North America, along the coast of southern Chile, and in New Zealand.

The rain forest of most concern in the United States and around the world for the past decade or so has been the Amazonia forest, located mostly in Brazil. The causes for the concern are many. The forest is the most extensive of its type left on the planet. Population growth and economic expansion have resulted in removal of the forest at an enormous rate. Parts of the forest are home to the last indigenous peoples in the Americas who continue to preserve most of their pre-Columbian way of life. The destruction of the forest appears to be permanent; forest that is cleared may never regenerate. Vast numbers of species in the forest have never been studied, and evidence has accumulated to show that species are becoming extinct faster than they can be named. Until recently the government of Brazil seemed to be promoting forest destruction. Widely publicized murders were committed over forest issues.

Probably as a result of worldwide attention, most of the news concerning the Brazilian rain forest was good in the early 1990s. In Mar 1991 satellite studies showed that the rate of burning in the Brazilian forest during the previous year was 27 percent less than in 1989, with 1990 forest clearing—not all necessarily rain forest—of 13,815 km² (5334 sq mi) as compared with comparable clearing in 1989 of 18,837 km² (7273 sq mi). On Jun 24 Brazilian president Fernando Collor de Mellor promised to end tax subsidies that promoted converting forest to ranches and farms. That month he also took other steps to change government policies to protect not only the forests but also the indigenous peoples of Brazil. In Oct the United States took steps toward joining an effort spearheaded by Germany to protect the Amazon forest and to provide the money to set up vast reserves.

Another story offering hopeful news concerns the Lacandona rain forest in Mexico near the Guatemalan border. The Lacandona forest is the largest tropical rain forest in North America, even after logging and slash-and-burn agriculture halved it to about 6500 km² (2500 sq mi) over the past 50 years. In 1991 Conservation International worked out a debt-for-nature swap in which $4 million of Mexican debt will be retired in return for expenditure of about $2.6 million from 1992 to 1994 on conservation projects, mostly to be spent on preserving the remaining Lacandona forest.

Forest Issues in the US West

A growing United States cut down most of its natural forests for farmland early in its history. In 1990 about 5 percent of the forest in the lower United States remained from the extensive forests of 1620, the year the pilgrims landed. A lot of this is second- (or third-) growth forest that has regrown. With the rise of more efficient agricultural practices, many lands once farmed have been allowed to revert to forest. As a result, there are more trees in the United States today than there were in the 1920s, but the new forest is not the same complex ecosystem as the old-growth forest it replaced. At least a hundred years is needed for trees to grow to full size.

The largest remaining old-growth forests are those in mountainous regions of the West, where logging has become a way of life for many. In recent years, environmental pressure to save the western forests has resulted in conflict at many levels between loggers and their allies and people opposed to cutting the forest. Activist environmentalists have taken many steps, legal and otherwise, to stop the cutting, especially clear-cutting, in which all the trees in a given region are removed and the debris burned. Over 200 sawmills closed in the 1980s, and the number of loggers and sawmill workers in Oregon dropped from about 42,000 in 1979 to about 33,000 in 1990. But economic changes caused most of the closings and lost jobs, not environmental pressure.

A lot of the skirmishing in the early 1990s concerned the old-growth rain forest of the Pacific Northwest. A surprise entry into the controversy came in the early 1990s when a small, otherwise undistinguished yew in old-growth forests became known as the only source of a potent anti-cancer drug. The drug, taxol, was first shown to be clinically useful in 1989, although it had been suspected as a potential treatment since its dis-covery in the 1960s. Taxol and the Pacific yew from which it is derived did not become well known, however, until the positive results of large-scale studies were announced in May 1991. On Jun 19 1991 the US government promised to direct nearly 40,000 Pacific yews, enough to treat about 6000 cancer patients, to the manufacturer of taxol, Bristol-Myers. Before this, the fate of most Pacific yews that were removed was to be burned with other "trash" trees during clear-cuts.

The Desert Is Gaining

Several reports in the early 1990s tell of serious losses to other parts of the environment in the American West, largely as a result of cattle ranching (but possibly aggravated by early signs of global warming). Land used by ranchers as range for animals, mostly cattle, is most affected. In 1990 the Environmental Protection Agency said that rangeland regions along water-ways through the region are in the worst condition in history. The US General Accounting Office issued a study of herds in public lands that noted that overstocking was common and that only 25 percent of over-stocked lands had been scheduled for reduction in herd size. A 1991 United Nations report, written by Harold Dregne of Texas Tech University, claims that 85 percent of the rangeland in the West is being degraded, largely as a result of overgrazing. Dregne believes that 15 percent of the rangeland in the West is "irreversibly degraded and cannot be economi-cally rehabilitated or improved," mostly as a result of salinization caused by irrigation.

Basically, the combination of overgrazing and salinization is turning the semidesert of the American West into actual desert as a result of lower groundwater tables, reduced surface water, and high loss of native vegetation, resulting in increased soil erosion. Cattle eat the native grasses, which are replaced by desert sagebrush, mesquite, and Russian thistle (tumbleweed).

(Periodical References and Additional Reading: *National Wildlife* 11/12-90, p 24; *New York Times* 2-26-91; *Science* 3-29-91, p 1559; *International Wildlife* 3/4-91, p 18; *New York Times* 3-26-91; *New York Times* 5-13-91, p A1; *New York Times* 6-4-91, p C4; *New York Times* 6-20-91; *New York Times* 6-26-91, p A9; *Science* 6-28-91, pp 1780 & 1784; *Science News* 7-21-90, p 40; *New York Times* 8-20-91, p C4; *Science News* 8-31-91, p 143; *Science* 9-6-91, p 1091; *International Wildlife* 9/10-91, p 12; *International Wildlife* p 1091; *International Wildlife* 9/10-91, p 12; *International Wildlife* 11/12-91, p 25; *New York Times* 11-3-91, pp E2 & E3; *New York Times* 11-7-91)

OUT OF EXTINCTION

Once in a while a species previously thought to be extinct is found to be still existing in small pockets. Among these in the early 1990s are Fender's blue butterfly and a small desert fish (that's right, a desert fish). A few years ago the list also included the black-footed ferret, thought for a time to be extinct. After discovery and as a result of an infectious disease, the ferret, like the California condor, was withdrawn from the wild for captive breeding; individuals were returned to the wild in 1991.

Fender's Blue Butterfly Found

Fender's blue butterfly (*Icaricia icarioides fenderi*), a subspecies of the generally endangered blue butterfly of the West Coast, was thought to have been extinct since 1937. Previously it was known to have been in existence since the end of the last ice age, some 11,000 years ago. The rare butterfly was spotted by Paul Hammond of Oregon State University while hiking through the Macdonald Forest near Corvallis OR.

The particular cause for the near extinction of Fender's blue butterfly was the near extinction of a plant, Kincaid's lupine. Just as the monarch butterfly is dependent on the common milkweed, the caterpillar stage of Fender's blue will eat only the leaves of Kincaid's lupine, a rare wildflower found almost exclusively in the Willamette Valley of Oregon.

Desert Fish Found

Almost all fish listed by the US Fish and Wildlife Service as endangered live in small bodies of water in the desert. If the water survives so, in general, do the fish. When people compete with the fish for the water, the fish are generally the losers. This situation occurred at Lake Magdalena, formerly about 150 km (100 mi) west of Guadalajara, Mexico. People used all the water in the lake for drinking and irrigation, drying it up completely in 1970. Any unique species living in the lake were presumed to have become extinct as a result.

Among the species thought to have been lost was *Opal allotoca*, a 5-cm (2-in.) fish. In Apr 1990, however, Michael L. Smith of the American Museum of Natural History and two graduate students found that *O. allotoca* individuals had been pumped into a small artificial pond near the original lake, where they had survived and reproduced. Smith and coworkers captured fifteen of the tiny fish, leaving five in the lake and taking ten back to the United States to breed.

(Periodical References and Additional Reading: *Science News* 6-9-90, p 359; *International Wildlife* 11/12-91, pp 25 & 26)

UNWANTED SPECIES

Although much of the concern about the environment centers on the loss of species or ecosystems, a related problem is the introduction of species into ecosystems where they were not formerly present. Much of the environmental damage on islands, for example, has come from the introduction of exotic goats, rats, snakes, pigs, and other species that either compete directly with existing species for a specific resource, eat the native species, or simply crowd them out. Few if any regions on Earth are left with the same mix of species that they had a few thousand years ago, before people began to move animals and plants from place to place, often with the best of intentions. Freed of natural restraints, exotic species can quickly become the dominant ones in a landscape. Today, neither an eastern meadow in the United States nor a hillside in California consists largely of native species

Even when we know an unwanted population has arrived, it is often difficult to eradicate it. The Mediterranean fruit fly (*Ceratitis capitata*), commonly called the medfly, has been known in California since 1975; the first major proliferation of the medfly occurred in 1986. Aerial spraying of the pesticide malathion was used to combat it. Another major outbreak started in Jul 1989 and continued through Nov 1990, when the California Department of Food and Agriculture (CDFA) declared that the imported fly had once again been eradicated by spraying it with malathion. Yet a controversial study by James R. Carey of the University of California at Davis published in *Science* on Sep 20 1991 says that malathion spraying has never succeeded in wiping out medflies completely. Instead of being reimported, which CDFA claims caused the second proliferation, the flies according to Davis had become endemic to southern California. Because medflies are difficult to trap, it is not clear which side is right, but most of the evidence, according to a University of California panel summoned in May 1990 to look at the data, favors Carey.

Arrival of the Killer Bees

Rarely do we know that an unwanted animal or plant is on its way. Such has been the case with the Africanized bees nicknamed "killer bees" because easily angered swarms sometimes sting victims to death. The bees started as an experiment in Brazil to see if bees from Africa could be more productive in the tropics than the European honeybee is. Some queen bees escaped in 1957 and founded wild colonies. The new colonies thrived and interbred with European honeybees. Gradually the hybrids moved north, extending their range some 300 to 500 km (200 to 300 mi) each year.

After much fanfare, the hybrid bees reached the United States in 1990, founding colonies in south Texas, just across the Rio Grande. By Aug 1991 a total of 137 swarms had been located in Texas, still moving north. They were expected to reach Houston by spring of 1992. It is not clear how far north the hybrids will thrive.

Africanized bees are less desirable than pure European bees not only because they are wilder and more aggressive, but also because they produce less honey. Some beekeepers have announced that they plan to abandon their hives if the Africanized bees are spotted in the vicinity. Bees are not only important for their honey but also for their role in pollinating various food crops. A reduction in the bee population because of fewer beekeepers would be bad for much of US agriculture.

And the Zebra Mussels

Unlike the Africanized bees, which moved up from Brazil, the invasion of zebra mussels (*Dressena polymorpha*) from Europe started in North America, specifically in Lake St Clair, a large lake between Canada and the United States that is part of the connection between lakes Erie and Huron. Zebra mussels were first observed in Jul 1988 by William P. Kovalak of Detroit Edison Company in Lake St Claire, where they probably had been brought in ballast by oceangoing ships using the St Lawrence Seaway and the Great Lakes. But, like the Africanized bees, the zebra mussels quickly began to extend their range, first throughout the Great Lakes and then into rivers connected with the Great Lakes. In the nineteenth century the zebra mussel, which originated in the Caspian and Black Sea regions, spread throughout Europe.

In Nov 1990 there was a zebra mussel scare in the Hudson River, but it turned out to be a false alarm caused by the misidentification of a native species of mussel. No matter. By May 15 1991 the mussel was definitely in the Hudson, found by fishers Everett Nack, Sr., and Steven Nack near Hudson NY.

The zebra mussel is so prolific that the small, 3-cm (1.5-in.) mussel clogs all sorts of underwater pipes, causing much economic damage. It also competes with native species for food and for space. On Nov 29 1990 US President Bush signed legislation setting up a program to control the expansion of the mussel.

(Periodical References and Additional Reading: *Science* 3-9-90, p 1168; *Science* 9-20-91, pp 1331, 1351, & 1369; *New York Times* 10-13-91, p 22; *New York Times* 12-3-90; *New York Times* 5-26-91; *Science* 9-21-90, p 1370; *Discover* 1-92, p 44)

POLLUTION
OUTDOORS AND IN

OVERALL AIR POLLUTANTS IN THE UNITED STATES

Type of Pollution	Change from 1980 to 1989	Estimated Metric Tons Emitted in 1989	Comments
Nitrogen dioxide	-5%	19,900,000	Contributes to acid rain, and is one of the principal catalysts that causes ozone formation near ground level
Sulfur dioxide	-10%	21,100,000	Considered to be the principal cause of acid rain
Total suspended particulates	-15%	7,200,000	Essentially this is dust; starting in 1987 a new measure used by the EPA concentrates on tiny particles that are thought to cause most respiratory ailments
Carbon monoxide	-23%	60,900,000	This gas is like the familiar carbon dioxide except that only a single oxygen atom is bonded to a single carbon atom; the carbon atom would prefer two oxygen atoms, and in the bloodstream it is able to bind to oxygen, making carbon monoxide an effective poison; adding extra oxygen to gasoline can help prevent carbon monoxide from forming.
Volatile organic compounds	-34%	18,500,000	Volatile organic compounds, along with nitrogen dioxide, are among the principal contibutors to the formation of ozone concentrations at ground level

Type of Pollution	Change from 1980 to 1989	Estimated Metric Tons Emitted in 1989	Comments
Lead	-90%	7,200,000	Lead tends to accumulate in tissues, where it interferes with various biochemical reactions, leading to mental retardation and various physical ailments; most lead in the air comes from burning leaded gasoline in automobiles; reductions in the use of leaded gasoline have resulted in substantial declines of lead in air

Source: US National Air Quality and Emissions Trends Report, 1989.

CURRENT INDUSTRIAL POLLUTION

Since 1987 many manufacturing corporations have reported to the US Environmental Protection Agency (EPA) the extent of their production and disposal of toxic wastes in the United States. Although these data are not compiled by the EPA, they are published. Using the *Toxics Release Inventory,* the Citizens Action arm of the Washington-based Citizen's Fund has produced rankings that are useful in identifying those corporations that are responsible for pollution and the locations of the most serious pollution problems.

Since 1989 the EPA has compiled similar data by state, based on mandatory reports of releases of about 330 hazardous wastes. The total reported for these kinds of wastes amounted to the following:

1987	7.000 billion pounds
1988	4.423 billion pounds
1989	5.700 billion pounds

These figures, however, are not comparable in various ways. For example, most of the reduction since 1987 came from changing the list of chemicals that were required to be reported. The most common pollutant reported was ammonium sulfate, which accounted for 750 million pounds of the 1989 total. (The most recent available data for both corporations and states are from 1989.)

Total Toxic Wastes in 1989 by Polluter

| Rank | Corporation | Mass of Pollution | | Short Tons | Percent of Total |
		Kilograms	Pounds		
Total		2,590,126,825	5,710,282,027	2,855,141	100
1.	Dupont	155,925,635	343,758,978	171,879	6.02
2.	Monsanto	133,391,531	294,079.534	147,040	5.15
3.	American Cyanamid	91,690,490	202,143,984	101,072	3.54
4.	BP America	56,205,752	123,913,120	61,957	2.17
5.	Renco Holdings	54,133,651	119,344,894	59,672	2.09
6.	3M	49,212,410	108,495,359	54,248	1.90
7.	Vulcan Materials	42,219,067	93,077,597	46,539	1.63
8.	General Motors	38,851,902	85,654,230	42,827	1.50
9.	Eastman Kodak	36,002,763	79,372,920	39,686	1.39
10.	Phelps Dodge	35,225,725	77,659,836	38,830	1.36

Total Toxic Wastes in 1989 by State

| Rank | State | Mass of Pollution | | Short Tons | Percent of Total |
		Kilograms	Pounds		
Total		2,590,126,82	5,710,282,027	2,855,141	100
1.	TX	359,600,000	792,800,000	396,400	13.9
2.	LA	214,800,000	473,500.000	236,750	8.3
3.	OH	162,700,000	358,700,000	179,350	6.3
4.	TN	119,900,000	264,300,000	132,150	4.6
5.	IN	115,700,000	255,000,000	127,500	4.5
6.	IL	112,200,000	248,000.000	124,000	4.3
7.	MI	99,800.000	220,100,000	110,050	3.9
8.	PA	88,100,000	194,200,000	97,100	3.4
9.	FL	87,100,000	192,000,000	96,000	3.3
10.	KS	84,000,000	185,100,000	92,550	3.2

Toxic Water Pollution by Corporation

While the top ten industrial polluters produce only 26.74 percent of all toxic wastes reported, the top ten *water* polluters produce 57.6 percent of the reported water pollution (air pollution by the top ten is only 21.64 percent of the total reported).

| | | *Mass of Pollution* | | | |
Rank	Corporation	Kilograms	Pounds	Short Tons	Percent of Total
Total		85,725,844	188,994,123	94,497	100
1.	Arcadian	10,670,085	22,414,703	11,207	11.86
2.	3M	7,029,519	15,497,518	7,749	8.20
3	Freeport McMoran	6,643,753	14,647,045	7,324	7.75
4	ITT	5,186,414	11,434,144	5,717	6.05
5	Allied Signal	4,449,171	9,808,795	4,904	5.19
6.	Louisiana Pacific	4,166,276	9,185,114	4,593	4.86
7.	Weyerhauser	3,831,945	8,448,037	4,224	4.47
8.	Strategic Minerals	3,540,477	7,805,457	3,903	4.13
9.	Monsanto	2,323,170	5,121,741	2,561	2.71
10.	Simpson Investment	2,048,848	4,516,960	2,258	2.39

Toxic Air Pollution by Corporation

| | | *Mass of Pollution* | | | |
Rank	Corporation	Kilograms	Pounds	Short Tons	Percent of Total
Total		1,101,178,614	2,427,695,968	1,213,848	100
1.	Renco Holdings	54,067,870	119,199,872	59,600	4.91
2.	3M	32,925,241	72,588,109	36,294	2.99
3.	Eastman Kodak	31,383,591	69,189,335	34,595	2.85
4.	Dupont	24,225,930	53,409,311	26,705	2.20
5.	General Motors	21,252,747	46,854,532	23,427	1.92
6.	Courtaulds Fibers	20,261,686	44,669,606	22,335	1.84
7.	Ford	14,425,440	31,820,817	15,901	1.31
8.	Hoechst Celanese	13,544,497	29,860,660	14,930	1.23
9.	BASF	3,104,026	28,889,582	14,445	1.19
10.	General Electric	13,104,026	28,889,582	14,445	1.19

Worst Toxic Sites

| | | | *Mass of Pollution* | | |
Rank	Location	Corporation	Kilograms	Pounds	Short Tons
1.	Alvin TX	Monsanto	93,200.000	205,500,0001	103,000
2.	Westwego LA	Cyanamid	87,300,000	192,400,000	96,000
3.	Tooele UT	Magnesium Corp of America	54,000,000	119,100,000	60,000
4.	Wichita KS	Vulcan	41,900,000	92,300,000	46,000

(Continued)

			Mass of Pollution		
Rank	Location	Corporation	Kilograms	Pounds	Short Tons
5.	Beaumont TX	Dupont	40,000,000	88,100,000	44,000
6.	Port Lavaca TX	BP America	29,700,000	65,500,000	33,000
7.	New Johnsonville TN	Dupont	26,000,000*	57,400,000*	29,000*
8.	East Chicago IN	Inland Steel	26,000,000*	57,300,000*	29,000*
9.	Lima OH	BP America	25,700,000	56,700,000	28,000
10.	Wichita KS	ATOCHEM	24,700,000	54,500,000	27,000

*Similarity in kilograms and short tons due to rounding. Figures are collected in pounds.

MAJOR INDOOR AIR POLLUTANTS IN THE HOME

Until recently, most air pollution was viewed as something that occurred outside the home and that could enter only when it drifted in through open windows. Today, with better sealed houses, it has become clear that the house itself, chemicals or activities within it, and even the ground the house is built on may contribute to air pollution indoors. Some of the pollutants found indoors are more dangerous to individuals than any outdoor air pollution. One estimate is that one-fifth to one-third of office buildings in the United States have polluted indoor air.

Pollutant	Sources	Effects	Levels in Homes	Steps to Reduce Exposure
Asbestos	Old or damaged insulation, fireproofing, or acoustical tile	After many years, chest and abdominal cancers and lung diseases	Elevated levels can occur where asbestos-containing materials are damaged or disturbed	Seek professional help from trained contractors; follow proper procedures for replacing wood stove gaskets that may contain asbestos
Biological pollutants	Bacteria, mold and mildew, viruses, animal dander and cat saliva, mites, cockroaches, and pollen	Eye, nose, and throat irritation; shortness of breath; dizziness; lethargy; fever; digestive problems; asthma; influenza and other infectious diseases	Higher levels occur in homes with wet or moist walls, ceilings, carpets, and furniture; poorly maintained humidifiers, dehumidifiers, or air conditioners; and household pets	Use fans vented to the outdoors in kitchens and bathrooms; vent clothes dryers outdoors; clean humidifiers daily; empty water trays in appliances frequently; clean and dry or remove water-damaged carpets; use basements as living areas only if they are leak-proof and have adequate ventilation, keeping the humidity between 30 and 50 percent

Pollutant	Sources	Effects	Levels in Homes	Steps to Reduce Exposure
Carbon monoxide	Unvented kerosene and gas heaters; leaking chimneys and furnaces; wood stoves and fireplaces; gas stoves; automobile exhausts from attached garages; tobacco smoke	At low levels: fatigue in healthy people and chest pain in people with heart disease; at higher levels, impaired vision and coordination; headaches; dizziness; confusion; nausea; fatal at very high concentrations	Homes without gas stoves contain from 0.5 to 5 parts per million (ppm); levels near properly adjusted gas stoves are often 5 to 15 ppm; near poorly adjusted gas stoves levels can be 30 ppm	Keep gas appliances properly adjusted; use vented gas space heaters and furnaces; use proper fuel in kerosene space heaters; install exhaust fans vented to outside over gas stoves; open flues when gas fireplaces are used; choose wood stoves that meet EPA emission standards; have annual inspection of home heating system; do not idle car inside garage
Formaldehyde	Plywood, wall paneling, particleboard, fiberboard; foam insulation; fire and tobacco smoke; durable-press drapes, textiles, and glues	Eye, nose, and throat irritation; wheezing and coughing; fatigue; skin rash; severe allergic reactions; possibly cancer	Average concentration in older homes without urea-formaldehyde foam insulation is generally below 0.1 ppm, but may be greater than 0.3 ppm in newer homes that use pressed wood products	Use interior grade wood products; use air conditioners and dehumidifiers to maintain moderate temperatures and reduce humidity levels; increase ventilation after bringing new formaldehyde sources into home
Lead	Automobile exhausts; sanding or burning of lead paint; soldering	Impaired mental and physical development in fetuses and children; decreased coordination and mental abilities; damage to kidneys, nervous system, and red blood cells; may raise blood pressure	Lead dust levels are 10 to 100 times greater in homes where sanding or burning of lead paint has occurred	Have paint tested before removing it in older homes; if it is lead based, cover it with wallpaper or other building material and replace moldings and woodwork; use no-lead solder; have drinking water tested for lead; if exposure is suspected, consult your health department
Nitrogen dioxide	Kerosene heaters, unvented gas stoves and heaters; tobacco smoke	Eye, nose, and throat irritation; may impair lung function and increase respiratory infections in young children	Average levels in homes without heaters is about half that of outdoors; homes with gas stoves or unvented gas or kerosene heaters often exceed outdoor levels	The same steps that prevent carbon monoxide should be taken to prevent buildup of nitrogen dioxide
Organic gases	Paints, paint strippers, solvents, wood preservatives; aerosol sprays; cleansers and disinfectants; moth repellents; air fresheners; stored fuels; hobby supplies; dry-cleaned clothing	Eye, nose, and throat irritation; headaches; loss of coordination; nausea; damage to liver, kidneys, nervous system; some organics cause cancer in animals and are suspected of causing cancer in humans	May average 2 to 5 times higher indoors than outdoors; activities such as paint stripping can raise levels to 1000 times outdoor levels	Follow manufacturers' instructions when using household products; use volatile products outdoors or in well-ventilated places; dispose of unused or little used products safely; buy volatiles in quantities you will soon use up

(Continued)

Pollutant	Sources	Effects	Levels in Homes	Steps to Reduce Exposure
Particles (soot)	Fireplaces, wood stoves, kerosene heaters; tobacco smoke	Eye, nose, and throat irritation; respiratory infections and bronchitis; lung cancer	Unless the home contains smokers or other strong particle sources, levels are the same as or lower than outdoors	Vent all furnaces outdoors; keep doors to rest of house open when using unvented heaters; choose wood stoves that meet EPA standards; have annual tune-up of heating system; change filters on central heating and cooling systems and air cleaners according to manufacturer's directions
Pesticides	Products used to kill household pests and products used on lawns or gardens that drift or are tracked inside the house	Irritation to eyes, nose, and throat; damage to the nervous system and kidneys; cancer	Preliminary research shows widespread presence of pesticide residues in homes	Use strictly according to manufacturer's instructions; mix or dilute outdoors; take plants or pets outside when possible; increase ventilation when using indoors; use other methods of pest controls when possible; do not store pesticides indoors; dispose of unwanted containers safely
Radon	Earth and rock beneath home; well water; building materials	No immediate symptoms; estimated to cause about 10 percent of lung cancer deaths; smokers at higher risk	Estimated national average is 1.5 picocuries per liter, but levels in homes have been found as high as 200 picocuries per liter; EPA believes levels in homes should be less than 4 picocuries per liter	Test your home for radon; get professional advice if radon reduction is indicated; seal cracks and other openings in basement floor; ventilate crawl space; install sub-slab ventilation or air-to-air heat exchangers; treat well water by aerating or filtering through granulated activated charcoal
Tobacco smoke	Cigarette, pipe, and cigar smoking	Eye, nose, and throat irritation; headaches; bronchitis; pneumonia; increased risk of respiratory and ear infections in children; lung cancer; contributes to heart disease; the American Heart Association claimed on Jan 9 1991 that passive tobacco smoke kills 53,000 nonsmoking people in the United States annually	Homes with one or more smokers may have levels several times higher than outdoor levels	Stop smoking and discourage others from smoking; if you do smoke, smoke outdoors

Life Science

THE STATE OF LIFE SCIENCE

As used in this volume, the term life science includes the various branches of biology and physical anthropology, with a small amount of paleoarchaeology (see below) included along with anthropology. Medicine is not treated in *Current Science* except for some aspects of medical technology (see "Biotechnology," p 589).

The study of biology itself has been divided throughout the twentieth century between the work of whole-organism biologists and the work of scientists who deal with organisms at the level of the cell, chemical, or molecule. While this dichotomy is not going to go away completely, the 1980s and early 1990s have seen a number of bridges built. The field of genetics has increasingly supplied a link as scientists trace the effects of a specific gene (part of a large molecule) on a protein (a whole molecule) that interacts with other molecules in chemical reactions, producing results at the cellular level that affect the whole organism. A major application of this technique, sometimes going up from gene to organism but more often traveling in the other direction, has been the elucidation of congenital diseases, such as cystic fibrosis. In pure biology, major investigations of this type concern efforts to reconstruct the cell-by-cell developmental history of some organisms (see *C. elegans* UPDATE," p 459) and studies of the cell or the chemical basis of memory. The driving force for research at all levels of organization continues to be evolution by natural selection.

A new tool for all sorts of purposes is the polymerase chain reaction (PCR), a technique for obtaining many copies of a gene (see "Who Owns PCR?," p 593). Large numbers of copies are needed to study genes in any detail. PCR has been vital in medical and forensic uses as well as in the Human Genome Project (see "Human Genome Project UPDATE," p 446).

New Recognition of Taxonomy

The glamour in biology since World War II has been with the molecular, chemical, and cell biologists, but there has also been a well-recognized

A PRIMER: DNA AND GENES

It is difficult to understand modern life science without some concept of genetics. The details of the subject are complex, but the basic ideas follow.

Most cells contain recognizable small pairs of bodies called chromosomes. Each chromosome is observed to double in normal cell division, but in the division that produces egg or sperm cells for plants or animals the pairs separate instead of doubling. Each chromosome consists mostly of a large molecule of deoxyribonucleic acid (DNA), which has the structure of a ladder that has been twisted. When chromosomes double, the DNA untwists and each new chromosome gets one of the uprights of the ladder. Each of the two uprights is then used to build a new chromosome that, barring mistakes, is exactly like the original before the doubling. By the 1940s it was already apparent that this process is used to transmit hereditary characteristics from one cell to its daughters, although the role of DNA and the mechanism involved were not understood until the 1950s. The unpairing process in egg and sperm cells is matched by the formation of new pairs in fertilization, which carries hereditary characteristics from one generation to the next. Since such characteristics are expressed in the body cells and ultimately in the whole organism, the DNA can be said to control cell development as well as heredity.

Each upright of the DNA twisted ladder, or double helix to use a more technical name, carries the information needed for this process in the form of strings of four different subunits called bases. The bases are adenine, thymine, guanine, and cytosine, usually abbreviated to A, T, G, and C. Although each DNA molecule contains the same four bases, the number and arrangement varies enormously. It is the exact arrangement that is preserved when a chromosome doubles itself.

The work and structure of all cells is mainly carried out by or consists of a group of highly variable complex polymers called proteins. Each protein is constructed from twenty-some subunits called amino acids. The information in the sequence of bases in DNA is a method of describing each protein. Nearly every combination of three of the four bases A, T, G, and C identifies exactly one amino acid. DNA is such a large molecule that the average chromosome can describe tens of thousands of proteins. One sequence of bases describing exactly one protein is called a gene, and the twenty-three pairs of human chromosomes carry the information for a couple of hundred thousand genes. Not all of these genes are active at any one time, but mechanisms based ultimately in the DNA turn genes on and off as needed. Indeed, in addition to the genes, DNA of plants and animals contains long sequences of bases that seem to do nothing.

role for evolutionary biology, ecology, and behavioral studies. Some of the older disciplines of biology, such as comparative anatomy, tended to be viewed as necessities for students but not as vital centers for new biological learning. Taxonomy, the science of classifying living organisms into groups according to scientific principles, mostly fell into the same dull category of a textbook necessity but a research dead end.

The 1980s saw the beginning of a new respectability for taxonomic studies. Taxonomy based on molecular studies, which had been proposed by Linus Pauling and Emile Zuckerkandl of CalTech in 1962, was at first viewed with suspicion, but it began to be vindicated by other research. Later, PCR enabled molecular taxonomists to improve the depth, accuracy, and application of molecular techniques based on DNA. By the 1990s the new approach was giving rise to laboratories and departments devoted to molecular taxonomy at the Smithsonian Institution's Museum of Natural History in Washington DC, the American Museum of Natural History in New York City, and the British Museum in London.

Another factor in the revival of taxonomy has been the popular writing—popular among scientists as well as the general public—of Harvard paleontologist Stephen Jay Gould. Gould, a fan of both traditional taxonomy and comparative anatomy as well as of the new molecular approaches, has analyzed and promoted taxonomy in several best-selling books as well as in regular magazine articles.

The realization that species are disappearing before they can even be properly classified (see "Is This a Mass Extinction?," p 390) has also generated new interest in taxonomy.

Taxonomy not only helps clarify evolutionary issues, it also is essential in the recognition of new species. There is widespread knowledge that we do not know anything about perhaps as many as half of all the species on Earth, mainly invertebrates. Surprisingly, however, the early 1990s brought discovery of two new species of large mammals, a primate (see "New and Rare Primate Species Found," p 483) and a small beaked whale about 3.5 m (12 ft) long. The whale was identified in 1991 on the basis of eleven carcasses found washed up on shore since Feb 2 1976. James Mead of the National Museum of Natural History, who spotted the first recognized carcass, has named the whale *Mesoplodon peruvianus*. Small whales are elusive, but this was still the first new species recognized since 1963. As these cases illustrate, taxonomy of newly discovered species remains vital.

Animal Rights

Animals rights became newsworthy in the 1980s, mainly as a result of protests and direct action against people seen as violating those rights.

Biologists, along with workers in medical research and various related pursuits (such as the testing of cosmetics or designing of safe vehicles), continue to run afoul of animal rights activists in the early 1990s. Incidents in the early 1990s may not have been deemed as newsworthy as those of the 1980s because the public has become more accustomed to animal rights demonstrations. Naturally, the first time that a protest over treatment of laboratory animals is mounted calls for larger headlines than the hundredth time. Also, scientists and others have revised many of their practices in the face of continuing agitation, so there are somewhat fewer occasions for protest. Nevertheless, the presidents of the National Academy of Sciences and the Institute of Medicine jointly reported in 1991 that something had to be done because animal rights activists have "resorted to violence, breaking into laboratories, setting fires, destroying records, and harassing researchers," a good summary of what had started happening in the late 1980s.

This trend toward quieter animal rights battles in the 1990s may accelerate after 1992 if legislation passed by the US Senate and House in 1991 becomes law that year. The Senate bill would protect scientists from people breaking into their laboratories in protest, while the House bill would also protect farmers or ranchers and food-processing facilities from animal rights activists. The separate bills still need to be merged by a conference, repassed, and signed before there is new law on the subject.

Paleoarchaeology

Reflecting trends that have accelerated through the 1980s, some people who used to think of themselves as physical anthropologists have now started to call themselves paleoarchaeologists. Anthropology has long been a diffuse label that covered many distantly related activities that centered on humans, their evolution, and their behavior. As it was viewed until recently, paleontology of early primates, a sub-branch of the earth science of paleontology, blended into physical anthropology, which had no clear boundary with archaeology, which treated the same subject matter as cultural anthropology, except at a different time. It was inevitable that there would be regrouping. The problem in classification is compounded because for our earliest ancestors, physical studies of teeth and bones are intimately involved with such behavioral questions as what the early ancestors ate and how they moved about.

One part of the current sorting out of disciplines seems to be happening largely by setting back the time boundary between archaeology and physical anthropology. Until the 1920s, that boundary had largely been drawn around the time of the agricultural revolution, roughly 10,000 years ago; if an event happened before farming, it was anthropology, afterward archaeology. By the 1980s the dividing line was becoming the arrival of *Homo sapiens* on the scene, roughly 40,000 years ago (or earlier, if you count archaic humans and Neanderthals). This change was not primarily caused by archaeologists encroaching on the subject matter previously studied by physical anthropologists. Instead it occurred largely because the anthropologists learned to apply techniques developed by archaeologists. These include the use of careful site maps that locate every fossil or artifact found, sampling through carefully planned trenches, and analysis of garbage.

In the 1990s the boundary is being pushed back further by the application of archaeological techniques to the study of prehuman ancestors. At a site in Kenya called Olorgesaile, Richard Potts of the Smithsonian Institution is applying archaeological methods to a *H. erectus* site thought to be about 980,000 years old. At the famous Oulduvai Gorge, Robert Blumenschine is using the same approach to study *H. habilis,* an earlier ancestor from about 2.4 to 1.7 mya, as well as various other hominids that lived in the region around that time, including the earliest known representatives of *H. erectus.*

Potts's project, funded by the Smithsonian, is an outgrowth of ideas of Glynn Isaac and Louis Binford that Binford called "landscape archaeology." Binford wanted to put fossils into the whole landscape that surrounds them. To accomplish this, Potts is running what amounts to a 5-km (3-mi) trench through the stratum that contains the fossils, although in reality there are more than twenty separate trenches. A major difference between this project and that at an ordinary archaeological site is that Potts has to use a bulldozer to uncover the stratum, which is buried as much as 3.5 m (12 ft) below a layer of sediment that contains no macrofossils. The goal is to determine how *H. erectus* 980,000 years ago behaved. Potts thinks that our view of *H. habilis* behaving one way for its species lifetime, followed by *H. erectus* with another fixed behavior pattern that persisted for over a million years, has to be revised. Even if *H. erectus's* life-style changed more slowly than that of modern humans, nevertheless it changed as our ancestor became a hunter, tamed fire, and explored the world beyond Africa.

New Tools for Anthropologists

Accompanying the landscape archaeology approach to physical anthropology has been the growth of a new subgroup of scientists, specialists in fossil formation, site dating, materials analysis with scanning electron microscopes, and related areas. The new methods supplement previous specializations in such physical characteristics as teeth, brain size, and limb bones.

Among the important, but still experimental, new techniques are electron-spin resonance and thermoluminescence, both designed to measure the dates of crystalline material by determining how many electrons are trapped in tiny holes in crystals. Such holes and the trapped electrons both increase with age on what is thought to be a regular basis.

Another new dating technique comes from changes in the handedness of molecules in ancient ostrich shells. Because such shells are found in many sites in Africa and also in Asia, the method, called racemization or diagenesis of proteins, has widespread applications. It was introduced in 1990 as a means of dating events too old for precise carbon-14 dating (typically more than 40,000 years old) and not old enough for potassium-argon dating (less than 200,000 years old). The same technique would also work with larger shells of nontropical birds, such as those of owls, extending its usefulness into Europe.

It is certain the new approaches and tools will enrich our understanding of the earliest hominids, and perhaps even the homonoids. It is somewhat less clear whether or not there is much of a future for pure physical anthropology as it was known for most of the twentieth century. As the border between anthropology and archaeology retreats in time, physical anthropologists are spending more of their time working with creatures that most people would call apes. While the work is very good science, it lacks a lot of the excitement of finding the "missing link."

(Periodical References and Additional Reading: *Science* 2-16-90, p 798; *Science* 3-23-90, p 1407; *Discover* 7-90, p 77; *Science* 2-22-91, p 872; *Science* 8-23-91, p 846; *International Wildlife* 9/10-91, p 32; *Discover* 11-91, p 9; *Science* 11-22-91, p 1099)

Timetable of Life Science to 1990

10,000 BC The dog becomes the first domesticated animal

9000 BC The agricultural revolution starts in the Near East with the domestication of sheep, goats, and wheat

8000 BC The agricultural revolution starts independently in what are now Latin America and Indochina

350 BC Aristotle [Greek: 384-322 BC] classifies the known animals using a system that will continue to be used for the next 2000 years

1648 Jan Baptista van Helmont's [Flemish: 1580-1635, or 1644] experiment showing that plants do not obtain the materials for their growth from the soil is published posthumously

1665 Robert Hooke [English: 1635-1703] describes and names the cell

1668 Francesco Redi [Italian: 1626-1697] shows that maggots in meat do not arise spontaneously, as most people then believed, but are hatched from flies' eggs

1669	Anton van Leeuwenhoek [Dutch: 1632-1723] discovers microorganisms, creatures too small to see with the naked eye, and recognizes that sperm are a part of reproduction
1683	Van Leeuwenhoek is the first to observe bacteria
1735	Carolus Linnaeus [Swedish: 1707-1778] introduces the system still used for classifying plants and animals
1779	Jan Ingenhousz [Dutch: 1730-1799] discovers that plants release oxygen when exposed to sunlight and that they consume carbon dioxide; this is the beginning of the understanding of photosynthesis
1827	John James Audubon [French-US: 1785-1851] starts publication of *Birds of America*
1839	Theodor Schwann [German: 1810-1882], building on the work of Matthias Schleiden [German: 1804-1881] in 1838, develops the cell theory of life (generally attributed to both Schleiden and Schwann)
1856	The first skeleton of what we now call the Neanderthals is found in a cave in the Neander valley, near Dusseldorf (Germany)
	Louis Pasteur [French: 1822-1895] discovers that fermentation is caused by microorganisms
1858	Charles Darwin's [English: 1809-1882] and Alfred Wallace's [English: 1823-1913] theory of evolution by natural selection is announced to the Linnaean Society
1859	Darwin's *Origin of Species* is published
1865	Gregor Mendel's [Austrian: 1822-1884] theory of dominant and recessive genes is published in an obscure local journal
1868	Workers building a road in France discover the skeletons of the first known Cro Magnons in a cave
1894	Marie-Eugène Dubois [Dutch: 1858-1940] announces discovery of the "Java ape-man," now known to be the first discovered specimen of *Homo erectus*
1898	Tobacco mosaic disease is recognized as being caused by a virus, resulting in the first identification of a virus, although viruses cannot yet be seen and are known only from their effects
1900	Three different biologists rediscover the laws of genetics originally found by Gregor Mendel
1901	The okapi is discovered
1919	Karl von Frisch [Austrian-German: 1886-1982] discovers that bees have a language that can be used to communicate how to find a good source of flower nectar
1924	Raymond Dart [Australian-South African: 1893-1988] identifies the first fossil of an australopithecine, a close relative of early humans
1938	The first known live coelacanth is captured; scientists had believed the species to have been extinct for 60 million years

1952   Eugene Aserinsky [US: 1921-] discovers that sleep with rapid eye movements (REM) is a specific stage of sleep, later found to be associated with dreams

1953   James Watson [US: 1928-] and Francis Crick [English: 1918-] determine the structure of DNA, the basis of heredity

1961   Louis Leakey [English: 1903-1972] and Mary Leakey [English: 1913-] discover a previously unknown ancestor of humans, *Homo habilis*

Marshall Nirenberg [US: 1927-] learns to read one of the "letters" of the genetic code

1962   Linus Pauling [US: 1901-] and Emile Zuckerkandl propose that the evolution of molecules can be used to determine how long one species has been separate from a related species.

1967   Allan Wilson [New Zealand-US: 1934-1991] and Vincent Sarich propose that humans and chimpanzees shared a common ancestor 4 to 6 million years ago

1968   Werner Arber [Swiss: 1929-] discovers restriction enzymes, a class of proteins that will make genetic engineering possible

1969   Jonathan Beckwith [US: 1935-] and coworkers are the first to isolate a single gene

1970   Har Gobin Khorana [Indian-US: 1922-] and coworkers produce the first artificial gene

Howard Temin [US: 1934-] and David Baltimore [US: 1938-] discover the enzyme that causes RNA to be transcribed to DNA, a key step in the development of genetic engineering

1973   Stanley Cohen [US: 1917-] and Herbert Boyer [US: 1936-] succeed in putting a specific gene into a bacterium, the first instance of true genetic engineering

1974   Don Johanson [US: 1943-] and coworkers discover Lucy, the nearly complete skeleton of *Australopithecus afarensis,* an early relative of humans (more than 3 million years old)

1975   César Milstein [British: 1927-] announces the discovery of how to produce monoclonal antibodies

1980   Martin Cline [US: 1934-] and coworkers succeed in transferring a functioning gene from one mouse to another

Chinese scientists succeed in cloning a fish

1988   New fossil finds indicate that modern *Homo sapiens* has existed for at least 92,000 years

GENES

FOUND GENES

The Human Genome Project (see below) is searching for all of the human genes, but finding all of them is going to take quite some time. Along the way, individual researchers are seeking and finding genes already recognized as of particular importance, whether in humans, mice, plants, or other organisms.

Gene for Maleness

In 1990 and 1991 a British group of researchers jointly led by Robin Lovell-Badge of the British Medical Research Council National Institute for Medical Research and Peter Goodfellow of the Imperial Cancer Research Fund claimed discovery of the gene on the Y chromosome that initiates maleness. They were not the first researchers to make this claim, since at least two different genes had in the past been touted as the key gene for maleness, only to be shown to be insufficient. Indeed, a gene found by David Page in 1987 was believed to be the one until as recently as Dec 1989. The British team, aware of the earlier misses, took special care to make sure they were right. They had originally announced probable success on Jul 11 1990 but were wary of making a definitive claim until almost a year later, on May 9 1991, after they had obtained clear evidence that their gene was absolutely necessary for maleness. (The popular press was not so cautious and the gene was declared found in 1990 and even made one of the 1990 "Top 50 Science Stories" by *Discover.*) The British claim so far has been universally accepted. The final piece of evidence is that the gene by itself causes maleness in mice that otherwise would be female.

Lovell-Badge and Goodfellow named the mouse maleness gene *Sry* and the human version *SRY* to mean sex-determining region on the Y chromosome. In humans, the *SRY* gene in the seventh week of pregnancy initiates changes that turn developing gonads into testes. Then the gene shuts down its operation. Surprisingly, the gene resumes activity in adult males (shown in mouse studies), suggesting that it has some additional effects after initiating maleness. The gene is quite tiny, however, so its main role may be to act as an on or off switch for other genes; it could work either by turning maleness on in some positive way or by turning female development off. The type of protein produced by the gene is used commonly to regulate other genes. The same gene appears in male mammals of various types, including chimpanzees, tigers, horses, rabbits, and, of course, mice.

Timetable for Genetics

Genetics really is *new* science. Almost all ideas that most people today take for granted were discovered within the lifetime of some people now living, and most genetic discoveries are well within the life spans of the generation Americans call the baby boomers, those born in the 1950s.

1900   Three different biologists—Hugo Marie De Vries [Dutch: 1848-1935], Karl Franz Joseph Correns [German: 1864-1933], and Erich Tschermak von Seysenegg [Austrian: 1871-1962]—rediscover the laws of heredity originally found by Gregor Mendel but published in an obscure local journal in 1865 and subsequently ignored until this rediscovery

1905   Clarence McClung finds that female mammals have two X chromosomes and males have an X paired with a Y

1907   Thomas Hunt Morgan [US: 1866-1945] starts experiments with the fruit fly *Drosophila melanogaster* that establish that the units of heredity are located on small bodies in cells called *chromosomes*

1909   Wilhelm Johannsen [Danish: 1857-1927] coins the word *gene* to describe the unit of heredity

1910   Thomas Hunt Morgan discovers that some genes are linked to a particular sex; in the fruit fly *D. melanogaster* a white-eyed mutation appears in males only

1911   Morgan starts mapping gene locations on chromosomes

1918   The number of human chromosomes is counted (incorrectly) for the first time; Herbert M. Evans counts forty-eight

1936   Andrei Nikolaevitch Belozersky isolates deoxyribonucleic acid (DNA) in the pure state for the first time

1941   George Wells Beadle [US: 1903-] and Edward Lawrie Tatum [US: 1909-1975] theorize that each enzyme is controlled by a single gene

1944   Oswald Theodore Avery [Canadian-US: 1877-1955], Colin MacLeod, and Maclyn McCarthy determine that DNA in chromosomes is the repository of genes, not proteins, as had been generally believed

1953   James Watson [US: 1928-] and Francis Crick [UK: 1918-] determine the structure of DNA, the basis of heredity; DNA is a double helix that carries information in combinations of the four bases that make up the two strands of the helix

1954   J. Lin Tjio and Albert Levan show that humans have forty-six chromosomes (arranged in twenty-three pairs) rather than forty-eight, as was previously believed

1959   Scientists establish that the chromosome known as Y confers maleness in humans; a human with two X chromosomes will normally be female, although there can be exceptions when a male gene crosses over; the condition of two X chromosomes and no Y is later found to happen about once in every 20,000 males

1961   Marshall Nirenberg [US: 1927-] and J.H. Matthaei of the US National Institutes of Health learn to read one of the "letters" of the genetic code when they find that a combination of three uridylic acid bases in a row (UUU—called a *codon*) in ribonucleic acid (RNA) codes for the amino acid phenylalanine

1967   Charles Yanofsky is the first to prove that the sequence of codons in a gene determines exactly the sequence of amino acids in a protein

1968   Werner Arber [Swiss: 1929-] discovers restriction enzymes, a class of proteins that will make genetic engineering possible

David Zipser discovers the meaning of the last remaining undeciphered codon of the genetic code, the pattern uracil-guanine-adenine (UGA) that means "stop making this protein"

1969   Jonathan Beckwith [US: 1935-] and coworkers are the first to isolate a single gene

Roger Donahue is the first to map a human gene to a chromosome other than the X and Y chromosomes; it is the gene for the "Duffy" blood type

1970   Har Gobin Khorana [Indian-US: 1922-] and coworkers produce the first artificial gene

Howard Temin [US: 1934-] and David Baltimore [US: 1938-] discover the enzyme that causes RNA to be transcribed to DNA, a key step in the development of genetic engineering

1973   Stanley N. Cohen [US: 1917-] and Herbert W. Boyer [US: 1936-] succeed in putting a specific gene into a bacterium, the first instance of true genetic engineering

1977   Phillip A. Sharp and, independently, Richard J. Roberts and coworkers discover that DNA in organisms more complex than bacteria contains long stretches of meaningless material that is not part of any gene; the meaningless material is labeled an *intron*

1980   Martin Cline and coworkers transfer genes from one mouse to another and succeed in having the gene function in its new organism

1981   The Chinese produce a genetic copy (clone) of a zebra fish

At Ohio University in Athens OH scientists transfer genes from other organisms into mice for the first time

1983   Walther J. Gehring and coworkers discover the homeobox gene (see below), essential for the development of a wide variety of organisms from yeasts to humans

1984   Alex Jeffreys develops "genetic fingerprinting," a method of identifying a specific person using any material from that person that contains DNA (for example, white blood cells, skin cells, or sperm)

1985   The first accounts of the DNA polymerase chain reaction, invented in 1983 by Kary B. Mullis, are published (see "Who Owns PCR?" p 593)

1986   The first genetically altered virus (for herpes in swine) is marketed and the first field trials of a genetically altered plant begin

1989   Francis Collins, Lap-Chee Tsui, and coworkers find and make copies of
        the gene that causes most cystic fibrosis

1990   On Jul 31 the Recombinant DNA Advisory Committee, a watchdog group
        for genetic engineering, approves the first serious attempt to insert cells
        that have been genetically altered into human patients to correct an
        inborn genetic defect

Nevertheless, all mysteries concerning sex determination are far from solved. Some humans with Y chromosomes become females for unknown reasons—perhaps something turns *SRY* off. Finding what controls *SRY* and what it controls is the next goal of the researchers.

Female Turn-Off Gene

Related to the gene for maleness, at least in terms of the problem it solves, is a gene that turns off most of the genes on an X chromosome. Although this might be termed a "female turn-off gene," it has nothing to do with being fat, bald, stupid, or any other condition that might turn off women. Instead, it seems to save the lives of normal female mammals by shutting down the production of what otherwise would be excess proteins.

For the most part, men and women need the same amount of most proteins; even where needs by sex are different, they are not as different as one might suspect. A normal female has two X chromosomes, each with a vast number of genes, one on each chromosome for each protein needed. A normal male has one of those big X chromosomes, but it is paired with a runt of a Y chromosome. The Y has very few genes. It would appear, therefore, that a normal female would have double amounts when compared with those of the normal male of most of the proteins produced by the X gene. But that does not happen. Instead, most of the genes on one of the female's X chromosomes are turned off while the genes on the other X chromosome produce about the same amount of proteins as would the male XY pair.

Late in 1990 Carolyn J. Brown of Stanford University and coworkers found the female turn-off gene. On one of the X chromosomes, this gene is active and switches off all but a half dozen genes. On the other X chromosome, the turn-off gene is itself shut down and the rest of the genes on the chromosome operate normally.

Gene for Marijuana Receptor

Every cell is enclosed by a membrane that keeps unwanted substances out and lets wanted ones in. The membrane is an active barrier, just as human skin is, but even more so. Most important substances that must enter cells to affect the functioning of organisms cannot penetrate the membranes by themselves. Instead, special molecules, called *receptors,* actively act to capture (bind to) molecules of a particular substance as the substance passes by. They then drag the substance into the cell, perhaps taking it to an appropriate body within. The importance of these receptors gradually became clear in the 1970s and 1980s, especially with regard to opiate and cholesterol receptors.

The opiate receptors, discovered in the early 1970s by Solomon Snyder of Johns Hopkins and coworkers, can be used as an example. Since there are receptors in the brain for such opiates as morphine and heroin, which cause the well-known effects when the receptors bind to the drugs, scientists realized that there should be some natural substances produced in humans with which the receptors would normally bind. By 1975 the first of these substances, enkephalin, was found. This was quickly followed by the discovery of similar chemicals called endorphins. This group of chemicals is involved in signaling the brain for such positive concepts as joy and such negative ones as boredom. Better understanding of how the endorphins and their receptors work led to new interpretations of the roles of neurotransmitters. These chemicals, used by nerve cells for communication, include serotonin and dopamine, now known to be actively involved in such disorders as Parkinson's disease.

Ten years before the first opiate receptors were found, researchers began to look for the active site of tetrahydrocannibinol (THC), the newly isolated active ingredient in marijuana. By the mid-1980s research by Allyn Howlett and coworkers at St. Louis Medical School indicated that THC must have a receptor, just as the opiates do, and that such a receptor could control important activities of the brain.

The receptor proved extremely elusive. But in the Aug 9 1990 *Nature* the team of Lisa Matsuda, Stephen Lolait, Michael Brownstein, Alice Young, and Tom Bonner from the US National Institute of Mental Health reported success in finding its gene, completing the quarter-century search. As expected, the gene is active in the regions of the brain that deal with movement, thinking, and memory, but not in regions that handle automatic functions such as respiration or heartbeat. The next search will be to find the natural brain substance that also binds to THC.

Recovering a Gene from Honest Abe

Three groups of researchers, including scientists ranging from Connecticut to Oregon, announced in Jul 1991 that they had discovered the gene whose defects cause Marfan syndrome, a relatively common genetic disease. Many Americans first learned of the disease when it caused the sudden and unexpected deaths of Olympic volleyball player Flo Hyman and

FIRST THE GENE OR FIRST THE PROTEIN?

The hypothesis that each gene makes (through several steps involving ribonucleic acid, or RNA) one protein and that each protein stems from a single gene goes back to 1941, when very little was understood about the physical nature of genes. At that time proteins were moderately well understood. By 1953 Frederick Sanger was able, in a heroic effort, to analyze completely for the first time the structure of a fairly small protein, insulin. That was the same year that James Watson and Francis Crick deduced the structure of DNA, the key step in unraveling the physical basis of genetics. Within a few years, the principal action had shifted from studying proteins to studying genes. The genetic code was deciphered and machines were developed that could find the sequence of bases on DNA that make up the code.

One result was to remove a big part of the difficult biochemical problem of describing the structures of proteins and replace it with the largely solved problem of deciphering a gene. Once the gene that accomplishes something is found, the sequence of amino acids in the protein follows automatically. This is not the complete description of a protein, since the shape of proteins determines a large part of their action, but it may be enough for many purposes. Thus, a biochemist may start with a result and proceed directly to trying to find the gene, leaving the question of the protein that actually accomplishes the result until later, since that is the easier problem. (Of course, if for some reason a protein's sequence of amino acids is already known, you can back into knowing a lot about the gene that makes the protein.)

A classic example of how the process works occurred with respect to muscular dystrophy in the late 1980s. By 1987 a team headed by Louis M. Kunkel had located the gene that causes a major form of the disease known as Duchenne muscular dystrophy. By May of the following year the protein that the gene produces became known. Soon the physiology of the disease was better understood and by the early 1990s it appeared that treatment based on the new knowledge was on the way.

University of Maryland basketball star Chris Patton. Others heard of it when speculation arose that Abraham Lincoln may have been affected by the syndrome. Despite its obscurity until recently, Marfan syndrome is thought to affect as many as one person in 10,000, including about 30,000 persons in the United States alone.

Marfan syndrome was named after a Frenchman who wrote a description of it in 1896, although a clear account of death from the syndrome was given by Giovanni Morgagni as early as 1761. As in the deaths of Hyman and Patton, Morgagni's case history involved a young person who died suddenly during physical activity—although Morgagni's example was a prostitute engaging in her trade. In all three cases, the lining of the aorta, the main artery of the body, split. More than half of the people with Marfan syndrome have enlarged or thickened aortas that can rupture in this way. Other symptoms include features we associate with Abraham Lincoln—elongated limbs and loose joints. In addition, the lens of the eye may easily become dislocated, while the chest can puff out or be sunken, and heart valves can function abnormally.

The main result of the discovery of the genetic basis for Marfan syndrome is that it is now clear that all the different symptoms relate to defects in a single protein, as one would expect for a purely genetic disease. The protein is fibrillin, the structural scaffolding in the connective tissues of the body—found in skin, ligaments, linings of blood vessels, and elsewhere. Different mutations of the gene that makes fibrillin cause the different manifestations of Marfan syndrome.

Fibrillin became suspect almost exactly a year before the gene was located. The research was reported on in the Jul 19 1990 *New England Journal of Medicine* by David W. Hollister of the University of Nebraska Medical Center and Reed Pyeritz of Johns Hopkins School of Medicine.

The now classical method of searching for a disease-causing gene is based on comparing the genetic makeup of members of families in which the syndrome is common. This method was used in conjunction with a search for the fibrillin gene. The first breakthrough in 1990 came when a Finnish team's family search showed that the defective Marfan gene was somewhere on chromosome 15. Final proof came when two patients with a severe form of the syndrome were found to have defects in the fibrillin gene. Because it is a large gene, mutations can occur in many different places, accounting for the variety of symptoms and the differences in severity among those with the syndrome.

It is hoped that knowledge of the gene will quickly lead to a simple test for defects in it. Early diagnosis of the syndrome could lead to surgical or medical treatments to remedy the changes in the aorta that lead to death.

Among the scientists who located the gene are Harry C. Dietz at Johns Hopkins University, Petros Tsipouras of the University of Connecticut Health Center at Farmington, Francisco Ramirez and Brendan Lee of Mount Sinai School of Medicine in New York City, and Lynn Y. Sakai of the Shriners Hospital for Crippled Children in Portland OR.

Knowing the gene will also enable researchers to determine whether Lincoln actually did have Marfan syndrome. If he did, the resulting publicity could help other sufferers feel better about their problem. A panel of experts was appointed by the National Museum of Health and Medicine on Feb 9 1991 to decide whether to attempt to recover Lincoln's DNA by the polymerase chain reaction method (see "Who Owns PCR?" p 593). About 280 g (10 oz) of Lincoln's tissues have been preserved, including blood, bone, and hair. Lincoln was fifty-six when he was assassinated, an age by which about three quarters of people with Marfan syndrome would have died from their disease under nineteenth-century care, but Lincoln's physical prowess as a young man was legendary. One expert, Jerold M. Lowenstein of the University of California at San Francisco, estimated in 1991 that the best efforts to recover some of Lincoln's genes from the existing material might locate 1 percent of his total genome. Whether that small amount would include the proper stretch for the Marfan gene is unknown.

Other genetic diseases of connective tissue may also be caused by defects in proteins related to fibrillin. Among these are congenital contractural arachnodactyly and mitral valve prolapse, the latter a common defect that affects one person in twenty. In 1990 Darwin J. Prokop and coworkers at Thomas Jefferson University in Philadelphia PA proposed that a faulty gene that should code for collagen III, another connective protein, can also predispose a person to rupture of the aorta.

Plant Homeobox

The homeobox has been recognized since 1983 as a key part of regulatory genes in all kinds of organisms. It was first observed in fruit flies as the core of genes whose mutations lead to gross developmental changes, such as feet growing instead of antennae. Soon the same core was found in many different animals as well as in yeast. When the same feature appears in various species, especially distantly related ones, biologists say that it is "conserved." The homeobox is highly conserved, which suggests that it arose early in the development of life and that it performs some essential function.

Because the homeobox is so highly conserved, biologists expected to find the gene structure in plants as well as in animals and yeast. Efforts to locate it in plants with a probe failed. A probe is a single strand of DNA that is labeled in some way (often with radioactive elements) and that contains a known gene. If the probe is put near single strands of DNA, it will wind itself about any part of the DNA that contains exactly, or even approximately, the same gene. The label can then be used to recover the probe and its attached DNA. But homeobox probes from animals or yeast failed to detect plant homeoboxes.

Nevertheless, Sarah Hake of the University of California at Berkeley and coworkers from the university and the US Department of Agriculture's Plant Gene Expression Center, also at Berkeley, reported in 1991 that they had found the plant homeobox by other means. Although it is too different from an animal homeobox to be recovered with a probe, the plant homeobox is enough like the animal one to be recognized both from its sequence of bases and from its effects. Using the first plant homeobox, found in a gene called *knotted* that causes parts of maize leaves to be misplaced, other plant homeoboxes were soon located in such plants as tomatoes and rice.

(Periodical References and Additional Reading: *Science News* 7-28-90, p 60; *Science News* 8-4-90, p 78; *Science* 8-10-90, p 624; *Science News* 11-10-90, p 293; *Discover* 1-91, pp 47 and 82; *New York Times* 2-10-91, p A1; *Science* 3-1-91, p 1030; *Science* 5-10-91, p 782; *Scientific American* 6-91, p 24; *New York Times* 7-25-91, p A18; *Science News* 7-27-91, p 55; *Discover* 8-91, p 18; *Science News* 8-3-91, p 70)

SURPRISES FROM GENES

Transcribing RNA

Scientists who speculate about the origin of life have generally reached the conclusion that before there was DNA, the other nucleic acid, ribonucleic acid (RNA), was on hand to get things started. In that view, DNA eventually arose and became central because it is structurally sounder than RNA and because its method for making copies of itself is superior. If RNA was present from the beginning, it would not be surprising to find that even today RNA is capable of performing all of the functions needed for life. In the past decade or so, RNA has been shown to perform almost all those functions, ranging from enzymelike catalysis to forming structural components. In 1991 yet another basic function was found to have—at least some of the time—an RNA component that was previously unsuspected.

When a strand of messenger RNA is produced, a large protein called RNA polymerase and several small helper proteins called simply A, B, C, and D (not all of which are present for a given situation) form what is known as the transcription machine. In the late 1980s Karen Sprague of the University of Oregon at Eugene observed that a silkworm (*Bombyx mori*) gene contained a larger-than-usual and more complex "on" switch, known technically as the promoter region. She wondered how it was possible for RNA polymerase and the B and C helpers, the only proteins found in the transcription machine for that particular gene, to operate such a switch by themselves.

Reasoning by analogy with some other cellular mechanisms, Sprague suspected that RNA might also be a part of the transcription machine and not just the medium of transcription. After initial results that suggested an RNA presence, she and coworkers hit a blank wall. Giving up on RNA, they tried to analyze the transcription machine in other ways, which did not work either. The team then returned to the concept of a small RNA molecule as part of the machine. This time, with other techniques, they found the RNA.

HOW TO MAKE A PROTEIN

DNA by itself does not do anything. It is a recording, like a CD, cassette, or floppy disk. You need a complex machine to turn tiny pits in a CD or magnetic patterns on a cassette into music or music and pictures. Most functions of living cells are accomplished by complex molecules that are the equivalent of sound and pictures if DNA is the equivalent of a CD or cassette. For the most part, the complex molecules are proteins, polymers built by connecting about twenty different amino acids in various configurations. Involved every step of the way, and sometimes acting for itself alone, is RNA, a polymer made from four different bases that are strung out along a backbone made from a simple sugar and phosphoric acid. The four bases are like those of DNA except that thymine (T) is replaced with the similar base uracil (U), so the four bases of RNA are cytosine (C), guanine (G), adenine (A), and uracil.

In a simplified view of how a protein is made, the main actors are various forms of RNA. The first step is transcription, in which a type of RNA called messenger RNA forms along the DNA and matches it base for base —that is, each C on the DNA matches a G on the RNA and vice versa while each A on the DNA matches a U on RNA and each T on DNA matches an A on RNA. Thus the RNA is just like the complementary strand of DNA except for the substitution of U for T. The messenger RNA then moves to a ribosome, a complex with some proteins, where the second process, translation, takes place. In the ribosome, the messenger RNA is pulled through the ribosome one codon at a time (remember that a codon is a pattern of three bases such as ACC or UAU that represents a single amino acid or the beginning or end of a gene). As each codon passes through, it is matched with a small molecule of RNA called a transfer RNA. The transfer RNA has brought along an amino acid from among those that are just floating around in the cell waiting to be picked up by a transfer RNA. Twenty-odd special ligase enzymes (proteins) match the appropriate transfer RNA and amino acid. The transfer RNAs drop off their amino acids in the ribosome, where they are added one at a time to the new protein chain that is forming there. When a stop codon is reached, the protein chain pops out and the messenger RNA leaves the ribosome.

Note that the above description is simplified. One simplification is that all the action of most of the already existing proteins that make this process take place has been omitted. Left to itself, DNA does nothing and RNA, versatile as it is, does not do much. A group of proteins is there to assemble the RNA on the DNA template to begin with, for example. This group is called a transcription factor or transcription machine. Its main component is a protein called RNA polymerase. Another group of about eighty proteins operates the ribosome.

Another simplification is that something needs to turn the gene on in the first place so that the various steps toward protein synthesis can take place. This is a different gene that encodes a different protein. Similarly, a third gene and concomitant protein is needed to turn the whole process off.

Other simplifications include leaving out the process of methylization in which methyl groups are attached to the messenger RNA to protect parts of it from enzymes that cut it to pieces; and omitting the processing that removes "junk" parts of RNA before it reaches the ribosome (for organisms other than bacteria, there is generally more junk than functioning RNA).

The transcription machine is a new place for RNA to be found working. Its presence there raises more questions than it answers. Among the questions are, What does it do? and Is there RNA in any other transcription machine? Continuing research is intended to solve these and related questions.

The Five Histones

Ever since Oswald Avery and coworkers discovered in 1944 that genes in chromosomes are on DNA and proteins in chromosomes are structural (previously the true roles were assumed to be reversed), not much attention has been paid to chromosomal proteins. In the early 1990s, however, the few scientists who took the chance of studying the proteins were able to show that some of them, at least, take an important and active role in how the chromosome works.

This should not have been a surprise because available evidence shows that life manages to get as much as possible out of every part of every cell—for example, bones are used to make blood and store calcium, while the heart produces a hormone, and skin helps teach immune cells how to behave.

The main chromosomal proteins of interest in the early 1990s were the five histones, proteins that often switch genes on and off along with their structural role of helping the meter-long (three-foot) DNA molecule coil up to fit into a chromosome that is small even when compared with a microscopic cell. Histones apparently act by allowing or failing to allow other proteins, called *transcription factors*, to get near a gene and turn it on. The histones may also push a gene out to the edge of the chromosome to allow it to operate or hide it deep inside the chromosome where nothing happens.

Some diseases, especially one version of a hereditary form of anemia called thalassemia, may be the result of histone malfunction. Mark Groudine of of the Fred Hutchinson Cancer Research Center in Seattle WA has found a rare thalassemia that is apparently not caused by genetic mutation directly, but by a defect in the histone that leaves part of the DNA in a chromosome looking like a tangled fishing line after a bad cast.

Mom's Genes, Pop's Genes

The basic idea about paired chromosomes and sex is that offspring receive a mix-and-match set of genes that partakes randomly of the genes from each parent. Biologists were startled a few years ago to discover that, as with so many ideas about living organisms, this view is oversimplified. A phenomenon known as *imprinting* can make a difference as to whether a given gene in an offspring originated in the mother or the father. The original meaning of imprinting in biology referred to learning from experiences shortly after birth; for example, geese think that the first thing that moves that they see after hatching is their mother. In the 1960s the meaning of imprinting was transferred to genes to refer to the effect on a gene of its position on a chromosome. In the 1980s the term began to be applied to parental influences on genes.

In the mid-1980s Bruce Cattanach and Anthony Searle, both of the Molecular Research Council of Radiology, showed that mice that received two copies of the same chromosome from a single parent died during development (depending on the chromosome). In 1988 Arthur Beaudet found the first known case of human uniparental disomy, a previously unknown type of genetic disease that is caused by the presence of two copies of one chromosome from one parent in the absence of any copies of the same chromosome from the other parent

Around the mid-1980s other researchers also showed differences in development that depended on the parental source of the chromosomes. Understanding of how this worked and the implications for ordinary development were notably lacking at the time. Progress was made by three groups of scientists in 1987 who reported that differences in maternal and paternal mouse genes were a result of methylization, a process frequently involved in turning off genes. Mother mice seemed to methylate a foreign gene, but father mice demethylated it. Similarly, daughters methylated and sons demethylated before passing the gene on to yet another generation.

In 1990 and 1991, however, confusion about the operation of the phenomenon came on the heels of the location of three different normal genes in mice that are affected by imprinting. Specifically, Denise Barlow of the Research Institute for Molecular Pathology in Vienna, Austria, and coworkers from the Max Planck Institute in Tübingen, Germany, and Vanderbilt University found imprinting for the gene for the receptor for insulin-like growth factor-2 (ILGF-2); Argiris Efstradiadis, Elizabeth Robertson, and coworkers at Columbia University found that the gene for ILGF-2 is itself subject to imprinting; and Shirley Tilghman and coworkers from Princeton University found imprinting in the gene known as H19, which has an unknown function. Unlike the foreign genes used in the 1987 studies, the normally occurring genes did not appear to be heavily methylated

MITOSIS AND MEIOSIS

Among the many unhappy memories of an American high-school education is the necessity for remembering (1) which is mitosis and which is
meiosis and (2) what do these words mean in any case. Walther Flem-
ming, who discovered chromosomes, coined the word <u>mitosis</u> in 1882
from the Greek <u>mitos</u> meaning "thread" and <u>-osis</u> meaning "a condition
or process." The word <u>meiosis</u> also comes from Greek, specifically from
<u>meioo,</u> which means "to make less," and the common ending <u>-osis.</u> It is
unfortunate for the memorization process in English that both the <u>i</u> in
<u>mitosis</u> and the <u>ei</u> in <u>meiosis</u> are pronounced with the same sound.

Mitosis is the process in which each strand of DNA in the double
helix unwinds and separates from its complementary strand. The cell
then builds a new chromosome on the basis of the single strand, resulting in two double helices where once there was one. Then the cell has
twice as many chromosomes as it needs or that would be good for it, so
it divides in two. The result is two identical cells, called daughter cells. If
all goes well, each daughter cell is exactly like its parent insofar as
genes are concerned. All cells have undergone mitosis somewhere
along the way, since each organism starts with a single cell (a zygote)
and every cell is the daughter, usually many generations removed, of
that original zygote.

Meiosis is a different process that occurs only in cells that give rise
to ova (eggs) and sperm. It is more complex than mitosis. As a result of
meiosis, four nonidentical daughter cells are produced, each with half
the chromosomes of the original cell. If all has gone well, each daughter
cell contains one and only one member of the original pairs. If all has not
gone well, serious genetic disease can result in offspring. In one of the
early stages of meiosis, however, genes are sometimes exchanged
between members of a pair of chromosomes, so the resulting chromosome in the daughter cell may not be exactly like either original chromosome in the parent.

A mnenomic device can help differentiate between mitosis and
meiosis. Think of the cell that divides asexually as <u>it</u> as in mITosis, while
the sexual one, mEiosis, has an <u>e,</u> just like <u>sEx.</u>

by either parent. Nevertheless, lack of a maternally derived H19 gene or of
either the maternal or paternal receptor for ILGF-2 results in mouse
embryos that die about the fifteenth day of development. Lack of a pater-
nally derived ILGF-2 gene results in offspring that are about half normal
size.

Imprinting is not just a phenomenon of mice. Since 1989 an instance of
imprinting in humans has been recognized and studied. The deletion of
part of one copy of chromosome 15, resulting in the loss of one set from

pairs of specific genes, is the cause of a serious hereditary disease. But the disease has two entirely different manifestations that depend on whether the chromosome with the deletion comes from the mother or the father. Loss of the genetic material in a maternal chromosome results in small feet and hands, a small penis if the child is a boy, a voracious appetite with concomitant weight gain, slowness of movement, shortness, and mental retardation—characteristics identified as Prader-Willi syndrome. On the other hand, exactly the same loss of genetic material in a chromosome derived from the father leads to a jerky gait, excessive smiling and laughing, large mouth and red cheeks, seizures, and profound mental retardation, a condition known as Angelman syndrome.

Genes are not the whole story of prenatal material influence. William Atchley and David Cowley of North Carolina State University worked with strains of mice for which inbreeding over many generations had fixed each strain's genome. The two strains they used could easily be distinguished by size. But when they transplanted embryos of small mice into large mothers or vice versa, they found that the resulting offspring reflected their surrogate mothers' bodies more than they did their genetic mothers'.

Mom's *and* Pop's (Mitochondrial) Genes

Mitochondria are small, energy-producing bodies within cells that have their own rapidly evolving DNA, separate from the main DNA of the cell. There is considerable reason to think that the mitochondria were once free-living. They moved into the ancestors of multicelled animals so long ago that they have become essential elements in the cell's workings instead of parasites. Until the early 1990s it was believed that all mitochondria in multicelled animals passed from mother to offspring. The father's mitochondria, while essential to the functioning of the father's cells, were always at an evolutionary dead end, since they were not inherited.

In 1988 and 1990, researchers found the first apparent exception to this rule. When different species of fruit fly interbred, the defenses against paternal mitochondria sometimes failed. The resulting hybrids had mitochondria from both parents. But this was viewed as an unusual artifact of controlled interbreeding, not something that could happen in a natural population.

Also in 1990, scientists from the University of Michigan studied the edible mussel of the North Atlantic and its possible relatives. Mussels are so much alike that no one was certain whether there are three species in the North Atlantic or some other number of somewhat different subspecies. The scientists—Walter R. Hoeh, Karen H. Blakley, and Wesley M. Brown—examined the mitochondrial DNA from 150 mussels taken from sixteen different populations. They were surprised to find that most of the mussels,

85 percent of them, had more than one kind of mitochondria. Eleven of those had three types and two of the mussels each had four kinds of mitochondria.

As early as 1915, studies of mussel cells showed that mussel egg cells do not effectively destroy or resist mitochondria from mussel sperm. Thus, the likelihood is that different types of mitochondria coming from both parents instead of just the mother accounts for the observed differences in mitochondrial DNA from individual mussels. In short, mussel mitochondria may not be inherited in the way that other mitochondria are. Perhaps this difference comes from interbreeding among the several putative species of North American mussels, similar to the way that different species of fruit flies can interbreed.

Jumping Genes Get Help from Mite

Barbara McClintock [US: 1902-] won the 1983 Nobel Prize for her discovery of transposons, popularly known as jumping genes, in maize (known in the United States as corn), but her fame and the prize had waited many years until after transposons had been found by other scientists and in species other than maize. Even today, transposons are not completely understood and those outside of maize are little known except to specialists. The story of jumping genes in the common laboratory fruit fly (*Drosophila melanogaster*) has been interesting, but also little known until recently except to specialists. Human jumping genes also exist, but it was not until 1991 that new discoveries brought the jumping genes of fruit flies, known as P elements, as well as human jumping genes, to the fore.

The unusual history of P elements is that they suddenly appeared in wild fruit fly populations about 1950, a short time ago for a set of genes. Around that time wild populations of *D. melanogaster* were found to have P elements, but those being bred in laboratories did not. Since *D. melanogaster* genes have been intensively studied since 1907, the appearance of a new gene did not go unnoticed. The new genes quickly spread through the wild population and even into unprotected laboratory populations as a result of sexual relations between fruit flies. But *D. melanogaster* does not breed with other species of fruit flies in the wild; so how did the P elements get into the *D. melanogaster* population to begin with?

A 1991 study of the problem by Marilyn Houck (a mite expert), Margaret Kidwell (a P element expert), Jonathan B. Clark, and Kenneth R. Peterson, all of the University of Arizona, suggests strongly that the P elements, common in another species, *D. willistoni*, since time immemorial, made the transition to *D. melanogaster* through mite bites, just as malaria parasites and others can be transferred from one human to another via mosquito bites. One piece of evidence that suggests that the P elements came from *D. willistoni* is that, although *D. melanogaster* and *D. willistoni* diverged some 50 million year ago, the P elements of each species are almost identical. Of course, this could also have occurred if both fruit flies had received the P elements from the same source, but a suitable source has not been found. Instead, there is just enough overlap in the ranges of the two fruit flies that it seems likely for genes from one to enter another if a vector could be found.

The Houck and Kidwell team thinks that they have found the vector, a mite (*Proctolaelaps regalis*) that is parasitic on both species of fruit fly. The mite feeds on all the growth stages of the fruit fly, and may have lifted the genetic material from eggs of *D. willistoni,* a method often deliberately used by human geneticists. The mite takes only a fraction of a second to extract its food from one source and move on to the next egg or larva (although feeding on the pupal stage of the fruit fly can take longer because of the thicker covering of the pupa). Such a pattern would make it easy to transfer material from one egg or larva to the next, even if they are different species. Furthermore, the mite has a range that includes the overlap of the ranges of the two species of fruit fly. Various experiments tended to confirm both the possibility and the likelihood of transfer of P elements by *P. regalis.* Now the researchers are trying to show an instance of the transfer occurring in a laboratory setting.

Geneticists are interested beyond the solution to a mystery. If genetic material can be transferred by mite bites, the mites could be harnessed deliberately to introduce new genes into species.

More work needs to be done in this area. If such transfers are relatively common, then the use of genes for extracting evolutionary history is endangered, for example. Or at least such use needs to be rethought to see what kinds of information can be obtained from genes that may or may not have come from some other species.

First Human Gene Caught Jumping

For a long time there has been evidence that some gene sequences in humans are transposons. Before 1991, however, the transposons were known only because they shared characteristics with other jumping genes; they had not been observed to have "jumped" from a location on a parental chromosome to a different location on the child's chromosome. In 1991 Haig H. Kazazian, Jr., and coworkers at Johns Hopkins observed a stretch of DNA in the middle of a gene that, when damaged, causes a type of hemophilia. They were able to track the gene back to a different location in both parents. Thus, the gene, long thought capable of jumping, had been proven to have done so. The transposon in question is a common DNA sequence of a type known as one of the Long INterspersed Elements or LINE. This LINE is found on chromosome 22 in all humans and also in chimpanzees and gorillas, meaning that it has been part of the human genome for at least the past 6 million years.

LINE transposons consist of a gene for the enzyme reverse transcriptase and various nonsense sequences. The reverse transcriptase holds the secret of jumping, since that enzyme causes RNA to make a DNA version of itself, called complementary DNA or cDNA (see "Human Genome Project UPDATE" below for further discussion of this process). The resulting cDNA then becomes inserted into the genome, not necessarily where it was before. The transposon has "jumped." Normally, not all of the original transposon makes the jump, since bits and pieces are easily lost when the gene is copied or when it becomes inserted into the genome.

The report by Kazazian and coworkers appeared in the Dec 20 *Science.* Earlier in 1991 there had been two reports of finding parts of another unused segment of the human genome being inserted within genes, just as the LINE sequence was inserted into the hemophilia gene. In both these cases the intervening material was identified as an *Alu* sequence, a type of DNA that does not possess the ability to jump. The possibility was raised that a LINE sequence could provide the reverse transcriptase needed to move the *Alu* sequences (and similar sequences) from place to place in the genome.

Although most jumps result in damage to a gene, some could produce important evolutionary changes. Thus, the role of the jumping genes may include both a cause of genetic diseases and a cause of mutations of many types. Further work will help sort these theories out.

Blue Genes in the Brain

It has become increasingly clear that there is a close relationship between the immune system and the nervous system. From a medical point of view, it appears that mental attitudes can influence control over disease that is mediated by elements of the immune system, notably the various kinds of cells that are often lumped together as white blood cells. Biologists have also begun to suspect that there are fundamental similarities in the way that immune system cells and nervous system cells function.

In 1991 a group of immunologists and neurobiologists from the University of California at Berkeley investigated whether the connections between the two systems include what is sometimes called gene shuffling, the rearrangement of genetic material used by the immune system to produce all the many types of antibodies. They inserted a backward gene as a marker into the genome of mice eggs. In adulthood, each cell in the bodies of the resulting mice would contain the marker.

If the marker gene were turned around and expressed, it would stain cells that contain it blue. At each end of the marker, the researchers put genes that are involved in genetic rearrangement in white blood cells. The reasoning was that as the mice developed, some cells that were rearranged would wind up with the marker gene turned around. When that happened, the cell would turn blue.

Sure enough, cells in seventy different locations in the brains of the genetically altered mice turned blue. The scientists think that this implies that such gene shuffling always happens as brain cells grow, but they cannot be certain. Perhaps the shuffling was caused by the genes from white blood cells that are known to be involved in such rearrangements. Further experiments may preclude this possibility.

If the brains cells do reshuffle genes as the brain develops, the process may have a role in determining the activities of different regions in the brain.

The scientists involved in the experiment were Hitoshi Sakano, Masao Matsuoka, Linda Kingsbury, Fumikiyo Nagawa, and Kazuya Yoshida from the Department of Immunology and David T. Larue and Jeffery A. Winer of the Department of Neurobiology, along with Kenko Okazaki, now of Kurume University in Japan, and Urs Müller of the Basel Institute for Immunology in Switzerland.

(Periodical References and Additional Reading: *Discover* 2-91, p 20; *Science* 3-22-91, pp 1403 & 1488; *Science* 4-26-91, pp 483, 506, & 542; *Discover* 5-91, p 14; *Science* 5-31-91, p 1250; *New York Times* 7-16-92, p C1; *Science* 9-6-91, pp 1092, 1110, & 1125; *New York Times* 9-17-91, p C1; *Scientific American* 10-91, p 30; *Science* 10-4-91, pp 7 & 81; *Science News* 10-5-91, p 212; *Science* 12-20-91, pp 1703, 1728, 1805, & 1808)

HUMAN GENOME PROJECT UPDATE

In the early 1990s the Human Genome Project continued to generate controversy, even as the actual sequencing of genes was just getting off the ground. Disputes over strategies that had seemed to be settled had a vampirelike way of arising again. And new controversies, such as the patenting issue described below, emerged.

In the spring of 1990 David Galas of the University of Southern California took over the DOE's portion of the Human Genome Project. He refocused research on finding expressed genes first (as Craig Venter's group is doing; see below), on genomes other than those of humans, and on cooperation with NIH instead of competition.

In May 1991 Richard Roberts of Cold Spring Harbor Laboratory announced that a trial of a new computer program he had devised

Background

The human being is a fairly complex animal that scientists believe is almost completely described by its genes. The idea behind the Human Genome Project is to uncover the complete sequence of about 3 billion bases in human DNA, to identify the genes encoded therein, and to recognize the immediate purpose for the genes. Because most of the DNA base sequence is tossed aside in the process of making proteins, it is expected that only 2 to 5 percent of the base sequence actually encodes human genes. There are thought to be about 50,000 to 100,000 human genes in all. Since 5 percent of 3 billion is 150 million, one can calculate that the average gene is thought to consist of a sequence of about 1500 to 3000 bases, although it is apparent that genes vary considerably in length.

A related project that is further along, sequencing the genome of the nematode Caenorhabditis elegans (see "C. elegans UPDATE," p 459), has uncovered indications that there are twice as many nematode genes as had been expected, perhaps a hint of what is yet to be learned about the human genome.

The genesis of the Human Genome Project can be traced to 1984. That year Robert Sinsheimer, chancellor of the University of California at Santa Cruz, was inspired by contemplating the high cost of major telescopes to think of a truly expensive project in biology. His notion was to make a map of the human genome. The ultimate goal that Sinsheimer had in mind would be a string of the letters T, G, C, and A that would fill 510 volumes of a standard-sized encyclopedia. The next year he arranged a workshop to present his idea.

Sinsheimer was not the only one looking for a big project. The US Department of Energy (DOE) was supporting national laboratories that were losing their funding. Sinsheimer's idea seemed to be just what the DOE was looking for, even though it had no obvious relationship to energy. The tenuous link claimed by DOE was that they had previously funded studies of radiation damage to genes. DOE sponsored a second conference on the idea, which led to others. The Office of Technology Assessment, the National Research Council, and, somewhat belatedly, the US National Institutes of Health (NIH) got into the act. Along the way the idea became the Human Genome Project, formed in response to many influences, from the DOE's desire to stay in existence after the fading cold war began to render much of its mission obsolete to the US Congress's natural tendency to back large and impressive-sounding scientific projects.

The Human Genome Project started formally on Jan 3 1989. The DOE continued to keep its hand in, but the NIH finally established leadership. By the fall of 1989, the basic nature of the project, involving many small groups of genetics researchers around the world cooperating under the leadership of James D. Watson [US: 1928-] and the NIH had taken shape.

(Continued)

Problems and progress mark the start of the 1990s, early days for the project, which may take until the late 1990s to get properly under way and another decade or two to complete.

The NIH and DOE sponsored a joint conference in the summer of 1989 to figure out how to get the job done. They developed a plan that would start in 1991, run for 15 years, and cost about a buck per nucleotide, or $3 billion. They also developed a strategy for doing the work, and put it out for public comment.

One basic idea was to begin by developing a map of the chromosomes, where the genes are located, by identifying a marker on the average of every 2 million bases. This was to be an interim goal that would be accomplished by 1996. A marker is any short section of DNA that can be recognized by a related section of cDNA whose sequence is known. Gene banks set up for other purposes already carry libraries of cDNA.

The problem with the interim goal for the Human Genome Project was that most specialists in the field thought it could not be accomplished in the way it was planned. So they convened a separate meeting in Mar 1990 to work out a new plan. The new plan, which was endorsed by the project, was based on spending 2 years finding "index" markers—common and useful sequences that come along about once every 10 to 15 million bases. An index map would be of immediate use not only in developing the more closely spaced map, but also in looking for a particular gene.

Scientists from the European Community enthusiastically joined the project and the Japanese reluctantly fell in line as well.

The project has been marked by a high degree of cooperation among research groups throughout the genetics community. Work has been parceled out so that geneticists are cooperating instead of competing in the way that Genome Project leader Watson described so vividly in his book The Double Helix. Each of the main groups has its own chromosome or set of chromosomes to study. Researchers also agreed on common languages and tools to make sharing data easier. The financial goal became to complete the project for half the originally estimated cost, or only $0.50 a base.

The project still has some unresolved questions on exactly how data will be shared and how money will be allocated, so the spirit of cooperation may not last the entire project. Many think that the only way to continue the early and needed emphasis on sharing data will be to institute tough rules that all researchers have to follow. Another result of massive cooperation is that the peer-review system is falling apart; since so many scientists in the same field are working on the project, there are few available to review the grants and papers of others.

The Human Genome Project has faced serious opposition from outside the genetics community as well, some of it well organized. Some socially conscious scientists and nonscientists fear that the knowledge gained will be used to discriminate against people with "bad genes."

showed 156 clear errors in 6000 sequences analyzed. If there are 50,000 genes in the human genome, that would mean that present methods are likely to result in errors in about 1300 of the genes. Other researchers showed, however, that there is considerable margin for error in that partial or incorrect knowledge of a sequence often does not damage its utility.

Among the questions facing the project is whether to use noncommercial sources exclusively or to turn to help from industry. In Aug 1991 the decision was made to turn to a commercial source. Collaborative Research Inc. in Bedford MA was awarded a $5 million grant for sequencing the genes of the two small microorganisms that cause leprosy and tuberculosis. Each mycobacterium has a genome about 4 million bases long, sixteen times as long as the most completely sequenced organism, the cytomegalovirus. This is fairly close to the current goal of $0.50 a base.

Patenting Our Genes

To the layperson the idea that human genes can be patented is patently absurd. Will scientists next will want to patent the human elbow or eye? Nevertheless, in the United States patent applications were filed by the hundreds in 1991 for something approximating human genes. The US Patent and Trade Office (PTO) had not yet decided what to do about them by early 1992.

The actual "devices" for which the patent applications were filed are sequences of what is known as complementary DNA (cDNA) from clones of human brain genes and other human genes. First, messenger RNA is collected from human brain tissue, with each molecule a template created from a specific gene. A new DNA strand is formed on each molecule of the messenger RNA, which is called *complementary DNA* since it is the complement, found by matching bases, of the messenger RNA; otherwise, cDNA is exactly the same as any single-stranded DNA. The sequence of bases in cDNA can be read by a gene analysis machine, although it takes time. A sequence of about 350 to 500 bases from a gene is enough, however, to be likely to be unique to that gene. Such a sequence can then be synthesized in other automatic machines. The synthetic cDNA fragment can then be used as a probe to locate the entire gene.

At that point, it may be helpful to determine what the gene does. Until recently at least, most patents for naturally occurring substances described the function rather than the chemical composition. The Harvard patent on protein GP120, issued in 1988, did not give a specific sequence of amino acids, for example.

When Craig Venter of the US National Institute of Neurological Disorders and Stroke announced that his project for locating and sequencing the bases on all human genes expressed in the brain would patent its results as the project proceeded, many scientists strongly objected. James Watson, director of the Human Genome Project, led the critics by saying that "virtually any monkey" could do the kind of work Venter and his group were doing. Most believed that the work failed to meet the requirements of the US patent law that the work be new, nonobvious, and useful—especially the nonobvious and useful parts. The methods used are common to much of genetic research and the applications are at this point unknown.

Venter's group first made its scientific intentions plain in an article in the Jun 21 1991 issue of *Science.* At that point, a few thousand human genes had been sequenced and listed in a central repository called Gen-Bank. The first Venter report noted that although his group was finding a lot of junk, they also had located a couple of hundred pieces of genes that controlled something unknown in human brains. Venter's project came to the attention of patent attorney Max Hensley, who proposed to the NIH technology transfer office that the pieces be patented to protect them from unauthorized use by commercial companies. The idea was passed on to Venter, who filed for patents not only for the 350 "expressed sequence tags," or base sequences of cDNA that can used as probes, but also for the entire (but still unknown) gene sequence and expressed protein. Ideas differ on whether any or all of this will be granted by the PTO. In the meantime, the NIH Office of Technology Transfer, headed by Reid Adler, has been the main backer of the scheme.

In Nov the White House Office of Science and Technology Policy joined the fracas; they announced that a policy of patenting genes would be delineated early in 1992.

Early in 1992 more than 2000 additional patents were applied for.

Timetable of Gene Patents

1789	The United States introduces its first patent law
1949	Because of an increase in patent applications involving specific organisms, the US Patent and Trade Office (PTO) requires that an example of the actual organism be submitted along with the patent
1969	Johnathan Beckworth and coworkers are the first to isolate a single gene; it is the bacterial gene for a step in the metabolism of sugar
1970	Har Gobind Khorana [US: 1922-] and coworkers at the University of Wisconsin announce the first complete synthesis of a gene, the gene for analine-transfer RNA; although previous workers had synthesized genes using natural genes as templates, this gene was assembled directly from its chemical components
	The PTO extends patent protection to plant seeds
1976	Khorana and coworkers announce construction of the first completely functional synthetic gene along with its regulators
1980	The US Supreme Court rules that a microbe developed by General Elec-

tric for oil cleanup can be patented, allowing the first patent for microorganisms produced by people

Chinese scientists clone a golden carp, the first genetic copy of a higher organism

Stanley N. Cohen of Stanford University and Herbert W. Boyer of the University of California at San Francisco receive the first patent issued for a method of genetic engineering; royalties on the patent will go to their universities, which applied for the patent in 1974 along with an application for a patent on the results of using the technique

1981   Bio Logicals of Toronto, Ontario, announces that the first fast and inexpensive gene synthesizer, a device for automatically assembling DNA or RNA sequences to order, will go on sale

1983   Monsanto scientists succeed in getting a bacterial gene to express itself in a petunia plant

1984   On Aug 28 Cohen and Boyer's patent on recombinant DNA molecules containing foreign genes in bacteria (plasmids), applied for in 1974, is granted; a request for a patent on the use of recombinant DNA molecules in yeasts is put off for later decision

1986   The US Department of Agriculture grants the Biologics Corporation of Omaha NE the world's first license to market a living organism produced by genetic engineering; it is a virus used as a vaccine to prevent herpes in swine

The US Federal Technology Transfer Act becomes law; some interpret this law to mean that any human genes found by US researchers must be patented

Leroy Hood and coworkers at CalTech announce development of an automated device that can determine the sequence of 10,000 bases along a DNA molecule in a 24-hour period

1987   On Apr 21 the PTO extends patent protection to all animals except humans; humans are deemed not eligible for patenting because of the Thirteenth Amendment to the US Constitution, which forbids slavery; unanswered is the question of how much of the human genome need be involved to invoke the Thirteenth Amendment

The Board of Patent Appeals and Interferences rules that genetically engineered oysters can be patented

More than 6000 patents related to genetic engineering are pending before the PTO, including patents for twenty-one genetically engineered animals

Researchers from MIT and Collaborative Research, Inc., announce that they have mapped over 400 markers that can be used to identify human genes

The US Department of Energy announces that its goal is to map the human genome

1988   On Feb 16 the PTO issues a patent to Harvard University for commercial use of the naturally occurring protein GP120, patentable because Harvard isolated the protein, a part of the HIV virus, in 1984

On Apr 12 the PTO issues patent No. 4,736,866 to Harvard Medical School for a mouse developed by Philip Leder and Timothy Y. Stewart by genetic engineering; it is the first US patent issued for a vertebrate

It agreed that the US National Institutes of Health (NIH) will take the lead in mapping the human genome with James D. Watson as head of the Human Genome Project

Speeding the Process

Cost and time are the main technical problems faced by the Human Genome Project. Genes can definitely be sequenced by the methods worked out in the 1980s, but with 3 billion bases to sequence, both time and money are formidable problems. Even if the current goal of $0.50 a base is met, the cost comes to $1.5 billion. If bases were identified at a rate of one per second, the project would take 95 calendar years. (Current sequencing machines produce about 2000 bases per day, or one every 40 seconds or so.) Ideas for speeding the process are attractive.

Among the ideas that are being tried out is one based on recognizing very small fragments of DNA. The computers employed for sorting out information use letters that are based on powers of two—older personal computers used eight binary digits (bits) for everything, while newer personal computers handle thirty-two bits at a time. One idea for sequencing genes depends on using all possible fragments of DNA that are eight bases long. This would need more than eight bits to describe, since there are 65,536 possible combinations of eight things taken four at a time with replacement—meaning that any one of four bases can be used to fill any one of the eight slots. For a binary system, such as a computer, the eight bases can be related to a code that is sixteen bits (sixteen things taken two at a time with replacement). Thus, a computer that handles sixteen or thirty-two bits at a time can work easily with eight bases at a time.

The sequencing idea, whose technical details are yet to be worked out, would be to have a grid of all 65,536 eight-base sequences, called octamers. The unknown stretch of DNA would hybridize with some subset of the octamers. These could then be matched by the computer to arrange them in the correct order for the previously unknown sequence.

There are still many bugs to work out. The grid has not been created in a practical form. Identical octamers pose problems, so that there may turn out to be several equally likely sequences to be studied by other means to find out which sequence is correct. The problems appear to be capable of resolution, however. If all works as planned, the octamer method should speed up the sequencing process and thus reduce costs.

(Periodical References and Additional Reading: *Science* 1-19-90, pp 270 & 281; *Science* 4-6-90, pp 44 & 49; *Science* 5-18-90, pp 804 & 805; *Science* 5-25-90, p 953; *Science* 6-29-90, p 1600; *Science* 7-27-90, p 342; *Science* 11-9-90, p 756; *Science* 2-22-91, p 854; *Science* 4-26-91, p 498; *Science* 5-31-91, p 1255; *Science* 6-21-91, pp 1618 & 1651; *Science News* 6-22-91, pp 389; *Science* 8-16-91, p 743; *Science* 9-27-91, p 1489; *Science* 10-11-91, pp 171, 173, 214, 221, & 293; *Science* 11-8-91, p 805; *Science* 11-22-91, p 1104)

THE THREAT OF ORGANISMS IN NEW GENES

On Feb 19 1991 the White House Council on Competitiveness, headed by US Vice President Dan Quayle, proposed new rules for evaluating the products of genetic engineering. Quayle's group asked that genetically engineered bacteria, plants, animals, and products of those organisms be treated just like any organisms or products produced by other means, such as selective breeding or chemical manufacture. The genetic means used to develop the organism or product should, in the council's view, be treated as irrelevant. Reactions were predictable, with Democratic representatives and spokespersons for such environmentally oriented groups as the National Wildlife Federation opposed to relaxing the rules and the Industrial Biotechnology Association applauding the council's work. The council deputy director, David McIntosh, explained that the Bush administration hoped that both the Agriculture Department and the Environmental Protection Agency would heed the council's call and write new regulations by the end of 1991. This did not occur.

Background

Genetic engineering had its tentative beginning in 1973. Almost immediately people began to worry about it. In Jul 1974 Paul Berg [US: 1926-] of Stanford chaired an eleven-scientist Committee on Recombinant DNA Molecules that wrote to *Science* and *Nature* about possible dangers from genetically engineered bacteria. The Berg committee proposed a voluntary ban on specific experiments with genetic engineering until their "potential hazards" could be evaluated. The following Feb a conference of 139 scientists from sixteen countries met at Asilomar CA and adopted a set of voluntary guidelines for genetic engineering. In 1975 there were six laboratories in the United States and a few outside the United States that conducted genetic engineering.

(Periodical References and Additional Reading: *Science* 7-13-90, p 124; *Science News* 8-25-90, p 116; *New York Times* 2-19-91, p D1; *Science* 2-22-91, p 878; *Science* 6-7-91, p 1368; *Science* 6-21-91, pp 1612 & 1613; *Science* 7-5-91, p 32; *Science* 10-4-91, p 35; *Science News* 12-14-91, p 390)

WHOLE ORGANISMS AND THE BRAIN

NEW KNOWLEDGE OF EVOLUTION

Creationists or others opposed to the theory of evolution like to point to changes in evolutionary dogma or to battles between evolution's supporters to show that evolution does not occur. Evolutionary theory continues to evolve. A concept of evolution existed before the theory of natural selection was put forward by Charles Darwin and Alfred Russel Wallace in 1858. Darwin himself continued to improve evolutionary theory during his lifetime, revising—not always for the better—his own earlier ideas.

Today the theory of evolution is a vital and changing part of science. Experiments are confirming parts of the theory that had not been previously tested. In addition, the new understanding of genes and how they work continues to add to our understanding of the mechanisms of evolution.

Rapid or Slow Maize Evolution

Among the most controversial topics of the 1980s was the evolution of maize (corn in American English and *Zea mays* in botany). Wild maize is very different from domesticated maize. Wild maize is a common Mexican weed that is a species of teosinte. Although it hybridizes easily with its tame relative, it does not have ears with easily edible kernels. Instead, teosinte has an ear that contains two intertwined rows of tapering kernels with each kernel wrapped in a shell as hard as that of a nut. When teosinte ripens, the whole ear falls apart, scattering the kernels.

The archaeological record shows popcorn with small maizelike ears appearing in what is now Mexico about 7000 years ago. There is no evidence of a form intermediate between teosinte and popcorn, although it is easy to trace the development of modern field and sweet maizes from popcorn. Thus, it could be argued that maize arose in a sudden step from teosinte, perhaps as the result of a single mutation. A detailed description of how this could have been accomplished was put forth by Hugh Iltis of the University of Wisconsin, Madison, in 1983. The Iltis theory, based on a change in the sex of the two different kind of flowers found on both maize and teosinte, was widely discussed, both by supporters and opponents of evolutionary theory.

The early 1990s brought new evidence. Some of it supports Iltis's theory and some indicates that his detractors may be correct. In particular, genetic studies show that maize, which has a male tassel and female ears, could easily have evolved this arrangement from teosinte by switching the sex roles of the two kinds of teosinte flowers. After that, however, the evidence, collected mostly by John Doebley and coworkers at the University of Minnesota, shows that other independent changes have to have taken place to get from teosinte to maize.

But not very many changes are needed. As few as five key single-gene mutations would be sufficient. That small a number of mutations could easily happen together in a short time. The most likely scenario to Doebley is that the first mutation changed the hard outer coat of the teosinte kernel to the thin flexible covering found on maize kernels. Such a change would in itself make the altered teosinte a desirable food crop, unlike wild teosinte. With people now paying attention to the suddenly edible plant, other changes that are beneficial from an agricultural point of view could have been encouraged to spread through the population. Within a short time—shorter than a thousand years, that is—*Z. mays* could have been bred in the form of popcorn.

Although the research deals only with maize, evolutionary theorists are interested in maize as an example of processes that occur in general. Scientists who favor the theory of sudden evolutionary steps, such as Stephen Jay Gould of Harvard University, like to point to the Iltis theory as the way evolution could proceed. Those who want to stay with original Darwinian theory find a lot wrong with the Iltis theory. If Doebley is right, neither side is completely wrong and both sides are partly correct: Evolution does proceed in jumps, but in little jumps, not big ones.

Guppy Experiments

The guppy (*Poecilia reticulata*) is a familiar pet fish. Because of its variability, frequent breeding, short pregnancy, and ease of observation, the guppy is also a favorite subject for studies of evolution.

In 1871 Darwin proposed his theory of sexual selection—mate choice by females leads to the evolution of secondary sexual traits in males. A guppy study released in the Jun 15 1990 *Science* by Anne E. Houde of Princeton University and John A. Endler of the University of California, Santa Barbara, establishes a good case for the corollary to Darwin's theory: Patterns in the genetic makeup of the females drive sexual selection. Houde and Endler studied differing female preferences for orange-spotted male guppies using seven populations of guppies collected from six different streams in Trinidad. The populations in the streams differed in the orange area in males.

Guppies were raised in sexual isolation and then brought together. Females from streams where males have large areas of orange prefer those kind of males; females from streams where males have little orange prefer males with little orange. The researchers think that this result implies that over long periods of time, the populations would grow into separate species. The new species with the orange-spotted males would not breed with the one with plain males even if change in the environment brought the two new species together in the same river.

David A. Reznick and Heather Bryga announced that their 11-year experiment with wild guppies in Trinidad demonstrates changes caused by natural selection that were predicted by current evolutionary theory. They moved populations of guppies from an environment with large predators that preferred to eat adult guppies to an environment with small predators that could only manage to consume very young guppies, or fry. (Guppies are born live, not hatched from eggs that have been laid.) Current evolutionary theory predicted that the transferred guppies would change their reproductive strategy when confronted with predation on their fry. This was observed, as the transferred guppies switched in a few generations to producing fewer and larger fry.

Mimicry Misidentified

Textbooks since the nineteenth century have used the nearly identical appearance of the monarch butterfly (*Danaus plexippus*) and the viceroy butterfly (*Limentis archippus*) to explain a phenomenon known as Batesian mimicry (after Henry Walter Bates who first described it). The basic idea is that viceroys look like monarchs because monarchs taste bad to birds. By disguising itself as a monarch, the viceroy, none of whose close relatives much resemble monarchs, evades being eaten.

Research in the early 1990s established that while the resemblance between the viceroys and monarchs probably helps keep the viceroy from being eaten, it is not because of Batesian mimicry. David B. Ritland and Lincoln P. Brower of the University of Florida in Gainesville reported in the Apr 11 1991 *Nature* their definitive study showing that viceroys taste as bad to birds—red-winged blackbirds, at least—as monarchs. They presented a panel of sixteen wild blackbirds with abdomens—minus identifying wings—of a number of butterflies, including not only monarchs and viceroys, but some butterflies known to be popular blackbird food. The blackbirds rejected viceroy bodies after a single peck 35 percent of the time. Monarchs from south Florida were rejected at around the same rate, although almost all of the butterflies thought to be tasty were eaten completely. Taking this study together with other research, the evidence suggests that all viceroys taste bad and only some monarchs do, making the situation more complicated than expected.

Monarchs taste bad and are even poisonous to birds or other predators primarily because their larvae feed on poisonous milkweeds, which contain the bitter substance called cardenolide. But milkweeds vary in how bitter or poisonous they are, and thus the monarchs do too. Some whole populations of monarchs taste reasonably good according to humans who have made the necessary trials. Furthermore, James Seiber and coworkers from the University of California at Davis showed that migrating monarchs lose their cardenolide slowly after they become adults. The adults begin to taste good unless they can recharge their poisons. They do this by collecting nectar that contains chemicals called pyrrolizidine alkaloids. The problem is that while the monarchs are migrating, the cardenolide is wearing thin and the new poisons have not yet been collected. At that point, monarchs taste better than viceroys.

Viceroy larvae feed on willows. The larvae may pick up some precursors of aspirin that way (the basic chemical in aspirin was discovered in willows), but no toxic chemicals. Instead, the adults appear to have evolved to produce their own toxins, a more reliable system than depending on what your larvae eat.

So why do monarchs and viceroys look alike? Even though a slight case might be made for Batesian mimicry, by sometimes tasty monarchs of always unpleasant viceroys, it is more likely that another kind of mimicry, called Mullerian after 19th-century German biologist Fritz Muller, is the cause. The difference between the two types of mimicry is that Batesian mimicry assumes that one species evolves to look like another to steal the other's advantage; but Mullerian mimicry is based on two species evolving to look like each other for mutual advantage. If all viceroys and most monarchs taste bad, then birds will more quickly learn to avoid any orange-and-black butterflies. Both butterfly species benefit.

Moving the Good Genes Around

A major concept of evolutionary theory developed by Sewall Wright [US: 1889-1988] is that evolution may begin by a happy mutation in an isolated subpopulation and then diffuse throughout a set of such subpopulations by migration and sexual recombination. In evolutionary terms this is called "interdemic selection by differential dispersion." Sometimes this is also called "island model evolution" because it describes what might happen if populations are initially isolated, as on islands. The island model assumes that the most successful populations send out members to other populations, spreading the good genes they developed back home. Although difficult to demonstrate and not based on experiment, but rather on reasoning and plausible analogy, Wright's ideas have been the core of much of evolutionary thought since they were first put forth in 1931.

Michael J. Wade of the University of Chicago and Charles J. Goodnight of the University of Vermont reported in the Aug 30 1991 *Science* that their 4-year experiment with flour beetles (*Tribolium castaneum*) constitutes experimental proof that Wright's theory is correct. Basically, they measured fitness in beetle populations by the number of progeny produced. Excess beetles from fit populations were sent to less fit groups, After twenty-four generations, all the populations were more fit. A control group in which beetles were allowed to join whatever population they pleased showed no such improvement in the average number of progeny.

Critics claimed that numbers and circumstances were special and that the flour beetle experiment showed that Wright's mechanisms *could* work, not that they always *did.* More experimental work will probably ensue.

(Periodical References and Additional Reading: *New York Times* 4-16-91; *Science* 6-15-90, p 1271 & 1405; *Discover* 3-91, p 14; *Science News* 6-1-91, p 348; *Science* 6-28-91, p 1792; *Science,* 8-30-91, pp 946, 973, & 1015)

HOW THE LEOPARD GOT ITS SPOTS

It is not often these days that a significant understanding of events in the development of living organisms comes from basic inorganic chemistry. But that seems to be the case with regard to knowledge of how living creatures develop spots, stripes and other patterns. It is not known at this time whether similar mechanisms account for more fundamental structures, but nature is parsimonious.

This story begins in 1952, when famed mathematical theorist, wartime code breaker, and computer designer Alan Turing [UK: 1912-1954] turned his attention to the question of how the stripes on a zebra are formed. He demonstrated mathematically that a zebra pattern could occur when two different, interactive chemicals diffuse through a zebra's skin in certain ways. A pattern developed by such a process came to be called a Turing structure by biologists and chemists. If Turing structures exist, they account not only for a zebra's stripes, but also for the bold patterns of some domestic animals and the leopard's spots. The problem until the early 1990s was that no one could demonstrate a physical system that behaved according to Turing's mathematics.

It is rare for anything mathematically possible to fail to exist somewhere in the real world (infinity, continuity, and perfection probably excepted). Turing's biological mechanism would be difficult to detect in an organism except by its effects unless the underlying chemistry were understood. Experiments, however, persistently failed to find systems of chemicals that formed stable stripes, spots, or other patterns.

Finally, in 1990, Patrick De Kepper and coworkers at Bordeaux University in France located a simple system that has the necessary complicated behavior. They published their results in the Jun 11 1990 *Physical Review Letters.* Their method involves using two reservoirs, one of chlorite and iodine and another of malonic acid and iodine. The reservoirs are connected by a strip of gel that permits both mixtures to diffuse toward the other end. At first the two chemicals form stripes of yellow and blue; after a time, however, the stripes break up to become yellow spots on a blue background.

Initial reaction from other chemists was excited but cautious. The problem was showing that the spots were the result of steady interactions among the chemicals and not an artifact from some other condition of the experiment. Nine months later, in the Feb 8 1991 *Science,* chemists Irving R. Epstein and Isvan Lengyel of Brandeis University in Waltham MA reported on their experiments that show how the Bordeaux reaction works. They first demonstrated that the chlorite diffuses through the gel faster than the iodine does by putting the two chemicals through separate gel columns. They showed that this happened because the iodine forms short-lived complexes with the gel that slow it down. From that premise they were able to calculate that the iodine would concentrate in dots, producing the pattern observed in Bordeaux.

The next steps are expected to include a search for similar situations in living creatures, probably occurring during early developmental stages. For example, theoretical considerations make it most likely that a spot of about a hundred cells would be about the right size to have been formed in this way. Spots that size would form in unborn leopards and later grow into the much larger spots observed in adult animals.

As a sidelight to the theory of Turing structures, it is interesting that as a result of the way such structures form and if they do account for color patterns in animals, it would be impossible for a striped animal to have a spotted tail, while it would be perfectly reasonable for a spotted animal to have a striped tail (and this does occur among various species of catlike civets and genets as well as in some of the wild cats).

(Periodical References and Additional Reading: *Scientific American* 3-88, p 80; *Science News* 8-11-90, p 88; *Science* 2-8-91, pp 627 & 650)

C. ELEGANS UPDATE

Biological convention has it that one should refer to a species with its genus abbreviated to its first letter for second or later references; but *Caenorhabditis elegans* has become so famous that more and more first references use the abbreviation *C. elegans*. Almost as often, this tiny nematode—only a millimeter (four hundredths of an inch) long—is simply called the worm.

The worm became famous originally because it was chosen to be the first multicellular organism for which the developmental fate of every cell was traced. This intimate knowledge of the worm's 959 cells is unprecedented, but many hope this feat will set a precedent for other animals and ultimately for humans. The worm is suitable for many projects because it has a life cycle of just 3 days; 100,000 individuals can be kept in a laboratory dish; and it is largely transparent.

A decade-long project is still under way in mapping the worm's genome. In addition to its purely scientific interest, the gene mapping has a practical application because so many genes are conserved by evolution; as a result, mammals, including humans, use the same genes to make the same proteins as the worm does. Mutations in worm genes that correspond to the human disease hypertrophic cardiomyopathy and to brittle bone disease have been already found. In Nov 1990 Paul Sternberg of Caltech described a worm gene that is similar to the *ras* gene that is implicated in many human cancers.

Timetable of the Worm

1963	Sydney Brenner of the Medical Research Council's Laboratory of Molecular Biology (MRC) in Cambridge, England, conceives of a project aimed at learning the complete development of a multicellular organism; he chooses *C. elegans* because of its transparency and its small number of cells
1983	John Sulston of MRC and coworkers on the Worm Project succeed in sequencing the development of each of the 969 cells that make up a single *C. elegans*
1986	John White, Eileen Southgate, and Nichol Thomson of MRC establish the location of every nerve connection among the 302 neurons of *C. elegans*, a result known as the wiring diagram
1988	The Worm Book, formally known as *The Nematode Caenorhabditis Elegans*, edited by William B. Wood and the Community of *C. elegans* Researchers, is published by Cold Spring Harbor Laboratory; it contains all the data collected on the Worm Project to that date
1990	In Apr Alan Coulson and John Sulston of MRC, working with Richard Wilson, Robert Waterson, and coworkers at Washington University in St. Louis MO, make available sheets of filter paper containing 95 percent of the *C. elegans* genome broken down by gene
1991	The Cambridge and Washington University laboratories have sequenced 200,000 bases out of the expected 100 million bases in the DNA of *C. elegans*
	Catherine H. Rankin of the University of British Columbia and Norman Kumar of the University of Toronto report that *C. elegans* can be used in learning studies because it can be conditioned
1994	Workers on the Worm Project plan to have 3 percent of the genome of *C. elegans* completed
2000	Worm Project workers are scheduled to complete the base-by-base map of the *C. elegans* genome

(Periodical References and Additional Reading: *Science* 6-15-90, p 1310; *New York Times* 1-8-91, p C1; *Science* 6-21-91, p 1619; *Discover* 8-91, p 22)

BRAIN NEWS

Human beings have a long tradition of valuing the brain, although the ancient Greeks thought that its main purpose was to act as an air conditioner for the blood. Humans most likely hold the brain in high regard because they believe the human brain to be superior to that of other organisms. In the early 1990s there was considerable activity regarding the scientific study of brains, both human and those of other animals, and how those brains react with the remainder of the organism.

Some would even go further. On Jun 27 1991 a panel from the Institute of Medicine, for example, proposed a project similar to the Human Genome Project that would map the brain's molecular biology, chemistry, and function as a set of computer data bases.

Growth and Regrowth

It has long been believed that unlike most other tissues in the body, mature nerve cells (neurons) do not and cannot grow or change. Loss of a mature neuron is a complete loss, as many unlucky people with nerve damage have learned.

But for years there have been hints that mature neurons, even brain cells, can sometimes grow, reproduce, or change function under certain conditions. Because of the implications for those with damaged or diseased nerve cells, there has been a significant effort to learn more about the seeming exceptions to the general rule that mature nerve cells of mammals cannot grow. A study by Fernando Nottebohm, reported in the Sep 21 1990 *Science*, showed that a male canary that changes its song in the autumn grows new neurons when it does so. While a canary is certainly not a mammal, it is a high form of life with a complex brain.

In 1991 a case of nerve reorganization in mammals, primates at that, surfaced. Ironically, this happened as a result of the activities of animal rights activists. Experiments being conducted on seventeen monkeys in Silver Spring MD were the target of action by the future founder of People for the Ethical Treatment of Animals (PETA) in 1981. Legal struggles between PETA and the owners of the monkeys, the Institute for Behavioral Research, successfully kept the monkeys from further experiments or study until 1987.

It was not until the Jun 28 1991 *Science,* however, that the results of new research were reported by a group from the National Institutes of Health (NIH) led by Tim P. Pons and Mortimer Mishkin (with aid from psychologists from Vanderbilt University in Nashville TN and the University of Alabama at Birmingham). The researchers obtained permission to study the brains of four of the surviving experimental monkeys using procedures that followed NIH guidelines for the treatment of laboratory animals, as was carefully pointed out in the report. The procedures involved putting the aging animals to death.

The original work on the monkeys in the late 1970s had involved removing from one or more limbs the nerve connection that carried messages from the limb to the brain. The Pons group was able to show that the part of the brain previously involved in analyzing information from the "deafferentiated" limb did not simply atrophy. Instead, for as much as 10 to 12 mm (0.4 to 0.5 in.)—much farther than any similar research had revealed—the affected section of the brain began to accept messages from facial nerves that were previously accepted only in regions adjacent to the one for nerves from the limb. This discovery shows that mature brain structures can change. It is thought that the results are of an order of magnitude greater than earlier results because of the size of the part of the body involved (earlier studies typically used a thumb instead of an entire limb). The long time period in which the brain could reorganize—a total of 12 years—may also have been a factor.

A related area of study in the early 1990s concerned humans with damaged brains who seemed to be able to use the undamaged portion of the brain to accomplish some task normally done by the portion of the brain that could no longer function. In early 1991 researchers from Western Ontario University in London, Ontario, reported on a woman who was able to behave as if she recognized the geometry of shapes presented to her for certain tasks, even though brain damage prevented her from actually recognizing the shapes. Similarly, researchers at Brown University reported late in 1990 on patients whose visual centers in the brain had been destroyed by accident or disease but who could still perceive certain stimuli such as bars of light. Whether the brain in these cases had adapted after damage or already used these capabilities in some way was not clear.

Timetable of Memory, Perception, and Learning

1973	Tim Bliss and coworkers at the National Institute for Medical Research in London find that learning occurs when connections between synapses are strengthened by repeated stimuli, a process called long-term potentiation (LTP)
1979	The substance protein kinase C is found to work by adding a phosphate group to molecules, which increases or decreases their activity
1980	Daniel L. Alkon of the US National Institute of Neurological Disorders and Stroke (NINDS) shows that electrochemical currents in a neuron change as an animal learns
	Robert Furchgott of Downstate Medical Center in Brooklyn NY, part of the State University of New York (SUNY), discovers a mysterious substance involved in telling the cells in the lining of blood vessels to relax

1981	Christoph von Malsburg of the University of Southern California proposes that different regions in the brain may be synchronized during visual perception
1983	It is discovered that LTP probably takes place when the NMDA receptor for the neurotransmitter glutamate is repeatedly activated
1986	Wolf Singer and Laurence Mioche from the Max Planck Institute for Brain Research in Frankfurt, Germany, discover the first experimental evidence that von Malsburg's theory of brain synchronization may be correct
	Joseph Farley of Princeton University shows that protein kinase C affects learning and memory by opening and closing ion channels in cells, causing the electrochemical changes observed by Alkon in 1980
	Furchgott and Louis Ignarro of the University of California in Los Angeles find that the mysterious chemical signal Furchgott found in blood vessel linings in 1980 is the common inorganic gas nitric oxide
1988	Barry Bank with coworkers from NINDS and Yale University finds that conditioning elevates levels of protein kinase C in the hippocampus of rabbits
	Salvador Moncada of Wellcome Research Laboratories in London, UK, proposes that nitric oxide is the universal substance used by the nervous system to convert messages into action
1989	Gary Lynch at the University of California, Irvine, and Roger Nicoll at the University of California, San Francisco, report that LTP is stimulated by glutamate at a different receptor, called the non-NMDA receptor
	Wolf Singer and coworkers show in experiments with cats that several different sets of neurons act in synchrony during perception, producing a recognizable set of brain waves that some think may be the cellular basis of consciousness
1990	Thomas J. Nelson of NINDS shows that messenger RNA is elevated in neurons from trained sea snails; messenger RNA is used by cells to make new proteins
	Richard Tsien and Roberto Malinow of Stanford University Medical Center and, independently, Charles Stevens of the Salk Institute in San Diego CA show that LTP takes place in the nerve transmitting a message, not in the receiving nerve; that is, LTP is presynaptic, not postsynaptic
	Terry J. Crow of the University of Texas Medical School at Houston finds that a chemical that inhibits protein synthesis prevents sea snails from developing long-term conditioning memory; in other experiments he shows that chemicals that block protein kinase C prevent short-term conditioning memory in the snails

(Continued)

Tsunao Saitoh and coworkers at the University of California, San Diego, report that the brains of eleven persons who died of Alzheimer's disease had half the amount of protein kinase C of the brains of seven people who had died from other causes

1991   Workers led by Tim Tully at Cold Spring Harbor Laboratory start an ambitious program to study learning in the fruit fly *Drosophila melanogaster*.

Solomon Snyder, David Bredt, and coworkers at Johns Hopkins School of Medicine show that nitric oxide is a neurotransmitter in the human brain and is involved in various interactions between nerves and muscles; it is also a factor in certain kinds of cell death, including cell death in the brain

Making Sense of Sense

New findings about visual perception, briefly described in the "Timetable of Memory, Perception, and Learning," are important in our overall understanding of consciousness. New findings about a less studied sense, that of smell, suggest the beginning of a whole new line of research on perception.

Evidence suggests that mammals can recognize about 10,000 different odors. Until the early 1990s most scientists assumed that this recognition occurs in a way somewhat similar to the way in which vision works. In vision there are a small number of different receptors and a large region of the brain that interprets the data from the receptors.

In the Apr 5 1991 issue of *Cell*, Linda Buck and Richard Axel of Columbia University reported that they had found evidence for a different system for smell than that of vision. Working with rats, they found the genes for many different families of odor receptor, suggesting that there may be anywhere from 200 to 1000 different odor receptors. Since their work is preliminary, many think that the higher number of about 1000 receptors will eventually turn up. If that is the case, then little processing in the brain needs to take place after an odor is detected by a small number of the receptors.

Can Homosexuality Be Detected in the Brain?

Darwin recognized that sexual preference was an important factor in evolution. In the early 1990s, geneticists learned a good deal about sex and development that was previously unknown or only suspected. Related to the gross facts of sexual differentiation, but much more subtle, is the question of development of sexual preference—combining in a sense Darwin's concerns with new understandings of anatomy and genetics. The 1990s saw progress on this front as well.

Homosexuality—the "love that dare not speak its name" of the nineteenth century—can be big news at the end of the twentieth century. Several results of 1990 and 1991 concerning a physical basis for homosexuality were treated as lead stories by many media outlets.

In 1990 D.F. "Dick" Swaab and M.A. Hofman of the Netherlands Institute for Brain Research in Amsterdam observed a difference in the brains of homosexual men, an enlarged suprachiasmatic nucleus (SCN). In the language of brain anatomists, a "nucleus" refers to a small identifiable region of the brain. The SCN is the nucleus that governs circadian (daily) rhythms of activity. Swaab's team found that homosexual men had SCNs twice as large as those of heterosexual men.

In the Aug 30 1991 *Science*, Simon LeVay of the Salk Institute in San Diego reported that he had also found observable differences between the brains of homosexual and heterosexual men. A nucleus in the brain's anterior hypothalamus called INAH-3 in homosexual men is more like the same nucleus in women than it is like that region in heterosexual men. LeVay had compared the INAH-3 region from the cadavers of nineteen male homosexuals who had succumbed to AIDS with those of sixteen presumed heterosexual men (of whom six died of AIDS) and six presumed heterosexual women (of whom one died of AIDS). He also compared the INAH-2 regions of the anterior hypothalamus, but found no differences there that he could attribute to homosexuality or to sex. Both the INAH-2 and INAH-3 regions had been found in 1989 to be larger in men than women, although the results were for older subjects than those in LeVay's study. LeVay observed that the INAH-3 region in homosexual men was half the size of that region in heterosexual men and similar in size to that region in women. LeVay's results were viewed as more significant than those of the Amsterdam group because animal studies have demonstrated that the anterior hypothalamus is the place in the brain that directs sexual activity. Male monkeys with damage to their anterior hypothalamus are no longer interested in female monkeys, for example, but they continue to masturbate.

Other studies revealing differences in the brains of homosexual and heterosexual people are forthcoming.

In Dec 1991 J. Michael Bailey of Northwestern University and Richard C. Pillard of Boston University reported on a study that Bailey says reveals that genes influence development of the INAH-3 region, leading ultimately to a predisposition toward homosexuality. Their study, like others before it, showed that about half of the identical twins of a homosexual man are also homosexual. They also investigated pairs of fraternal twins and adopted brothers. The fraternal twins of homosexual men were found to be homosexual 22 percent of the time and the adoptive brothers only about 11 percent of the time. The study included 161 homosexual men in the primary group (who were interviewed) and 170 of their brothers (sampled by questionnaire). If one assumes that only 4 percent of the total population of male humans are homosexual, these results would imply that genes account for half of male homosexuality. Nontwin biological brothers were not questioned, but the interviews with the primary group revealed 9.2 percent of the brothers to be homosexual, a figure so low as to cast doubt on the idea that there is a genetic component to homosexuality. In an interview with *Science*, Bailey stated that he assumes that the entire basis of homosexuality is biological, with half coming from heredity and the other half from other environmental and random biological factors. It is difficult to find reliable studies showing an environmental influence on sexual orientation. The results by Bailey and Pillard were published in *Archives of General Psychiatry*.

(Periodical References and Additional Reading: *Science* 6-29-90, p 1603; *Science* 8-24-90, p 896; *Science* 9-21-90, p 1444; *Science News* 9-22-90, p 182; *New York Times* 1-15-91, p C1; *New York Times* 4-5-91, p A1; *Science* 4-12-91, p 209; *Science News* 5-25-91, p 328; *New York Times* 6-28-91, p A12; *Science* 6-28-91, pp 1763, 1789, & 1857; *New York Times* 7-2-91, p C1; *Science News* 7-13-91, p 23; *New York Times* 8-30-91, p A1; *Science,* 8-30-91, pp 956 & 1034; *New York Times* 9-1-91, p 1; *Science* 9-27-91, p 1486; *Science* 1-3-92, p 33)

NEWS ABOUT EARLY HUMANS AND OUR OTHER ANCESTORS

BONES OF OUR ANCESTORS

Progress in understanding human evolution since 1856 has depended largely, although not completely, on discovery and interpretation of extremely rare fossils. Early humans and their ancestors often failed to die in places where bones turned easily into stone. Furthermore, human hard parts are not so large or hard as those of some other creatures. In addition, the early population of our ancestors was not very large. Despite these impediments, the early 1990s brought forth reports of several new fossils that may lead to important new understandings.

Return to Ethiopia

On Nov 30 1974, Donald C. Johanson, fossil-hunting in the desert of the Afar triangle in Ethiopia, spotted part of small arm sticking out of the sand. Unearthed, it turned out to be Lucy, the most nearly complete specimen of a prehuman hominoid ever discovered and one of the oldest as well. The next year, working in the Afar once again, Johanson's expedition, co-led with French geologist Maurice Taieb, made another unique find, a collection of fossils from thirteen *Australopithecine afarensis* individuals—about 2000 bones from the same site. This group of 3-million-year-old humans became known as the First Family. The following year, as another Johanson expedition to the Afar triangle was winding down, a coup put new rulers in place in Ethiopia. Johanson barely managed to take his current box of fossils and early hominid tools out of the country.

But then he could not get back in. He tried in 1982, making it as far as Addis Ababa, the capital city; but a ban on foreign expeditions was declared before he could get out of the city to the site. All efforts failed until 1990, when the political situation, still complicated by several independent groups controlling different parts of the country, finally cooled down enough to permit a new expedition.

The expedition was led by Johanson, Walter H. Kimbel, and Robert C. Walter, all of Johanson's Institute of Human Origins in Berkeley CA. As with the expeditions in the 1970s, this one headed for a site in the desert called Hadar. At Hadar, wind and water erodes the loose desert soil, exposing fossil bones on the surface. There is very little digging, but a lot of searching with sharp eyes, especially in the morning and early evening. Not only is the desert heat less intense then, but the lower angle of the sun gives the fossils shadows that help make the bones visible. In the 13 years since Johanson's last visit, some precious fossils may have been lost to the forces of erosion, but far more should have been uncovered.

This indeed proved the case. Hunting in 1990 during the preferred months of Oct and Nov, the ten members of the team and their Afar helpers located fossils of eighteen individuals. Fifteen of them appear to be *Australopithecus afarensis,* like Lucy, but the other three may be something entirely different, although Kimbel says he does not believe this to be the case. At least one of the fossils may indicate that Lucy and her relatives spent more time living in trees than previously believed. Another surprise is that one jaw and some of the facial bones show hallmarks of *A. africanus,* a species generally thought to come later than *A. afarensis.* This would make the jawbone similar to the Malawi jaw (see "The Malawi Jawbone" below).

Johanson's team was not alone in Ethiopa in 1990. South of Hadar, Johanson's former colleagues Tim White and Desmond Clark, both of the University of California at Berkeley, successfully searched the Middle Awash valley.

The third team, led by John Fleagel and Solomon Yirga of the State University of New York (SUNY), at Stony Brook and partially funded by the National Geographic Society, worked farther south yet, at a site called Fejej. There the team mainly collected teeth from *A. afarensis,* but the teeth were from earlier individuals than any found elsewhere. The teeth are at least 3.7 million years old, but may date back as far as 4.4 million years, approaching the time when *A. afarensis* was just beginning to branch from its still unknown ancestor.

All fossils, especially those of hominids, require careful study after fieldwork is over to interpret their meaning—and hominid fossils generally provoke fights among anthropologists that last for years before consensus is reached. Clearly, the more fossils the better. Future expeditions to Ethiopia can help sort out what the new fossils suggest about the history of our species.

The Fayum Skull

Elwyn L. Simons of Duke University has long been one of the leaders in early primate studies, especially primates that are the earliest known sprouts on the branch that leads to modern humans. In 1989, on the basis of a few teeth and a jawbone, Simons described a previously unknown species, *Catopithecus browni,* that he thought might be the most ancient yet. The fossil was found in a bed in the Fayum badlands of Egypt that was first dis-

Background

Understanding of the various relatives of modern human beings and how they are connected undergoes continual change. Currently, there is little consensus among paleoanthropologists, the people who study early humans and their relatives, about the details of who is related to whom.

Human beings are part of an order of mammals called <u>primates</u> (see "Taxonomy of Living Organisms", page 484, and "Primate Evolution" below). Primates more closely related to humans are called <u>hominoids</u>; these include the great apes (gibbons, orangutans, gorillas, chimpanzees, and pygmy chimpanzees), humans, and their closest ancestors. Until recently, our relatives that are not great apes were classed along with us as <u>hominids</u>, while the great apes were classed as <u>pongids.</u> Evidence from DNA studies and other recent work suggests that gorillas and both species of chimpanzee are more closely related to humans than either the gibbons or the orangutan. Currently, many anthropologists identify three families— for living species, these are Hylobatidae for the gibbons, Pongidae for the orangutan, and Hominidae for the human, the gorilla, and both chimpanzees. Although there is only a single species of human alive today, in the past there are thought to have been at least two others. In the old classification, the Hominidae, or hominids, had two main branches over the course of time: australopithecine and human. In the new system there would be three branches, to account for the gorilla and the chimpanzees.

The first column below gives scientific and common names (with "man" used only because of traditional terminology; modern paleoanthropologists shun the use of <u>man</u> to mean "human") along with the time and place in which the animal is known to have flourished. The abbreviation mya means "million years ago."

<u>Catopithecus browni</u> c 40 mya (Egypt)	Earliest known representative of higher primates, which include Old World monkeys, gibbons and apes, and all hominids
<u>Aegyptopithecus</u> c 30 mya (Egypt)	This monkeylike creature may be the earliest known ancestor of hominoids
<u>Proconsul</u>; three known species c 20 mya (East Africa)	Oddly named after Consul, a popular chimpanzee in the London Zoo (<u>Proconsul</u> means "before Consul"); <u>Proconsul</u> is generally recognized as an ancestor of all hominoids
<u>Otavipthecus</u> <u>namibiensis</u> c 13 mya (Nambia)	See "The Nambian Jawbone" below
<u>Sivapithecus</u> c 10 mya	(India, Pakistan, and Turkey) Once believed to be directly on the line to humans, today <u>Sivapithecus</u> is thought to be in the Pongidae family and the ancestor of the orangutan (but see "The Pakistani Arm" on p 471)

(Continued)

Pan troglodytes and Pan paniscus Chimpanzee c 7 mya to present (Sub-Saharan African forests)	Several studies of proteins and DNA suggest that the chimpanzee is our closest living relative; there is some evidence that we are more closely related to the pygmy chimpanzee or bonobo, P. paniscus, than to the common chimp
Australopithecus afarensis c 4 mya (Ethiopia, Kenya, and Tanzania)	Many believe that A. afarensis is the ultimate ancestor of humans, but it is not clear who is the ultimate ancestor of A. afarensis; the famous fossil known as "Lucy" is a member of this species, as well as "the First Family"
Australopithecus aethiopicus c 2.8 to 2.2 mya (East Africa)	One fossil, known familiarly as "the black skull" and more formally as KNT-WT 17000, and discovered in 1985 by Alan Walker, is thought to be a previously unknown species of australopithecine
Australopithecus africanus; "The Taung Child" c 2.5 mya (South Africa)	A. africanus was the first nonhuman hominid to be discovered (in 1924) and was widely doubted at first; like all australopithecines, it walked upright and had a brain much smaller than members of Homo of the same size
Homo habilis; "Handy Man" c 2 mya (East Africa)	The first known member of our own genus, and—if classified correctly— probably a direct ancestor; it is widely believed that stone tools that date from the same time were made by H. habilis; it is not clear whether H. habilis was a hunter or scavenger or both
Australopithecus bosei; "Zinj"; "Nutcracker Man" c 2 mya (East Africa)	A still controversial classification that some would merely label the East African subspecies of A. robustus; when the first specimen was found by L.S.B. Leakey in 1959, he thought it was a new genus, Zinjanthropus, hence the nickname
Homo erectus; "Java Ape Man" "Peking Man" from 2 mya to c 90 thousand years ago (Africa, Asia, and Europe)	Depending on how one classifies the Neanderthals, this is the first nonhuman hominid to have been discovered (in 1890 by Eugene Dubois); it was a successful creature who could make fire and was probably a good hunter; the fact that it did not change its basic tool kit for 1.5 million years suggests that it was not very bright; H. erectus is generally thought to be the immediate ancestor of Homo sapiens; the other hominids became extinct and by at least a million years ago, H. erectus was the only hominid on Earth with the possible exception of very early archaic H. sapiens
Australopithecus robustus c 1.5 mya (South Africa)	Despite the name, A. robustus was not especially robust by modern human standards; however, compared with the gracile A. afarensis and A. africanus, this is a larger and stronger species; it coexisted with Homo, so it cannot be an ancestor of modern humans

Archaic Homo sapiens c 1 mya (Africa and possibly Europe)	Today many anthropologists recognize that our species evolved about 90,000 years ago from Homo erectus, but the first humans still show some primitive features that modern H. sapiens have lost; some classify the Neanderthals (see below) with the archaic H. sapiens; often the archaic H. sapiens are called "anatomically modern humans" (AMH) to show that, while we can see no difference in fossils from these creatures, unpreserved soft parts or important connections between such soft parts (the brain is on anthropologists' collective mind) could be quite different from those of modern humans
"Neanderthal Man" 200,000 to 35,000 years ago (Mostly European, but some fossils from Africa and the Near East)	While paleoanthropologists since World War II have classified this well-known group as Homo sapiens neandertalis, a subspecies of modern humans, current thinking is that this "cave man" of the ice ages may be a separate species; Neanderthals share certain traits with modern humans, such as a large brain and such customs as burial of the dead; anatomical differences between Neanderthals and modern humans, while not pronounced, are clear; some scientists have taken to spelling the name of these hominids "Neandertal" (without the h) since the German for "Neander Valley" uses tal for valley, not the anglicized thal
Homo sapiens; "Cro Magnon Man" 90,000 years ago to present (worldwide)	Modern humans may have started with archaic H. sapiens (see above) and we may have evolved—although most think it is not likely—from the Neanderthals around 35,000 years ago; in any case, Homo sapiens replaced all other hominids about 35,000 years ago

covered in 1983 and that is thought to be more than 38 million years old, or of the Eocene Epoch. *Catopithecus* means "below the apes," but *C. browni* is also just above the prosimians (such as lemurs and lorises).

In Mar 1990 Simons described a much more complete specimen of *C. browni*, a whole skull that is sometimes simply called the Fayum skull. Possession of a whole skull clarifies the relationship of the prosimians with the other early apes and monkeys. *C. browni* is the earliest known creature on the Old World monkey line. Although the skull has been crushed during its long change from bone to rock, it is possible to derive a lot of useful technical information from it. One result of the crushed state is that the size and configuration of the brain cannot be determined.

The Pakistani Arm

Sivapithecus is an early ape whose remains have been found in India, Pakistan, Turkey, and Greece. The reinterpretation of *Sivapithecus* from an ancestor of humans to an ancestor of orangutans was one of the main changes in our view of human evolution during the 1980s. However, bones found in Pakistan from a *Sivapithecus* arm have recently confused the simple picture of *Sivapithecus* as a proto-orangutan.

David Pilbeam of Harvard University, principal architect of the orangutan view, and coworkers reported in the Nov 15 1990 issue of *Nature* that analysis of the two upper arm bones showed that *Sivapithecus,* although known from other fossils to have a face like an orangutan, walked about on all fours like a gorilla or chimpanzee.

Orangutan limbs are specialized for swinging through trees rather than for knuckle-walking on the ground. Of course, *Sivapithecus,* which lived between 13 and 7 million years ago (mya), may have evolved from a knuckle-walker to a tree dweller as it became an orangutan. The other possibility is that *Sivapithecus* was unrelated to any of the modern apes.

The Namibian Jawbone

On Jun 4 1991 Martin Pickford, a British paleontologist working for the Collège de France in Paris, while on an expedition led by Glenn Conroy of Washington University in St. Louis MO to the Otavi hills in Namibia, found a fossil jawbone (lower mandible) of a primate from about 13 mya. Although the purpose of the expedition was to locate such fossils, Conroy and his team of three were astonished when Pickford made his discovery 15 minutes into the first day of searching.

While the significance of the find and better dating will be worked out over a 3-year period of study in the United States (after which the fossil returns to Namibia), it appears from early indications that the jawbone represents a creature with characteristics between those of apes and humans. The arch of a human jawbone is distinctly different from that of an ape. The sides of ape jawbones are nearly parallel, unlike human jawbones, which spread farther and farther apart, like a parabola.

Also, the jawbone contains four well-preserved teeth with hominoid characteristics, although described as a bit "chunky and thick" by Phillip V. Tobias of the University of Wittwatersrand in Johannesburg, South Africa. The teeth also suggest that the creature, named *Otavipthecus namibiensis,* was apelike and the size of a young modern chimpanzee. The Otavi ape was also a juvenile, judging from the wear on the teeth. Tobias thinks that the ape was a bit more than a meter (3 1/2 ft) tall.

This is the first hominoid fossil of any kind from southwestern Africa.

The Malawi Jawbone

In Jul 1991 Tim Bromage of Hunter College in New York found a fossil now known as UR-501 in the Chiwondo Beds of Malawi in southeastern Africa.

Because UR-501 is another lower mandible, it may be called the Malawi jaw-bone. The Chiwondo Beds are in a branch of the Great African Rift system, a split that goes from the Red Sea through a series of great lakes into southern Africa. The Chiwondo Beds are near Lake Nyasa, the northern end of the southernmost part of the large African lakes of the Rift. This locates Chiwondo approximately halfway between the Hadar fossil bed of Ethiopia and the famous Swartkrans, Sterkfontein, and Taung sites of South Africa.

The possible significance of the Malawi jawbone is still being explored, although its location suggests that it may be a link between the *A. africanus* of South Africa and the *A. afarensis* found farther north in the Rift valley (see above). Another early idea is that the jawbone could come from a link between the australopithecines and *Homo habilis,* whose fossils also have been found to the north of the Chiwondo Beds.

The Georgian Jawbone

In Dec 1991 Leo Gabunia of the Georgian Academy of Sciences (in the former USSR) reported to a scientific conference in Frankfurt, Germany, that a team of Georgian and German archaeologists led by Vachtang Dzarparidee of the Georgian academy and Gerhard Bosinski of the University of Cologne had discovered the earliest hominid fossil known from either western Asia or Europe. It is a jawbone of either *Homo erectus* or archaic *Homo sapiens* that may be as old as 1.6 million years or date from around 90,000 years ago. It is between two magnetic reversals (see "Reversals of Magnetism," p 241) that bracket it with those ages. If it is associated with the volcanic layer below, the jawbone with its sixteen teeth is almost certainly that of *H. erectus*. If it dates to 1.6 mya, this would cast doubt on indications from studies of DNA that hominids remained in Africa until about a million years ago (see "The African Eve and the Pygmy Adam," p 476). Such a date would help explain the presence in Pakistan of stone tools thought also to be from about 1.6 mya.

However, the fossil may be much younger, since some authorities who have seen it think that the jawbone could be from archaic *H. sapiens*. The earliest other fossils suspected of being archaic *H. sapiens* date from about 250,000 years ago. The best evidence is for archaic humans emerging a little less than 100,000 years ago.

The jawbone was found along with the remains of creatures such as saber-toothed cats while researchers were investigating the cellar of a medieval house near Tbilsi, Georgia.

(Periodical References and Additional Reading: *Science* 3-30-90, pp 1527 & 1567; *Science* 4-6-90, pp 7 & 60; *Science News* 11-17-90, p 311; *Science* 5-22-91, p 1428; *New York Times* 6-22-91, p 2; *Science News* 6-29-91, p 405; *Science* 8-23-91, p 846; *Discover* 1-92, p 42; *New York Times* 2-3-92, p B7)

NEANDERTHAL NEWS

The first truly early hominids to be recognized were the Neanderthals, with skeletons first found in 1856 (3 years before publication of Darwin's *Origin of Species,* but Darwin apparently did not know of the Neanderthal finds). Although the first hominids found, they have been arguably the most controversial from that time to this. In the early 1990s considerable attention was focused on trying to determine their abilities, life-style, and relationship to people living today. It is hotly debated whether Neanderthals are our literal ancestors or a branch of the hominid bush that was pruned after the most recent glacial advance. Champions of the ancestor concept include Milford Wolpoff of the University of Michigan, while the driven-to-extinction school is led by Christopher B. Stringer of the Natural History Museum in London. People who believe interbreeding took place between Neanderthals and anatomically modern humans (AMH) find evidence in present-day bones of Europeans, evidence dismissed by the other school. David Frayer of the University of Kansas has suggested that comparisons of Neanderthal bones with bones of European AMH from more than 300 years ago show more evidence of interbreeding than comparisons between Neanderthal bones and those of contemporary Europeans. People who think Neanderthals did not or could not interbreed with AMH get solace from mitochondrial DNA studies that point to an African origin for AMH that would suggest no contribution from early European mitochondria (see "The African Eve and the Pygmy Adam," p 476).

Neanderthal Dates

One of the main problems with explaining the Neanderthals is the apparent proximity in both time and space of Neanderthals and AMH in the Mount Carmel region of northern Israel. Careful dating by several researchers puts the Neanderthals in that region between 60,000 and 48,000 years ago. Equally careful dating from about 20 km (12.5 mi) away shows that AMH were around at least 92,000 years ago. Since AMH are still in that region today, it would appear that at some point the Neanderthals and the AMH were living where they could interact and compete for the same resources. Wolpoff believes that the two were living in contact with each other, as different races do in the same region today, and that the period of togetherness lasted 50,000 years, plenty of time for interbreeding.

On Jun 27 1991 a team of French scientists led by Norbert Mercier of the *Centre des Faibles Radioactivités* in Gif-sur-Yvette, France, reported that the dating of flint tools found with a Neanderthal skeleton near St-Césaire, France, established the find as the most recent Neanderthal site known. The thermoluminescence date for the tools was 36,200 years, with a margin of error of 2700 years in either direction; that is, the site could be as young as 33,500 years before present (bp). The AMH culture in this part of Europe, known as the Cro Magnon culture, has been established by modern dating of bones to about 30,000 years bp, but tools of the type associated with Cro Magnons are known from sites as old as 40,000 bp, suggesting that Neanderthals and AMH coexisted in Europe as well as in the Near East.

Better dating of Neanderthal sites in Spain by James Bischoff of the US Geological Survey and others has suggested that Neanderthal toolmaking underwent a sharp and dramatic change around 40,000 years ago—several thousand years earlier than previously thought. This change could also have been caused by interactions with AMH, so it tends to confirm the implications of the site dates.

Neanderthal Life-styles

Neanderthals were physically well adapted to living in ice age conditions—stocky with a respiratory tract designed for living in a cold, dry climate. Neanderthal mothers produced large babies that could survive harsh conditions early in life. Some think that the climatic changes when the ice retreated caused the Neanderthal demise by making such adaptations unnecessary. According to Ezra Zubrow of the State University of New York (SUNY) at Buffalo, an increased mortality rate of only 2 percent could cause extinction in thirty generations, about a thousand years.

Not everyone agrees that the Neanderthals were well-adapted to the ice age, superb physical specimens though they were. John J. Shea and Ofer Bar-Yosef of Harvard University proposed in the spring of 1990 that the first Neanderthals appeared in Europe before the ice arrived. Their main diet was fruit, nuts, and other plant foods (hominoid staples, as for the African apes of today). When the ice descended on Europe from the north and from the mountains of the south, the Neanderthals mostly moved to the Middle East, where conditions were closer to what they were already used to. But the Middle East was already occupied by AMH. Because the physically strong Neanderthals could muscle out the AMH, the AMH were forced to live mainly by hunting (the Neanderthals took over places with fruit and nut trees and so forth). Then the climate got even worse; conifers with no fruit and only tiny "nuts" became the dominant vegetation. Now the hunters had the edge. The Neanderthals became extinct and the hunters became our ancestors.

In this argument Shea and Bar-Yosef assume that neither species—they consider AMH and Neanderthals to be different species—had any intellectual advantage because the evidence is that at that time AMH and Neanderthals used similar tools and had similar customs, such as burial of the dead.

Since 1939 it has appeared that the Neanderthals survived the ice and snow 45,000 years ago much the way some members of the notorious Donner Party survived a winter trapped in similar conditions 150 years ago—by cannibalism. This conclusion was based on a skull found in 1939 in Italy's Guattari cave. The skull was interpreted to be that of the victim of a ritual murder in which the brains of the victim were eaten. This interpretation by Alberto Carlo Blanc was based upon a comparison of the damaged Neanderthal skull with that of skulls remaining after brains had been consumed by recent Melanesian headhunters. Reexamination in 1991 of the same skulls, however, by Tim White of the University of California at Berkeley and Nicholas Toth of Indiana University suggested that the cannibal interpretation is wrong. A closer look showed significant differences between the Neanderthal wound and those made by the headhunters. The hole in the base of the Neanderthal skull and other marks on it are just the kind that a scavenging hyena makes on animal skulls. This leaves no evidence that the Neanderthals were cannibals at all.

(Periodical References and Additional Reading: *Science* 2-16-90, p 798; *Science News* 3-24-90, p 189; *Science News* 10-13-90, p 235; *Science* 4-19-91, p 376; *Science News* 6-1-91, p 341; *Science News* 6-8-91, p 360; *New York Times* 6-27-91, p A6; *Science* 8-23-91, p 834; *Scientific American* 9-91, p 40; *Smithsonian* 12-91, p 114; *Discover* 1-92, p 42; *New York Times* 2-4-92, p C1)

THE AFRICAN EVE AND THE PYGMY ADAM

Although past progress in physical anthropology relied almost entirely on fossils, modern biochemists claim that they can obtain better information of certain kinds by examining the genes or other molecules of living people and apes. Early work along these lines that showed a recent common ancestor for African apes and *Homo sapiens* was originally doubted by traditional anthropologists, but since 1980 has been largely accepted. More recent results from biochemists are still controversial, however.

African Eve

In 1987 a report in *Nature* by Allan C. Wilson of the University of California at Berkeley and colleagues led to a theory that has been described as the African Eve or the Mitochondrial Eve hypothesis. Their analysis of the DNA found in mitochondria of a sample of modern humans indicates that all 5.384 billion humans living today stem from a single source in Africa dating from some 100,000 to 200,000 years ago, much more recently than anyone suspected. The "Eve" label was given because mitochondria are tiny bodies within cells that are passed from generation to generaton in the cytoplasm of ova, and thus go from mother to daughter unaffected by male genes. Thus, the study could be described as tracing all modern humans to a single ancestral African woman.

A second analysis, answering criticisms of the first in detail, appeared in *Science* on Sep 23 1991. The second study used a more carefully selected population of 189 modern humans, a better method for comparison of mitochondrial DNA, and an improved method for anchoring the time frame that was unavailable when the earlier study was made. It confirmed the African origin and refined the date to around 200,000 years ago. The main finding was that there is more genetic variation in African mitochondrial genes than in those of other populations. The new anchor was specific genetic information about differences between humans and chimpanzees.

The second study was published 2 months after the death of Wilson from leukemia; it listed Linda Vigilant and Mark Stoneking of the University of California at Berkeley, Henry Harpending of Pennsylvania State University, and Kristen Hawkes of the University of Utah as coauthors along with Wilson.

Many traditional anthropologists found both Wilson studies unbelievable. For one thing, the predecessors of modern humans had left Africa and traveled through Asia and Europe before 200,000 years ago. Also, fossils suggest that some *Homo erectus* populations, notably those of eastern Asia, separately evolved into modern humans; that is, anthropologists studying *H. erectus* remains from that region discern affinities with the skeletons of modern humans from the same region. A few months after the second study was published, various molecular biologists and computer experts found new difficulties in the methods Wilson and Vigilant had used. Among this round of critics was Stoneking, an author on the study.

Pygmy Adam

One way to confirm or reject the African Eve hypothesis would be to find a molecule or other entity that is transmitted only from father to son, just as mitochondria are transmitted only from mother to daughter. Then, if biochemical analysis reveals the same African origin for that entity within the same time frame, the hypothesis that our ancestors came from Africa about 200,000 years ago to replace totally all extant populations in the rest of the world would be strengthened, if still not completely proven. At first thought, the Y chromosome would appear to be such an entity. Not only is the Y chromosome transmitted from father to son, but it is not as involved in genetic recombination as some other chromosomes are. Especially, the gene that induces maleness can be presumed not to be recombined with genes from the X chromosome.

Alas, the situation is not that simple. The part of the Y that does not recombine is mostly junk, with only three still poorly understood genes on it—the main gene that initiates maleness was announced as found only as recently as May 9 1991. Variations in these genes, called polymorphisms, can be studied in different populations and charted to show evolutionary relationships, but very few polymorphisms are known.

Despite the meager evidence, what there is of it tends to confirm an African origin for modern humans, a result with which even many bone anthropologists concur. Not enough data can be extracted so far, however, to reveal a time scale, which is where the real controversy lies.

One biochemist, Gérard Lucotte of the Collège de France in Paris, has used polymorphisms on the Y chromosome of modern humans and a computer algorithm to take the process further than anyone else. Lucotte claims to know the exact region in Africa where the modern expansion began and to have identified the remnant of the population that started it all. The region is the southern tip of the Central African Republic and its native population is made of up Aka pygmies. Lucotte has claimed a "pygmy Adam" of about 200,000 years ago, although he has no evidence to support that date other than it is the date Wilson's group established for African Eve.

Other biochemists think that Lucotte's method is not good enough to make such a fine determination. Fossil evidence suggests that humans of about 200,000 years ago were more likely to have been far to the east of the Central African Republic. Finally, even if Lucotte were correct about the pygmy Adam's location, the independence of male and female lines means that there is no reason to assume that this Adam even knew that Eve.

Africa in the Genes

One other massive study has used genetic information from nuclear DNA and other biochemical data to try to establish a human "family tree." Work by Luigi Luca Cavilli-Sforza of Stanford University with Paolo Menozzi of the University of Parma, Alberto Piazza of the University of Turin, and Kenneth K. and Judith R. Kidd of Yale University has also established an African origin for modern humans. The team found that the genes of Africans are twice as distant from those of non-Africans as Australian genes are distant from Asian genes. They interpret this to mean that Asians separated from Africans twice as long ago as Australians separated from Asians. Thus, if Australians settled their continent 60,000 years ago, our ancestors left Africa 120,000 years ago. This is near to the most recent possible date for the African Eve according to the Wilson and Vigilant study of 1991 (the widely quoted 200,000 years bp is the midpoint of a possible range that has 166,000 years bp as the near extremity). Furthermore, the Cavilli-Sforza study shows the Australian-Asian distance to be twice that of the European-Asian distance, suggesting a separation about 30,000 years ago. These dates are all roughly consonant with the fossil record. They also provide an independent verification of the African Eve hypothesis.

(Periodical References and Additional Reading: *Science* 1-25-91, p 378-9; *Science* 9-27-91, pp 1463 & 1503; *Science News* 9-28-91, p 197; *Scientific American* 10-91, p 30; *New York Times* 10-1-91, p C1; *Scientific American* 11-91, p 104)

BRONZE-AGE MAN FOUND IN ICE AND ACHEULEAN TOOL FOUND IN GREECE

Two of the science stories that excited people the most in the early 1990s might properly be termed archaeology rather than life science.

Bronze-Age Hunter Found as Glacier Melts

On Sep 19 1991 a pair of German hikers on the Smilaun Glacier in the Tyrolean Italian Alps near the Austrian border discovered the mummified and frozen remains of a body that had apparently been trapped in the ice for more than 5000 years. Newspapers soon dubbed the remains the "Ice Man." With the Ice Man were typical tools of the time, including a fire flint and kindling, a bow and fourteen arrows, a knife with a stone blade, and an ax with a copper head. The Ice Man was tattooed on his back and behind his knee. In addition to remnants of leather clothing, he was wearing boots insulated with straw stuffing. Evidence of killed small game and picked berries was found nearby.

Further studies will be conducted on such potential sources of information as the contents of the Ice Man's stomach and his DNA, along with a search for any surviving microbes or parasites. At the moment, the main concern is how to thaw out the body without destroying any important evidence.

Melting of the glacier is taken to indicate that the climate in the Tyrol is warmer now than it has been in the past 5000 years. Hikers are keeping their eyes peeled for the emergence of other ancient bodies.

An Ancient Tool

One Jun day in 1991, Curtis Runnels of Boston University discovered a typical Acheulean "hand ax" in an ancient lake bed near Nikopolis, Greece, while he was on the way back to the car to grab a bit of lunch. It was an excellent example of the art and created quite a stir.

The 23-cm (9-in.) flint tool is thought to be between 200,000 and ·

VERSATILE TOOL WITH A LONG LIFE

Acheulean tools were manufactured in essentially the same way for well over a million years, from about 1.6 mya to 200,000 years ago. The typical Acheulean tool is traditionally called a hand ax, although there is no consistent understanding of what purpose these very common tools served. Consequently, careful anthropologists refer to such tools by the more neutral name of <u>biface.</u> While one study of European bifaces shows wear patterns that suggest they were used for butchering, another study, based on locations in which bifaces are often found, proposes that they were intended to be thrown at game. Another suggested use is digging up underground tubers and roots. Modern copies of bifaces have been made and tested on these and various other tasks. Bifaces have generally been found to be useful for anything tried. Thus, the "hand ax" may have been an all-purpose tool, the <u>H. erectus</u> version of the Swiss army knife.

400,000 years old, with 200,000 more likely. Part of the significance of the find is that it is the first Acheulean tool known from that part of the world. Such tools were first found in quantity at St. Acheul, France (hence the name). Since then, they have been shown to be common throughout Africa and western Europe and present as well from Turkey through India. Their absence, until now, from southeastern Europe—Italy, the Balkan States, and Greece—was difficult to understand. Also, for other reasons, it is surprising that they have not been found in the Far East.

Acheulean tools were long thought to be principally connected with *Homo erectus,* the precursor of both *Homo sapiens* and the Neanderthals that existed from about 2 to about 0.9 mya. Recently, the discovery of fossils attributed to archaic *H. sapiens* that overlap slightly in time with those of *H. erectus* suggests that Acheulean tools may also have been produced by the archaic humans. Indeed, Runnels thinks this particular tool was probably made by archaic *H. sapiens,* although proof is lacking so far. Some other anthropologists, such as F. Clark Howell of the University of California at Berkeley, think that the tool may be the work of Neanderthals.

The site where the tool was found is comprised of a type of clay that is unlikely to preserve bones; thus, although further excavation was planned for 1992, completely identifying the maker will probably never be accomplished. Other sites, however, may provide evidence that could be strongly suggestive of either *H. erectus* or *H. sapiens.*

(Periodical References and Additional Reading: *New York Times* 7-25-91, p A11; *Science* 8-2-91, p 505; *Science News* 8-3-91, p 68; *New York Times* 9-26-91, p A15; *Science* 10-11-91, p 187; *Discover* 1-92, p 34; *New York Times* 2-22-92, p 5)

PRIMATE EVOLUTION

Before the mysteries of our closer relatives among the hominoids can be completely resolved, it is necessary to solve some of the less well known problems connected with the evolution of our whole order, the primates.

The Primate Split

Today there are three branches of the primate order, the order that consists of mammals that use sight more than scent, have nails instead of claws on grasping hands and feet, are mostly active in daylight, and have relatively large brains. The branches are the tarsiers, the lemurs, and the simians (monkeys, apes, and humans). At some time in the past, according to evolutionary theory, there was only a single type of ancestral primate. It is thought that this time was toward the end of the Cretaceous period or shortly thereafter, about 65 million years ago (mya). At some later date the split into three branches occurred.

In the Jan 3 1991 issue of *Nature* Leonard Krishtalka and Christopher Beard of the Carnegie Museum of Natural History in Pittsburgh PA and Richard Stucky of the Denver (CO) Museum of Natural History announced that they had identified four fossils of a very early representative of the tarsier line in the Wind River Basin near Shoshoni WY. Geological evidence puts the fossils at 50 mya. The species has been named *Shoshonius cooperi.* The fossils had been found during the mid-1980s. Fossils found in 1990, but still under study, tend to confirm the designation as an early tarsier.

If *S. cooperi* is a tarsier, then that line had split off from the ancestral primate nearly 10 million years earlier than previously believed.

Bernard Sigé and coworkers at the University of Montpellier in France suggest that they have evidence of an even earlier split. Their primate consists of ten scattered teeth from about 60 mya, which they think are all the known remains of an ancestral lemur, which they have named *Altiatlasius* after the High Atlas mountains of Morocco, where the teeth were found.

Essential differences in the three branches of the primate order suggest that if any one of the more specialized suborders existed, then the other two did as well.

The Flying Lemur Connection

But what was the ancestral primate, anyway? Until 1990 it was thought that early mammals called plesiadapiforms who lived shortly after the dinosaur extinction of 65 mya (a time called the Paleocene Epoch) filled the bill. Separate studies of plesiadapiform fossils by Christopher Beard and by Richard Kay of Duke University in Durham NC and coworkers showed that this role did not fit the observed facts of plesiadapiform anatomy—the middle fingers were too long, the artery to the brain was in the wrong place for a primate, and the middle ear was also misplaced.

Although these three details were wrong for a primate, they just fit a small order known only from southeastern Asia and the Philippines, the Demoptera (earlier Demopterans also lived in North America until about 57 mya). The ones living today are called flying lemurs or colugos. Thus, the plesiadapiforms are not the ancestors of primates, but of colugos.

Plesiadapiforms are still more like primates than other mammals from the Paleocene. It would appear that the colugo, not the tree shrew as was previously believed, is the closest living relative of the primates.

Beard thinks that he has found an alternative primate ancestor, an early mammal from the Wyoming digs of 60 mya that is called a microsyopid. He argues that mycrosyopid limbs are designed like primate limbs, specialized for tree climbing. This would push back primate origins even earlier, for if mycrosyopids are that specialized, they must have an ancient ancestor, possibly from even before the dinosaur extinctions.

The Flying Fox Connection?

Another candidate for a close relative of the primates is the flying fox, along with other members of the order or suborder Megachiroptera. The Megachiroptera, or "big bats," all either eat fruits or drink flower nectar, while the Microchiroptera, or "small bats," mostly eat insects, although some drink blood, eat fish, or eat fruit. Unlike flying lemurs, which really glide instead of fly, the flying foxes have wings similar to those of other bats and engage in true flight. It is the structure of the wings that has in the past caused taxonomists to classify both the Megachiroptera and the Microchiroptera in the same order.

In the Mar 14 1986 issue of *Science* John Pettigrew of the University of Queensland in Australia began a long-lasting crusade to take the Megachiroptera out of the bat order and establish them as a separate order closely related to the primates. His evidence has largely been based on similarities in the nervous system between flying foxes and primates.

A number of studies since 1986 have used DNA or other molecular evidence to investigate the relationships between the Megachiroptera and the Microchiroptera. They have all concluded that the two groups are more closely related to each other than to any other mammals, although the two kinds of bat diverged long ago. Pettigrew remained unconvinced as recently as an article in the Jun 1991 *Systematic Zoology,* but the weight of evidence was increasingly against him. Everyone agrees, however, that the controversy has been useful in making people look more closely at the bats, which constitute the most diverse group of mammals.

In the Feb 22 1991 *Science,* J.G.M. Thewissen and S.K. Babcock of Duke

University reported that their studies of wing nerves suggested that the Megachiroptera were more closely related to the Demoptera than to the Microchiroptera. This would tie together the flying foxes and the flying lemurs, with the flying lemurs interposed between flying foxes and primates.

Evolution of Handedness

Another long-standing controversy from the late 1980s also remains at issue in the early 1990s. In 1987 linguist Peter MacNeilage of the University of Texas in Austin, Michael G. Studdert-Kennedy of Yale University, and Bjorn Lindblom of the University of Stockholm proposed that, contrary to long-standing belief, handedness exists among primates other than humans. About 90 percent of humans preferentially use their right hands, while nearly 10 percent are left handed and only a few show no handedness. Other primates, even our close relatives, have been found not to prefer one hand (or foot) over the other according to conventional wisdom. But Mac-Neilage and coworkers found forty-five studies that showed handedness in all sorts of primates, ranging from primitive lemurs to chimpanzees.

Their 1987 article generated many new studies of the problem. The preponderance of evidence from the new studies is that lemurs and some monkeys are left handed or left footed, but great apes are not handed at all. MacNeilage thinks that the early simians and monkeys all held onto trees with their right hands while reaching for fruit with the left. Because it was more important to hang onto the tree than to get the food, higher primates, including humans, became right handed and left brained. As recently as the spring 1991 issue of *Behavioral and Brain Science*, Mac-Neilage continued to argue for these beliefs.

NEW AND RARE PRIMATE SPECIES FOUND

Maria Lúcia Lorini, Vanessa Guerra Persson, and Dante Martins Texeira discovered in 1990 a monkey previously unknown to science. The monkey, named the black-faced lion tamarin, lives on an island off the coast of Brazil. This makes four species of lion tamarins in all. The other three live along the coast of Brazil. The researchers spotted more than a dozen of the new species. However, this is probably a large percentage of all living individuals of the species. Plans are being laid to try to increase the number through a captive breeding program.

(Periodical References and Additional Reading: *Science News* 6-30-90, p 406; *Science* 7-6-90, p 20; *International Wildlife* 7/8-90, p 28; *Discover* 1-91, pp 57 & 64; *New York Times* 1-9-91; *Science News* 1-12-91, p 20; *Science* 2-22-91, pp 851 & 934; *Science* 7-5-91, p 36; *Science* 10-4-91, p 33)

TAXONOMY OF LIVING ORGANISMS

Biologists who classify organisms are called taxonomists. They use a system first introduced by Linnaeus in 1735.

KINGDOMS

At that time, Linnaeus and other scientists divided all life forms into two kingdoms—plants and animals. Since then, taxonomists have learned that there are fundamental differences among organisms that go beyond the differences between plants and animals. The following kingdoms are generally used today, although in this, as in all other matters, taxonomists disagree.

To give some dimension to the new thinking in taxonomy, this table is arranged in two columns. The column on the left represents the more traditional system, one that many taxonomists still prefer. The column on the right represents newer thinking, ideas that are accepted by some and rejected by others. It is not possible at this time to determine which, if any, of the newer ideas will stand the test of time. The five-kingdom scheme was first proposed in 1969 by R.H. Whittaker, and was not accepted by most taxonomists until the 1980s.

Five Kingdoms		*Six Kingdoms*	
Monerans	Bacteria and certain one-celled algae	**Archaebacteria**	Mostly microbes metabolizing chemicals
Protists	More complex one-celled organisms	**Monerans**	Bacteria and certain one-celled algae
Fungi	Mushrooms, molds, and yeasts	**Protists**	More complex one-celled organisms
Plants	Mosses, ferns, and higher plants	**Fungi**	Mushrooms, molds, and yeasts
Animals	Sponges to humans	**Plants**	Mosses, ferns, and higher plants
		Animals	Sponges to humans

Scientists believe that organisms in a given kingdom are more closely related to each other than they are to organisms from a different kingdom.

VIRUSES

Viruses are not included among the kingdoms because it is not clear that they are alive in the same sense as the organisms in the five or six kingdoms. Nevertheless, they can be classified into eight groups that have related structures. Most viruses consist of a strand or two of genetic material (the genome) along with a protein envelope to enclose the genetic material. A basic division is between viruses based on RNA and those based on DNA, the two different types of nucleic acid that are used as genetic material. Here is a brief classification of viruses with examples of human diseases that they cause.

> Unenveloped plus-strand RNA (many colds, poliomyelitis)
> Enveloped plus-strand RNA (some cancers, AIDS, yellow fever)
> Minus-strand RNA (flu, mumps, rabies)
> Viroids—single-stranded unenveloped RNA (infect plants only)
> Double-stranded RNA (Colorado tick fever)
> Small-genome DNA (viral hepatitis, warts)
> Medium- and large-genome DNA (herpes, shingles, some cancers, poxes)
> Bacteriophages (infect bacteria only)

OTHER TAXONS

Any set of organisms that is grouped together in a classification scheme is called a taxon. The kingdoms are the largest taxons. Within the kingdoms there are smaller taxons nested within one another.

Each kingdom is divided into two or more phyla. Organisms within a phylum are supposed to be descended from a common ancestor. The phyla are also divided into parts, which are then further divided, each time on the basis of presumed descent from common ancestors. In descending order of size, here are the main divisions:

> Kingdom
>   Phylum
>   Class
>   Order
>   Family
>   Genus
>   Species

(The mnemonic is "**K**ing **P**hil **C**ame **O**ver **F**or **G**ene's **S**pecial.")

Many taxonomists add to this list by classifying groups of species with *sub-*, *super-*, or *infra-*, as in *subphylum, subspecies, superfamily,* or *infra-class.* Plants, for historical reasons, are often placed into large groups called *divisions.* When divisions are used, they generally take the place of phyla. Sometimes fungi are also grouped into divisions instead of phyla. Although divisions seem to be beloved by botanists, other biologists do not see the point of using one system for plants and a different one for animals.

The interpretation of taxons as derived from a single ancestor has encountered many difficulties. Many taxonomists, for example, doubt that all Animalia have a single ancestor. Improved classifications are frequently suggested. Taxonomists who adhere most closely to evolutionary history are called cladists. Taxonomists who put shared characteristics ahead of strict evolutionary lines of descent are pheneticists. The list below reflects common current usage, which tends to be more phenetic than cladistic. Usually, a more cladistic interpretation is the one in the second column where a second column is used, mostly for reptliles and birds on pp 497-503 and for hominoids on p 505.

TAXONOMIC CONVENTIONS

By convention, Latin names except for genus and species are given in Roman type, while genus and species are in *italics.* All Latin names except for species have their initial capitalized, although their English counterparts do not. Family names are formed by adding the suffix *-idae* to the genus name for what is called the "type species"; this is supposed to represent the typical genus for that family.

In writing, spell out completely the genus and species the first time a reference is made to it in a paragraph or short passage, but abbreviate the genus to its first letter in later references in the passage. Thus, a discussion of human beings would first refer to *Homo sapiens* and thereafter to *H. sapiens.* If a species is not known, but its genus is, spell out the generic name in italics and add an appropriate abbreviation for the word *species* in roman. One unknown hominid of genus *Homo* is *Homo* sp., while two or more such unknowns are *Homo* spp.

MAJOR TAXONS

KINGDOM: MONERA

(monerans) One-celled organisms with simple cells that lack a membrane around the genetic material. Bacteria do not produce their own food; blue-green algae do.

Phylum: Archaebacteria
Phylum: Omnibacteria or Eubacteria (most commonly recognized bacteria)
Phylum: Cyanobacteria (blue-green algae, also called blue-green bacteria)
Phylum: Chloroxybacteria (symbiots of sea squirts)
Phylum: Mycoplasma (parasites on plants and animals)
Phylum: Spiroplasma (parasites on plants and animals)
Phylum: Spirochetae (decomposers and animal pathogens)
Phylum: Pseudomonads (decomposers and plant pathogens)
Phylum: Actinomycetes (funguslike monerans that live in soils)
Phylum: Myxobacteria (found in soils and some animals)
Phylum: Nitrogen-fixing aerobes (convert nitrogen from air into organic compounds)
Phylum: Chemoautotrophs (oxidize various inorganic chemicals)

KINGDOM: ARCHAEBACTERIA

This is a controversial but increasingly accepted kingdom of primitive microbes that mostly metabolize inorganic chemicals, such as sulfur compounds, or other difficult-to-digest materials, such as cellulose. Most are anaerobic, so oxygen is a poison. When not classified as a kingdom, archaebacteria are classed as a phylum among the bacteria.

KINGDOM: PROTISTA (protists) One-celled or colonial; complex cells that have a membrane around their genetic material.

Phylum: Ciliophora (ciliated protozoans such as *Paramecium*)
Phylum: Dinoflagellata (dinoflagellates, most with two flagella; famous as the cause of red tides.)
Phylum: Zoomastigina (protozoans with many flagella, such as *Trypanosome,* the cause of sleeping sickness)
Phylum: Rhizopoda (amoebas without cells walls or sexuality)
Phylum: Heliozoa (another group of amoebas)
Phylum: Sarcodina (protozoans that move by flowing, such as amoebas)
Phylum: Sporozoa (parasitic protozoans with no means of motion during most of their lives, such as *Plasmodium,* the cause of malaria)
Phylum: Euglenophyta (one-celled organisms, many of which are photosynthetic)
Phylum: Foraminifera (forams, one-celled creatures with shells)
Phylum: Bacillariophyta (golden algae and diatoms)
Phylum: Chlorophyta (green algae)
Phylum: Phaeophyta (brown algae)
Phylum: Rhodophyta (red algae)
Phylum: Acrasiomycota (cellular slime molds)
Phylum: Myxomycota (plasmodial slime molds)
Phylum: Oomycota (water molds, white rusts, and downy mildews)

KINGDOM: FUNGI One-celled or multicelled; cells have nuclei; the nuclei stream between cells giving the appearance that cells have many nuclei; do not produce their own food. There is also a group known as *Fungi imperfecti,* which includes the fungi that cause athlete's foot and ringworm, as well as molds used in cheese production. These are members of the other phyla that do not appear to reproduce sexually. Finally, there is a large group of lichens, which consist of associations between a fungus and photosynthetic partner.

(Continued)

Phylum: Zygomycetes (black bread mold is an example)
Phylum: Ascomycetes (includes Penicillium, truffles, yeasts)
Phylum: Basidiomycetes (includes mushrooms)

KINGDOM: PLANTA (Plants) Multicellular land-living organisms that carry out photosynthesis; cells have nuclei.
Phylum: Bryophyta (mosses and liverworts)
Phylum: Psilophyta (whisk ferns)
Phylum: Lycophyta (club mosses, quillworts, and relatives)
Phylum: Spenophyta (horsetails)
Phylum: Pterophyta (ferns)
Phylum: Spermatophyta (seed-bearing plants)
  Class: Cycadales (cycads)
  Class: Ginkgoales (the ginkgo)
  Class: Gnetales (Mormon teas)
  Class: Coniferales (conifers)
      Family: Taxacae (yews)
      Family: Pinacae (pines)
      Family: Taxodiaceae (redwoods)
      Family: Cupressaceae (cypresses)
      Family: Araucariacae (monkey-puzzle trees, Norfolk Island pines, and relatives)
  Class: Anthophyta (flowering plants, or angiosperms)
    Subclass: Dicotyledonae (plants with two seed leaves—for example, most fruits and vegetables, common flowers, and trees)
    Subclass: Monocotyledonae (plants with a single seed leaf—for example, onions, lilies, and grasses)

KINGDOM: ANIMALIA (Animals) Multicellular organisms that get their food by ingestion; most are able to move from place to place; cells have nuclei
Phylum: Porifera (sponges)
  Class: Calcarea (calcareous sponges)
  Class: Hexactinellida (glass sponges)
  Class: Demospongiae (siliceous sponges)
  Class: Sclerospongiae (coralline sponges)
Phylum: Cnidaria (coelenterates)
  Class: Anthozoa (sea anemones and corals)
  Class: Hydrozoa (hydras)
  Class: Scyphozoa (jellyfish)
Phylum: Ctenophora (comb jellies)
Phylum: Platyhelminthes (flatworms)
  Class: Turbellaria (free-living flatworms)
  Class: Monogenea (monogenetic flukes)
  Class: Trematoda (digenetic flukes)
  Class: Cestoda (tapeworms)
Phylum: Nemertea (ribbon worms)
  Class: Palaonemertea (unarmed nemerteans)
  Class: Enopla (armed nemerteans)
Phylum: Nematoda (roundworms)
  Class: Adenophorea
  Class: Secernentea
Phylum: Nematomorpha (horsehair worms)
  Class: Nectonematoida (marine nematomorphs)
  Class: Gordioda (freshwater and terrestrial nematomorphs)

Phylum: Rotifera (microscopic wormlike or spherical animals)
   Class: Seisonidea
   Class: Bdelloidea
   Class: Monogononta
Phylum: Gastrotricha (microscopic animals with spiky cuticles)
Phylum: Kinorhyncha (microscopic animals with retractable oral cones)
Phylum: Priapula (priapulan worms)
Phylum: Acanthocephala (spiny-headed worms)
   Class: Palaeacanthocephala
   Class: Archiacanthocephala
   Class: Eoacanthecephala
Phylum: Entoprocta (tiny sessile water animals)
Phylum: Gnathostomulida (tiny wormlike animals)
Phylum: Loricifera (tiny animals of tidal sands)
Phylum: Bryozoa (moss animals)
Phylum: Brachiopoda (lampshells)
Phylum: Phoronida (tube worms)
Phylum: Sipuncula (peanut worms)
   Class: Phascolosomida (without tentacles around its mouth)
   Class: Sipunculida (with tentacles around its mouth)
Phylum: Echiura (echiuran worms)
Phylum: Pogonophora (pogonophoran beard worms)
Phylum: Vestimentifera (vestimentiferan beard worms)
Phylum: Ectoprocta (sessile, water-dwelling animals)
Phylum: Annelida (segmented worms)
   Class: Polychaeta (sand, tube, and clam worms)
   Class: Oligochaeta (earthworms and relatives)
   Class: Hirudinida (leeches)
Phylum: Mollusca (soft-bodied animals with a mantle and foot)
   Class: Aplacophora (aplacophorans)
   Class: Monoplacophora (monoplacophorans)
   Class: Polyplacophora (chitons)
   Class: Bivalvia (clams, oysters, mussels)
   Class: Scaphopoda (tooth or tusk shells)
   Class: Gastropoda (slugs and snails)
   Class: Cephalopoda (the nautilus, octopuses, cuttlefish, and squids)
Phylum: Arthropoda (segmented animals with an external skeleton)
   Class: Merostomata (horseshoe crabs)
   Class: Crustacea (hard-shelled arthropods with jointed appendages)
      Order: Nectiopoda (living remipedes)
      Order: Anostraca (fairy shrimps)
      Order: Notoscraca (tadpole shrimps)
      Order: Cladocera (water fleas)
      Order: Conchostraca (clam shrimps)
      Order: Calanoida (calanoidian copepods)
      Order: Harpacticoida (harpacticoidan copepods)
      Order: Cyclopoida (cyclopoidan copepods)
      Order: Poecilostomatoida (poecilostomatoidan copepods)
      Order: Siphonostomatoida (siphonostomatoidan copepods)
      Order: Monstrilloida (monstrilloidan copepods)
      Order: Misophrioida (misophrioidan copepods)
      Order: Mormonilloida (*Mormonilla*, a copepod genus)
      Order: Thoracica (goose and acorn barnacles)

(Continued)

Order: Ascothoracia (ascothoracican parasitic barnacles)
Order: Acrothoracica (boring barnacles)
Order: Rhizocephala (rhizocephalan parasitic barnacles)
Order: Leptostraca (nebaliids)
Order: Stomatopoda (mantis shrimps)
Order: Bathynellacea (bathynellaceans)
Order: Anaspidacea (anaspidaceans)
Order: Euphausiacea (krill)
Order: Amphionidacea (amphionidaceans)
Order: Decapoda (crabs, shrimps, lobsters, and relatives)
Order: Mysida (opossum shrimps)
Order: Lophogastrida (lophogastrids)
Order: Cumacea (cumaceans)
Order: Tanaidacea (tanaids)
Order: Mictacea (mictaceans)
Order: Spelaeogriphacea (*Potiicoara braziliense* and *Spelaeogriphus lepidops*)
Order: Thermosbaenacea (thermosbaenaceans)
Order: Isopoda (isopods, such a sea slaters, rock lice, and pill bugs)
Order: Amphipoda (amphipods, such as beach hoppers, sand fleas, scuds, and skeleton shrimps)
Class: Arachnida (scorpions, spiders, mites, ticks)
  Order: Scorpiones (true scorpions)
  Order: Uropygi (whip scorpions and vinegaroons)
  Order: Schizomids (schizomids)
  Order: Amblypygi (tailless whip scorpions and whip spiders)
  Order: Palpigradi (palpigradi)
  Order: Araneae (spiders)
  Order: Ricinulei (ricinuleids)
  Order: Pseudoscorpionida (false scorpions)
  Order: Solpugida (sun spiders, or wind spiders)
  Order: Opiliones (daddy longlegs, or harvestmen)
  Order: Acari (mites and ticks)
Class: Pycnogonida (sea spiders)
Class: Myriapoda (millipedes and centipedes)
Class: Insecta (insects)
  Order: Thysanura (silverfish and bristletails)
  Order: Archaeognatha (rockhoppers)
  Order: Ephemeroptera (mayflies)
  Order: Odonata (dragonflies and damselflies)
  Order: Blattodea (cockroaches)
  Order: Mantodea (mantids and praying mantids)
  Order: Isoptera (termites)
  Order: Plecoptera (stone flies)
  Order: Orthoptera (locusts, katydids, crickets, and grasshoppers)
  Order: Dermaptera (earwigs)
  Order: Grylloblattodea (grylloblattids)
  Order: Embioptera (embiopterans)
  Order: Phasmida (stick and leaf insects)
  Order: Zoraptera (zorapterans)
  Order: Psiciotera (book and bark lice)
  Order: Phthiraptera (phthirapterans)
  Order: Hemiptera (true bugs)
  Order: Thysanoptera (thrips)

Order: Anoplura (sucking lice)
Order: Mallophaga (biting lice and bird lice)
Order: Homoptera (plant bugs such as white flies, aphids, scale insects, and cicadas)
Order: Stresiptera (twisted-wing insects)
Order: Coleoptera (beetles)
Order: Neuroptera (lacewings, ant lions, snake flies, dobsonflies, fish flies, alder flies, and owl flies)
Order: Raphidioptera (raphidiopterans)
Order: Megaloptera (megalopterans)
Order: Hymenoptera (ants, bees, and wasps)
Order: Mecoptera (scorpion flies and snow flies)
Order: Siphonaptera (fleas)
Order: Diptera (true flies, mosquitoes, and gnats)
Order: Trichoptera (caddisflies)
Order: Lepidoptera (butterflies and moths)
Phylum: Onychophora ("slugs" with legs)
Phylum: Tardigrada (water bears)
Phylum: Pentastomida (tongue worms)
Phylum: Echinodermata (animals with pentamerous radial symmetry)
　Class: Crinoidea (sea lilies and feather stars)
　Class: Asteroidea (sea stars—often called starfish)
　Class: Concentricycloidea (sea daisies)
　Class: Ophiuroidea (brittle stars and basket stars)
　Class: Echinoidea (sea urchins and sand dollars)
　Class: Holothuroidea (sea cucumbers)
Phylum: Chaetognatha (arrow worms)
Phylum: Hemichordata (acorn worms)
Phylum: Chordata
　Subphylum: Urochordata (tunicates)
　Subphylum: Cephalochordata (lancelets)
　Subphylum: Vertebrata (animals with backbones)
　Class: Agnatha (jawless fishes)
　　Order: Myxiniformes (hagfishes)
　　　Family: Myxinidae (hagfishes)
　　Order: Hyperoartia (lampreys)
　　　Family: Petromyzontidae (lampreys)
　Class: Chondrichthyes (sharks, rays, and relatives)
　　Order: Hexanchiformes (comb-toothed sharks)
　　　Family: Hexanchidae (cow sharks)
　　Order: Chlamydoselachiformes
　　　Family: Chlamydoselachidae (frilled sharks)
　　Order: Heterodontiformes (horn sharks)
　　　Family: Heterodontidae (bullhead sharks)
　　Order: Squaliformes (sharks with 2 dorsal fins and 5 pairs of gill openings)
　　　Family: Orectolobidae (carpet sharks)
　　　Family: Scapanorhynchidae (goblin sharks)
　　　Family: Rhincodontidae (whale sharks)
　　　Family: Odontaspididae (sand tigers)
　　　Family: Alopiidae (thresher sharks)
　　　Family: Lamnidae (mackerel sharks)
　　　Family: Scyliorhinidae (cat sharks)
　　　Family: Carcharhinidae (gray sharks)

(Continued)

Family: Sphyrnidae (hammerhead sharks)
Family: Squalidae (dogfish sharks)
Family: Squatinidae (angel sharks)
Family: Pristiphoridae (saw sharks)
Order: Rajiformes (skates and rays)
  Family: Pristidae (sawfishes)
  Family: Rhinobatidae (guitarfishes)
  Family: Torpedinidae (electric rays)
  Family: Narkidae (other electric rays)
  Family: Rajidae (skates)
  Family: Anachantobatidae (skate relatives)
  Family: Arynchobatidae (other skate relatives)
  Family: Dasyatidae (stingrays)
  Family: Gymnuridae (butterfly rays)
  Family: Myliobatidae (eagle rays)
  Family: Urolophidae (eagle ray relatives)
  Family: Potamotrygonidae (other eagle ray relatives)
  Family: Mobulidae (mantas)
Order: Chimaeriformes (ratfishes)
  Family: Chimaeridae (chimaeras)
  Family: Rhinochimaeridae
  Family: Callorynchus
Class: Osteichthyes (bony fishes)
  Order: Acipenseriformes (sturgeons and paddlefishes)
    Family: Acipenseridae (sturgeons)
    Family: Polyodontidae (paddlefishes)
  Order: Semionotiformes (gars)
    Family: Lepisteidae (gars)
  Order: Amiiformes (the bowfin)
    Family: Amiidae (the bowfin)
  Order: Elopiformes (tarpons and bonefishes)
    Family: Elopidae (tarpons)
    Family: Albulidae (bonefishes)
    Family: Pterothrissidae (deep-sea bonefishes)
  Order: Anguilliformes (eels)
    Family: Anguillidae (freshwater eels)
    Family: Moringuidae
    Family: Myrocongridae
    Family: Muraenidae (morays)
    Family: Muraenesocidae (pike eels)
    Family: Congridae (conger eels)
    Family: Hetercongridae (garden eels)
    Family: Ophichthidae (snake eels)
    Family: Synaphobranchidae (cutthroat eels)
    Family: Ilyophidae
    Family: Simenchelyidae (slime eels)
    Family: Nemichthyidae (snipe eels)
    Family: Serrivomeridae
    Family: Cyemidae
    Family: Saccopharyngidae (swallowers)
    Family: Eupharyngidae (gulpers)
    Family: Monognathidae
    Family: Notacanthidae (spiny eels)

Family: Halosauridae
Order: Clupeiformes (herringlike fishes)
  Family: Culpeidae (herrings and shads)
  Family: Engraulidae (anchovies)
  Family: Chirocentridae (wolf herrings)
Order: Osteoglossiformes (mooneyes and relatives)
  Family: Osteoglossidae (bonytongues)
  Family: Pantodontidae (butterfly fishes)
  Family: Hiodontidae (mooneyes)
Order: Mormyriformes (mormyrids)
  Family: Mormyridae (elephant fish)
  Family: Gymnarchidae (the gymnarchid)
Order: Salmoniformees (salmonlike fish)
  Family: Salmonidae (trouts and salmons)
  Family: Osmeridae (smelts)
  Family: Umbridae (mudminnows)
  Family: Esocidae (pikes and pickerels)
Order: Myctophiformes (light-emitting deepwater fish)
  Family: Synodontidae (lizardfishes)
Order: Cypriniformes (carplike fish)
  Family: Characidae (characins)
  Family: Cyprinidae (carps and minnows)
  Family: Catostomidae (suckers)
Order: Siluriformes (catfishes)
  Family: Ictaluridae (bullhead catfishes)
  Family: Clariidae (walking catfishes)
  Family: Ariidae (sea catfishes)
Order: Percopsiformes (cavefishes, trout-perches, and relatives)
  Family: Amblyopsidae (cavefishes)
  Family: Aphredoderidae (the pirate perch)
  Family: Percopsidae (trout-perches)
Order: Batrachoidiformes (toadfishes)
  Family: Batrachoididae (toadfishes)
Order: Gobiesociformes (clingfishes and relatives)
  Family: Gobiesocidae (clingfishes)
Order: Lophiiformes (anglerfishes)
  Family: Lophiidae (goosefishes)
  Family: Antennariidae (frogfishes)
  Family: Ogcocephalidae (batfishes)
Order: Gadiformes (codfishlike fishes)
  Family: Gadidae (codfish, the haddock, and relatives)
  Family: Ophidiidae (cusk-eels)
  Family: Zoarcidae (eelpouts)
Order: Atheriniformes (killifishes and relatives)
  Family: Ecocoetidae (flyingfishes)
  Family: Belonidae (needlefishes)
  Family: Cyprinodontidae (killifishes)
  Family: Poeciliidae (livebearers)
  Family: Atherinidae (silversides)
Order: Beryciformes (squirrelfishes and relatives)
  Family: Holocentridae (squirrelfishes)
Order: Gasterosteiformes (sticklebacks and relatives)
  Family: Gasterosteidae (sticklebacks)

(Continued)

Family: Aulostomidae (trumpetfishes)
Family: Fistulariidae (cornetfishes)
Family: Syngnathidae (pipefishes)
Order: Perciformes (perchlike or basslike fishes)
Family: Centropomidae (snooks)
Family: Percichthyidae (temperate basses)
Family: Serranidae (sea basses)
Family: Grammistidae (soapfishes)
Family: Centrarchidae (sunfishes)
Family: Percidae (perches)
Family: Priacanthidae (bigeyes)
Family: Apogonidae (cardinalfishes)
Family: Malacanthidae (tilefishes)
Family: Pomatomidae (bluefishes)
Family: Rachycentridae (the cobia)
Family: Echeneidae (remoras)
Family: Carangidae (jacks)
Family: Coryphaenidae (dolphins)
Family: Lutjanidae (snappers)
Family: Lobotidae (tripletails)
Family: Gerreidae (mojarras)
Family: Haemulidae (grunts)
Family: Sparidae (porgies)
Family: Sciaenidae (drums)
Family: Mullidae (goatfishes)
Family: Kyphosidae (sea chubs)
Family: Ephippidae (spadefishes)
Family: Chaetodontidae (butterflyfishes)
Family: Pomacanthidae (angelfishes)
Family: Cichlidae (chiclids)
Family: Embiotocidae (surfperches)
Family: Pomacentridae (damselfishes)
Family: Labridae (wrasses)
Family: Scaridae (parrotfishes)
Family: Mugilidae (mullets)
Family: Sphyraenidae (barracudas)
Family: Clinidae (clinids)
Family: Blenniidae (combtooth blennies)
Family: Stichaeidae (pricklebacks)
Family: Pholidae (gunnels)
Family: Anarhichadidae (wolffishes)
Family: Ammodytidae (sand lances)
Family: Eleotridae (sleepers)
Family: Gobiidae (gobies)
Family: Acanthuridae (surgeonfishes)
Family: Trichiuridae (cutlassfishes)
Family: Scombridae (mackerels and tunas)
Family: Xiphiidae (swordfishes)
Family: Istiophoridae (billfishes)
Family: Stromateidae (butterfishes)
Family: Scorpaenidae (scorpionfishes)
Family: Triglidae (searobins)
Family: Anoplopomatidae (sablefishes)

        Family: Hexagrammidae (greenlings)
        Family: Sottidae (sculpins)
        Family: Agonidae (poachers)
        Family: Cyclopteridae (snailfishes)
     Order: Pleuronectiformes (flatfishes)
        Family: Bothidae (lefteye flounders)
        Family: Pleuronectidae (righteye flounders)
        Family: Soleidae (soles)
        Family: Cynoglossidae (tonguefishes)
     Order: Tetraodontiformes (filefish, puffers, and relatives)
        Family: Balistidae (leatherjackets)
        Family: Ostraciidae (boxfishes)
        Family: Diodontidae (porcupinefishes)
        Family: Molidae (molas)
Class: Amphibia (amphibians)
     Order: Gymnophiona (caecilians)
        Family: Caeciliidae
        Family: Ichthyophiidae
        Family: Rhinatrematidae
        Family: Scolecomorphidae
        Family: Typhonectidae
     Order: Urodela (salamanders and newts)
        Family: Cryptobranchidae (giant salamanders)
        Family: Hynobiidae (Asian salamanders)
        Family: Salamandridae (newts, brook salamanders, and fire salamanders)
        Family: Amphiumidae (Congo eels)
        Family: Proteidae (the olm, mudpuppies, and waterdogs)
        Family: Ambystomatidae (mole salamanders)
        Family: Dicamptodontidae (Pacific mole salamanders)
        Family: Plethodontidae (lungless salamanders)
        Family: Sirenidae (sirens)
     Order: Anura (frogs and toads)
        Family: Leipelmatidae (the tailed frog and New Zealand frogs)
        Family: Discoglossidae (disc-tongued toads)
        Family: Pipidae (clawed and Surinam toads)
        Family: Rhinophrynidae (the burrowing toad)
        Family: Pelobatidae (spadefoot toads, horned toads, and parsley frogs)
        Family: Bufonidae (true toads)
        Family: Brachycephalidae (gold frogs)
        Family: Rhinodermatidae (mouth-brooding frogs)
        Family: Heleophrynidae (ghost frogs)
        Family: Myobactrachidae (myobactrachid frogs)
        Family: Leptodactylidae (leptodactylid frogs)
        Family: Dendrobatidae (poison-arrow frogs)
        Family: Hylidae (true tree frogs)
        Family: Centrolenidae (glass frogs)
        Family: Pseudidae (pseudid frogs)
        Family: Ranidae (true frogs)
        Family: Sooglossidae (Seychelles frogs)
        Family: Hyperoliidae (sedge and bush frogs)
        Family: Rhacophoridae (Old World tree frogs)
        Family: Microhylidae (narrow-mouthed frogs)

(Continued)

Class: Reptilia (reptiles)
  Order: Chelonia (turtles and tortoises)
    Suborder: Cryptodira (hidden-necked turtles)
      Family: Carettochelyidae (the pig-nosed softshell turtle)
      Family: Chelydridae (snapping turtles)
      Family: Dermatemydidae (the Central American river turtle)
      Family: Cheloniidae (sea turtles)
      Family: Dermochelyidae (the leatherback sea turtle)
      Family: Emydidae (pond and river turtles)
      Family: Kinosternidae (American mud and musk turtles)
      Family: Staurotypidae (Mexican musk turtles)
      Family: Testudinidae (tortoises)
      Family: Trionychidae (softshell turtles)
      Family: Platysternidae (the big-headed turtle)
    Suborder: Pleurodira (side-necked turtles)
      Family: Chelidae (snake-necked turtles)
      Family: Pelomedusidae (hidden-necked side-necked turtles)
  Order: Squamata (lizards and snakes)
    Suborder: Lacertila (lizards)
      Family: Agamidae (chisel-teeth lizards)
      Family: Chamaeleontidae (chameleons)
      Family: Iguanidae (iguanas)
      Family: Gekkonidae (geckos)
      Family: Pygopodidae (snake, or flap-footed, lizards)
      Family: Teiidae (whiptails and racerunners)
      Family: Lacertidae (wall and sand lizards)
      Family: Xantusiidae (night lizards)
      Family: Scinidae (skinks)
      Family: Corylidae (girdle-tailed lizards)
      Family: Dibamidae (blind lizards)
      Family: Xenosauridae (xenosaurs)
      Family: Anguidae (anguids)
      Family: Helodermatidae (beaded lizards)
      Family: Lanthanotidae (the Bornean earless lizard)
    Suborder: Amphisbaenia (worm lizards)
      Family: Amphisbaenidae (true worm lizards)
      Family: Rhineuridae (the Florida worm lizard)
      Family: Trogonophidae (edge-snouted worm lizards)
      Family: Bipedidae (Mexican worm lizards)
    Suborder: Serpentes (snakes)
      Family: Boidae (boas)
      Family: Pythonidae (pythons)
      Family: Aniliidae (pipesnakes)
      Family: Tropidophiidae (proctocolubroids)
      Family: Uropeltidea (shieldtail snakes)
      Family: Leptotyphlopidae (thread snakes)
      Family: Anomalepidae (dawn blind snakes)
      Family: Typhlopidae (typical blind snakes)
      Family: Colubridae (harmless snakes)
      Family: Elapidae (front-fanged snakes)
      Family: Viperidae (vipers)

Order: Rhynchocephalia (the tuatara)
  Family: Spenodontidae (the tuatara)
Order: Crocodylia (crocodilians)
  Family: Alligatoridae (alligators and caimans)
  Family: Crocodylidae (crocodiles)
  Family: Gavialidae (the gharial)
Class: Aves (birds)
Order: Struthioniformes (the ostrich)
  Family: Struthionidae (the ostrich)
Order: Rheiformes (rheas)
  Family: Rheidae (rheas)
Order: Casuariiformes (emus and cassowaries)
  Family: Dromaiidae (the emu)
  Family: Casuariidae (cassowaries)
Order: Apterygiformes (kiwis)
  Family: Apterygidae (kiwis)
Order: Tinamiformes (tinamous)
  Family: Tinamidae (tinamous)
Order: Sphenisciformes (penguins)
  Family: Spheniscidae (penguins)
Order: Gaviformes (loons)
  Family: Gaviidae (loons)
Order: Podicipediformes (grebes)
  Family: Podicipedidae (grebes)
Order: Procellariiformes (albatrosses and petrels)
  Family: Diomedeidae (albatrosses)
  Family: Procellariidae (shearwater and petrels)
  Family: Hydrobatidae (storm petrels)
  Family: Pelecanoididae (diving petrels)
Order: Pelicaniformes (pelicans, gannets, and cormorants)

Class: Archosauria (a new class suggested by revisionist dinosaur experts)
  Order: Thecodonta (thecodonts)
  Order: Crocodylia (crocodilians)
  Order: Pterosaura (pterosaurs)
  Order: Dinosaura (dinosaurs)
  Order: Aves (birds)

Class: Aves (a new system for classifying birds based on the DNA studies of Charles G. Sibley and Jon Ahlquist)
  Order: Struthioniformes (ratites)
    Family: Struthionidae (the ostrich)
    Family: Rheidae (rheas)
    Family: Casuariidae (cassowaries and the emu)
    Family: Apterygidae (kiwis)
  Order: Tinamiformes (tinamous)
    Family: Tinamidae (tinamous)
  Order: Craciformes (guans, megapodes, and relatives)
    Family: Cracidae (guans, chachalacas, and relatives)
    Family: Megopodiiae (megapods)
  Order: Galliformes (game birds)
    Family: Phasianidae (grouse, turkeys, pheasants, partridges, and relatives)
    Family: Numididae (guineafowls)
    Family: Odontophoridae (New World quails)
  Order: Anseriformes (waterfowl)
    Family: Anhimidae (screamers)
    Family: Anseranatidae (the magpie goose)
    Family: Dendrocygnidae (whistling-ducks)
    Family: Anatidae (swans, geese, and ducks)

(Continued)

Family: Pelecanidae (pelicans)

Family: Sulidae (gannets and boobies)

Family: Phaethontidae (tropicbirds)

Family: Phalacrocoracidae (cormorants)

Family: Fregatidae (frigates)

Family: Anhingidae (darters)

Order: Ciconiiformes (stilt-legged birds)

Family: Areidae (herons and bitterns)

Family: Ciconiidae (storks)

Family: Threskionrnithidae (spoonbills and ibises)

Family: Scopidae (the hammerhead)

Family: Balaenicipitidae (the whale-headed stork)

Family: Phoenicopteridae (flamingos)

Order: Anseriformes (waterfowl)

Family: Anatidae (swans, geese, and ducks)

Family: Anhimidae (screamers)

Order: Falconiformes (birds of prey)

Family: Cathartidae (New World vultures)

Family: Sagittariidae (the secretary bird)

Family: Pandionidae (the osprey)

Family: Falconidae (falcons, falconets, and caracaras)

Family: Accipitridae (kites, fish eagles, Old World vultures, sparrowhawks, buzzards, etc.)

Order: Galliformes (game birds)

Family: Phasianidae (pheasants and quails)

Family: Tetraonidae (grouse)

Family: Meleagrididae (turkeys)

Order: Turniciformes (button-quails)

Family: Turnicidae (button-quails)

Order: Piciformes (woodpeckers, barbets, toucans, and honeyguides)

Family: Indicatoridae (honeyguides)

Family: Picidae (woodpeckers and wrynecks)

Family: Megalaimidae (Asian barbets)

Family: Lybiidae (African barbets)

Family: Ramphastidae (toucans and New World barbets)

Order: Galbuliformes (puffbirds and jacamars)

Family: Galbulidae (jacamars)

Family: Bucconidae (puffbirds)

Order: Bucerotiformes (hornbills)

Family: Bucerotidae (typical hornbills)

Family: Bucorvidae (ground-hornbills)

Order: Upupiformes (hoopoes and relatives)

Family: Upupidae (hoopoes)

Family: Phoeniculidae (woodhoopoes)

Family: Rhinopomastidae (scimitarbills)

Order: Trogoniformes (trogons)

Family: Trogonidae (trogons)

Order: Coraciiformes (rollers, kingfishers, and relatives)

Family: Coraciidae (typical rollers)

Family: Brachytperaciidae (ground-rollers)

Family: Leptosomidae (cuckoo-rollers)

Family: Momotidae (motmots)

Family: Numididae (guinea-fowls)
Family: Megapodiidae (megapodes)
Family: Cracidae (guans and curassows)
Family: Opisthocomidae (the hoatzin)
Order: Gruiformes (cranes, rails, and bustards)
    Family: Gruidae (cranes)
    Family: Aramidae (the limp-kin)
    Family: Psophiidae (trum-peters)
    Family: Rallidae (rails)
    Family: Otididae (bustards)
    Family: Turnicidae (button-quails)
    Family: Rhynochetidae (the kagu)
    Family: Eurypygidae (the sun bittern)
    Family: Cariamidae (seriemas)
    Family: Pedionomidae (the plains wanderer)
    Family: Mesitornithdae (mesites)
    Family: Heliornithidae (fin-foots)
Order: Charadriiformes (shore-birds, gulls, terns, and auks)
    Family: Charadriidae (plovers)
    Family: Scolopacidae (sand-pipers)
    Family: Recurvirostridae (avocets and stilts)
    Family: Phalaropodidae (phalaropes)
    Family: Jacanidae (jacanas)
    Family: Rostratulidae (painted snipes)
    Family: Haematopodidae (oystercatchers)
    Family: Dromadidae (the crab plover)
    Family: Burhinidae (stone curlews)
    Family: Glareolidae (pratin-coles and coursers)

    Family: Todidae (todies)
    Family: Alcedinidae (alce-dinid kingfishers)
    Family: Facelonidae (dacelonid kingfishers)
    Family: Ceryidae (cerylid kingfishers)
    Family: Meropidae (bee-eaters)
Order: Coliiformes (mousebirds)
    Family: Coliidae (mouse-birds)
Order: Cuculiformes (cuckoos and the hoatzin)
    Family: Cuclidae (Old World cuckoos)
    Family: Centropodidae (coucals)
    Family: Opisthocomidae (the hoatzin)
    Family: Crotophagidae (anises and the guira cuckoo)
    Family: Neomorphidae (roadrunners and ground cuckoos)
Order: Psittaciformes (parrots and relatives)
    Family: Psittacidae (parrots and relatives)
Order: Apodiformes (swifts)
    Family: Apodidae (typical swifts)
    Family: Hemiprocnidae (crested swifts)
Order: Trochiliformes (hum-mingbirds)
    Family: Trochilidae (hum-mingbirds)
Order: Musophagiformes (tura-cos)
    Family: Musophagidae (turacos)
Order: Strigiformes (owls, nightjars, and relatives)
    Family: Tytonidae (barn and grass owls)
    Family: Strigidae (typical owls)
    Family: Aegothelidae (owlet-nightjars)

(Continued)

Family: Thinocoridae (seed snipes)
Family: Chionididae (sheathbills)
Family: Laridae (gulls)
Family: Sternidae (terns)
Family: Stercorariidae (skuas)
Family: Rynchopidae (skimmers)
Family: Alcidae (auks)
Order: Pteroclidiformes (sandgrouses)
Family: Pteroclididae (sandgrouses)
Order: Columbiformes (pigeons)
Family: Columbidae (pigeons)
Order: Psittaciformes (parrots, lories, and cockatoos)
Family: Psittacidae (parrots, lories, and cockatoos)
Order: Cuculiformes (cuckoos and turacos)
Family: Cuculidae (cuckoos)
Family: Musophagidae (turacos)
Order: Strigiformes (owls)
Family: Strigidae (typical owls)
Family: Tytonidae (barn and bay owls)
Order: Caprimulgiformes (nightjars and frogmouths)
Family: Caprimulgidae (nightjars or goatsuckers)
Family: Podargidae (frogmouths)
Family: Aegothelidae (owlet-nightjars)
Family: Nyctibiidae (potoos)
Family: Steatornithidae (the oilbird)
Order: Apodiformes (swifts and hummingbirds)
Family: Apodidae (swifts)
Family: Hemiprocnidae (crested swifts)
Family: Trochilidae (hummingbirds)
Order: Trogoniformes (trogons)
Family: Trogonidae (trogons)

Family: Podargidae (Australian frogmouths)
Family: Batrachostomidae (Asian frogmouths)
Family: Steatornithidae (the oilbird)
Family: Nyctibiidae (potoos)
Family: Eurostopodidae (eared nightjars)
Family: Caprimulgidae (nighthawks and nightjars)
Order: Columbiformes (pigeons and relatives)
Family: Columbidae (pigeons)
Order: Gruiformes (cranes, rails, and bustards)
Family: Eurypygidae (the sunbittern)
Family: Otididae (bustards)
Family: Gruidae (cranes)
Family: Heliornithidae (sungrebes and the limpkin)
Family: Psophiidae (trumpeters)
Family: Cariamidae (seriemas)
Family: Rhynochetidae (the kagu)
Family: Rallidae (rails)
Family: Mesitornithidae (mesites)
Order: Ciconiiformes (stilt-legged birds, shorebirds, and birds of prey)
Family: Pteroclidae (sandgrouses)
Family: Thinocoridae (seed-snipes)
Family: Pedionomidae (the plains wanderer)
Family: Scolopacidae (sandpipers, phalaropes, snipes, and relatives)
Family: Rostratulidae (painted snipes)
Family: Jacanidae (jacanas)
Family: Chionididae (sheathbills)
Family: Burhinidae (thick-knees)

Order: Coliiformes (mouse-birds)
　Family: Coliidae (mouse-birds)
Order: Coraciiformes (kingfishers, bee-eaters, and hoopoes)
　Family: Alcedinidae (kingfishers)
　Family: Motmotidae (motmots)
　Family: Todidae (todies)
　Family: Meropidae (bee-eaters)
　Family: Coraciidae (rollers)
　Family: Leptosomatidae (the cuckoo-roller)
　Family: Phoeniculidae (wood-hoopoes)
　Family: Upupidae (the hoopoe)
　Family: Bucerotidae (hornbills)
Order: Piciformes (woodpeckers, toucans, and barbets)
　Family: Ramphastidae (toucans)
　Family: Indicatoridae (honeyguides)
　Family: Capitonidae (barbets)
　Family: Galbulidae (jacamars)
　Family: Bucconidae (puffbirds)
　Family: Picidae (woodpeckers)
Order: Passeriformes (perching birds)
　Family: Eurylaimidae (broadbills)
　Family: Menuridae (lyrebirds)
　Family: Atrichornithidae (scrub-birds)
　Family: Furnariidae (ovenbirds)
　Family: Dendrocolaptidae (woodcreepers)
　Family: Formicariidae (antbirds)
　Family: Tyrannidae (tyrant flycatchers)
　Family: Pittidae (pittas)

Family: Charadriidae (plovers, avocets, stilts, and oystercatchers)
Family: Glareolidae (pratincoles, coursers, and the crab plover)
Family: Laridae (gulls, terns, skuas, and skimmers)
Family: Accipitridae (hawks, eagles, and the osprey)
Family: Sagittariidae (the secretary bird)
Family: Falconidae (falcons and caracaras)
Family: Podicipedidae (grebes)
Family: Phaethontidae (tropicbirds)
Family: Sulidae (boobies and gannets)
Family: Anhingidae (anhingas)
Family: Phalarcrocoracidae (cormorants)
Family: Ardeidae (herons, bitterns, and egrets)
Family: Scopidae (the hammerhead)
Family: Phoenicopteridae (flamingos)
Family: Threskiornithidae (spoonbills and ibises)
Family: Pelecanidae (pelicans and the shoebill)
Family: Ciconiidae (New World vultures and storks)
Family: Fregatidae (frigates)
Family: Spheniscidae (penguins)
Family: Gaviidae (loons)
Family: Procellariidae (petrels, shearwaters, and albatrosses)
Order: Passeriformes (perching birds)
　Family: Acanthisittdae (New Zealand wrens)
　Family: Pittidae (pittas)
　Family: Eurylaimidae (broadbills)
　Family: Philepittdae (asities)
　Family: Tyrannidae (tyrant flycatchers, cotingas, manakins, and relatives)

(Continued)

Family: Pipridae (manakins)
Family: Cotingidae (cotingas)
Family: Conopophagidae (gnateaters)
Family: Rhinocryptidae (tapaculos)
Family: Oxyrunicidae (the sharpbill)
Family: Phytotomidae (plantcutters)
Family: Xenicidae (New Zealand wrens)
Family: Philepittidae (sunbird asitys)
Family: Hirundinidae (swallows)
Family: Alaudidae (larks)
Family: Motacillidae (wagtails and pipits)
Family: Pycnonotidae (bulbuls)
Family: Laniidae (shrikes)
Family: Campephagidae (cuckoo-shrikes)
Family: Irenidae (leafbirds)
Family: Prionopidae (helmet shrikes)
Family: Vangidae (vanga shrikes)
Family: Bombycillidae (waxwings and relatives)
Family: Dulidae (the palmchat)
Family: Cinclidae (dippers)
Family: Troglodytidae (wrens)
Family: Mimidae (mockingbirds)
Family: Prunellidae (accentors)
Family: Muscicapidae (thrushes and relatives)
Family: Paridae (true tits)
Family: Aegithalidae (long-tailed tits)
Family: Remizidae (penduline tits)
Family: Sittidae (nuthatches)
Family: Climacteridae (Australian treecreepers)
Family: Certhiidae (holarctic treecreepers)

Family: Thamnophilidae (typical antbirds)
Family: Furnariidae (ovenbirds and woodcreepers)
Family: Formicariidae (ground antbirds)
Family: Conopophagidae (gnateaters)
Family: Rhinocryptidae (tapaculos)
Family: Climacteridae (Australo-Papuan treecreepers)
Family: Menuridae (lyrebirds and scrub-birds)
Family: Ptilonorhynchidae (bowerbirds)
Family: Meliphagidae (honeyeaters)
Family: Pardalotidae (pardalotes, bristlebirds, scrubhens, thornbills, whitefaces, and relatives)
Family: Eopsaltriidae (Australo-Paupuan robins and relatives)
Family: Irenidae (fairy-bluebirds and leafbirds)
Family: Orthonychidae (logrunners and chowchillas)
Family: Pomatostomidae (Australo-Papuan babblers)
Family: Laniidae (true shrikes)
Family: Corvidae (quailthrushes, shrike-thrushes, crows, birds-of-paradise, orioles, and relatives)
Family: Callaeatidae (New Zealand wattlebirds)
Family: Bombycullidae (palmchats and waxwings)
Family: Cinclidae (dippers)
Family: Muscocapidae (true thrushes, old-world flycatchers, and chats)
Family: Sturnidae (starlings, mynas, catbirds, mockingbirds, and thrashers)
Family: Sittidae (nuthatches and the wallcreeper)
Family: Certhiidae (creepers, wrens, gnatcatchers, and relatives)

Family: Rhabdornithidae (Philippine creepers)
Family: Zosteropidae (white-eyes and relatives)
Family: Dicaeidae (flower-peckers and relatives)
Family: Pardalotidae (diamond birds)
Family: Nectariniidae (sunbirds and spider-hunters)
Family: Meliphagidae (honeyeaters)
Family: Ephthianuridae (Australian chats)
Family: Emberizidae (buntings and tanagers)
Family: Parulidae (wood warblers)
Family: Vireonidae (vireos)
Family: Icteridae (American blackbirds)
Family: Fringillidae (finches)
Family: Estrildidae (waxbills)
Family: Ploceidae (weavers)
Family: Sturnidae (starlings)
Family: Oriolidae (orioles and figbirds)
Family: Dicruridae (drongos)
Family: Callaeidae (New Zealand wattlebirds)
Family: Grallinidae (magpie-larks)
Family: Corcoracidae (Australian mud-nesters)
Family: Artamidae (wood swallows)
Family: Cractididae (bell magpies)
Family: Ptilonorhynchidae (bowerbirds)
Family: Paradisaeidae (birds of paradise)
Family: Corvidae (crows)
Class: Mammals
Subclass: Ornithodelphia (egg-laying mammals)
Order: Monotremata (egg-laying mammals)
Family: Tachyglossidae (echidnas)
Family: Orinthorhynchidae (the platypus)

Family: Paridae (penduline-tits, titmice, and chickadees)
Family: Aegithalidae (long-tailed tits and bushmice)
Family: Hirundinidae (river-martins and swallows)
Family: Regulidae (kinglets)
Family: Pycnonotidae (bulbuls)
Family: Hypocolidae (no common name)
Family: Cisticolidae (African warblers)
Family: Zosteropidae (white-eyes)
Family: Sylviidae (leaf-warblers, grass-warblers, babblers, and relatives)
Family: Alaudidae (larks)
Family: Nectariniidae (sugarbirds, flowerpeckers, sunbirds, and relatives)
Family: Melanocharitidae (no common names)
Family: Paramythiidae (no common names)
Family: Passeridae (sparrows, wagtails, accentors, weavers, whydahs, and relatives)
Family: Fringillidae (chaffinches, goldfinches, crossbills, Hawaiian honey-creepers, buntings, wood warblers, tanagers, cardinals, meadowlarks, new-world blackbirds, and relatives)

(Continued)

Subclass: Metatheria (nonplacental mammals)
  Order: Marsupialia (nonplacental mammals)
    Family: Didelphidae (opossums)
    Family: Dasyuridae (marsupial mice, dasyures, numbats, and the Tasmanian wolf)
    Family: Notorcyctidae (marsupial moles)
    Family: Peramelidae (bandicoots)
    Family: Caenolestidae (rat opposums)
    Family: Phalangeridae (phalangers)
    Family: Phascolarctidae (the koala)
    Family: Vombatidae (wombats)
    Family: Macropodidae (wallabies and kangaroos)
Subclass: Eutheria (placental mammals)
  Order: Insectivores (shrews, moles, and hedgehogs)
    Family: Solenodontidae (solenodons)
    Family: Tenrecidae (tenrecs)
    Family: Potamogalidae (otter shrews)
    Family: Chrysochloridae (golden moles)
    Family: Erinaceidae (hedgehogs)
    Family: Macroscelididae (elephant shrews)
    Family: Soricidae (shrews)
    Family: Talipidae (moles)
  Order: Dermoptera (flying lemurs)
    Family: Cynocephalidae (flying lemurs)
  Order: Chirotpera (bats)
   Suborder: Microchiroptera (insect-eating bats)
    Family: Rhinopomatidae (mouse-tailed bats)
    Family: Emballonuridae (sac-winged bats)
    Family: Noctilionidae (bulldog bats)
    Family: Nycteridae (slit-faced bats and relatives)
    Family: Megadermatidae (false vampires)
    Family: Rhinolophidae (horseshoe bats)
    Family: Hipposideridae (leaf-nosed bats)
    Family: Phyllostomatidae (large-eared and long-tongued bats and relatives)
    Family: Desmodontidae (true vampires)
    Family: Natalidae (funnel-eared bats)
    Family: Furipteridae (smoky bats and relatives)
    Family: Thyropteridae (disc-winged bats)
    Family: Myzopodidae (the golden bat)
    Family: Vespertilionidae (common bats and relatives)
    Family: Mystacinidae (the New Zealand short-tailed bat)
    Family: Molossidae (free-tailed bats and relatives)
   Suborder: Megachiroptera (fruit-eating bats)*
    Family: Pteropidae (fruit-eating bats)
  Order: Primates (lemurs, monkeys, apes, and humans)
   Suborder: Prosimii (prosimians)
    Family: Tupaiidae (tree shrews)
    Family: Lemuridae (lemurs)
    Family: Daubentoniidae (the aye-aye)

---

* Some authorities now consider fruit-eating bats to be more closely related to the primates than to insectivorous bats.

Family: Lorisidae (lorises and bushbabies)
Family: Tarsiidae (tarsiers)
Suborder: Anthropoidea
Family: Cebidae (New World monkeys)
Family: Callithricidae (marmosets and tamarins)
Family: Cercopithecidae (Old World monkeys)
Family: Pongidae (apes and gibbons)
Family: Hominidae (humans)
Order: Edentata (mammals without teeth)
Family: Myrmecophagidae (anteaters)
Family: Bradypodidae (sloths)
Family: Dasypodidae (armadillos)
Order: Pholidota (scaly ant-eaters)
Family: Manidae (pangolins)
Order: Lagomorpha (pikas, rabbits, and hares)
Family: Ochotonidae (pikas)
Family: Leporidae (rabbits and hares)
Order: Rodentia (squirrels, rats, mice, porcupines)
Suborder: Sciuromorpha (squirrels and gophers)
Family: Sciuridae (squirrels)
Family: Geomyidae (pocket gophers)
Family: Heteromyidae (pocket mice and kangaroo mice)
Family: Castoridae (the beaver)
Family: Anomaluridae (scaly-tailed squirrels)
Suborder: Myomorpha (rats and mice)
Family: Cricetidae (New World mice, hamsters, lemmings, voles, gerbils, and relatives)
Family: Spalacidae (palearctic mole rats)
Family: Rhizomyidae (east African mole rats and bamboo rats)
Family: Muridae (true mice and rats)
Family: Gliridae (dormice)
Family: Platacanthomyidae (spiny dormice)
Family: Zapodidae (jumping mice)
Family: Dipodidae (jerboas)
Suborder: Hystricomorpha (mostly larger rodents)
Family: Hystricidae (Old World porcupines)
Family: Erethizontidae (New World porcupines)
Family: Caviidae (cavies)
Family: Hydrochoeridae (the capybara)
Family: Dinomyidae (the paca-rana)
Family: Dasyproctidae (pacas and agoutis)
Family: Chinchillidae (chinchillas)
Family: Capromyidae (hutias and the coypu)
Family: Octodontidae (bush rats, the cururo, and relatives)
Family: Ctenomyidae (tucotucos)
Family: Abrocomidae (rat chinchillas)
Family: Echimyidae (spiny rats and arboreal rats)
Family: Thryonomyidae (African cane rats)
Family: Petromyidae (the rock rat)

Superfamily: Hominoidea (system of classifying human relatives based on DNA matching)
Family: Hylobatidae (gibbons)
Family: Hominidae (apes and humans)
Subfamily: Ponginae (the orangutan)
Subfamily: Homininae (the gorilla, chimpanzees, and the human)

(Continued)

Family: Bathyergidae (African mole rats)
Family: Ctenodactylidae (gundis)
Order: Cetacea (whales and dolphins)
  Suborder: Odontoceti (toothed whales)
  Family: Platanistidae (river dolphins)
  Family: Ziphiidae (beaked whales)
  Family: Physeteridae (sperm whales)
  Family: Monodontidae (the white whale and the narwhal)
  Family: Stenidae (white dolphins and relatives)
  Family: Delphinidae (dolphins)
  Family: Phocaenidae (porpoises)
  Suborder: Mysticeta (whalebone whales)
  Family: Eschrichtidae (the California gray whale)
  Family: Balaenopteridae (rorquals)
  Family: Balaenidae (right whales)
Order: Carnivora (carnivores)
  Family: Canidae (dogs)
  Family: Ursidae (bears and the giant panda)
  Family: Procyonidae (raccoons and the red panda)
  Family: Mustelidae (weasels, skunks, and otters)
  Family: Viverridae (civets, mongooses, and relatives)
  Family: Hyaenidae (hyenas and the aardwolf)
  Family: Felidae (cats)
Order: Pinnipedia (seals)
  Family: Otariidae (eared seals)
  Family: Odobenidae (the walrus)
  Family: Phocidae (seals)
Order: Tubulidentata (the aardvark)
  Family: Orycteropodidae (the aardvark)
Order: Proboscidea (elephants)
  Family: Elephantidae (elephants)
Order: Hyracoidea (hyraxes)
  Family: Procaviidae, (hyraxes or dassies)
Order: Sirenia (sirenians)
  Family: Dugongidae (dugongs)
  Family: Trichechidae (manatees)
Order: Perissodactyla (odd-toed ungulates)
  Family: Equidae (horses)
  Family: Tapiridae (tapirs)
  Family: Rhinocerotidae (rhinoceroses)
Order: Artiodactyla (even-toed ungulates)
  Suborder: Suiformes (pigs)
  Family: Suidae (Old World pigs)
  Family: Tayassuidae (peccaries)
  Suborder: Tylopoda (camels)
  Family: Camelidae (camels)
  Suborder: Ruminantia (ruminants)
  Family: Tragulidae (chevrotains)
  Family: Cervidae (deer)
  Family: Giraffidae (the giraffe and the okapi)
  Family: Antilocaprinae (the pronghorn)
  Family: Bovidae (bovines, including cattle, antelopes, goats, and sheep)

A human being in this scheme could be classified as follows:

Kingdom Animals: Organisms that use other organisms for food and that often move rapidly

Phylum Chordates: Animals that are partially supported by a rod of cartilage or bone vertebrae and an internal skeleton

Subphylum Vertebrates: Chordates that have vertebrae, such as fish, amphibians, reptiles, birds, mammals

Subclass Mammals: Vertebrates that have hair and suckle their young

Order Primates: Mammals that use sight more than scent, have nails instead of claws on grasping hands and feet, are mostly active in daylight, and have relatively large brains

Suborder Anthropoids: New and Old World monkeys, apes, and hominoids

Superfamily Hominoids: Anthropoids that are tailless, generally large in size, can climb trees, and have relatively flat faces; specifically, the great apes, australopithecines, and human beings

Family Hominids: Classically, hominoids that walk upright, have small canines, and large brains; specifically, the australopithecines and human beings; an alternate scheme based on DNA matching includes the gorilla and chimpanzees in this group

Genus *Homo:* Hominids with especially large brains that make tools and show other signs of culture; specifically, *Homo habilis, Homo erectus,* and *Homo sapiens*

Species *Homo sapiens:* Anatomically modern human beings and possibly the Neanderthals

Subspecies *Homo sapiens sapiens:* This is only necessary if one considers the Neanderthals a subspecies of *Homo Sapiens*

LIFE SCIENCE RECORD HOLDERS

By Size

Largest structure made by living creatures	Great Barrier Reef off coast of Australia	Length: 2030 km (1260 mi) Area: 200,000 km^2 (80,000 sq mi)
Largest living creature	"General Sherman," a sequoia in Sequoia National Park CA	Height: 83.8 m (274.9 ft) Circumference: 34.9 m (114.6 ft)
Largest living animal	Blue whale	Length: 34 m (110 ft) Weight: 150 m tons (136 sht tons)
Largest living land animal	African elephant	Height: 4 m (13 ft) Weight: 13 m tons (13 sht tons)
Longest animal	Bootlace worm, a sea worm	Length: 55 m (180 ft)
Largest land animal ever	*Brachiosaurus*, a dinosaur from 150 million years ago	Weight: 90 m tons (80 sht tons)
Smallest free-living creature	Pleuro-pneumonialike organisms of the genus *Mycoplasma*	Diameter: 0.0001 mm (0.000004 in.)
Smallest creatures of any kind	Viroids, viruslike plant pathogens without coats	Diameter: 0.000000002 mm (0.00000000007 in.)
Largest invertebrate	Giant squid	Length: 17 m (55 ft)
Most teeth in a mammal	Toothed whales	Amount: 260 teeth
Longest trip by a shorebird	Semipalmated sandpiper	Length: 4500 km (2800 mi) in four days
Largest egg ever	Laid by extinct elephant bird of Madagascar	Capacity: 8.9 L (2.35 gal) Weight: 12 kg (27 lb)
Largest nest	Built by bald eagle	Weight: 3000 kg (6700 lb)

Time

Oldest living creature	King Clone, a creosote plant in California desert	Age: 11,700 years
Longest recorded age of an animal	Ocean quahog (clam)	Age: 220 years
Oldest living bird	Wandering albatross	Age: Banded birds recovered after 40 years; estimated at 80 years in the wild
Longest gestation	Alpine black salamander	Period: Up to 38 months

Speed

Fastest flight	Peregrine falcon	While diving, 340 km/hr (212 mph)
Fastest swimmer	Cosmopolitan sailfish	Speed: 110 km/hr (68 mph)
Slowest mammal	Three-toed sloth	Speed 6 km (4 mi) a day on the ground

The Nobel Prize for Physiology or Medicine

Date	Name [Nationality]	Achievement
1991	Erwin Neher [German: 1944-] Bert Sakmann [German: 1942-]	Analysis of ion channels in cells
1990	Joseph E. Murray [US: 1919-] E. Donnall Thomas [US: 1920-]	First kidney transplant (Murray) and first successful bone-marrow transplant (Thomas)
1989	J. Michael Bishop [US: 1936-] Harold E. Varmus [US: 1939-]	Discovery of the cellular origin of cancer-causing genes found in retroviruses
1988	Sir James W. Black [UK: 1924-] Gertrude B. Elion [US: 1918-] George H. Hitchings [US: 1905-]	Development of artificial variations on DNA that block cell replication, the basis for many new drugs
1987	Susumu Tonegawa [US: 1939-]	Discovery of the genetic principle of antibody diversity
1986	Stanley Cohen [US: 1922-] Rita Levi-Montalcini [Italian-US: 1909-]	Discovery of growth factors
1985	Michael S. Brown [US: 1941-] Joseph L. Goldstein [US: 1940-]	Discovery and analysis of cholesterol receptors
1984	Niels K. Jerne [Danish: 1911-] Georges J.F. Köhler [German: 1946-] César Milstein [UK-Argentinian: 1927-]	Pioneering work in immunology (Jerne) and the invention of monoclonal antibodies (Köhler and Milstein)
1983	Barbara McClintock [US: 1902-]	Discovery of mobile genes in chromosomes of corn
1982	John R. Vane [US: 1927-] Sune K. Bergstrom [Swedish: 1916-] Bengt I. Samuelsson [Swedish: 1934-]	Studies on formation and function of prostaglandins, hormonelike substances that combat disease
1981	Roger W. Sperry [US: 1913-] David H. Hubel [US: 1926-] Torsten N. Wiesel [US: 1924-]	Studies on the organization and local functions of brain areas
1980	George D. Snell [US: 1903-] Baruj Benacerraf [US: 1920-] Jean Dausset [French: 1916-]	Discovery of antigens useful for making the immune system accept transplanted organs
1979	Allan McLeod Cormack [US: 1924-] Godfrey N. Hounsfield [English: 1919-]	Invention of computedized axial tomography, or CAT scan
1978	Daniel Nathans [US: 1928-] Hamilton O. Smith [US: 1931-] Werner Arber [Swiss: 1929-]	Use of restrictive enzymes in gene splicing to produce mutants in molecular genetics
1977	Rosalyn S. Yalow [US: 1921-] Roger C.L. Guillemin [US: 1924-] Andrew V. Schally [US: 1926-]	Advances in the synthesis and measurement of hormones
1976	Baruch S. Blumberg [US: 1925-] D. Carleton Gajdusek [US: 1923-]	Identification of and tests for different infectious viruses
1975	David Baltimore [US: 1938-] Howard M. Temin [US: 1934-] Renato Dulbecco [US: 1914-]	Discovery of the interaction between tumor viruses and the genetic material of host cells
1974	Albert Claude [US: 1898-1983] George E. Palade [US: 1912-] C. René de Duve [Belgian: 1917-]	Advances in cell biology, electron microscopy, and structural knowledge of cells

(Continued)

Date	Name [Nationality]	Achievement
1973	Karl von Frisch [German: 1886-1982] Konrad Lorenz [German: 1903-1989] Nikolaas Tinbergen [Dutch: 1907-1988]	Study of individual and social behavior patterns of animal species for survival and natural selection
1972	Gerald M. Edelman [US: 1929-] Rodney R. Porter [English: 1917-1985]	Determination of the chemical structure of antibodies
1971	Earl W. Sutherland, Jr. [US: 1915-1974]	Work on the action of hormones
1970	Julius Axelrod [US: 1912-] Ulf von Euler [Swedish: 1905-1983] Sir Bernard Katz [English: 1911-]	Discoveries in the chemical transmission of nerve impulses
1969	Max Delbrück [US: 1906-1981] Alfred D. Hershey [US: 1908-] Salvador E. Luria [US: 1912-1991]	Discoveries in the workings and reproduction of viruses in human cells
1968	Robert Holley [US: 1922-] Har Gobind Khorana [US: 1922-] Marshall W. Nirenberg [US: 1927-]	Understanding and deciphering the genetic code that determines cell function
1967	Haldan K. Hartline [US: 1903-1983] George Wald [US: 1906-] Ragnar A. Granit [Swedish: 1900-]	Advanced discoveries in the physiology and chemistry of the human eye
1966	Charles B. Huggins [US: 1901-] Francis Peyton Rous [US: 1879-1970]	Research on causes and treatment of cancer
1965	François Jacob [French: 1920-] André Lwoff [French: 1902-] Jacques Monod [French: 1910-1976]	Studies and discoveries on the regulatory activities of human body cells
1964	Konrad Bloch [US: 1912-] Feodor Lynen [German: 1911-1979]	Work on cholesterol and fatty acid metabolism
1963	Sir John C. Eccles [Australian: 1903-] Alan Lloyd Hodgkin [English: 1914-] Andrew F. Huxley [English: 1917-]	Study of mechanism of transmission of neural impulses along a single nerve fiber
1962	Francis H.C. Crick [English: 1916-] James D. Watson [US: 1928-] Maurice Wilkins [English: 1916-]	Determination of the molecular structure of DNA
1961	Georg von Békésy [US: 1899-1972]	Study of auditory mechanisms
1960	Sir Macfarlane Burnet [Australian: 1899-1985] Peter Brian Medawar [English: 1915-]	Study of immunity reactions to tissue transplants
1959	Severo Ochoa [US: 1905-] Arthur Kornberg [US: 1918-]	Artificial production of nucleic acids with enzymes
1958	George Wells Beadle [US: 1903-1989] Edward Lawrie Tatum [US: 1909-1975] Joshua Lederberg [US: 1925-]	Beadle and Tatum for genetic regulation of body chemistry, Lederberg for genetic recombination
1957	Daniel Bovet [Italian: 1907-]	Synthesis of curare
1956	Werner Forssmann [German: 1904-1979] Dickinson Richards [US: 1895-1973] André F. Cournand [US: 1895-1988]	Use of catheter for study of the interior of the heart and circulatory system
1955	Hugo Theorell [Swedish: 1903-1982]	Study of oxidation enzymes
1954	John F. Enders [US: 1897-1985] Thomas H. Weller [US: 1915-] Frederick C. Robbins [US: 1916-]	Discovery of a method for cultivating poliomyelitis virus in tissue culture

Date	Name [Nationality]	Achievement
1953	Fritz A. Lipmann [US: 1899-1986] Hans Adolph Krebs [English: 1900-1981]	Discovery by Lipmann of coenzyme A and by Krebs of citric acid cycle
1952	Selman A. Waksman [US: 1888-1973]	Discovery of streptomycin
1951	Max Theiler [S. African: 1899-1972]	Development of 17-D yellow fever vaccine
1950	Philip S. Hench [US: 1896-1965] Edward C. Kendall [US: 1886-1972] Tadeusz Reichstein [Swiss: 1897-]	Discovery of cortisone and other hormones of the adrenal cortex and their functions
1949	Walter Rudolf Hess [Swiss: 1881-1973] Antonio Egas Moniz [Portuguese: 1874-1955]	Hess for studies of middle brain function, Moniz for prefrontal lobotomy
1948	Paul Müller [Swiss: 1899-1965]	Discovery of effect of DDT on insects
1947	Carl F. Cori [US: 1896-1984] Gerty T. Cori [US: 1896-1957] Bernardo A. Houssay [Argentinian: 1887-1971]	Coris for discovery of catalytic metabolism of starch, Houssay for pituitary study
1946	Hermann J. Muller [US: 1890-1967]	Discovery of X-ray mutation of genes
1945	Sir Alexander Fleming [English: 1881-1955] Sir Howard W. Florey [English: 1898-1968] Ernst Boris Chain [English: 1906-1979]	Discovery of penicillin and research into its value as a weapon against infectious disease
1944	Joseph Erlanger [US: 1874-1965] Herbert Spencer Gasser [US: 1888-1963]	Work on different functions of a single nerve fiber
1943	Henrik Dam [Danish: 1895-1976] Edward A. Doisy [US: 1893-1986]	Dam for discovery and Doisy for synthesis of vitamin K
1942	No award	———————
1941	No award	———————
1940	No award	———————
1939	Gerhard Domagk [German: 1895-1964]	Discovery of first sulfa drug, prontosil (declined award)
1938	Corneille Heymans [Belgian: 1892-1968]	Discoveries in respiratory regulation
1937	Albert Szent-Györgyi [Hungarian: 1893-1986]	Study of biological combustion
1936	Sir Henry Dale [English: 1875-1968] Otto Loewi [Austria: 1873-1961]	Work on chemical transmission of nerve impulses
1935	Hans Spemann [German: 1869-1941]	Discovery of the "organizer effect" in embryonic development
1934	George R. Minot [US: 1885-1950] William P. Murphy [US: 1892-] George H. Whipple [US: 1878-1976]	Discovery and development of liver treatment for anemia
1933	Thomas H. Morgan [US: 1866-1945]	Discovery of chromosomal heredity
1932	Sir Charles Sherrington [English: 1857-1952] Edgar D. Adrian [US: 1889-1977]	Multiple discoveries in the function of neurons
1931	Otto H. Warburg [German: 1883-1970]	Discovery of respiratory enzymes

(Continued)

Date	Name [Nationality]	Achievement
1930	Karl Landsteiner [US: 1868-1943]	Definition of four human blood groups
1929	Christiaan Eijkman [Dutch: 1858-1930] Sir Frederick G. Hopkins [English: 1861-1947]	Eijkman for antineuritic vitamins, Hopkins for growth vitamins
1928	Charles Nicolle [French: 1866-1936]	Research on typhus
1927	Julius Wagner von Jauregg [Austria: 1857-1940]	Fever treatment, with malaria inoculation, of some paralyses
1926	Johannes Fibiger [Danish: 1867-1928]	Discovery of Spiroptera carcinoma
1925	No award	————————
1924	Willem Einthoven [Dutch: 1860-1927]	Invention of electrocardiograph
1923	Sir Frederick Banting [Canadian: 1891-1941] John J.R. MacLeod [English: 1876-1935]	Discovery of insulin
1922	Archibald V. Hill [English: 1886-1977] Otto Meyerhof [German: 1884-1951]	Hill for discovery of muscle heat production, Meyerhoff for oxygen-lactic acid metabolism
1921	No award	————————
1920	Shack August Krogh [Danish: 1874-1949]	Discovery of motor mechanism of blood capillaries
1919	Jules Bordet [Belgian: 1870-1961]	Studies in immunology
1918	No award	————————
1917	No award	————————
1916	No award	————————
1915	No award	————————
1914	Robert Bárány [Austrian: 1876-1936]	Studies of inner ear function and pathology
1913	Charles Robert Richet [French: 1850-1935]	Work on anaphylaxis allergy
1912	Alexis Carrel [French: 1873-1944]	Vascular grafting of blood vessels and organs
1911	Allvar Gullstrand [Swedish: 1862-1930]	Work on dioptrics, refraction of light in the eye
1910	Albrecht Kossel [German: 1853-1927]	Study of cell chemistry
1909	Emil Theodor Kocher [Swiss: 1841-1917]	Work on the thyroid gland
1908	Paul Ehrlich [German: 1854-1915] Elie Metchnikoff [Russian: 1845-1916]	Pioneering research into the mechanics of immunology
1907	Charles L.A. Laveran [French: 1845-1922]	Discovery of the role of protozoa in disease generation
1906	Camillo Golgi [Italian: 1843-1926] Santiago Ramón y Cajal [Spanish: 1852-1934]	Study of structure of nervous system and nerve tissue
1905	Robert Koch [German: 1843-1910]	Tuberculosis research
1904	Ivan P. Pavlov [Russian: 1849-1936]	Study of physiology of digestion
1903	Niels Ryberg Finsen [Danish: 1860-1904]	Light ray treatment of skin disease
1902	Sir Ronald Ross [English: 1857-1932]	Work on malaria infections
1901	Emil von Behring [German: 1854-1917]	Discovery of diphtheria antitoxin

Mathematics

THE STATE OF MATHEMATICS

Mathematics is not necessarily a science in the modern sense of the word, but it is increasingly essential for all the branches of science covered in this book. Furthermore, it is organized sociologically like a science and progresses like one in many ways.

One problem with mathematics is that, even more than for the advanced sciences, such as theoretical physics, people not actually involved with current developments have a difficult time in understanding the subject matter. As a result, popular accounts of mathematical research tend to be limited to advances in number theory and the solution of computational problems—topics that are often easy to explain. Only when newer ideas, such as chaos theory or fractal advances, become widespread do most accounts for the nonspecialist get written and published. *Current Science* is operating under the same constraints as other publications offering reports on higher mathematics, so the discussions here perforce are also limited. In other words, more is going on in higher mathematics than can be explained to nonmathematicians. Even mathematicians may know little about parts of the subject that are not closely related to their own work.

Mathematics and the Computer

It probably seems obvious that the advent of the computer would have a great impact on mathematics. However, this was far from obvious to mathematicians in the first heady days of the computer in the 1940s and 1950s. Nearly all of modern mathematics has little to do with computation and the parts that do involve computation tend to use clever tricks rather than brute force.

It was not until 1976, when three mathematicians announced that they had proved a conjecture that dated from 1850 using a computer, that mathematicians began to take notice. The conjecture was that any map on

COMMON TERMS

Often people say that mathematics is not a science but a language. If so, it is a language that is foreign to the vast majority of people. The following deals with terms that are used in the mathematics section of <u>Current Science</u> that you may have forgotten from high school or college or that may have been passed over without notice or definition during that time.

<u>Algorithm</u> A sequence of steps or a procedure that always solves a problem if followed correctly is called an algorithm. The rules learned in elementary school for adding or multiplying two numbers written with more than a single digit are typical algorithms.

<u>Factor</u> As a noun, a factor is any one of the numbers that make up a product; for example, 3 is a factor of 12 because 3(4) = 12. As a verb, to factor is to find the counting numbers that can form a product; for example, one pair of factors for 60 is 15(4). To factor completely is to find the prime factors. For 60, they are 3, 5, 2, and a repeat of 2.

<u>Factorial</u> The product of all the consecutive counting numbers (1, 2, 3, ...) up to a given number $\underline{n}$ is written as $\underline{n}!$ and read as $\underline{n}$ factorial. The order of the numbers used as factors does not matter, but it is convenient notationally to list $\underline{n}$ first, so the technical definition is usually given as $\underline{n}! = \underline{n}(\underline{n} - 1)(\underline{n} - 2)(\underline{n} - 3) ... 1$. Thus, $1! = 1$, $2! = 2(1) = 2$; $3! = 3(2)(1) = 6$; $4! = 4(3)(2)(1) = 24$; and so forth. Notice that factorials increase rapidly in size as $\underline{n}$ gets larger: $6! = 720$; $7! = 5040$; and $8! = 40,320$. By the time you get to 70! you are dealing with a number that has over a hundred digits.

<u>Number</u> There are various kinds of numbers that are used in mathematics. These types can be organized in an ever-ascending pattern for which each successive level includes all of the numbers in the levels that precede it.

<u>Counting or natural numbers</u> 1, 2, 3, ... (where "..." means that the pattern keeps repeating in the same way)

<u>Whole numbers</u> The counting numbers and 0, or 0, 1, 2, 3, ...

<u>Integers</u> The whole numbers and their negatives:
$$... -3, -2, -1, 0, 1, 2, 3, ...$$

<u>Rational numbers</u> The ratio of one integer to another; equivalently, the number found by dividing one integer by another (division by 0 excluded). These include all of what we think of as positive or negative fractions or as finite decimals.

<u>Real numbers</u> Numbers that can be represented by infinite decimals, which include the <u>irrational numbers</u> such as $\sqrt{2}$ and π as well as the rational numbers.

<u>Complex numbers</u> The sums of real numbers and numbers formed as the product of a real number and $\sqrt{-1}$, with $\sqrt{-1}$ usually written as $\underline{i}$. Any mathematical entity more complicated than this (such as a vector or a matrix that does not reduce to a complex number) is not considered a

number. If the real number part of a complex number is 0, the complex number is called <u>imaginary.</u>

<u>Order of magnitude</u> This phrase is often used as if its meaning were self-evident, but it is seldom defined formally; it refers to the powers of 10 or nearly equivalently in our decimal system of notation to the number of digits involved. For example, a bowling score is often an order of magnitude greater than a football score (that is, while 17 points by one team is often a football score, a decent bowler might score 170 points), but often a baseball score is an order of magnitude less than a football score (for example, the football score might be 42 while the baseball score is 5). Digits alone may not tell the tale. A score of 9 is not an order of magnitude less than one of 10. For more striking differences, you refer to the number of orders of magnitude. The bowling score of 170 is roughly two orders of magnitude greater than the baseball score of 5; although a Scrabble score of 400 would be more accurately described as two orders greater than a baseball score of 5. Order of magnitude is usually used as a rough comparison, not as an exact one.

<u>Polygon</u> In a plane, any geometric figure with straight sides that can be traced completely once with a pencil without crossing any of the sides and still return to the starting point. Polygons with small numbers of sides include triangles, squares, rectangles, kites, stars, and so forth. Letters such as B and D are not polygons because not all the sides are straight; A and E are not polygons because the pencil cannot return to the starting point without retracing. The common method for drawing a six-pointed star does not produce a polygon because lines intersect, but the outline of the star with the inner lines erased is a polygon.

<u>Polynomial</u> A polynomial is any finite sum or difference of the products of powers of a variable with numbers, such as $3x^5 + 2x^4 + x^2 - 9$. The important thing to remember is what is not involved: Entities ruled out include division in an expression involving the variable; radicals, such as square or cube roots; use of a variable as an exponent; factorials; and—in general—any mathematical operation more complicated than addition, subtraction, multiplication, and raising to a whole-number power.

<u>Prime number</u> A counting number whose only factors are itself and 1 (except for 1) is prime. The first few prime numbers are 2, 3, 5, 7, 11, 13, 17, 19, 23, 29, 31, 37, 41, 43, and 47. The numbers 39 and 51 are not prime, for example, because 39 = 13(3) and 51 = 17(3). Euclid proved that there is no greatest prime; in effect, this means that there are as many prime numbers as there are counting numbers.

---

a plane could be colored with four colors in such a way that no two regions with the same color would share a common boundary.

One problem with verifying computer proofs is that the only way to validate a proof is to rerun the program or program a different computer to check it. This is because computer proofs of the type used thus far in mathematics consist of checking all of a vast number of possibilities to make sure that the theorem is true for each and every one. Often the number of steps is so great that if all the working mathematicians in the world did nothing else for the rest of their lives but check the computer's "math," only a small fraction could be checked over hundreds of years. Thus, a few holdouts refuse to accept computer results as proofs. Despite this limitation, computer results have become more and more a part of mathematics and probably will continue to be so, especially as increasing numbers of mathematicians have access to very powerful computers.

At the beginning of the computer era, however, the impact was in the other direction. The people who early on figured out how computers could work and what their powers and limitations would be were mathematicians, most of them little known outside the mathematics community (some of their less-than-famous names are Emil L. Post [US: 1897-1954], Alonzo Church, Stephen Cole Kleene, and Abraham Robinson). A few, for one reason or another, however, did achieve fame, notably John von Neumann [Hungarian-US; 1903-1957], Claude Shannon [US: 1916-], and Alan Turing [UK: 1912-1954]. Since those early days, mathematicians have continued to develop the basic ideas behind computation and information processing. Some new findings are discussed later in the mathematics section.

Other recent contributions of mathematics to computer science follow.

Fuzzy logic was developed by Lofti Zadeh in 1965 at the University of California at Berkeley, as he considered sets that might not be rigidly defined. A classic definition of a set in mathematics is a collection for which, when presented with any entity whatsoever, one can tell whether the entity belongs to the collection. Using that definition, the counting numbers form a set but handsome men do not, because it is not certain whether any given man would correctly be called handsome. Zadeh found a precise mathematical way to describe the degree of handsomeness and other fuzzy characteristics so that they could be used to define what he named "fuzzy sets." Thinking with such fuzzy sets then followed the rules of fuzzy logic.

Like many great ideas, Zadeh's took a little time to work its way into practical use, but by 1985 the first fuzzy computer chips were being made, including one developed at AT&T Bell Laboratories (Bell Labs) in Murray Hill NJ by Masaki Togai and Hiroyuki Watanabe. In 1989 Japan's Ministry of Trade and Industry was sufficiently impressed to commit $36 million to a Laboratory for International Fuzzy Research. By 1990 the first consumer products using fuzzy logic began to appear: a washing machine that can tell how dirty clothes are, a vacuum cleaner that knows how thick the carpet is, and a car with a fuzzy transmission.

Interactive proof of problems once thought to require brute force is a new result for computers that could lift some of the unease mathematicians have about many computer proofs. Mathematicians have been able to classify problems into types in several ways; often they can show that all problems of a given kind can be reduced to a single type

For example, almost all simple probability problems, such as tossing dice or dealing cards, can be changed into problems involving drawing colored marbles from an urn (for example, the six faces of a die can be modeled as six different colors of marble; throwing two dice is the same as randomly drawing two of the marbles from the urn, replacing the first marble after the first draw). In another example, a very large class of computational problems has been shown to be equivalent to finding compatible matches between members of a group of women and a different sized group of men; it is known as the matchmaker's problem.

At the end of 1989 a group of mathematicians in the United States and Israel, working over computer networks, showed that all the problems reducible to the matchmaker's problem can be solved by a much simpler method in which the computer asks whether or not a relatively small set of apparent solutions found by some method actually works. If it does, then the basic method will always work. The mathematicians involved were Richard Lipton of Princeton University; Noam Nisan at MIT; Carsten Lund, Lance Fortnow, and Howard Karloff at the University of Chicago; and Adi Shamir of the Weismann Institute in Israel. Although a superfast computer is still needed to use the new method, the existence of interactive proofs means that some difficult solutions can be checked. Interactive proofs also point to the possibility that much more useful methods of solution can be developed in the future, a likelihood that did not seem to be even a remote possibility before the scope of interactive proofs was expanded in 1989.

Fusion trees are another way to get computers to solve problems faster. Many problems involve comparing numbers to find out which one is the least or the greatest, or whether the numbers are equal. Computers are designed to do this in a pairwise fashion; that is, two numbers at a time. In 1991 Michael L. Freedman of the University of California, San Diego, and Dan E. Willard of the State University of New York (SUNY) at Albany, figured out a way for computers to do comparisons for whole groups of numbers at a time; they called these groups fusion trees. The basic idea is simple: The computer does not need to use all the data in a number to make comparisons. People use the same method when they compare such numbers as 23 and 2308 or such numbers as 4085 and 5085: In comparing 23 and 2308, one need not think of the digits involved, only how many there are; while in comparing 4085 and 5085 one need only note the first digit. Getting a computer to behave this way and to do so for large groups of numbers at a time, however, was quite a feat. Applications are still in the future, but the most obvious would be improvements in handling large databases. Fusion trees may also be useful in solving the Traveling Salesman Problem (see following article).

Polygon decomposing is not directly related to computer science, but many problems are solved by having a computer find all the triangles that can be put together to make a given polygon, a process known as decomposing the polygon into triangles or, more simply, as decomposing the polygon. The problem of finding the triangles so that they cover the polygon and intersect with each other is easy for people dealing with small polygons. It is difficult for computers and especially difficult when large polygons are involved. In 1991, after many years of development, mathematicians were able to show that ways exist for computers to solve the problem in a reasonable number of steps, although the actual methods so far are not especially practical.

Timetable of Polygon Decomposing

BC

c 400   Greek mathematicians use decomposition of polygons into triangles as a method of proof.

c 300   Euclid uses decomposition of triangles to prove and to organize propositions concerning area in the *Elements.*

CE

1911   N.J. Lennes shows that any polygon can be decomposed into a finite number of triangles.

1978   Michael Garey and David Johnson at Bell Labs, Franco Preparata at the University of Illinois, and Robert Tarjan at Stanford University find a way to decompose a polygon into triangles that uses a number of steps that is about the product of the number of vertices and the logarithm of the number of vertices. This is a big improvement over previous algorithms that required a number of steps on the order of the square of the number of vertices. Note that the logarithm of a number expressed in decimal form is roughly one less than the number of digits in the number, so the logarithm of 1289 is roughly 3. Instead of about 1000000(1000000) = 1 trillion steps to decompose a million-vertex polygon the old way, the new method uses about 6 million steps.

1988   Tarjan and Christopher Van Wyk of Bell Labs find an algorithm for reducing the number of steps in decomposing a polygon to the order of the product of the number of vertices by the logarithm of the logarithm of the number of vertices. This reduces the number of steps for decomposing a million-vertex polygon to a little more than a million steps and the number for a trillion-vertex polygon from about 12 trillion steps by the 1978 algorithm to about 2 trillion.

1991   Bernard Chazelle of Princeton University finds an algorithm that reduces the number of steps in decomposing a polygon to about the number of vertices, although the algorithm is too difficult for computers to use to make it worthwhile except on exceptionally large polygons.

Conjecture Progress

Many mathematicians believe to some degree that progress in mathematics comes largely through intuition. The best evidence that this belief is true is the large number of conjectures in mathematics that have eventually been proved, ranging from vague ideas of the early Greek mathematicians to modern guesses in the higher realms of mathematical complexity. The recent probable proof of a conjecture from 1611, for example, is dealt with in greater detail in the following section. Here is yet another account of a proof of a conjecture.

It has been known since early times that the sum of the lengths of the line segments connecting a point in the interior of a triangle can be shorter than the sum of two of the sides of the triangle. A simple illustration, if you remember some trigonometry, uses an equilateral triangle with sides each $2\sqrt{3}$ units long and the interior point that is equidistant from the vertices of the triangle. In that case, the sum of any pair of sides is $4\sqrt{3}$ units. The interior point is 2 units from each of the vertices. Thus, the sum of the three distances from a vertex to the interior point is 6. Since $\sqrt{3}$ is about 1.7, the sum of two sides, or $4\sqrt{3}$, is about 6.8, or about 0.8 unit longer than the sum of the three distances to the interior point. More exactly, the ratio of the shorter path to the longer is $\sqrt{3}/2$, or about 0.866. Put another way, the saving would be slightly more than 13 percent.

This difference in length can be significant when you want to connect three points by the shortest possible path that does not need to return to the starting place. For example, telephone lines between cities forming an equilateral triangle 100 km on a side would be 200 km long by going along two of the sides of the triangle but only about 173 km long if all three cities were connected to a central point equidistant from the cities. This is also the strategy used when airlines route everyone through a hub city, although more than three cities are generally involved.

In 1968 Henry O. Pollak and Edgar N. Gilbert, both of Bell Labs, conjectured that the saving of about 13 percent that could be computed for an equilateral triangle is the largest amount of saving possible by connecting points in this way. Proving this conjecture for connecting an arbitrarily large number of points by adding intermediates was found to be difficult. Using various means, mathematicians showed that the Pollak-Gilbert conjecture is true for four, five, and six points. Finally, in 1990, Frank K. Hwang of Bell Labs and Dingzhu Du of Academia Sinica in Beijing, China, were able to prove the Pollak-Gilbert conjecture for an arbitrary number of points. Essentially they showed that placing an added point inside any triangle of three points produces at most a saving of about 13 percent; there is no better strategy.

Not all conjectures work out, however. Sometime in the seventeenth century amateur mathematician Pierre de Fermat [French: 1601-1665] conjectured that all numbers formed by raising two to a power of two and adding one to the result would be prime. The first such Fermat number is simply $2^2 + 1 = 1$; the second is $2^4 + 1$ or 17; the third, $2^8 + 1 = 257$; the fourth, $2^{16} + 1 = 65,537$; and so forth. As you can see, the Fermat numbers increase rapidly in size, with concomitant difficulty in showing that they are prime or that they can be factored. As it happens, Fermat was wrong and the fifth Fermat number, $2^{32} + 1$, was factored by Leonhard Euler [Swiss: 1703-1783] in 1732. Subsequently, no one has ever found a higher Fermat number that is prime. It was eventually learned that the sixth Fermat number, seventh (factored in 1970), and eighth (factored in 1980) are not prime either.

On Jun 15 1990 Arjen Lenstra of Belicore in Morristown NJ and Mark Manasse of Digital Equipment Corporation's Systems Research Center in Palo Alto CA succeeded in factoring the ninth Fermat number, $2^{512} + 1$, a number of 155 digits, into three primes. This makes the ninth Fermat number the largest difficult number factored so far by any means. Next in line for Fermat numbers is $2^{1024} + 1$, which has 309 digits.

Factoring some numbers named by large numerals is not difficult. For example, 2^{1024} is also 309 digits long, but it is an even number that easily factors into $2(2^{1023})$ along with more than a thousand other obvious possibilities.

Lenstra and Manasse had the help of a new approach to factoring devised in the 1970s by British mathematician John Pollard, a Connection Machine supercomputer, and a large number of collaborators who did bits and pieces of the algorithm on their own computers around the world. The algorithm works for numbers known to be of the form $an + b$, which of course includes the class of Fermat numbers. Even so, prospects for factoring the tenth Fermat number with the algorithm are small because it would take a vast amount of computer time to accomplish.

(Periodical References and Additional Reading: *Science* 6-1-90, p 1079; *Science* 6-29-90, p 1608; *Science News* 12-22&29-90, p 389; *Discover* 1-91, pp 60 & 67; *Science News* 6-29-91, p 406; *Science* 7-19-91, p 270)

Timetable of Mathematics to 1990

20,000 BC	People in the Near East begin using notches to record numbers
3500 BC	Egyptians develop a numeral system that can record very large numbers, with different symbols for ones, tens, hundreds, and so forth
2400 BC	In Mesopotamia (now Iraq, Syria, and Turkey), a numeration system based on place value (similar to the Hindu-Arabic system) is introduced
2000 BC	Mesopotamian mathematicians learn to solve quadratic equations (equations for which the highest power of the variable is 2, called degree 2)
1900 BC	Mesopotamian mathematicians discover what we now call the Pythagorean theorem: The sum of the squares of the legs of a right triangle equals the square of the hypotenuse

876 BC	The first known use of a symbol for zero occurs in India
470 BC	Greek mathematician Hippasus discovers the dodecahedron, a regular solid with twelve faces
450 BC	Greek mathematician Hippacos of Metapontum shows that some measures, such as $\sqrt{2}$, cannot be described in ratios of units; today we call these measures irrational numbers
300 BC	The *Elements* of Euclid [Greek: c 325-c 270 BC] shows that virtually all known aspects of mathematics can be proved from a short list of assumptions
260 BC	In Central America, the Maya develop a numeration system based on place value and zero
230 BC	Hellenic mathematician Apollonius of Perga [c 262-c 190 BC] writes *Conics,* an analysis of such curves as the parabola, ellipse, and hyperbola
1100	Persian poet, mathematician, and astronomer Omar Khayyám [1048-1131] develops geometric methods for solving cubic equations (polynomial equations of degree 3)
1321	Levi ben Gerson [French: 1288-1344] is the first to use mathematical induction in a proof
1515	Scipione del Ferro [Italian: 1465-1526] discovers an algebraic method for solving one form of cubic equations
1536	Niccolò Tartaglia [Italian: 1499-1557] announces that he can solve most cubic equations
1545	The *Ars Magna* of Girolamo Cardano [Italian: 1501-1576] contains the complete solution by Luigi Ferrari [Italian: 1522-1565] of the quartic (the polynomial equation of degree 4) as well as a complete solution of the cubic based on Tartaglia's work
1572	Rafael Bombelli [Italian: 1526-c 1572] uses complex numbers (numbers that include $\sqrt{-1}$) to solve equations
1614	John Napier [Scottish: 1550-1617] describes logarithms
1637	René Descartes [French: 1596-1650] publishes the first account of analytical geometry (also discovered by Pierre de Fermat [French: 1601-1665]), the link between algebra and geometry
1639	Gerard Desargues [French: 1593-1662] develops projective geometry
1654	Blaise Pascal [French: 1623-1662] and Pierre de Fermat develop the basic laws of probability
1666	Isaac Newton [English: 1642-1727] describes his invention of the calculus, but does not have it published at this time
1684	Gottfried Wilhelm Leibniz [German: 1646-1716] publishes the first account of his independent discovery of the calculus, the first description to reach print
1742	Christian Goldbach [German: 1690-1764] conjectures that every even number greater than two is the sum of two primes
1763	Gaspard Monge [French: 1746-1818] invents descriptive geometry, the mathematical techniques that are the basis of mechanical drawing and most architects' plans

1799	Karl Friedrich Gauss [German: 1777-1855] proves the fundamental theorem of algebra, which is that every polynomial equation has a solution
	Paolo Ruffini [Italian: 1765-1822] proves that not all polynomial equations of the fifth degree can be solved by algebraic methods
1801	Gauss publishes *Disquistiones arithmeticae,* greatly extending number theory
1822	Jean-Victor Poncelet [French: 1788-1867] develops projective geometry
1826	Nikolai Ivanovich Lobachevski [Russian: 1793-1856] gives the first public address concerning non-Euclidean geometry
1837	Pierre Wantzel [French: 1814-1848] proves that an angle cannot be trisected with a compass and straightedge
1854	Bernhard Riemann [German: 1826-1866] shows that several non-Euclidean geometries are possible (including the one that Albert Einstein later demonstrated to be the most likely geometry of the universe)
1873	Charles Hermite [French: 1822-1901] shows that the number *e* is not the solution to any algebraic equation with rational coefficients; *e* therefore is a transcendental number
1877	Georg Cantor [German: 1845-1918] shows that the number of points in a line segment is the same as the number in the interior of a square
1881	Willard Gibbs [US: 1839-1903] introduces vector analysis
1882	Ferdinand Lindemann [German: 1852-1939] proves that π is transcendental, implying that the circle cannot be squared with straightedge and compass
1892	Cantor shows that there are at least two types of infinities—specifically that the infinity of real numbers (including all infinite decimals) is bigger than the infinity of counting numbers (1, 2, 3, ...)
1900	David Hilbert [German: 1862-1943] proposes his famous list of twenty-three unsolved problems
1931	Kurt Gödel [Austrian-US: 1906-1978] shows that any formal system strong enough to include the laws of arithmetic is either incomplete or inconsistent: *incomplete* if not all true theorems can be proved or *inconsistent* if two contradictory theorems can be proved
1936	Independently, Alan Turing [English: 1912-1954] and Alonzo Church [US: 1903-] discover that there is no single method for proving whether a statement in mathematics is true or false
1949	Claude Shannon [US: 1916-] publishes his work on information theory, a general approach to handling communications
1976	In the first major computer-assisted proof, it is shown that any map can be colored with four colors in such a way that no two regions of the same color share a common border
1980	The classification of all finite simple groups, started in 1830, is completed, perhaps the longest proof in the history of mathematics
1989	In Apr Hungarian Miklós Laczkovich of the Eötvös Loránd University in Budapest shows that it is possible to cut a circular region into a finite number of pieces and rearrange the pieces to form a square; the finite number, however, is very large—about one followed by fifty zeros!

PROBLEM SOLVING

PROGRESS ON THE TRAVELING SALESMAN PROBLEM

A sales representative regularly travels from point to point calling on prospective clients, finally returning to home base. The best route for such a business is a minimum of some kind, usually the least cost or time. In many cases, locating such a minimum reduces to finding the shortest distance; however, if the travel is from city to city by air, minimum distance may not equal least cost or time. The problem is further complicated when travel is asymmetric; for example, when travel from city A to city B has one cost, while travel from B to A has a different cost.

Practical Travelers

The above is the essential situation that goes by the old sexist name of Traveling Salesman Problem (TSP). Not only does the problem apply to saleswomen, but it also has a host of analogues in which the traveler is a machine part or even a process. Many of these versions of TSP are extremely practical.

Consider the Traveling Laser Problem, for which a laser must drill 1.2 million holes in a particular pattern on a very large-scale integrated chip for use in a computer, a real situation in the manufacture of these chips. In practice, the laser stays still and the chip is moved about to position each hole the laser drills.

One direct and simple way to program the computer to position the chip would be to have the first hole drilled at random and then to always move to the nearest place another hole is needed (with some simple method for breaking ties). Normally, however, this procedure will not produce a chip in anything like the least possible time. Indeed, this procedure can produce a result that is what mathematicians call "arbitrarily bad," meaning that a bad choice of a starting hole and some early bad tie-breaking can result in the longest possible route instead of the shortest.

On the other hand, a fairly good approximate solution to the Traveling Laser Problem was developed by Martin Grötschel and coworkers at the Technische Universität and Konrad Zuse Sentrum in Berlin. Grötschel reported on his work and related scheduling problems at the Jul 1991 International Conference in Industrial and Applied Mathematics in Washington DC. His solution to the Traveling Laser Problem cuts manufacturing time in half.

A "fairly good approximate solution" is one that is known to get close rapidly to an exact solution after the approximate solution has been calculated over a reasonable number of steps. The exact solution to the Traveling Laser Problem, however, remains unknown. But if the approximate solution is good enough, then for practical purposes it is not necessary to know the exact one. The difference between fairly good approximate and exact can be astonishing. When Manfred Padberg of New York University and Giovanni Rinaldi of the Institute for Systems Analysis in Rome solved TSP for 532 cities, they used 7 percent of their computer time to get within 1.7 percent of the best possible solution; getting the best solution took the remaining 93 percent of the computer time.

A general method of solving all versions of TSP would be an enormous benefit in many industries. Finding shortest routes for long-distance calls is an important application, for instance. A less obvious example is setting up the best arrangement of schedules in chemical manufacturing. The order of operations on an assembly line is another problem that can be viewed as a Traveling Salesman Problem.

Theoretical Travelers

TSP also has considerable mathematical interest. It is the archetypical problem of a type known as NP complete, short for nondeterministic polynomial-time complete. (Definitions of necessary common mathematical terms can be found in "The State of Mathematics," p 513.) A problem is NP complete if the number of steps needed to solve any possible example of the type is greater than any polynomial expression that describes the number of steps needed to solve the problem. Multiplying two integers using arithmetic's common multiplication algorithm is not NP complete. If you multiply an n-digit number by an m-digit number, the total number of operations (both multiplications and additions) is always less than or equal to $2nm$, a simple polynomial. Similarly, multiplication of two n by n matrices of integers always requires at most $2n^3 - n^2$ operations.

The numbers represented by the polynomials for these problems, even when n or m are large, are not vastly greater than n or m themselves. For example, if n is 100, $2n^3 - n^2$ is only 1,090,000, four orders of magnitude greater than 100. On the other hand, the number of operations that occur in TSP if one were to check every possible tour of n cities is $n!$, which is not a polynomial. 100! is 100 x 99 x 98 x ... x 1, which is a staggering product on the order of 1 with 200 zeros after it, or 10^{200}. This is 198 orders of magnitude greater than 100.

A mere 1,090,000 operations is easy work for a computer. Thus, a computer can solve a problem in which the algorithm is not NP complete in a reasonable time. But even at a billion operations per second, it would take 10^{191} seconds, or about 10^{182} years, to process an algorithm that required 10^{200} operations. So the trick for solving an NP complete problem, such as TSP, is to find an algorithm that takes many fewer steps. Furthermore, a theorem has been proved that says that if TSP can be solved in less than the time that it would take to compute the value of the polynomial formed when a reasonable measure of problem size (such as number of cities, holes, vertices, and so forth) is plugged into the polynomial, then ways exist to solve all NP complete problems by such efficient methods.

In 1991 Donald L. Miller of E.I. du Pont de Nemours and Company in Wilmington DE and Joseph F. Pekny of Purdue University in Lafayette IN described their method for solving TSP exactly for cases in which the direction of travel makes a difference; that is, when going from A to B is counted as different from going from B to A. This is known as the Asymmetric Traveling Salesman Problem (ATSP). The Miller-Pekny method for solving ASTP is also useful for some classes of TSP, although often not as good as methods known earlier.

Timetable of the Traveling Salesman Problem (TSP)

1954	TSP is solved for 49 cities
1963	The first branch and bound algorithms for pruning the number of possibilities to be tested in solving TSP are introduced
1971	Stephen Cook of the University of Toronto discovers the class of nondeterministic polynomial, or NP complete, problems
1972	Richard Karp of the University of California at Berkeley calculates how many problems fall into the NP complete class
1973	S. Lin and B.W. Kernighan develop a general method for getting approximate solutions to TSP
1980	TSP is solved for 318 cities by Manfred Padberg of New York University and Giovanni Rinaldi of the Institute for Systems Analysis in Rome; it takes 3.4 hours of computer time
1985	*The Traveling Salesman Problem: A Guided Tour of Combinatorial Organization* by E.J. Lawler, J.K. Lenstra, A.H.G. Rinooy Kan, and D.B. Shmoys summarizes known instances of TSP
1986	TSP is solved by Padberg and Rinaldi for connecting the 532 cities in the United States that have AT&T central telephone offices, taking 6 hours of computer time
1988	Martin Grötschel, then of Augsburg University in Germany, and Padberg solve TSP for 2392 cities, the current record; it takes only 2.6 hours of computer time
1990	Gerhard Reinelt of Augsburg University develops TSPLIB (the Traveling Salesman Problem Library), a computer list of outstanding problems available on electronic mail

(Periodical References and Additional Reading: *Science* 2-15-91, p 754; *New York Times* 3-12-91, p C1; *New York Times* 3-29-91; *New York Times* 5-24-91; *Science* 7-26-91, p 384)

SPHERE-PACKING PROBLEM SOLVED

The year 1991 saw the probable proof of a long-standing geometric conjecture of considerable importance. Early in the year Wu-Yi Hsiang of the University of California at Berkeley began to send out preliminary versions of his 100-page proof demonstrating the correctness of Johannes Kepler's 1611 conjecture that the closest possible packing of three-dimensional spheres is by what we now call the face-centered cubic lattice method. Packing spheres is roughly the same as stacking them. The face-centered cubic method of sphere packing is the way a neat grocer stacks oranges in pyramids. Official publication and scrutiny of the final version of the proof by Wu-Yi's colleagues will not be complete until sometime in 1992.

Higher-Dimensional Sphere Packing

There has been a lot of interest in sphere packing in the past few years, partly because the problem in higher dimensions is an analogue of finding the best locations for stations in a communications system. Another reason for recent work on sphere packing has been the inspiration of a major book, published in 1988, on the subject by John H. Conway of Princeton University and Neil J.A. Sloane of Bell Labs.

Reading Conway and Sloane's *Sphere Packings, Lattices and Groups* was the immediate inspiration for Noam D. Elkies of Harvard University to apply his specialty, elliptic curves, to the problem. Independently, Tetsuji Shioda of Rikkyo University in Tokyo ran into the connection between sphere packing and elliptic curves while working on another problem altogether. Both soon found that solutions to equations for elliptic curves give very good sphere packings in all dimensions up to 1024 (beyond that there are known packings that are better than the ones obtained from elliptic curves). The elliptic curve method revealed specific ways of sphere packing in dimensions up to twenty-four that had all been previously laboriously worked out by hand. This was a totally surprising result, since there was no reason to expect that elliptic curves would do this well, although Elkies has said the connection is "natural enough...once you have the two problems in mind."

Very dense sphere packings, although not necessarily the best, have been known for dimensions greater than three for some years. In 1970 John Leech and Sloane found a packing in ten dimensions in which each sphere is touched by 500 other spheres. The most remarkable is Leech's lattice in 24-dimensional space in which each sphere touches 196,560 other spheres, which is the closest possible packing in twenty-four dimensions.

Background

Experience tells us that for any kind of object there is more than one way to fit several into a box. A jumble usually results in a lot of empty space, but a careful packer can arrange the objects so that the empty space seems to be at a minimum. The trick is proving that the new arrangement actually is the best way to pack the objects. This is true even for the comparatively simple situation in which all of the objects are identical spheres. Part of the difficulty is handling any volume-related problem in three dimensions.

Dimension It helps to have a rudimentary idea of the meaning of dimension. The most common mathematical idea of what dimension means for a space is based on considering the number of numbers needed to locate a point in the space. Thus, a line or a curve can be considered to be a 1-dimensional space. A plane or a surface, such as the surface of Earth, is a 2-dimensional space. You can tell that Earth's surface is 2-dimensional because you need only two numbers—latitude and longitude—to locate any point. The conventional concept of the space we live in is the ordinary 3-dimensional space that needs three numbers. A more advanced concept, using relativity theory, uses time as a fourth dimension for the space we live in. In this view, any construct that uses more numbers, say eleven, to locate a point, is 11-dimensional space. Picture a situation in which you want to describe an ordinary location on Earth, the time, the temperature, the wind speed, the relative humidity, and the percentage of cloud cover. That is seven numbers in all. A computer modeling the weather that assigned seven such numbers to each point would be working in an 7-dimensional space.

Mathematicians have learned the hard way that the number of dimensions involved affects geometry in almost unpredictable ways. In the 1980s a series of mathematicians established that an uncountable infinity of 4-dimensional spaces behave very badly, for example. Except for situations in which the centers of spheres are all spaced regularly, no one knows the best sphere packing for four dimensions. The only spaces in which the exact solution is known are the trivial result in one dimension, two dimensions (see below), eight dimensions, and twenty-four. In three dimensions, many truly strange results and difficult problems have been connected in some way with volume, which is a part of the sphere-packing problem.

Spheres and circles A sphere as used in a packing problem is really a spherical region—all points in three dimensions that are within a set radius of a given point. The equivalent in two dimensions is a circular region, similar to a coin. In two dimensions the analogous problem to sphere packing is easily solved, as can be demonstrated with seven or more identical coins. When you push the coins together on a table, they naturally arrange themselves into a figure that has each coin touching six others. It has been easy for mathematicians to prove that this arrangement covers the plane with the most coins and leaves the least empty space

(Continued)

between them. Approximately 9 percent of the area is unoccupied by circular regions.

For three dimensions Isaac Newton conjectured in 1694 that the greatest number of spheres that can touch a given sphere is twelve. This conjecture was proved by R. Hoppe in 1874, but unlike the problem for coins, this packing leaves a very large gap. You can <u>almost</u> fit in a thirteenth sphere.

If you stack identical balls, you can demonstrate that there are various ways to make the stack. The first layer on a flat surface is arranged the same way as the close packing for coins, since the balls touch each other at their equators. This leaves a series of hollows into which the second layer of balls can go, but it is easy to discover that the hollows are too close to put a ball into each one. The best you can do is to put a ball in every other hollow. Now make the third layer so that each ball is directly over the still empty hollows of the first layer; and the fourth layer above the empty hollows of the second layer; and so forth. This procedure results in a packing that has a face-centered cubic lattice as the centers of the spheres. The empty space for such a packing is still about 26 percent of the total space, but you cannot do any better.

<u>A face-centered cubic lattice</u> The technical minded may wonder what the expression "face-centered cubic lattice" means, which is described in the remainder of this otherwise skippable paragraph. A lattice is a set of points that in two dimensions has been compared to the trees in an infinite orchard. Wherever you stand in such an orchard, you can observe rows of trees that are aligned. If all the rows meet at right angles and are the same distance apart in each direction along those angles, the lattice points form a square. If you make a lattice of this type in three dimensions, the lattice points are the vertices of cubes, hence a "cubic lattice." This is the most familiar method of identifying points in three dimensions. Each point can be identified by three numbers called coordinates that describe how far it is from three mutually perpendicular lines. If the distance between any two points is the unit used, all three of the coordinates will be integers. These integers can be used to identify two sublattices. When the sum of the integers is even, for example (0, 0, 0), (1, 1, 0), (2, 4, 8), and so forth, the sublattice is face centered. When the sum is odd, as in (1, 0, 0), (1, 1, 1), or (3, 6, 8), then the sublattice is body centered. These names come from a different lattice in which all of the coordinates are even, such as (0, 0, 0), (0, 0, 2), or (2, 4, 8). These points also produce a cubic lattice. In such a system, the center of a face of one of the cubes has two odd coordinates, and therefore the sum is even; but the center of the cube (or "body") has three odd coordinates and has a sum that is odd.

<u>Elliptic curves</u> An ordinary curve in two dimensions is the set of all solutions to an equation in x and y. For example, you learn in elementary algebra that the curve that is formed as solutions to an equation such as $2x + 3y = 4$, or indeed any equation of the form $ax + by = c$, is a straight line, considered a form of curve in mathematics. Square one of the variables x or y and the resulting curve, say $2x^2 + 3y = 4$, is a parabola. Ellip-

tic curves are not, as one might expect, ellipses. A typical ellipse is of the form $\underline{a}x^2 + \underline{b}y^2 + \underline{c}x + \underline{d}y = \underline{e}$, where $\underline{a}$ and $\underline{b}$ have the same sign. An elliptic curve, on the other hand, is of the form $y^2 = x^3 + \underline{A}x + \underline{B}$. Here $\underline{A}$ and $\underline{B}$ are capitalized to indicate that they could be algebraic expressions as well as numbers.

(Periodical References and Additional Reading: *Science News* 5-19-90, p 316; *Science* 3-1-91, p 1028)

SHUFFLING ENOUGH

Along with progress in the Traveling Salesman Problem and the apparent proof of the sphere-packing conjecture, solution of a third problem of an entirely different type received widespread notice in the early 1990s. Until 1990 no one knew how many times you have to shuffle a deck of cards to randomize it properly.

The first part of the problem is to define what you mean by a random arrangement of cards. Although the concept of randomness seems intuitively obvious, mathematicians have had a difficult time in producing truly random numbers or even an unambiguous way to recognize a random sequence of numbers.

The basic idea, of course, is that all random outcomes should be equally likely. For a deck of 52 cards, there are 52! or about 10^{68} possible outcomes, rather more than the number of seconds that have elapsed since the universe began (the universe is about 10^{16} seconds old according to a commonly used estimate). Starting by examining all of these possible outcomes proved a difficult task that had previously baffled mathematicians.

Persi Diaconis of Harvard University and David Bayer of Columbia University sidestepped this problem at first. They developed their own measure for a well-shuffled deck that corresponds to intuitive ideas of randomness and also makes mathematical sense. This definition, based on what they call rising sequences, could then be used as one way to determine how many shuffles are needed to achieve randomness.

Rising Sequences and a Well-Shuffled Deck

A deck that has a large number of consecutive cards, such as straights in poker, is suspicious. Suits complicate things, so to work with just numbers, assume that the whole unshuffled deck is numbered from 1 to 52. If you cut the deck at a point such as between cards 23 and 24 and interchange the partial decks, you will have a deck that runs 24, 25, 26, ..., 51, 52, 1, 2, 3, ..., 21, 22, 23. Such a deck is barely more "random" than the original.

Cutting the deck a second time at some random point, say between 36 and 37, and placing the short stack on top, produces 37, 38, ..., 52, 1, ..., 23, 24, ..., 35, 36. This is no more random than the deck after the first cut, but cutting between 5 and 6 is a little more random.

But only a young child tries to shuffle cards by repeated cutting. Experienced adults use a riffle shuffle. After the deck is cut, the cut halves are interlaced in such a way that the number of cards from each partial deck varies slightly. If one partial deck is 24, ..., 52 and the other is 1, ..., 23, the result of a riffle shuffle by a moderately good practioner of the art might be 1, 2, 24, 26, 27, 3, 4, 28, 29, 30, 5, 6, 31, 32, 33, 7, 8, 9, 34, 35, 10, 36, 37, 11, 38, 39, 12, 13, 14, 40, 41, 42, 15, 16, 43, 44, 45, 17, 18, 19, 46, 47, 20, 21, 48, 49, 22, 23, 50, 51, 52. At first glance this appears to be more random than the result of the single cut or even more random than the result of two cuts. To achieve the same effect by single cuts, you need to cut the deck about twenty-six times. But notice that this is not the pattern you would get from *randomly* cutting the deck twenty-six times, since the two basic sequences that are apparent in the single cut above are preserved. Yet if you cut the deck again between 36 and 37 and do a second riffle shuffle, the basic sequences begin to be scrambled.

Diaconis and Bayer decided to count such basic sequences to determine randomness. They chose to look at any sequence found by proceeding to the next higher card, which they call a rising sequence, as basic. Thus, the deck after a single cut has two rising sequences and after two cuts three rising sequences; but after a cut and a riffle shuffle there are only two rising sequences. A second cut and riffle shuffle, however, produces in general more rising sequences than just the three from two cuts without the riffles. For example, a possible arrangement after the second cut and riffle might begin 37, 1, 11, 2, 24, 38, 25, 39, 12, 26, 27, 3, 13, 4, ... The few cards just listed give four rising sequences:

$$37, 38, 39$$
$$1, 2, 3, 4$$
$$11, 12, 13$$
$$24, 25, 26, 27$$

A longer list from after two shuffles would turn up additional rising sequences.

Diaconis and Bayer decided that a perfectly shuffled deck would contain exactly twenty-six rising sequences after starting with just one such sequence. In other words, for each card in the deck there would be just one number that directly followed it in numerical sequence when the cards were dealt. Note that this is not the same as a random arrangement since such a well-shuffled deck would represent only a small fraction of the possible 52! different decks. Also, the deck that started out 51, 52, 49, 50, 47, 48, 45, 46,

... and followed that same pattern would have twenty-six rising sequences, but might seem like an arranged deck instead of an unshuffled one to a card player. Despite these difficulties, the Diaconis-Bayer definition does suggest what one would like to happen after a number of cut-and-riffle shuffles.

Using this definition of a well-shuffled deck for their calculations, and also frequenting casinos to observe what actually occurs when decks are shuffled, the two mathematicians concluded that only seven decent shuffles are needed to produce a well-shuffled deck.

Kneading 52-Dimensional Space

What Diaconis and Bayer really wanted to show, however, was that seven shuffles would produce a random deck. As noted, the concept of rising sequences does not result in true randomness. For this they had to leave the casino environment and enter the environment of 52-dimensional space, not an unfamiliar realm for mathematicians.

A regular feature of mathematical operations consists of transformations of a space. These can be very simple: A transformation called translation simply moves everything in the space the same amount in a single direction; while rotation turns the whole space and everything in it around a point. More complicated transformations do more. Dilation expands the whole space in the way that rising bread expands. Diaconis and Bayer found that a transformation called the baker's transformation corresponds to shuffling a deck of cards by the cut-and-riffle method. This transformation stretches the space in one direction, cuts it in half, and then mashes the two halves together. It is called the baker's transformation because it does to space what a baker kneading bread dough does to the dough.

A cube in 52-dimensional space is called a 52-dimensional hypercube. All the cards in a deck can be represented as points along the edge of the hypercube with the distance of each point from a chosen vertex of the hypercube indicating the position of the card within the deck.

Diaconis and Bayer were able to show that each baker's transformation of the hypercube produces a figure that represents a number of possible outcomes of a cut-and-riffle shuffle. That number increases as the transformation is performed additional times. By the time seven transformations have taken place, the number of possible decks represented will have risen from one to approximately 52!.

In other words, seven good cut-and-riffle shuffles are all that is needed to achieve a true random sequence (not just the twenty-six rising sequences described above). Casino operators probably already know this result from empirical studies, having performed billions of trials.

(Periodical Reference and Additional Reading: *Discover* 1-91, p 66)

COMPUTERS
AND MATHEMATICS

ENCRYPTION DEVICES AND DESIRES

Mathematicians have long been interested in problems relating to codes and code breaking. A well-known example of mathematical code breaking concerns a precursor of the computer developed during World War II by Alan Turing. The Enigma Device was used successfully to break the German military code, a major help in ending the war. With the advent of general-purpose electronic computers in the 1950s, code making and breaking became computerized. As a result, codes became both easier and harder—easier in application, but usually more difficult to break.

The early 1990s saw activity in codes move from the secret corridors of the spies to the front pages of newspapers.

In 1990 the US National Security Agency took strong steps—unnecessarily strong according to many manufacturers—to restrict access to chips and software used for keeping computer-held information secret. Notable were government refusals to allow IBM to sell its encryption system for the System 390 Mainframe computers that were introduced in Sep 1990; the weakening of Lotus Development Corporation's Notes security system for foreign sales of the Notes program; and restrictions on sales of various coding systems for desktop computers.

Most coding devices are based on the Data Encryption Standard, or DES, used by the US government for unclassified information and by banks and other financial institutions. DES was originally developed in the 1970s by mathematicians at IBM's Thomas J. Watson Research Laboratory in Yorktown NY under the code name Lucifer. DES is a public key code based on using prime numbers and their products to keep information secret from those who are not supposed to know about it and allowing those who are the intended recipients to have access. In the United States DES has been a national standard since 1977.

In 1991, however, the security of DES became suspect. Mathematicians revealed that Adi Shamir of the Weizmann Institute in Israel and his student Eli Biham had let it be known via computer networks that they had found a way to compromise the code. The new method is based on analyzing statistical variations in large samples of coded messages. This development tends to extend the view expressed by Edgar Allan Poe in "The Gold Bug", that all codes concocted by human ingenuity can be broken by human ingenuity, to include computer ingenuity. However, some follow-up stories minimized the Shamir-Biham achievement, claiming that it was only a slight improvement on methods already known.

Background

The basic concept of DES and other public key codes is that a secret key number is used as the basis of a set of mathematical operations that scramble a message. To unscramble the message, one needs to reverse the mathematical operations using the same secret key number. A disadvantage is that possession of a single original message and its scrambled form is sufficient for determining what the secret key number is. An advantage is that there are 2^{56} possible keys, or about 7.2×10^{15}. Martin Hellman of Stanford University has calculated that a $10 million supercomputer could crack DES in about 2 hours, but such a computer is not routinely available to people who might want to crack the code.

(Periodical References and Additional Reading: *Science News* 6-2-90, p 343; *Science News* 9-7-91, p 148; *Science* 9-20-91, p 1343; *New York Times* 10-3-91, p A18; *New York Times* 10-18-91, p E4)

TURING TEST TOURNEY

In 1950 mathematician Alan Turing predicted that by the year 2000 computers would be programmed so well that after a 5-minute "conversation" with the computer via a remote terminal, an ordinary person would have only a 70 percent chance of detecting that a computer was at the other end. One version or another of this idea has come to be known as the Turing Test.

Businessman and philanthropist Hugh Loebner offered a prize of $1500 for a competition based on a limited version of the Turing Test. Six computer programs competed by conversing via terminal with ten human judges, who communicated in the same way with half a dozen humans. Each conversation was limited to a predefined topic and timed to be 14 minutes long. After the discussions, which were staged in Nov 1991 at the Computer Museum in Boston MA and at the Cambridge Center for Behavioral Studies, each judge cast a ballot for either a computer or a human. The winning program, called PC Therapist, was a psychology program similar to the popular program ELIZA of the early days of the personal computer. PC Therapist had been programmed by Joseph Weinstein of Woodside NY. Half of the judges thought PC Therapist was a person.

Loebner has now offered to award $100,000 to the computer programmer who can produce an artificial intelligence program that meets full Turing Test requirements—convincing a panel engaged in unrestricted conversation that it is human.

(Periodical References and Additional Reading: *Science* 11-29-91, p 1291; *Scientific American* 1-92, p 30)

THE CURSE OF DIMENSIONALITY

A difficulty in solving many problems in mathematics is that known algorithms take too many steps, reminiscent of the old fairy tale task of emptying a large lake with a thimble. This happens with integration. A typical problem in integration consists of finding (in two dimensions) the area under a curve or (in three dimensions) the volume of an oddly shaped region. Indeed, all integration problems in two or three dimensions can be reduced to area or volume calculations.

The same applies in higher dimensions, but that is where the difficulty with too many steps comes in. If the "area" to be calculated is in 50-dimensional space, then the method of approximating a solution that is within a reasonable distance from exactness, say 0.1 units, requires about $(1/0.1)^{49}$, or 10^{49} steps. In general, if the dimension of the problem is n and the answer needs to be within a margin of error of ε, then the number of calculations is about $(1/\varepsilon)^{n-1}$. Even for accuracy to four places for a problem with ten variables, a modest and common goal, about a billion calculations are required for the solution. Each additional dimension makes the problem 10,000 times as hard.

Because this difficulty increases dramatically with the dimension of the "area" being calculated, it has come to be known among mathematicians as the curse of dimensionality. The curse applies not only to integration but also to many other problems for which a good approximation requires calculating a large number of sample points.

In 1990 Henryk Wasniakowski of Columbia University and the University of Warsaw made a big start toward lifting the curse of dimensionality. He devised what some have called the silver dagger algorithm. The silver dagger is a version of the much used Monte Carlo method of approximations, but one that reduces the number of calculations to about $1/\varepsilon$. Thus, four-place accuracy for a ten-variable problem would require only at most a few thousand steps instead of about a billion.

Wasniakowski's work was reported in the Jan 1991 *Bulletin of the American Mathematical Society* and is based on the 1954 solution to a related set of problems by Klaus F. Roth of Imperial College in London.

Wasniakowski followed this achievement up with an even more dramatic breakthrough in 1991. The 1990 silver dagger was an improvement on the Monte Carlo technique, but the one from 1991, aimed at a set of calculations called approximation problems, broke completely new ground. The new method reduces the number of calculations from the power of $n - 1$ where n is the dimension involved to the power of 2. For a ten-variable approximation problem to four-place accuracy, this would reduce the number of calculations from about a billion to about a million.

(Periodical References and Additional Reading: *Science* 1-11-91, p 165; *Scientific American* 7-91, p 29; *Science* 7-26-91, p 384)

Physics

THE STATE OF PHYSICS

To understand the state of physics in the early 1990s, a good place to start is the list of member societies of the American Institute of Physics, representing somewhat more than 90,000 scientists. These societies include, in addition to the general physics societies, groups devoted to optics, acoustics, rheology (the deformation and flow of matter), crystallography, astronomy, the vacuum, and geophysics. Notice that there is no separate group for particle physics, nuclear physics, cosmology, or materials science, the topics that often seem to dominate news accounts of physics, although many members of the American Physical Society work in those subdivisions of the subject. Several of the topics of considerable concern to physicists have been dealt with earlier in *Current Science,* notably cosmology (p 83), clusters of atoms (p 154), and geophysics (p 256).

Every two years the American Physical Society and the American Institute of Physics polls physics departments to find out what specialties are being sought—that is, what kind of job openings there are. The survey in 1990 revealed that more departments were looking for condensed matter physicists than any other category, although the number was sharply down from recent years. There was slightly more call for condensed matter theorists than for experimentalists, as was also the case in the second highest category, particle physics. In the third largest group, however, which lumps together atomic, molecular, and optical physicists, the call was more than twice as high for experimentalists as for theorists. The other groups, in order of demand, were astrophysicists, nuclear physicists (much more for experimentalists), and plasma physicists.

Unsolved Mysteries

Although some scientists near the end of the nineteenth century believed that everything that could be known about physics had been discovered, scientists in the twentieth century have learned that that is far from the case. Indeed, physics is fascinating today precisely because it abounds in unsolved mysteries. Here are two from the early 1990s that are looking for solution in the next few years.

Why Do Superdeformed Nuclei Radiate Alike? When two atomic nuclei collide they can fuse to form a rotating cigar-shaped complex known as a deformed nucleus; if the shape is more than twice as long as its diameter, the nucleus complex is superdeformed. As such a superdeformed nucleus slows down after the collision, it emits electromagnetic radiation at gamma-ray frequencies. The mystery is that very different superdeformed nuclei tend to emit exactly the same spectra of gamma rays. This phenomenon was first discovered in 1989 by Peter J. Twin, now at the University of Liverpool. Theoretical explanations of why nuclei with 150 neutrons and protons (nucleons when thought of collectively) should have the same spectra as those that contain 190 nucleons generally invoke an arcane concept of quantum theory called pseudospin, but no one is certain how it might produce the observed effect. It is hoped that a new detector, called Gammasphere, scheduled for Lawrence Berkeley Laboratory in 1993, will provide enough data for an explanation to be worked out that is complete.

Do Wigner Crystals Exist? We picture electrons as moving about rapidly most of the time, although those involved in static electricity do not move very far even in terms of subatomic particle distances. Thus electrons are often pictured as a gas, since gases consist of particles (molecules) that move around rapidly. Eugene Wigner [Hungarian-US: 1902-   ] suggested that electrons could also form solids by becoming immobilized, as in a crystal; hence such solids, if found, are called Wigner crystals.

As early as 1979, Charles Grimes and Gregory Adams at AT&T Bell Laboratories (Bell Labs) in Murray Hill NJ found a peculiar form of electron crystal (essentially a two-dimensional solid) above the surface of liquid helium. In 1990 two large groups found evidence of other two-dimensional Wigner crystals in metal sheets in strong magnetic fields. Some of the evidence, however, did not seem sufficiently ordered to be a crystal; perhaps the scientists are finding a two-dimensional Wigner glass. (In a glass, molecules do not form the ordered arrays found in crystals.)

And a Solved Mystery

On Aug 6 1990 Udi Meirav and Marc Kastner of MIT working with Shalom Wind at IBM described a new low-temperature transistor they had developed that works one electron at a time. They created a 1-dimensional version of a gallium-arsenide field-effect transistor that changes the flow of electrons back and forth in a single cycle when one electron is added or subtracted from the region that contains the free electrons—in effect, this means that half an electron turns the current on or off, even though no such thing as half an electron exists. They were able to use the device to amplify current by a factor of 100, which could have important applications if the engineering details, which are formidable, could be worked out.

Of theoretical interest is the peculiar feature of the transistor that the amplitude of the current varies up and down periodically in a totally unexpected manner. This was explained in 1991 by a group of theorists from MIT consisting of Patrick Lee, Yigal Meir, and Ned Wingreen, who reported on their theory at the Cincinnati meeting of the American Physical Society in Mar. The effect is a result of the complications induced by the Pauli exclusion principle that two electrons cannot occupy the same state. Because only a few (fewer than a hundred) electrons are cooped up in the one-dimensional part of the transistor, a region named the Coulomb island, each one that enters the region has to find a different state from the ones already there. The accessibility of the state varies by whether the electron makes the total even or odd and in other ways that depend on how many electrons are already on the island. It is as if the island has complicated immigration laws that require a different set of papers for each immigrant, stamped by different authorities.

Making Light

A distinct trend of the early 1990s was the rush to use synchrotron radiation for theoretical and practical purposes. A synchrotron is a particular design for a particle accelerator. Synchrotron radiation is, except for its synchrotron source, plain old electromagnetic radiation. Although physicists often call synchrotron radiation light, the most useful frequencies are much higher that those of visible light, in the realm of X rays. The synchrotron produces a continuous spectrum of electromagnetic radiation, but it can be tuned so that desired frequencies are dominant and narrow bands of a particular frequency can be picked out, just as a filter can pick out red light in the visible spectrum.

The source of the light is the energy that particles, in practice almost always electrons or positrons, give off when they are accelerated (one Danish device called ASTRID, which started operation in 1990, uses heavy ions instead of electrons). In a synchrotron the particles travel in a circle, which means that they are in constant acceleration. The particles also travel at speeds near that of light, so their mass increases as predicted in relativity theory. Some of the energy involved is given off as electromagnetic energy.

When synchrotron radiation was first discovered in 1947 by physicists at General Electric Corporation, it was considered at best a nuisance, since it robbed energy from the particles. Physicists eventually began to find uses for the energy, however. By the 1980s synchrotron energy was seen as an important source of energy for the manufacture of semiconductor chips as well as for other purposes requiring very bright electromagnetic radiation, such as a version of the X-ray microscope. By the early 1990s, plans were in place to build various new synchrotron sources. These include:

Name	Location	Year
Laboratorio National de Luz Sincrotron	Campenas, Brazil	1992
The Advanced Light Source	Berkeley CA	1993
European Synchrotron Radiation Facility	Grenoble, France	1994
The Advanced Photon Source	Argonne National Laboratory	1995
SPring-8	Harima Science Garden City, Japan	1998

Atom Interferometers

The basic idea of an interferometer is discussed on p 35. Interferometry is based on the interactions of waves, and it does not matter whether the waves are those of light, water, or matter of any kind. As long ago as 1923 Louis de Broglie [French: 1892-1987] deduced that particles of all sorts behave as waves as well as particles. The wave-particle duality has come to be a cornerstone of physics, but it is still hard to remember that *all* bodies have waves associated with them. A small body such as a photon or electron has a large wave, so it is easy to picture the photon as a wave, which people do most of the time; or the electron as a wave, which is the basis of such devices as electron microscopes. A large body has a small wave, so it is very difficult to picture the wave associated with the Sun or with a human being. In between, although at the small end of things from our human point of view, is the atom. Each atom can be thought of either as a particle or as a wave. When pictured as a wave it is common to call the wave a matter wave.

At least five groups during the early 1990s built interferometers based on matter waves. Such interferometers are generally known as atom interferometers. Because atoms are so much larger than photons or electrons, atom interferometers use waves with very small wavelengths, which is helpful in many applications. Previously the best available interferometers for measuring fundamental constants in physics were based on matter waves of neutrons, which are smaller particles that have necessarily larger waves. A possible practical use for atom interferometers is in very accurate inertial guidance systems.

Although atom interferometers became a major news story in 1991, physicists Saul Altshuler and Lee M. Frantz from TRW Corporation took out a US patent on the idea in 1973. So far no fights over rights have developed, although that might happen in the future if atom interferometers move out of the laboratory and into practical applications.

Supersymmetry Anyone?

There were many attempts through the 1980s to go beyond the standard model with various new theories, including theories with such names as strings, GUTs (Grand Unified Theories), superstrings, supersymmetry, and Theories of Everything. None have adduced any experimental evidence. Newer particle accelerators, such as the Large Electron Positron collider (LEP) at CERN have produced a profusion of neutral Z particles, however. At least one set of measurements of neutral Z decay suggests supersymmetry, but that result is still far from a ringing endorsement of the theory.

(Periodical References and Additional Reading: *Science News* 7-28-90, p 53; *Science* 8-10-90, p 629; *Physics Today* 12-90, p 17; *Physics Today* 4-91, p 17; *Science* 4-12-91, p 215; *Science* 5-10-91, p 778; *Science* 5-17-91, p 921; *Physics Today* 7-91, p 17; *Science* 7-9-91, p 272; *Science News* 9-7-91, p 157; *Scientific American* 10-91, p 26; *Science* 10-18-91, p 357; *Physics Today* 2-92, p 13)

Timetable of Physics to 1990

1586	Simon Stevinus [Belgian-Dutch: 1548-1620] shows that two different weights that are dropped will reach the ground at the same time
1604	Galileo [Italian: 1564-1642] discovers that a body falling freely will increase its distance as the square of the time
	Johannes Kepler [German: 1571-1630] shows that light diminishes as the square of the distance from the source
1663	Pascal's law by Blaise Pascal [French: 1623-1662] is published posthumously (probably discovered around 1648): Pressure in a fluid is transmitted equally in all directions
1675	Ole Römer [Danish: 1644-1710] measures the speed of light
1676	Robert Hooke [English: 1635-1703] formulates Hooke's law: The amount a spring stretches varies directly with its tension
1678	Christiaan Huygens [Dutch: 1629-1695] develops the wave theory of light
1687	The *Principia* of Isaac Newton [English: 1642-1727] is published; it contains his laws of motion and the theory of gravity
1746	At least two experimenters in Holland invent a method for storing static electricity; the device used becomes known as the Leyden jar after the city of invention
1752	Benjamin Franklin [US: 1706-1790] performs his kite experiment, demonstrating that lightning is a form of electricity
1787	Jacques Charles [French: 1746-1823] discovers Charles' law: All gases expand the same amount with a given rise in temperature
1791	Luigi Galvani [Italian: 1737-1798] announces his discovery that two different metals that touch produce an effect similar to that of an electric charge
1798	Count Rumford [US-British-German-French: 1753-1814] shows that heat is a form of motion

1798   Henry Cavendish [English: 1731-1810] determines the gravitational constant and the mass of Earth

1800   Alessandro Volta [Italian: 1745-1827] announces his invention, made the previous year, of the electric battery

1801   Johann Ritter [German: 1776-1810] discovers ultraviolet light

1802   Thomas Young [English: 1773-1829] develops his wave theory of light

1819   Hans Christian Oersted [Danish: 1777-1851] discovers that magnetism and electricity are two different manifestations of the same force (not published until 1820)

1820   André-Marie Ampère [French: 1775-1836] formulates the first laws of electromagnetism

1830   Michael Faraday [English: 1791-1867] in England and Joseph Henry [US: 1797-1878] in the United States independently discover the principle of the electrical dynamo

1842   Julius Robert Mayer [German: 1814-1878] is the first scientist to state the law of conservation of energy

1850   Rudolf Clausius [German: 1822-1888] makes the first clear statement of the second law of thermodynamics: Energy in a closed system tends to degrade into heat

1851   William Thompson, later Lord Kelvin [Scottish: 1824-1907], proposes the concept of absolute zero, the lowest theoretically possible temperature (roughly -273° C or -460° F)

1873   James Clerk Maxwell [Scottish: 1831-1879] publishes the complete theory of electromagnetism, which includes his prediction of radio waves

1887   Albert Michelson [German-US: 1852-1931] and Edward Morley [US: 1838-1923] attempt to measure changes in the velocity of light produced by the motion of Earth through space; the negative result of the Michelson-Morley experiment is later interpreted as helping to establish Einstein's special theory of relativity

1888   Heinrich Hertz [German: 1857-1894] produces and detects radio waves

1895   Wilhelm Konrad Roentgen [German: 1845-1923] discovers X rays

1896   Antoine-Henri Becquerel [French: 1852-1908] discovers natural radioactivity

1897   Joseph John Thomson [English: 1856-1940] discovers the electron

1900   Max Planck [German: 1858-1947] explains the behavior of light by proposing the quantum, the smallest step a physical process can take

1905   Albert Einstein [German-Swiss-US: 1879-1955] shows that the photoelectric effect can be explained if light has a particle nature as well as a wave nature

Einstein shows that the motion of small particles in a liquid (Brownian motion) can be explained by assuming that the liquid is made of molecules

Einstein develops his theory of relativity and the law $E = mc^2$ (energy equals mass times the square of the speed of light)

1911   Heike Kamerlingh Onnes [Dutch: 1853-1926] discovers superconductivity

1914   Ernest Rutherford [New Zealander-Canadian-English: 1871-1937] discovers the proton

1915   Einstein completes his general theory of relativity, a theory of gravity more accurate than that of Sir Isaac Newton

1919   An expedition led by Arthur Eddington [English: 1882-1944] confirms that Einstein's theory of gravity is more accurate than Newton's in predicting the effect of gravity on light

1923   Louis-Victor de Broglie [French: 1892-1987] theorizes that particles, such as the electron, also have a wave nature

1925   Wolfgang Pauli [Austrian-US: 1900-1958] discovers the exclusion principle

Werner Heisenberg [German: 1901-1976] develops the matrix version of quantum mechanics

1926   Erwin Schrödinger [Austrian: 1887-1961] develops the wave version of quantum mechanics

1927   Heisenberg develops his uncertainty principle

1932   James Chadwick [English: 1891-1974] discovers the neutron

Carl Anderson [US: 1905-1991] discovers the positron

John Cockcroft [English: 1897-1967] and Ernest Walton [Irish: 1903-] develop the first particle accelerator

1937   Anderson discovers the muon

1938   Otto Hahn [German: 1879-1968] splits the uranium atom, opening the way for nuclear bombs and nuclear power

1947   Quantum electrodynamics (QED) is born, with many parents, notably Richard Feynman [US: 1918-1988], Julian Schwinger [US: 1918-], Schin'ichiro Tomonaga [Japanese: 1906-1979], Willis Lamb, Jr. [US: 1913-] (all of whom got Nobel Prizes), and Hans Bethe [German-US: 1906-] (whose Nobel Prize was for other work)

Cecil Powell [English: 1903-1969] and coworkers discover the pion

1952   A group led by Edward Teller [Hungarian-US: 1908-] develops the hydrogen bomb

1955   Owen Chamberlain [US: 1920-] and Emilio Segrè [Italian-US: 1905-] produce the first known antiprotons

Clyde Cowan, Jr. [US: 1919-] and Frederick Reines [US: 1918-] are the first to observe neutrinos

1957   Experiments by a group led by Chien-Shiung Wu [Chinese-US: 1912-] and quickly confirmed by others show that the law of conservation of parity does not hold for the weak interaction

John Bardeen [US: 1908-1991], Leon Cooper [US: 1930-], and John Schrieffer [US: 1931-] develop a theory that explains superconductivity

1961   Murray Gell-Mann [US: 1929-] and, independently, Yu'val Ne'eman [Israeli: 1925-] and others develop a method of classifying heavy subatomic particles that comes to be known as the eightfold way

1964   Gell-Mann introduces the concept of quarks as components of heavy subatomic particles, such as protons and neutrons

1967   Steven Weinberg [US: 1933-], Abdus Salam [Pakistani-English: 1926-], and Sheldon Glashow [US: 1932-] independently develop a theory that combines the electromagnetic force with the weak force

1986   Alex Muller [Swiss: 1927-] and Georg Bednorz [German: 1950-] discover the first warm-temperature superconductor

P A R T I C L E   P H Y S I C S

WHAT'S GOING ON AT 17 KEV?

Ever since 1985 some experiments have consistently reported evidence for an uncommon neutrino that has a mass of 17 kilo electron volts (keV). Such evidence was first detected in that year by John Simpson of the University of Guelph, Canada. Other experiments have almost as consistently put the upper limit on the mass of common neutrinos at about 1/2000th of this amount, and have found no evidence for the heavy uncommon neutrinos. This would be old news if it were not the case that a number of particularly accurate experiments in 1989 (by Simpson again), in 1990, and in 1991 did find the 17-keV neutrino, pushing it to the fore.

One experiment, noted in part for the high number of events counted, was conducted by Andrew Hime and Nick Jelley of Oxford University. They used sulfur-35 absorbed on a thin film of plastic as a source of neutrinos. Sulfur-35 has 16 protons and 19 neutrons, too many neutrons to be stable. One of the neutrons has to change into a proton, so one of them emits an electron (beta particle) and an antineutrino, the neutron changes into a proton, and the atom goes from sulfur-35 to chlorine-35. Chlorine-35 has 17 protons and 18 neutrons, close enough to equal numbers to make the atom stable. Hime and Jelley wanted to confirm Simpson's measurements of 1989 (Hime had been a graduate student of Simpson's), but it is impossible to measure neutrinos directly because they interact extremely weakly with other forms of matter. So they measured the electrons and inferred the neutrinos. They caught electrons that were nearly perpendicular to the film of sulfur-impregnated plastic in a semiconductor diode that reported the electrons' energies. The curve describing how these energies were distributed indicates that about one in every hundred antineutrinos produced by the process has a mass of 17.2 keV, while the other 99 electrons are essentially massless.

About the same time, Eric Norman and coworkers Bhaskar Sur, Kevin Lesko, Munther Hindi, Ruth Mary Larimer, Paul Luke, and William Hansen at the Lawrence Berkeley Laboratory (LBL) were performing an experiment with the same aim, but a different design. Norman's group used a crystal that had been grown by Eugene Haller of LBL and the University of California. The crystal was grown with radioactive carbon-14 embedded in germanium. Norman's experiment, which took 11 months to complete, used a matched crystal of pure germanium as a control. The main idea, similar to that of Simpson's 1989 work, was to combine the source of antineutrinos and the detector in a single crystal, eliminating "noise" and enabling the experimenters to measure very small effects. They found evidence of a neutrino between 15 and 19 keV. Norman is planning to rerun the experiment with a better crystal that will give enough antineutrinos to fix the mass more closely.

Norman's group during this time period also looked for the heavy neutrino with another experiment based on an unrelated phenomenon. In this experiment the neutrinos were produced when electrons were captured in atomic nuclei, producing a positron and a neutrino. While the results were not so clear-cut as those of the carbon-14 decay experiment, they tended to support the 17-keV neutrino. Yet another experiment that found the heavy neutrino was conducted by a group from the Ruder Boskovic Institute in Zagreb, Yugoslavia, led by Igor Zlimen and Ante Lujubicic in collaboration with Brian Loan from the University of Ottawa, also using electron capture. Both electron-capture experiments used the decay of iron-55 as a source of neutrinos.

Taken together, experimenters have found evidence for the 17-keV neutrino in tritium (Simpson's original 1985 experiment), sulfur, carbon, and iron, sufficiently diverse a group of atoms to impress many. Yet all of these experiments used semiconductors to detect the electrons. Another kind of detector is the magnetic spectrometer, which was used in eight different hunts for the heavy neutrino in the late 1980s. The conventional view is that none of those eight experiments shows any evidence for 17-keV neutrinos, although Simpson says that when he reanalyzed the data for a negative 1985 experiment, he could show that a heavy neutrino was found in the vicinity. Thus, there is some suspicion that the heavy neutrino could be an artifact caused by the semiconductor detectors. The indefatigable Norman is also planning to rerun his crystal experiment without a crystal to determine whether that could be true. In this version the carbon-14 will be part of carbon dioxide mixed into xenon, which will serve as the detector material in an ionization or scintillation chamber. But the most convincing experiment planned has been worked out by Wolfgang Stoeffl of Lawrence Livermore National Laboratory. He would also use a magnetic spectrometer, but would avoid distortions that occurred in earlier experiments. The experiment also would run without a solid source involved, using neutrinos produced in the decay of the gas tritium. The result should be about ten times as sensitive as previous magnetic spectrometer experiments. Others will run similar experiments designed to eliminate specific problems seen in earlier versions.

Results of these experiments should convince the physics community by the end of 1992 either that the heavy neutrino exists or that something else is causing the observed results. If it does exist, it will pose a number of problems for theorists, some of whom call it "the neutrino from hell." First of all, the big bang theory of the creation of the universe was not worked out with heavy neutrinos in mind. If they exist as about one in a hundred, then the universe would have collapsed long ago, following a standard big bang. Not only would heavy neutrinos account for the "missing mass" in the universe, but they would produce too much of it! Furthermore, no one has the least idea how to account for masses of such lepton particles as the electron and the muon. There is no known reason why muons should be exactly like electrons, except about 200 times heavier. So why should we think it will be easy to find a reason why some neutrinos are about 2000 times heavier than the more common neutrinos?

Background

When the structure of atoms first began to be understood, atoms were thought to be made of two components, negatively charged electrons and something positive that balances out the charge so that atoms can be electrically neutral. Electrons are easy to spot because they can rub off (producing static electricity), steam off when substances are heated (the Edison effect), or be produced continuously by various elements (the beta decay of many radioactive elements). The positive charge was harder to identify and more complicated to understand. Historians generally credit Ernest Rutherford [New Zealand-UK: 1871-1937] with discovery of the particle of positive charge, the proton, in 1914. But he did not produce individual protons until 1919, 22 years after recognition of the electron. It was another 13 years before a scientist demonstrated the existence of another particle in the atom, the neutron, although some scientists predicted neutrons soon after the proton was isolated.

All atoms except the simplest form of hydrogen consist of a core that is made of shells of protons and neutrons surrounded by shells of electrons. Atoms cannot have cores of just protons because the force that holds the core together, known as the strong force, depends on the exchange of particles between neutrons and protons. A nucleus of just protons would not work because the particle exchange of the strong force changes protons to neutrons and vice versa (most of the time). Neutrons also need to be changed into protons to stay stable. Although the decay of isolated neutrons was not observed until 1948, it was clear soon after their discovery that neutrons in some atoms decay spontaneously after varying periods of time, losing some energy by emitting electrons as they become neutrons. However, both the amount of energy lost and the lack of change in spin (the particle equivalent to angular momentum) between the original neutron and new proton does not correspond to expectations; that is, the electron does not carry away enough energy, but at the same time it carries away too much spin—all of it, in fact. Wolfgang Pauli [Austrian-US: 1900-1958] proposed that there must be another particle involved that has an opposite spin and that carries small amounts of energy. The new particle would have no mass since there is no inexplicable loss of mass in the transition from the neutron to the proton. This idea was accepted when Enrico Fermi [Italian-US: 1901-1954] worked out the details of what has come to be known as the weak interaction, coining the name neutrino for the Pauli particle in the process. Later, when it was discovered that all particles have near twins called antiparticles, the Pauli particle was classed as an antineutrino.

A neutrino or its antineutrino twin, which has an opposite spin, is essentially spin and energy, rather like a dust devil, traveling at about the speed of light. The spin of an electron, a particle that can travel at any speed, can be reversed by stopping the electron and turning it around, just as you can change the spin of a top from clockwise to counterclockwise by turning it upside down. But a neutrino is always found with the same spin and an antineutrino with the opposite spin. This implies that they cannot be stopped, which many physicists in the past assumed meant that they always travel at exactly the speed of light. In turn, by Einstein's theory of special relativity, this would mean that neutrinos have no mass.

More recently, particle physicists have discovered that particles and their antiparticles can come in mixed states. This idea is hard to picture when you think of particles as particles but is easier to see when you think of particles as waves. Mixed states can impart mass to an otherwise massless particle. Using this idea, some theoretical physicists have suggested that the neutrino may have a small mass. Increasingly sophisticated experiments (except for the 17-keV experiments) have ruled out any mass greater than 9 eV, however. Some still think that the mass of the ordinary neutrino and its antiparticle is exactly 0 as Pauli first proposed.

Complicating this whole picture is the mystery of the three families of matter. For reasons that are not understood at all, in addition to ordinary matter there is a heavy family of matter and an even heavier one. Also, all three families are made from two different kinds of stuff, the quark stuff that interacts with a strong interaction and the lepton stuff that interacts with Fermi's weak interaction. Protons and neutrons are made from quark stuff combined in various ways. Electrons and neutrinos are particles of lepton stuff, which do not combine. The heavier lepton stuff consists of the muon family (a muon, an antimuon, a muon neutrino, and a muon antineutrino), while the heaviest consists of the similarly constituted tauon family. There is no fourth family.

The muon, discovered in 1937, is about 200 times as massive as an electron, but there is no evidence that the muon neutrino has any measurable mass. Experiments in particle accelerators actually rule out much of a mass for the muon neutrino. The tauon is about 3500 times as massive an an electron. It has been known only since 1975 and its properties are not fully investigated. There is considerable speculation that the tauon neutrino could be the 17-keV particle. The other possibility, assuming that the heavy neutrino exists, would be a fourth kind of matter, contradicting other particle-accelerator experiments.

(Periodical References and Additional Reading: *Science* 3-22-91, p 1426; *Physics Today* 5-91, p 17; *Science* 11-29-91, p 1298)

PROGRESS IN HOT FUSION

Nuclear fusion remains a major but seemingly ever-receding hope for cheap and abundant energy (see "Alternative Sources of Energy," p 359). In the early 1990s, funding for fusion research in the United States was down, but there was progress in Europe and a confirmation of an unconventional surprise that had been detected by accident in 1989.

Conventional Break, But No Break-even

The conventional way to achieve fusion is to put very fast light atomic nuclei into very close confinement using some form of magnetic field as the means of containment. The heat involved in such fast moving atomic nuclei would break ordinary matter containers, but magnetic fields are not affected by heat.

Until 1991 the conventional light nuclei to use were either protons or combinations of one proton and one neutron called deutrons, the nuclei of atoms of deuterium, also known as heavy hydrogen. Theory suggested that a combination of two neutrons with a proton, the nucleus of really heavy hydrogen or tritium, would work better than regular hydrogen or deuterium nuclei. But the catch is that tritium atoms and their nuclei are radioactive. Plasma physicists, the scientists who study this form of nuclear fusion, did not want to risk contaminating millions of dollars worth of equipment with radioactivity.

In a break with tradition, the scientists of the Joint European Torus (JET) laboratory in Culham, UK, took the step. They injected a small amount of tritium into already heated deuterium (the experiments are carried out at such high temperatures that the atoms quickly separated into nuclei and electrons, forming a plasma). The JET experiment did produce more power than previously obtained, although far less than the break-even point. The researchers had to use ten times more power to produce the reaction than they obtained from it.

It was still unknown at the end of 1991 how badly they had contaminated their equipment. Despite this, competitors in the United States sounded envious when interviewed about the experiment.

Cluster Impact Fusion

In the Sep 18 1989 *Physical Review Letters* and in a follow-up report in Jul 1991, Robert J. Beuhler, Lewis Friedman, and Gerhart Friedlander of Brookhaven National Laboratory announced yet another route to fusion. As in the Pons-Fleischmann experiments (see "Cold Fusion?" p 162), the Brookhaven method uses heavy water and a metal that absorbs hydrogen. Also, like "Cold Fusion," the Brookhaven process is disputed by theorists who say that fusion cannot be the explanation for the observations.

In the Brookhaven experiments, the heavy water is in the form of tiny droplets that carry electric charges. Because of the electric charges, an electric field can be used to speed the droplets through a vacuum chamber, at the end of which is a target consisting of titanium that has absorbed heavy hydrogen. When the droplets hit the target at several hundred thousand miles an hour, the pressure and temperature at the target both become very great. This classic combination of high pressure and temperature is enough to cause fusion. This method has been deemed cluster impact fusion because it works only when the droplets are clusters of between 25 and 1300 molecules of heavy water (see "Clusters of Atoms," p 154).

The evidence for cluster impact fusion is much solider than that of Pons and Fleischmann. The fusion route is a likely one, with two heavy hydrogen atoms joining to form the even heavier form of hydrogen called tritium, releasing protons in the process. Both the tritium and the protons, which quickly capture an electron to become ordinary hydrogen, have been observed. Changing the recipe by substituting ordinary hydrogen produces no evidence of fusion—which is another sign that the Brookhaven process really works. On the other hand, a detailed physical explanation of what is happening is hard to come by, which is why some theorists still believe that something other than cluster impact fusion is actually causing the effect.

The Brookhaven investigators will now use larger accelerators to see whether they can get even more definite evidence of the reaction. Any practical application would seem at this time to be far in the future .

(Periodical References and Additional Reading: *Science* 10-25-91, p 515; *New York Times* 11-11-91, p A1; *Science News* 11-16-91, p 309)

PARTICLE PHYSICS UPDATE

At first thought it would appear that nothing of any consequence has happened in particle physics since Carlo Rubbia [Italian-US: 1934-   ] and his cohorts detected the W and Z particles at the *Centre Européen de Recherche Nucléaire* (CERN) near Geneva, Switzerland, in 1983. And furthermore, not much can be expected to happen until the Superconducting Supercollider (SSC) begins operations at Waxahachie TX in 1999. While this impression is not entirely true, there is something to it. The theorists are proceeding, but not much new in the way of really significant experiments have appeared.

Searching for Higgs

When the SSC does go on-line, or perhaps even before that if the Europeans proceed with their plans for a giant synchrotron at CERN (decision scheduled for 1992 or 1993), the announced goal of the large particle accelerators will be to seek a particular large particle, known as the Higgs particle. A large accelerator is needed because the Higgs particle is expected to be found at about 600 giga-electron-volts (GeV).

Finding the existence of the Higgs particle seems to be an essential step in continuing to establish that the currently used standard model of particle physics—called the "standard model"—is valid. While that would be pleasant, it would not contribute much to advancing knowledge. Not finding the Higgs particle, or finding something quite different, would be far more interesting.

There are physicists, however, who think it unlikely that the Higgs particle exists. For one thing, it is calculated to have 0 spin, unlike any known fundamental particle except for some versions of the graviton, which is also an unobserved particle. The Higgs is needed, however, to account for the existence of mass, which is otherwise unexplained.

Furthermore, even if the Higgs explains where mass comes from, its existence, if it does exist, does not tell why different particles have the masses they do. Some other theory, with no doubt some other undiscovered particle, is needed for that.

Factory Fever

In particle accelerator parlance, a factory is an accelerator that produces a great many of some kind of particle so that the particle and its descendants can be studied. Generally, the particle is the result of an impact of two other particles, which produce together a short-lived complex that decays into a shower of particles. The factory is one that makes a lot of a particular particle among the various droplets in the shower.

In Canada, the excitement revolves around a plan to produce a kaon factory at the Tri-University Meson Facility (TRIUMF) near Vancouver, British Columbia. On Sep 19 1991 the Canadian government agreed to fund about a third of the new facility at a cost of $236 million. The TRIUMF backers are looking for additional funding from British Columbia and from sources outside of Canada.

In the United States hopes have been high for so-called B factories that would produce many B mesons. Both Stanford and Cornell universities want to build B factories.

Both kaons and B mesons are thought to be involved in processes that violate certain rules that other particles obey, called CP symmetry (for charge-parity symmetry), but kaons are the only particles so far actually caught in the act of such a violation (in 1964 by Val Fitch and James Cronin). The Fitch-Cronin, or garden-variety, violation is not what the factory fans have in mind, however. They are looking for exotic interactions that violate some parts of the standard model.

(Periodical References and Additional Reading: *Science* 10-4-91, p 36; *Science* 10-25-91, p 522)

BASIC   THEORIES

UNCOVERING A NAKED SINGULARITY

Stuart Shapiro and Saul Teukolsky from Cornell University reported in the Feb 25 1991 *Physical Review Letters* that according to Einstein's equations for general relativity, certain situations result in naked singularities (see Background). A supercomputer simulated the behavior of sets of particles representing stars as they followed Einstein's gravitational rules.

Inputting into the computer certain three-dimensional configurations of particles led to the singularities. The typical configuration that resulted in the problem was the collapse under gravity of a sufficiently large football-shaped population, which produced singularities near the two pointed ends of the region the population occupied. Smaller numbers of particles in the same configuration failed to produce naked singularities. In some cases the singularities were in empty space, beyond the population of particles in the region.

Shapiro and Teukolsky are continuing to investigate whether adding spin to the configuration or treating the population as a fluid instead of as individual points eliminates the singularities.

Since the Shapiro-Teukolsky result suggests what still seems impossible in reality, some way around it needs to be found. Even though naked singularities are unacceptable, the many triumphs of the general theory suggest that the theory should not be abandoned. The most likely possibility is that some mathematical juggling will have to be found to eliminate the naked singularities from the theory.

Background

A problem that has bedeviled modern physics throughout the twentieth century is that theories that seem perfectly good in many ways also show results that are totally impossible in the real world as we believe it to be. This happens with such regularity that theoretical physicists have developed sneaky ways to get over the difficulty so that the rest of the theory can be retained. The most famous of these is the renormalization process of quantum electrodynamics, in which countervailing infinities are balanced against each other to make theoretical electrons behave like real ones. Mathematically, it seems like an ugly procedure, but it is extremely successful in predicting particle behavior.

The generic name for the kind of problem that often emerges in mathematical theories is singularity. A singularity is something like the unpleasant result of dividing by 0 (try it on your calculator).

People who have studied calculus know that, even though division by zero is strictly prohibited, there are ways to get around it to find the limit of a function. Indeed, it was this discovery by Newton and Leibnitz (and

(Continued)

their immediate predecessors) that made calculus possible. A singularity is a situation in which even the methods that work for calculus fail, so you are confronted with what amounts to the forbidden division by zero. The result in that situation is beyond calculation, although this idea is commonly expressed by saying, somewhat misleadingly, that a quantity becomes infinite. It would be more nearly correct to say that calculations break down and can no longer predict the behavior of the system

The singularity problems in physics arise when the mathematics of a theory describes as infinite something that cannot be infinite, such as the charge of an electron or the gravitational pull of an object. If the mathematics says that you have a singularity and if infinite amounts do not exist in nature, then something must be wrong. This was the situation in the early 1940s in particle physics, but the renormalization procedure of quantum electrodynamics saved the day.

A similar problem with infinity emerged in Albert Einstein's theory of gravitation, known as the general theory of relativity, shortly after it was developed in 1915. The mathematics of this theory is daunting, so it was not clear at first that the singularities were really there, even though Einstein himself had encountered them in his early work on gravity in the period between 1908 and 1927. Karl Schwarzschild [German: 1873-1916] used Einstein's theory shortly after it was fully formed to predict in 1917 that gravitational singularities would result in what we now call black holes—but at that time no one believed that black holes could really exist, since they were singularities.

A study of black holes in 1939 by J. Robert Oppenheimer [US: 1904-1967] and Hartland S. Snyder at Caltech showed that black holes formed by the collapse of a sphere would be surrounded by a boundary outside of the singularity itself. This boundary, called the event horizon, prevents any information about the black hole from reaching the rest of the universe.

Following the discovery in 1963 of quasars—objects that certainly exist but are poorly understood even today—physicists began to think more seriously about gravitational singularities. In 1969 Roger Penrose pointed out that a singularity could exist without causing the universe to go haywire if the singularity occurred inside a black hole. Such a singularity is protected, or "clothed," by the black hole. This satisfied theory by allowing the singularity and it satisfied reality by keeping it where nothing could get to it and where it could not affect the rest of the universe. Unclothed, or "naked," singularities were presumed not to exist. Penrose's idea came to be known as "cosmic censorship."

(Periodical References and Additional Reading: *Science News* 3-9-91, p 148; *New York Times* 3-10-91; *Science* 3-29-91, p 1566; *American Scientist* 7/8-91, p 330)

THE BASIC LAWS OF PHYSICS

Key Terms

Mass is a measure of the amount of matter. Near the surface of Earth it is roughly equivalent to weight.
Velocity measures how an object changes position with time.
Acceleration expresses how an object changes velocity with time.
Momentum is the product of mass and velocity.
Energy is the ability to do work.

Law of Gravity

The gravitational force between any two objects is proportional to the of their masses and inversely proportional to the square of the distance between them. If F is the force, G is the number that represents the ratio (the *gravitational constant*), m and M are the two masses, and r is the distance between the objects:

$$F = \frac{GmM}{r^2}$$

In metric measure, the gravitational constant is 0.0000000667 (6.67×10^{-8}) dyne cm^2/g^2, so another way of writing the basic law of gravity is

$$F = \frac{0.0000000667\,mM}{r^2}$$

This law implies that objects falling near the surface of Earth will fall with the same rate of acceleration (ignoring drag caused by air). This rate is 32.174 feet per second per second (ft/sec^2) or 980.665 cm/sec^2, and is conventionally labeled g. Applying this rate to falling objects gives the velocity, v, and distance, d, after any amount of time, t, in seconds. If the object starts at rest and we use 32 ft/sec^2 as an approximation for g:

$$v = 32t$$
$$d = 16t^2$$

For example, after three seconds a dropped object that is still falling will have a velocity of 32 x 3 = 96 feet per second and have fallen a distance of $16 \times 3^2 = 144$ feet.

If the object has an initial velocity v_0 and an initial height above the ground of a, the equations describing the velocity and the distance d

above the ground (a positive velocity is *up* and a negative velocity is *down*) become

$$v = v_0 - 32t$$

and

$$d = -16t^2 + v_0 t + a$$

After three seconds an object tossed in the air from a height of 6 feet with a velocity of 88 feet per second will reach a speed of 88 - 96 = -8 feet per second, meaning that it has begun to descend, and will have a height of (-16 × 9) + (88 × 3) + 6 = -144 + 264 + 6 = 126 feet above the ground.

The maximum height, *H*, reached by the object with an initial velocity v_0 and initial height *a* is

$$H = a + \frac{v_0^2}{64}$$

For the object tossed upward at 88 feet per second from a height of 6 feet, the maximum height reached would be 6 + 88²/64 = 6 + 121 = 127 feet. Therefore, after three seconds, the object has just reached its peak and has fallen back only one foot.

Albert Einstein's general theory of relativity introduced laws of gravity that are more accurate than the one just given, which is owed to Sir Isaac Newton. Newton's gravitational theory is extremely accurate for most practical situations, however. For example, Newton's theory is used to determine how to launch satellites into proper orbits.

Newton's Laws of Motion

Newton's laws of motion apply to objects in a vacuum, and are not easily observed in the real world, where forces such as friction tend to overwhelm the natural motion of objects. To obtain realistic solutions to problems, however, physicists and engineers begin with Newton's laws and then add in the various forces that also affect motion.

1. *Any object at rest tends to stay at rest. A body in motion moves at the same velocity in a straight line unless acted upon by a force.* This is also known as the law of *inertia*.

Note that this law implies that an object will travel in a curved path only so long as a force is acting on it. When the force is released, the object will travel in a straight line. A weight on a string that is swung in a circle will travel in a straight line when the string is released, for the string was supplying the force that caused circular motion.

2. *The acceleration of an object is directly proportional to the force acting on it and inversely proportional to the mass of the object.* This law, for an acceleration a, a force F, and a mass m, is more commonly expressed in terms of finding the force when you know the mass and the acceleration. In this form it is written as

$$F = ma$$

The implication of this law is that a constant force will produce acceleration, which is an increase in velocity. Thus, a rocket, which is propelled by a constant force as long as its fuel is burning, steadily increases in velocity. If there were enough fuel, the rocket would eventually cease to increase in velocity, however, because Einstein's special theory of relativity tells us that no object can exceed the speed of light in a vacuum (see "Conservation of mass-energy" below). Nevertheless, even a small force, constantly applied, can cause a large mass to reach velocities near the speed of light if enough time is allowed.

3. *For every action there is an equal and opposite reaction.*

Conservation laws

Many results in physics come from various other *conservation laws:* Important effects result from a rule that a certain quality must not change during a certain class of operations. All such conservation laws treat closed systems, of course. Anything added from outside the system could affect the amount of the entity being consumed.

Conservation of momentum. *In a closed system, momentum stays the same.* This law is equivalent to Newton's third law.

Since momentum is the product of mass and velocity, if the mass of a system changes, then the velocity must change. For example, consider a person holding a heavy weight, such as an anchor, in a canoe that is still in the water. The momentum of the system is zero, since the masses have no velocity. Now the person in the canoe tosses the anchor toward the shore. The momentum of the anchor is now a positive number if velocity toward the shore is measured as positive. To conserve momentum, the canoe has to be accelerated in the opposite direction, away from the shore. The positive momentum of the anchor is balanced by the negative momentum of the canoe and its cargo. In terms of two masses, m and M, and matching velocities v and V

$$mv = MV$$

Conservation of angular momentum. An object that is moving in a circle has a special kind of momentum, called *angular momentum*. As noted above, motion in a circle requires some force. Angular momentum combines mass, velocity, and acceleration (produced by the force). For a body moving in a circle, the acceleration depends both on the speed of the body in its path and the square of the radius of the circle. The product of this speed, the mass, and the square of the radius is the angular momentum of the mass.

In a closed system, angular momentum is conserved. This effect is used by skaters to change their velocity of spinning. When skaters bring their arms close to their body, this tends to reduce the angular momentum because the center of mass is closer to the body. But, since angular momentum is conserved, the rate of rotation has to increase to compensate for the decreased radius. Because the rate depends on the square of the rotation, the rate increases dramatically.

Conservation of mass.   *In a closed system, the total amount of mass is conserved in all but nuclear reactions and other extreme conditions.*

Conservation of energy.   *In a closed system, energy is conserved in all but nuclear reactions and other extreme conditions.*

Energy comes in very many forms: mechanical, chemical, electrical, heat, and so forth. As one form is changed into another (except in nuclear reactions and under extreme conditions), this law guarantees that the total amount of energy remains the same. Thus, when you change the chemical energy of a dry cell into electrical energy and use that to turn a motor, the total amount of energy does not change (although some becomes heat energy—see Laws of Thermodynamics below).

Conservation of mass-energy.   Einstein discovered that his special theory of relativity implied that energy and mass are related. Consequently, in some situations, mass and energy by themselves are not conserved, since one can be converted to the other. The more general law, then, is the law of conservation of mass-energy: *The total amount of mass and energy must be conserved.* Einstein found the following equation that links mass and energy.

$$E = mc^2$$

In this equation, E is the amount of energy, m is the mass, and c is the speed of light in a vacuum.

One instance of energy changing to mass can be seen in Einstein's equation for how mass increases with velocity. If m_0 is the mass of the object when it is not moving, v is the velocity of the object in relation to an observer who is considered to be at rest, and c is the speed of light in a vacuum, then the mass, m, is given by the equation

$$m = \frac{m_0}{\sqrt{1 - \frac{v^2}{c^2}}}$$

This accounts for the rule that no object can exceed the speed of light in a vacuum. As the object approaches this speed, so much of the energy is converted to mass that it cannot continue to accelerate.

In both nuclear fission (splitting of the atomic nucleus) and nuclear fusion (the joining of atomic nuclei that produces the energy of a hydrogen bomb), mass is converted to energy.

Conservation for particles. Many properties associated with atoms and subatomic particles are also conserved. Among them are charge, spin, isospin, and a combination known as CPT for *charge conjugation, parity,* and *time.*

First and Second Laws of Thermodynamics

First law. This is the same as the law of conservation of energy. It is a law of thermodynamics, or the movement of heat, because heat must be treated as a form of energy to keep the total amount of energy constant.

Second law. *Heat in a closed system can never travel from a low-temperature region to one of higher temperature in a self-sustaining process.* "Self-sustaining" in this case means a process that does not need energy from outside the system to keep it going. In a refrigerator, heat from the cold inside of the refrigerator is transferred to a warmer room, but energy from outside is required to make that happen.

The second law has many implications. One of them is that no perpetual motion machine can be constructed. Another is that all energy in a closed system eventually becomes heat that is diffused equally throughout the system, so that one can no longer obtain work from the system.

The equations that describe the behavior of heat also can be applied to order and therefore to information. The word *entropy* refers to diffused heat, disorder, or lack of information. Another form of the second law of thermodynamics states that in a closed system, entropy always increases.

Laws of Current Electricity

Key terms. When electrons flow in a conductor, the result is electric *current.* The amount of current is based on an amount of electric charge called the *coulomb,* which is the charge of about 6,250,000,000,000,000,000 (6.25 $\times 10^{18}$) electrons. When one columb of charge moves past a point in one second, it creates a current of one ampere. Just as a stream can carry the same amount of water swiftly through a narrow channel or slowly through a broad channel, the energy of an electric current varies depending on the difference in charge between places along the conductor. This *potential difference* is measured in *volts.* The voltage is affected by the nature of the conductors. Some substances conduct an electric current much more easily than others. This *resistance* to the current is also measured in volts. Electric *power* is the rate at which electricity is used.

Ohm's law. *Electric current is directly proportional to the potential difference and inversely proportional to resistance.* If you measure current (I) in amperes, potential difference (V) in volts, and resistance (R) in volts, then the current is equal to the potential difference divided by the resistance.

$$I = \frac{V}{R}$$

Law of electric power. *If electric power* (P) *is measured in watts, then the power is equal to the current measured in amperes and the potential difference measured in volts.*

$$P = IV$$

Laws of Waves, Light, and Electromagnetic Radiation

Key terms. Light is a part of a general form of radiation known as *electromagnetic waves,* or, when thought of as particles, *photons.* Here we will treat electromagnetic radiation as a wave phenomenon for the most part. The *velocity of a wave* is how fast the wave travels as a whole. The *wavelength* is the distance between one crest of the wave and the next. The *frequency* is how many crests pass a particular location in a unit of time.

Law of wave motion. All waves (including water waves and sound waves) obey the *wave equation* that relates the velocity of the wave to its frequency and wavelength. *For all waves, the velocity is equal to the product of the frequency and wavelength.* The letters traditionally used in this equation have already been used in the equations above to mean something else, so we will use W for the velocity of the wave, f for the frequency, and l for the wavelength.

$$W = fl$$

Law of electromagnetic energy. The energy of an electromagnetic wave depends on a small number that is known as *Planck's constant.* Measured in ergs (a unit of energy) per second, Planck's constant is 6.6×10^{-27}, a number with 26 zeros after the decimal point before getting to a nonzero digit. *The energy is equal to the product of Planck's constant and the frequency.* Using E for energy, h for Planck's constant, and f for frequency,

$$E = hf$$

When thought of in terms of the particle called a photon, the energy of a photon obeys the same law. This law can be combined with the speed of light in a vacuum (c) to give

$$E = \frac{hc}{l}$$

Inverse-square law. All radiation obeys an inverse-square law, which is similar to the law of gravity. *The intensity of the radiation decreases as the inverse of the square of the distance from the source of the radiation.*

Two Basic Laws of Quantum Physics

Very small masses at very small distances behave differently from masses at the sizes and distances we can observe directly. Effects occur in discrete steps based on Planck's quantum, which is the size by which energy changes in steps (instead of continuously). The science of such effects is called *quantum physics*. Small masses act sometimes like particles, sometimes like waves, and sometimes like nothing that we know about on the scale at which we live. Two laws that describe their behavior in particular are basic and easily stated.

Heisenberg's uncertainty principle. *It is impossible to specify completely the position and momentum of a particle, such as an electron.*

Pauli's exclusion principle. *Two particles of a certain class that are essentially the same cannot be in the same exact state.* This class, the *fermions,* includes such particles as the electron, neutron, and proton. Particles of a different class, the *bosons,* do not obey Pauli's exclusion principle. (See "Subatomic Particles" on the next page.)

S U B A T O M I C   P A R T I C L E S

During the nineteenth century, most scientists came to believe that every-thing was made from atoms, even though there was no way then to detect atoms. We now know that they were right, and there are even "pho-tographs" of individual atoms available, images made with the scanning tunneling microscope. Early in the twentieth century, physicists discovered that atoms themselves are made from smaller pieces—*subatomic particles*. At first, these smaller pieces seemed the ultimate limit of matter, although more and more of them kept appearing. Then, in the 1960s, physicists pro-posed that many subatomic particles are themselves made from smaller particles that are not detectable. Like the undetectable atoms of the nine-teenth century, quarks, the undetectable smaller particles, have come to be accepted. Perhaps in the next century there will be "photographs" of quarks.

Beside quarks, various other groups of particles are thought not to be made up of smaller pieces. Among these are the leptons. Together, the quarks and leptons form what we think of as matter. They are character-ized by a spin of ½, as are the particles made from three quarks, the baryons. *Spin* is a number for each particle that behaves something like the number describing the rotation of an ordinary object. Other subatomic particles are similar to the photon, the particle that makes up light. Parti-cles like the photon and other particles with a spin of either 0 or 1 are called bosons, and they act as the "glue" that holds matter together. They produce the four known forces: gravity, electromagnetism, the strong force, and the weak force. Under certain conditions, electromagnetism and the weak force become a single force, the *electroweak* force.

Many subatomic particles carry a charge, which is a unit of the force of electromagnetism. All charges are counted as either -1 or +1 or 0 (no charge), except for the quarks, which have charges in multiples of ⅓. No one knows why these charges come only in these particular amounts.

The masses of subatomic particles are measured in terms of the energy of the particle, for energy E is related to mass m and the speed of light c by Albert Einstein's famous equation $E = mc^2$. Since c is a very large number, the energy of a particle is much larger than its mass. Even so, the energy is expressed in a very small unit, the MeV, which is a mil-lion *electron volts*. An electron volt is the energy measured as 0.0000000000000000001602 (1.602×10^{-19}) joule (see "Units of Measure," p 638).

Recently, many physicists have proposed various other undetectable or at least undetected particles, called WIMPs (for weakly interacting massive particles); these are far from being fully accepted at this time, and are omitted from the table. Every particle mentioned has an antiparticle whose charge is the opposite (negative particles have positive antiparticles) and whose spin is in the opposite direction. Only a few antiparticles are listed below, mainly as examples.

BASIC PARTICLES OF MATTER (FERMIONS—ALL SPIN 1/2)

Leptons

The electron is generally considered to have been discovered in 1897 by British physicist Joseph John (J.J.) Thomson. Movement of electrons is the source of current electricity, while an excess or deficit of electrons causes static electricity. The properties of the electron form the basis of electronic devices such as computer chips. Electrons are found in all atoms, where they occupy several shells around the outside of the atom. Interactions between electrons account for all chemical reactions. Like all subatomic particles, the electron has a related particle called an antiparticle. The antiparticle of the electron is known as the positron. It is a mirror image of the electron and has a charge of +1. The positron was the first antiparticle to be discovered. It was proposed in 1931 by Paul Adrien Maurice Dirac and discovered, accidentally, in 1932 by Carl David Anderson.

Muons and tauons are very poorly understood particles. The muon is often described as a "fat electron," since it has all the properties of an electron except that its mass is 200 times as great. Similarly, the tau is a "fat muon." No one predicted these particles and no one knows what their role in the universe is. The muon was discovered in cosmic ray radiation by Carl David Anderson in 1937. The tauon was not discovered until 1975.

	Charge	Mass (MeV/c²)	Average Lifetime (sec)	Spin
Electron (e)	-1	0.511	Stable	$\frac{1}{2}$
Positron ($\bar{e}$)	+1	0.511	Stable	$\frac{1}{2}$
Muon (μ)	-1	105.7	2.2×10^{-6}	$\frac{1}{2}$
Tauon (τ)	-1	1750	NA	$\frac{1}{2}$

The neutrino family is a group of particles associated with electrons, muons, and tauons (see "What's Going on at 17 KeV?," p 542). For each of the three charged particles there is a neutrino and an antineutrino. Aside from the fact of their associations, there would seem to be no difference among the three neutrinos. They do not interact strongly with anything. Neutrinos are passing through your body all the time and nearly all of them go on to pass through Earth and out the other side. Neutrinos were predicted by Wolfgang Pauli in 1930 and named neutrino ("little neutral one" in Italian) by Enrico Fermi in 1932. They were first found experimentally by Clyde Lorrain Cowan and Frederick Reines in 1955.

	Charge	Mass (MeV/c²)	Average Lifetime (sec)	Spin
Electron neutrino	0	Less than 2×10^{-5}	Stable*	$\frac{1}{2}$
Muon neutrino	0	Less than 0.3	Stable*	$\frac{1}{2}$
Tau neutrino	0	Less than 40	Stable*	$\frac{1}{2}$

*One theory suggests that one type of neutrino may change into another type.

Quarks

Quarks were first proposed by Murray Gell-Mann in 1964 to account for the relationships between various kinds of baryons. Each baryon is composed of three quarks and each meson of two quarks. Common baryons and mesons are composed of quarks that are known as up and down. Even though their individual masses are small, the binding energy between quarks in a baryon produces most of the mass in the universe. Other baryons or mesons have a quality known as *strangeness,* which is conferred by the strange quark, or a quality known as *charm,* conferred by the charm quark. Two other quarks, known as top and bottom (once also called truth and beauty) complete the list of six "flavors." Each quark also comes in one of three "colors," although physicists disagree on what to call the colors (red, blue, and green is one of the popular choices). In a meson or a baryon, the colors must be combined so as to produce absence of color, so none can be detected directly. Furthermore, quarks are confined within the particles they make up, and cannot be directly detected.

	Charge	Mass (MeV/c²)	Average Lifetime (sec)	Spin
Up (u)	$+\frac{2}{3}$	40	Stable	$\frac{1}{2}$
Down (d)	$-\frac{1}{3}$	70	Stable	$\frac{1}{2}$
Strange (s)	$-\frac{1}{3}$	150	Stable	$\frac{1}{2}$
Charm (c)	$\frac{2}{3}$	1500	N.A.	$\frac{1}{2}$
Top (t)*	$\frac{2}{3}$	Greater than 89,000 but less than 250,000	N.A	$\frac{1}{2}$
Bottom (b)	$-\frac{1}{3}$	4700	Less than 5×10^{-12}	$\frac{1}{2}$

*Not yet observed, but predicted by theory.

BARYONS (ALL SPIN 1/2)

Ordinary Baryons

The proton was the first baryon to be discovered (by Ernest Rutherford in 1914). It is found in the nucleus of the atom. For about 20 years it was assumed that atoms consist of a core of protons surrounded by electrons, although the actual situation is somewhat more complex. The proton appears to be stable, although some recent theories hold that protons may decay into pure energy after about 10^{31} years. So far, although much watched, no one has seen a proton decay. Every atom contains an equal number of protons and electrons, giving a total charge to the atom of 0.

The neutron is almost exactly like a proton, but it has no charge so it was much harder to detect. James Chadwick discovered the neutron, which was not predicted, early in 1932. Neutrons in the nuclei of atoms are usually stable, but neutrons left to themselves soon decay into a proton, an electron, and an electron antineutrino. All atoms except hydrogen must have neutrons in their nuclei to be stable. Each neutron has a mass slightly greater than its fellow nucleon, the proton.

Strange Baryons

Other baryons include two hyperons, three sigmas, two xis, and an omega. Except for the omega, none was predicted. Instead, the baryons were found in cosmic-ray and particle-accelerator experiments in the late 1940s and 1950s. Murray Gell-Mann predicted the omega on the basis of a theory that was preliminary to the quark theory in 1961. The omega was discovered in 1964. None of these particles is stable, decaying after much less than a second into other particles. They are not constituents of ordinary matter and are "strange" because each contains a strange quark.

In physics, a bar over the symbol for a particle indicates the symbol for the corresponding antiparticle, so $\bar{p}$ means antiproton and $\bar{u}$ refers to the anti-up quark.

	Quark content	Charge	Mass (MeV/c²)	Average Lifetime (sec)	Spin
Proton (p)	uud	+1	938.3	Stable or 10^{31}	½
Antiproton ($\bar{p}$)	$\overline{uud}$	-1	938.3	Stable or 10^{31}	½
Neutron (n)	udd	0	939.6	9.18×10^2	½
Antineutron ($\bar{n}$)	$\overline{uud}$	0	939.6	8.18×10^2	½
Lambda (L)	uds	0	1116	3×10^{-10}	½
Positive sigma (Σ+)		+1	1189.4	8×10^{-9}	½
Negative xi (Ξ)		-1	1315	1.7×10^{-19}	½
Omega (Ω)	sss	-1	1672	1.3×10^{-10}	½

MESONS (SPIN 0 OR 1)

Pions were predicted in 1935 by Hideki Yukawa. They are the bosons that hold the nucleus of atoms together and come in positive, negative, and neutral forms. When a pion is exchanged between a proton and a neutron, each nucleon can change each into the other. In the process, the exchange produces the strong force, which is needed to keep the positively charged protons from rushing apart due to the electromagnetic force (positive charges repel each other). When the muon was first discovered in 1937, scientists thought the muon was the particle Yukawa predicted, but by 1945 it was known that the properties of the muon were wrong for that role. In 1947 Cecil Frank Powell and coworkers located the pion in cosmic rays. Pions, like all mesons, are composite particles made from two quarks. Other bosons are thought to be elementary particles, not made up of other particles.

Other mesons. Various short-lived mesons heavier than the pion incorporate such quarks as strange, top, and bottom. None of them are constituents of ordinary matter, but the neutral K mesons, or kaons, have been very important in experiments that extended basic physical theories.

	Charge	Mass (MeV/c^2)	Average Lifetime (sec)	Spin
Positive pion (π^+)	+1	140	2.6×10^{-8}	0
Negative pion (π^-)	-1	189.6	2.6×10^{-8}	0
Neutral pion (π^0)	0	135	8×10^{-17}	0
Positive kaon (K^+)	+1	493.7	1.2×10^{-8}	0
Negative kaon (K^-)	-1	493.7	1.2×10^{-8}	0
K-zero-short (K^0_S)	0	497.7	9×10^{-11}	0
K-zero-long (K^0_L)	0	497.7	5.2×10^{-8}	0

OTHER BOSONS

The photon is the agent of electromagnetic radiation, including ordinary light, radio waves, microwaves, X rays, and gamma rays. In the nineteenth century Thomas Young demonstrated the wave nature of electromagnetic radiation, but in 1905 Albert Einstein showed that light also has a particle nature. The particle came to be called the photon. Exchanging photons causes charged particles to be attracted (if the charges are unalike) or repelled (if the charges are alike). When an electron absorbs or emits a photon of sufficient energy, it can change into a positron. If a positron and an electron meet, they disappear, leaving an energetic photon. Electrons can emit and absorb photons in other ways as well.

Gluons. Although the Yukawa theory of the pion seemed to explain the strong force, the quark theory soon led to the understanding that pions and the strong force are both side effects of a more essential strong force, one carried by eight neutral particles called gluons. Exchanging gluons between quarks usually causes quarks to change from one color to another, keeping the quarks attracted to each other.

	Charge	Mass (MeV/c²)	Average Lifetime (sec)	Spin
Red to blue	0	0	Stable	1
Red to green	0	0	Stable	1
Green to red	0	0	Stable	1
Green to blue	0	0	Stable	1
Blue to red	0	0	Stable	1
Blue to green	0	0	Stable	1
Neutral (1)	0	0	Stable	1
Neutral (2)	0	0	Stable	1

Other Vector Bosons

Like the photon, the positive and negative W particles and the neutral Z particle are vector bosons. These three were predicted by the electroweak theory, and produced and detected by Carlo Rubbia and coworkers in 1983. They are very massive. Two other particles of this class have been predicted, but not detected. They are the Higgs particle, a massive particle that helps give mass to the W and Z particles, and the massless graviton, which should have the same relation to gravity as the photon does to electromagnetism.

	Charge	Mass (MeV/c²)	Average Lifetime (sec)	Spin
Photon	0	0	Stable	1
Positive W	+1	80600	10^{-20}	1
Negative W	-1	80600	10^{-20}	1
Neutral Z	0	91160	10^{-20}	1
Higgs particle*	0	1000000	N.A.	0
Graviton*	0	0	Stable	0

* Not yet observed, but predicted by theory.

The Nobel Prize for Physics

Date	Name [Nationality]	Achievement
1991	Pierre-Gilles de Gennes [French: 1933-]	Studies of phase transitions, especially in liquid crystals
1990	Richard E. Taylor [US: 1929-] I. Friedman [US: 1930-] Henry W. Kendall [US: 1926-]	Confirmation of the existence of quarks
1989	Norman F. Ramsey [US: 1915-] Hans G. Dehmelt [US: 1922-] Wolfgang Paul [German: 1913-]	Development of the separated oscillatory field method and its use in atomic clocks (Ramsey) and the ion trap technique (Dehmelt and Paul)
1988	Leon M. Lederman [US: 1922-] Melvin Schwartz [US: 1932-] Jack Steinberger [US: 1921-]	Development of neutrino-beam methods and discovery of the muon neutrino
1987	Georg J. Bednorz [Swiss: 1950-] K. Alex Müller [Swiss: 1927-]	Discovery of warm-temperature superconductivity
1986	Ernst Ruska [German: 1906-1988] Gerd Binnig [German: 1947-] Heinrich Rohrer [Swiss: 1933-]	Development of the electron microscope (Ruska) and the scanning tunneling microscope (Binnig and Rohrer)
1985	Klaus von Klitsing [German: 1943-]	Discovery of the quantized Hall effect
1984	Carlo Rubbia [Italian: 1934-] Simon van der Meer [Dutch: 1925-]	Detection of the W and Z intermediate vector bosons
1983	William A. Fowler [US: 1911-] Subrahmanyan Chandrasekhar [US: 1910-]	Investigations into the aging and ultimate collapse of stars
1982	Kenneth G. Wilson [US: 1936-]	Theory of phase transitions
1981	Nicolaas Bloembergen [US: 1920-] Arthur L. Schawlow [US: 1921-] Kai M. Siegbahn [Swedish: 1918-]	Advances in technological applications of lasers for the study of matter
1980	James W. Cronin [US: 1931-] Val L. Fitch [US: 1923-]	Studies in the asymmetry of subatomic particles
1979	Steven Weinberg [US: 1933-] Sheldon L. Glashow [US: 1932-] Abdus Salam [Pakistani: 1926-]	Link between electromagnetism and the weak force of radioactive decay
1978	Arno A. Penzias [US: 1933-] Robert W. Wilson [US: 1936-] Pyotr L. Kapitsa [USSR: 1894-1984]	Penzias and Wilson for study of microwave radiation and Kapitsa for work in low-temperature physics
1977	Philip W. Anderson [US: 1923-] Nevill F. Mott [UK: 1905-] John H. Van Vleck [US: 1899-1980]	Electronic research in computer memories
1976	Burton Richter [US: 1931-] Samuel C.C. Ting [US: 1936-]	Discovery of J/psi subatomic-particle

Date	Name [Nationality]	Achievement
1975	James Rainwater [US: 1917-1986] Ben Mottelson [Danish: 1926-] Aage Bohr [Danish: 1922-]	Studies proving asymmetrical structure of the atomic nucleus
1974	Anthony Hewish [UK: 1924-] Martin Ryle [UK: 1918-1984]	Hewish for discovery of pulsars and Ryle for radiotelescopy
1973	Leo Esaki [Japanese: 1925-] Ivar Giaever [US: 1929-] Brian Josephson [UK: 1940-]	Theories on superconductors and semiconductors important to microelectronics
1972	John Bardeen [US: 1908-1991] Leon N. Cooper [US: 1930-] John R. Schrieffer [US: 1931-]	Theory of superconductivity without electrical resistance at temperature of absolute zero
1971	Dennis Gabor [UK: 1900-1979]	Invention of holography
1970	Hannes Alfvén [Swedish: 1908-] Louis Néel [French: 1904-]	Alfvén for plasma physics and Néel for antiferromagnetism
1969	Murray Gell-Mann [US: 1929-]	Classification of elementary particles
1968	Luis W. Alvarez [US: 1911-1988]	Work with elementary particles
1967	Hans A. Bethe [US: 1906-]	Study of energy production of stars
1966	Alfred Kastler [French: 1902-1984]	Optical study of subatomic energy
1965	Julian S. Schwinger [US: 1918-] Richard P. Feynman [US: 1918-1988] Shin'ichero Tomonaga [Japanese: 1906-1979]	Research into the basic principles of quantum electrodynamics
1964	Charles H. Townes [US: 1915-] Nikolai G. Basov [USSR: 1922-] Alexander Prokhorov [USSR: 1916-]	Development of maser and laser principles in quantum mechanics
1963	Eugene Paul Wigner [US: 1902-] Maria Goeppert-Mayer [US: 1906-1972] J. Hans D. Jensen [German: 1907-1973]	Wigner and Goeppert-Mayer for mechanics of proton-neutron interaction and Jensen for theory of nucleic structure
1962	Lev D. Landau [USSR: 1908-1968]	Superfluidity in liquid helium
1961	Robert Hofstadter [US: 1915-1990] Rudolf Mössbauer [German: 1929-]	Hofstadter for measurement of nucleons and Mössbauer for work on gamma rays
1960	Donald Glaser [US: 1926-]	Bubble chamber for subatomic study
1959	Emilio Segrè [US: 1905-1989] Owen Chamberlain [US: 1920-]	Demonstration of the existence of the antiproton
1958	Pavel A. Cherenkov [USSR: 1904-1990] Ilya M. Frank [USSR: 1908-1990] Igor Tamm [USSR: 1895-1971]	Discovery of principle of light emission by electrically charged particles
1957	Tsung-Dao Lee [Chinese: 1926-] Chen Ning Yang [Chinese: 1922-]	Prediction of violations of law of conservation of parity

(Continued)

Date	Name [Nationality]	Achievement
1956	William Shockley [US: 1910-1989] Walter H. Brattain [US: 1902-1987] John Bardeen [US: 1908-1991]	Studies on semiconductors and invention of the electronic transistor
1955	Willis Lamb, Jr. [US: 1913-] Polykarp Kusch [US: 1911-]	Lamb for measurement of hydrogen spectrum and Kusch for magnetic momentum of electron
1954	Max Born [UK: 1882-1970] Walther Bothe [German: 1891-1957]	Born for work in quantum mechanics and Bothe in cosmic radiation
1953	Fritz Zernicke [Dutch: 1888-1966]	Phase-contrast microscope
1952	Felix Bloch [US: 1905-1983] Edward Mills Purcell [US: 1912-]	Measurement of magnetic fields of atomic nuclei
1951	Sir John Cockcroft [UK: 1897-1967] Ernest T.S. Walton [Irish: 1903-]	Transmutation of atomic nuclei in accelerated particles
1950	Cecil Powell [UK: 1903-1969]	Photographic study of atomic nuclei
1949	Hideki Yukawa [Japanese: 1907-1981]	Prediction of mesons
1948	Patrick Blackett [UK: 1897-1974]	Discoveries in cosmic radiation
1947	Sir Edward Appleton [UK: 1892-1965]	Discovery of ionic layer in atmosphere
1946	Percy W. Bridgman [US: 1882-1961]	Laws of high-pressure physics
1945	Wolfgang Pauli [Austrian: 1900-1958]	Pauli principle of exclusion
1944	Isidor Rabi [US: 1898-1988]	Magnetic properties of nuclei
1943	Otto Stern [US: 1888-1969]	Magnetic momentum of protons
1942	No award	————————
1941	No award	————————
1940	No award	————————
1939	Ernest Lawrence [US: 1901-1958]	Invention of the cyclotron
1938	Enrico Fermi [Italian: 1901-1954]	Discovery of new radioactive elements
1937	Clinton Davisson [US: 1881-1958] George P. Thomson [UK: 1892-1975]	Discovery of electron diffraction by crystals
1936	Victor F. Hess [Austrian: 1883-1964] Carl D. Anderson [US: 1905-1991]	Hess for discovery of cosmic radiation and Anderson for the positron
1935	Sir James Chadwick [UK: 1891-1974]	Discovery of the neutron
1934	No award	————————
1933	Paul Dirac [UK: 1902-1984] Erwin Schrödinger [Austrian: 1887-1961]	Discovery of new equations for atomic theory
1932	Werner Heisenberg [German: 1901-1976]	Discovery of quantum mechanics
1931	No award	————————

Date	Name [Nationality]	Achievement
1930	Sir Chandrasekhara Raman [Indian: 1888-1970]	Laws of light diffusion
1929	Prince Louis de Broglie [French: 1892-1987]	Discovery of wave character of electrons
1928	Sir Owen Richardson [UK: 1879-1959]	Studies of effect of heat on electron emission
1927	Arthur Holly Compton [US: 1892-1962] Charles Wilson [UK: 1869-1959]	Compton for effect of wavelength change in X rays and Wilson for visible ion tracings
1926	Jean Baptiste Perrin [French: 1870-1942]	Discovery of discontinuous structure of matter and equilibrium of sedimentation
1925	James Franck [German: 1882-1964] Gustav Hertz [German: 1887-1975]	Discovery of the laws of electron impact on atoms
1924	Karl Siegbahn [Swedish: 1886-1978]	X-ray spectroscopy
1923	Robert Millikan [US: 1868-1953]	Calculation of elementary electrical charge
1922	Niels Bohr [Danish: 1885-1962]	Discovery of atomic structure and radiation
1921	Albert Einstein [German: 1879-1955]	Law of photoelectric effect
1920	Charles Guillaume [Swiss: 1861-1938]	Special nickel-steel alloys
1919	Johannes Stark [German: 1874-1957]	Doppler effect in canal rays and spectral lines in electrical fields
1918	Max Planck [German: 1858-1947]	Quantum theory of light
1917	Charles Barkla [UK: 1877-1944]	X-ray diffusion of elements
1916	No award	————
1915	Sir William Bragg [UK: 1862-1942] Sir Lawrence Bragg [UK: 1890-1971]	Study of crystals with X rays
1914	Max von Laue [German: 1879-1960]	Crystal diffraction of X rays
1913	Heike Kamerlingh Onnes [Dutch: 1853-1926]	Low temperature physics and liquefaction of helium
1912	Nils Gustaf Dalén [Swedish: 1869-1937]	Automatic gas regulators for lighthouses and sea buoys
1911	Wilhelm Wien [German: 1864-1928]	Laws of heat radiation
1910	Johannes Van der Waals [Dutch: 1837-1923]	Equation of state, liquids, and gases
1909	Guglielmo Marconi [Italian: 1874-1937] Karl Ferdinand Braun [German: 1850-1918]	Wireless telegraphy
1908	Gabriel Lippmann [French: 1845-1921]	Color photography
1907	Albert A. Michelson [US: 1852-1931]	Spectroscopic and metrologic study of light

(Continued)

Date	Name [Nationality]	Achievement
1906	Sir Joseph Thomson [UK: 1856-1940]	Electrical conductivity of gases
1905	Philipp Lenard [German: 1862-1947]	Work on cathode rays
1904	John Strutt, Lord Rayleigh [UK: 1842-1919]	Discovery of argon
1903	Antoine Henri Becquerel [French: 1852-1908] Pierre Curie [French: 1859-1906] Marie Curie [French: 1867-1934]	Becquerel for discovery of spontaneous radioactivity and Curies for later study of radiation
1902	Hendrik A. Lorentz [Dutch: 1853-1928] Pieter Zeeman [Dutch: 1865-1943]	Discovery of effect of magnetism on radiation
1901	Wilhelm K. Roentgen [German: 1845-1923]	Discovery of X rays

Technology

THE STATE OF TECHNOLOGY

Until Galileo's time (the early seventeenth century), science as we know it today did not really exist, but technology of a type not totally unlike what we have today had been a part of human experience for over 2 million years, even preceding the evolution of modern humans. Galileo himself was among the first to connect technology and science, but the flow was from technology to the newly emerging science. The technological advance of the telescope, although not invented by Galileo, became in Galileo's hands a scientific tool. If a needed tool had not been developed by technology, however, Galileo was not above inventing it himself, as was the case with the first crude thermometer.

Since the seventeenth century, the relationship between technology and science has remained close. By the middle of the nineteenth century the flow of information was often in the other direction, from science to technology. In our century it is clear that the two disciplines reinforce each other in many ways. The phrase "science and technology" today seems a single idea, like horse and buggy or bagels and lox. As in those famous pairs, the two ideas remain separate, but work much better together than apart.

Computer Progress and Change

A great deal of technology in the early 1990s that affects not only large organizations but also many individuals is tied into the computer. A whole book would be needed to handle the details, but here are some of the main trends.

It became apparent toward the end of 1991 that the massively parallel version of the supercomputer would replace the vector supercomputer, familiar from various Cray supercomputer models of the 1980s, for most data-intensive purposes. A massively parallel computer uses a large array of different chips to enable it to work on many small aspects of a problem at once. Previous supercomputers, like almost all computers since the first, were designed to perform all operations sequentially. On Jun 4 1991 Thinking Machines Inc. of Cambridge MA claimed that its latest version of a massively parallel supercomputer had set a record for speed in calculation at 9.03 billion mathematical operations a second. A supercomputing trade show in Albuquerque NM on Nov 18 1991 showcased several of the new machines, notably the CM-5 (or Connection Machine 5) from Thinking Machines. Cray Research announced in 1991 that it would build its own massively parallel computer with a target of 1993 or 1994.

On a more prosaic level, the early 1990s saw the first widespread introduction of computers for which the input device was a form of pen or pencil instead of a keyboard. Early "pentop" computers tried to recognize letters, and act as the functional equivalent of a keyboard. Human handwriting defeated them, and they generally missed about 5 to 10 percent of the letters' true identities. Some pentops introduced toward the end of 1991, however, simply capture the strokes and save them as graphics, leaving decipherment up to the human mind, which has considerable experience in the field (although not with a 100 percent success rate either). While solving some of the error-rate problem, the method of capturing pen strokes uses considerably more of the computer's storage memory than capturing the letters themselves. New pen-based software was promised for 1992 from the Go Corporation of Foster City CA and from Microsoft Corporation.

Another new development in personal computers is the ability to show realistic pictures (as realistic as color television, at least) in an environment in which the user can interact directly with sight and sound. Apple Computer's Quicktime, due out in the middle of 1992, will add movielike pictures and sound to existing Macintosh programming. Microsoft is planning similar capabilities for the next version of the Windows operating system. Because of the interactive nature of such programs, existing videotapes can be easily edited into whatever the operator desires. At a simpler level, Nintendo is introducing interactive games with real actors instead of the simplified cartoon characters seen in the past.

Somewhat related to the new pictorial capacity for computers is the increasing use of combined digital-analog chips, which are used among other things for image compression. Other uses include transferring between digital data and analog instructions in the tiny hard drives of laptop computers and working with sound recognition and production.

Unexpected Computer Problems

The computer moved from being an exciting new technology for individual use and a necessity for large corporations to being a necessity for most kinds of office workers and for all sorts of science in the 1980s. In the early 1990s the personal computer for ordinary tasks and the supercomputer for special purposes had become ubiquitous. But among the results of widespread use were several unexpected problems.

The most imaginative science fiction writers of the 1940s—a time when such writers as Frederic Brown and Arthur C. Clarke had already described one computer that became God and another that erased the universe one star at a time—failed to predict the computer virus, a wisp of memory that lodges in some part of the workings of a computer, hidden from defenses against it, and reproduces itself in that computer and in other computers that are exposed to the infected computer. The analogy with a biological virus is startlingly exact. Although computer viruses became familiar to the general public during a major epidemic in 1988, they have been around much longer—the term was created in 1983 by Fred Cohen, and primitive forms existed even earlier. Computer viruses were fought with little public fanfare in the early 1990s, but their menace was greater than ever because more and more computers found themselves tied together in networks, many of them worldwide.

Furthermore, the computer programmers who create viruses—now universally known as hackers, a corruption of the original meaning of *hacker*, first used to describe people who devoted virtually their total lives to programming—are learning new tricks from each other. By 1991 the common virus was the "stealth" type, so called because it is especially difficult to detect.

A typical stealth virus, which came to be called *stoned*, emerged in mid-1991. It took over the computer monitor and displayed the message "Your computer is now stoned. Legalize marijuana." By Dec 6 a version of *stoned* had infected the game Spacewrecked, produced by Konami Inc. of Buffalo Grove IL, although the virus was caught before many diseased disks were shipped. Of more concern, the variant called *stoned III* infected a reference disk for a program called Netware shipped to about 3000 customers of Novell Inc. of Provo UT, the largest supplier of office-network software.

The main problem with *stoned III* is that it can write itself over random portions of a hard disk, sometimes erasing vital control portions in the process. The computer can then misplace large sections of recent memory; in this way the virus can cause any problem from mild forgetfulness through total amnesia.

An even more virulent virus, *Michelangelo*, infected Leading Edge computers shipped between Dec 10 and Dec 27 1991. *Michelangelo*, first observed in Europe in Apr 1991, copies itself onto diskettes that then copy the virus onto hard disks in any computer that uses them. On Mar 6, the birthday of artist Michelangelo, the virus writes random information over the hard disk, effectively destroying any information stored in the computer.

When the first computers were being designed, no one expected this problem. The US National Academy of Sciences urged on Dec 5 1990 that the government adopt new security measures for computers to protect against viruses and against penetration by unauthorized people trying to obtain information or to subvert programs to undesirable purposes.

Computer viruses were not the headline stories in 1990 and 1991 that they were in the late 1980s, but *forgotten* does not necessarily mean *gone*. A 1991 survey released Nov 25 of 606 large companies and government agencies in the United States revealed that 63 percent had had problems with viruses in the past, with 19 percent reporting the presence of a virus in Sep 1991. All the companies or agencies involved in the survey had at least 300 computers in use. During the first 9 months of 1991, 9 percent of the companies reported virus attacks that affected as many as 25 percent of their computer force. There were about a thousand known viruses active in 1991. They attack via computer networks or by sharing of disks with programs.

A survey conducted by Dataquest of San Jose CA of 600,000 business computer users in North America also revealed that 63 percent claimed to have had trouble with computer viruses, causing data loss in 38 percent of the instances.

Viruses returned to the headlines early in 1992 as *Michelangelo* was scheduled to strike on Michelangelo's birthday anniversary of Mar 6, presumably to wipe out all the data on hard disks of infected personal computers. When the fateful day arrived, virtually no one was infected. In preparing for the virus to strike, many had carefully backed up their hard drives—which all go out eventually—and a few had found other viruses while looking out for *Michelangelo*. All in all, the *Michelangelo* virus scare probably did some good and very little harm.

The struggle between the legitimate software programmers and the hackers can be likened to the nineteenth-century confrontation between armor manufacturers and bomb designers. Each improvement by one side leads to corresponding improvements on the other. A few computer theorists think that this is a good thing, that ultimately beneficial programs will incorporate ideas and techniques from the viruses. Most, however, are more alarmed than encouraged. Although there are no known cases so far of malevolent computer terrorism, the possibility remains ever present. As we become more and more dependent on computers that are connected in more ways than ever, the possibility of an Andromeda strain virus destroying a main part of our current technology base is alarming indeed.

Another set of unexpected problems is related to the physical use of a keyboard and video display monitor, the most common appurtenances for the desktop computer. The keyboard had long been used by secretaries and clerk-typists. When executives began to use keyboards extensively, the difficulty known as carpal tunnel syndrome—a hand disorder—appeared. Until pentop systems are perfected, or the even better voice-operated computer, many people have solved their keyboard problems with various versions of the device called the mouse, which can be helpful in reducing dependence upon the obnoxious keyboard to a few strokes here and there.

Many have suspected that video monitors cause at least eyestrain and possibly other harmful health effects. People heard that some radiation leaked from monitors and presumed incorrectly that the monitors were radioactive. Various US legislative bodies from coast to coast (at least from Suffolk County, Long Island, in New York in 1988 to San Francisco in 1990) have tried to legislate hours and other conditions to protect workers from video monitors. The San Francisco rules, passed Dec 17 1990, suggest the nature of such regulations in general. Lighting, keyboards, and chairs were regulated and breaks every couple of hours were to be provided. A committee was set up to look for video-display-caused cancer or miscarriages. Compliance for the ordinance was scheduled for Jan 1992, but will be phased in over a four-year period.

Despite such concerns, a major study released Mar 14 1991 in *The New England Journal of Medicine* showed no additional risk from miscarriages in telephone workers who use computer terminals.

Workers were affected by the unwanted way that supervisors could use computer networks to look over employees' shoulders. Trans World Airlines is just one of many companies that uses the computers to keep track of how fast and efficient employees are. One study in Oct 1990 showed that monitored workers are more apt to be depressed or to be affected by physical side-effects while using a computer.

Engineering Advances

On Mar 6 1990 Ed Yeilding and J.T. Vida set a record for a flight from California to Maryland of 1 hour 8 minutes 17 seconds. They flew an SR-71 Blackbird spy plane on its way from its California base, where its use is being discontinued, to the Washington DC area so that the Blackbird could be shown at the Smithsonian Institution.

One breakthrough in engineering in the early 1990s was a literal breakthrough. At 11:21 a.m. on Dec 1 1990 Greenwich mean time the English and French workers on each side of the Channel Tunnel under the English Channel, known as the Chunnel, broke through and shook hands. They were 22.4 km (13.9 mi) from the English coast and 20.3 km (9.7 mi) from the French. Digging had begun in 1987 and the Chunnel was scheduled for operation as a railroad tunnel starting in Jun 1993.

Materials Advances

A lot of the progress in technology in the early 1990s had to do with the introduction of either new materials, such as the sol-gels that are incredibly light rigid structures, or new versions of old materials, such as diamond (not to mention buckyballs, discussed on p 157 and zeolites, p 171).

Sol-gels, also known as aerogels or solid smoke, are only four times as dense as air at sea level. They were discovered 60 years ago but began to be put to practical use for the first time in the early 1990s. The first uses were for insulation, but new uses of all sorts were proposed, including such exotic ideas as traps for micrometeorites.

A sol-gel is solid smoke, but a electrorheological (ER) fluid is temporarily solid goo. An ER fluid, such as cornstarch and water, becomes an ER solid for as long as it is in an electric field. When the field is removed, the solid becomes liquid again. Although known since the 1940s, the ER effect is still not explained by theory. Without waiting for an explanation, however, scientists in the early 1990s turned to ER for faster switches and for mechanical devices of extreme accuracy. No consumer products exist yet, but predictions are for ER components in automobiles before the 1990s are over.

While brides-to-be think of diamonds (cut stones), electronics engineers and materials scientists in general think of diamond (a particular arrangement of carbon atoms into crystal). Diamond has long been known as the hardest material available, resulting in industrial use as an abrasive and for cutting other hard materials. Diamond also has other characteristics that make it especially valuable. It is an extremely good insulator for electricity, but at the same time it conducts heat better than any other material. In addition, diamond is exceptionally transparent to most electromagnetic radiation. Such a combination of qualities is rarely found in the same substance.

These useful characteristics make diamond attractive to manufacturers of electronic devices. For such devices, the diamond should not be in the form of a lump, however; instead, carbon atoms should form a thin coating over other materials while still maintaining the characteristic diamond crystalline structure. Slightly thicker diamond chips could replace silicon in applications where heat would destroy the kind of chips now used. A properly doped diamond chip would not only resist heat, but it would be faster and use less electricity than a silicon chip.

On Apr 19 1991 Jagdish Narayan and Vijay Godbole of North Carolina State University and Carl White of the US Oak Ridge National Laboratory announced in *Science* that they had succeeded in growing thin single-crystal diamond films on metal surfaces. Although their first efforts resulted in films only 100 square microns in area, the researchers say that there is no reason why the same technique could not be used to produce films as large as the palm of a person's hand.

Timetable of Artificial Diamonds

Antoine Lavoisier [French: 1743-1794] had demonstrated in 1772 that diamond was nothing but an expensive form of carbon. The virtues of diamond are such, however, that the first attempts by modern chemists to make artificial diamond started in the nineteenth century. Despite early failures, the ability to grow diamonds in various forms has existed for a long time.

1893	Henri Moissan [French: 1852-1907] announces that he has used heat and pressure to produce artificial diamond; later studies show, however, that he did not have sufficient heat or pressure to accomplish this and that the diamonds he exhibited had probably been placed in his equipment as a hoax by perpetrators unknown.
1905	Percy Bridgman [US: 1882-1961] develops the first devices to produce pressures higher than 100 atmospheres.
1955	After 50 years' experience, Bridgman and coworkers at General Electric use his high-pressure devices to produce artificial diamonds, which are immediately offered for sale
1958	A method for producing diamond from methane, later called chemical vapor deposition, is patented in the United States, but the diamond is mixed with graphite, making it of limited use for commercial purposes.
1977	Researchers in the Soviet Union find a way to make diamond from methane without the graphite impurities.
1981	The Soviet Union diamond researchers produce both single-crystal diamond films on previously existing diamonds and multiple-crystal diamonds on metal; single-crystal diamonds are much more desirable for many purposes
1990	In Jul, workers in the United States report that they can grow pure carbon-12 diamond films that conduct heat 50 percent better than natural diamond, which contains 1 percent carbon-13; carbon-12 diamond also can withstand laser radiation much better than natural diamond

Narayan's team achieved success by abandoning the promising method called chemical vapor deposition (CVD) that had been used for most diamond growing since it was first invented in 1958. CVD requires producing a mist of a material containing carbon (such as methane) from which the carbon atoms are liberated by chemical or thermal techniques, where they fall onto heated surfaces and spontaneously arrange themselves into diamond islands. The main problem with this technique is that those islands do not merge into a single crystal.

Narayan puts the carbon directly on a copper surface by blasting the atoms onto the surface. Then he and his team use a laser to liquefy the layer of carbon-suffused copper, melting it. Essentially, the carbon atoms float to the top and form a thin layer. When the temperature returns to normal, the diamond is left like a layer of ice on a lake in winter.

(Periodical References and Additional Reading: *Science News* 1-13-90, p 21; *Science* 2-16-90, p 807; *Science* 3-9-90, p 1180; *Science News* 5-5-90, p 287; *Science* 8-10-90, p 627; *Discover* 8-90, p 26; *Science News* 11-17-90, p 316; *New York Times* 12-11-90; *New York Times* 12-18-90, p A22; *Science* 12-21-90, p 1641; *New York Times* 12-23-90, p A1; *New York Times* 12-25-90; *New York Times* 3-24-91, p A22; *Science* 4-19-91, pp 351, 375, & 416; *New York Times* 6-5-91, p D1; *New York Times* 6-24-91, p D2; *New York Times* 6-30-91, p F9; *New York Times* 10-13-91, p F11; *New York Times* 10-27-91, p F11; *New York Times* 10-29-91, p D9; *New York Times* 11-25-91, p L1; *New York Times* 12-2-91, p L16; *New York Times* 1-22-92, p D1)

Timetable of Technology and Invention to 1990

BC

2,400,000	Ancestors of human beings begin to manufacture stone tools
750,000	Ancestors of human beings learn to control fire
23,000	People in the Mediterranean regions of Europe and Africa invent the bow and arrow
7000	People in what is now Turkey begin to make pottery and cloth
5000	Egyptians start mining copper ores and smelting the metal
3500	The potters' wheel and, soon after, wheeled vehicles appear in Mesopotamia (now Iraq, Syria, and Turkey)
2900	The Great Pyramid of Giza and the first form of Stonehenge (with only three stones) are built
2000	Interior bathrooms are built in palaces in Crete
522	Eupalinus of Megara constructs a 1100-m (3600-ft) tunnel on the Greek isle of Samos to supply water from one side of Mt Castro to the other
290	The Pharos lighthouse at Alexandria is built
260	Archimedes [Greek: c 287-c 212] develops a mathematical description of the lever and other simple machines
200	The Romans develop concrete
140	The Chinese start making paper, but do not use it to write upon
100	Water-powered mills in Ilyria (now Yugoslavia and Albania) are introduced

CE

1	About this time the Chinese invent the ship's rudder
190	The Chinese develop porcelain
600	The first windmills are built in what is now Iran
704	Between 704 and 751 the Chinese start printing with woodblocks

1040	The Chinese develop gunpowder
1041	Between 1041 and 1048 Pi Sheng invents movable type in China
1070	The Chinese begin to use the magnetic compass for navigation
1190	The first known reference to a compass is made in Europe
1267	A book written by Roger Bacon [English: c 1220-1292] in 1267 and 1268 mentions eyeglasses to correct farsightedness
1288	The first known gun, a small cannon, is made in China
1310	Mechanical clocks driven by weights begin to appear in Europe
1440	Johann Gutenberg [German: c 1398-c 1468] reinvents printing with movable type about this time
1450	Nicholas Krebs develops eyeglasses for the nearsighted
1590	About this time the compound microscope (using two lenses) is invented in Holland, probably by Zacharias Janssen
1608	The telescope is invented in Holland, probably by Hans Lippershey [German-Dutch: c 1570-c 1619]
1620	Cornelius Drebbel [Dutch: 1572-1633] builds the first navigable submarine
1642	Blaise Pascal [French: 1623-1662] invents the first adding machine
1643	Evangelista Torricelli [Italian: 1608-1647] makes the first barometer, in the process producing the first vacuum known to science
1654	Christiaan Huygens [Dutch: 1629-1695] develops the pendulum clock
1658	Robert Hooke [English: 1635-1703] invents the balance spring for watches
1671	Gottfried Wilhelm Leibniz [German: 1646-1716] builds a calculating machine that can multiply and divide as well as add and subtract
1698	Thomas Savery [English: c 1650-1715] patents "the Miner's Friend," the first practical steam engine
1709	Gabriel Daniel Fahrenheit [German-Dutch: 1686-1736] invents the first accurate thermometer
1733	John Kay [English: 1704-1764] invents the flying-shuttle loom, which, along with improvements in making iron and the steam engine, is a key to the start of the Industrial Revolution
1751	Benjamin Huntsman [English: 1704-1776] invents the crucible process for casting steel
1761	A clock called a marine chronometer, built by John Harrison [English: 1693-1776], is shown to be accurate within seconds over a long sea voyage, making modern navigation possible
1764	James Hargreaves [English: 1720-1778] introduces the spinning jenny, a machine that spins from 8 to 120 threads at once
1765	James Watt [Scottish: 1736-1819] builds a model of his improved steam engine
1769	Richard Arkwright [English: 1732-1792] patents the water frame, a spinning machine that complements the spinning jenny

1783	Joseph-Michael Montgolfier [French: 1740-1810] and Jacques-Etienne Montgolfier [French: 1745-1799] develop the first hot-air balloon
	Later that year, Jacques Charles [French: 1746-1823] builds the first hydrogen balloon
1792	William Murdock [Scottish: 1754-1839] is the first to use coal gas for lighting
1793	Eli Whitney [US: 1765-1825] invents the cotton gin, a machine for separating cotton fibers from seeds
1807	Robert Fulton [US: 1765-1815] introduces the first commercially successful steamboat
1822	Joseph Niepce [French: 1765-1833] produces the earliest form of photograph
1825	George Stephenson [English: 1781-1848] develops the first steam-powered locomotive to carry both passengers and freight
1837	Samuel Finley Breese Morse [US: 1791-1872] patents the version of the telegraph that will be commercially successful
1839	Louis-Jacques Daguerre [French: 1789-1851] announces his process for making photographs, which come to be called daguerrotypes
	Charles Goodyear [US: 1800-1860] discovers how to make rubber resistant to heat and cold, a process called vulcanization
1856	Henry Bessemer [English: 1813-1898] develops the way of making inexpensive steel now known as the Bessemer process
1859	Edwin Drake [US: 1819-1880] drills the first oil well in Titusville PA
1876	Alexander Graham Bell [Scottish-US: 1847-1922] invents the telephone
	Karl von Linde [German: 1842-1934] invents a practical refrigerator
1877	Nikolaus Otto [German: 1832-1891] invents the type of internal combustion engine still used in most automobiles
1878	Louis-Marie-Hilaire Bernigaud [French: 1839-1924] develops rayon
1879	Thomas Edison [US: 1847-1931] and Joseph Swan [English: 1828-1914] independently discover how to make practical electric lights
1885	Karl Benz [German: 1844-1929] builds the precursor of the modern automobile
1889	Gustave Eiffel [French: 1832-1923] builds his famous tower in Paris; at 303 m (993 ft) it is the tallest free-standing structure of its time
1890	Herman Hollerith [US: 1860-1929] develops an electronic system based on punched cards, which is used in counting the US census; later he founds the company that will become IBM
1893	Rudolf Diesel [German: 1858-1913] describes the Diesel engine
1903	Orville Wright [US: 1871-1948] and Wilbur Wright [US: 1867-1912] fly the first successful airplane
1904	John Fleming [English: 1849-1945] develops the first electronic vacuum tube
1909	Leo Baekeland [Belgian-US: 1863-1944] patents Bakelite, the first truly successful plastic

1929   Robert Goddard [US: 1882-1945] launches the first instrumented, liquid-fueled rocket

1930   Frank Whittle [English: 1907-] patents the jet engine

1931   Ernst Ruska [German: 1906-] builds the first electron microscope

1937   John V. Atanasoff [US: 1903-] starts work on the first electronic computer

        Chester Carlson [US: 1906-1968] invents xerography, the first method of photocopying

1939   Paul Müller [Swiss: 1899-1965] discovers that DDT is a potent and long-lasting insecticide

1941   John Rex Whinfield [English: 1901-1966] invents Dacron

1948   William Shockley [English-US: 1910-], Walter Brattain [US: 1902-1987], and John Bardeen [US: 1908-1991] invent the transistor

1957   Gordon Gould [US: 1920-] develops the basic idea for the laser, which he succeeds in patenting in 1986, after a long struggle

1959   Richard Feynman [US: 1918-1988] proposes in a speech on Dec 29 that computers can be built from tiny transistors made by evaporating material away, that new products can be made by combining single atoms, and that circuits can be made from as few as seven atoms

1965   John Kemeny [US: 1926-] and Thomas Kurtz [US: 1928-] develop the computer language BASIC, which later becomes the main programming language for personal computers

1970   The floppy disk is introduced for storing data on computers

1971   The first computer chip is produced in the United States

        The first pocket calculator is introduced in the United States; it weighs about 2.5 pounds and costs about $150

1975   The first personal computer (in kit form) is introduced in the United States; it has 256 bytes of memory

1982   Compact disc players are introduced

1986   The Chernobyl nuclear reactor number 4 explodes, killing dozens of people and making a large area around it uninhabitable

1989   On Jan 3 Japan initiates daily broadcasts of High Definition Television (HDTV) with a one-hour program featuring the Statue of Liberty and New York Harbor

LIGHT AND COLD

OPTICAL COMPUTERS

On Jan 29 1990 a team of scientists at AT&T Bell Laboratories (Bell Labs) led by Alan Huang and Michael Prise demonstrated the first crude all-optical computer. Although practical optical computers are still not available, the various components are being developed. Light technology, as the compact disc (CD) and fiber optics have demonstrated, has various advantages over electronic technology.

The Bell Labs computer is based upon an invention of physicist David Miller, also of Bell Labs. Miller and other physicists learned in 1984 that a thin layer of gallium arsenide could be changed from opaque to transparent by an electric charge or by a laser light tuned to an appropriate frequency. By 1987 Miller had developed the first "optical transistor" using this property. Called the Symmetric Self-Electro-optic Effect Device, or S-SEED, the "optical transistor" can, like the true transistor, be part of a team of many such devices on a single chip. The 1990 computer used four arrays of S-SEEDs, each with thirty-two of the tiny devices. A single array is about as large as a capital letter produced by a typewriter. While the four arrays with the total of 128 S-SEEDs was small, the first optical computer was much larger (as is true of electronic computers as well, even laptops, compared to their chips). The board with the optical computer is about the size of a card-table top and is covered with lasers and lenses.

To operate the computer, the laser beams go on and off, some tuned to change the state of the S-SEED they are directed at and some tuned to be reflected from the S-SEED. The pattern of beams through the array of S-SEEDs can be arranged to follow the same pattern as wires take in connecting transistors in an electronic computer.

As development of the optical computer has continued, a principal ongoing effort is to replace any remaining electronic parts with optical ones. Here is a status report on some of the components being developed for future all-optical systems, including devices other than those from Bell Labs. In reality, however, the all-optical computer when finally developed will no doubt be an *almost all*-optical system; for example, a system that can be plugged into a wall outlet or that runs on batteries seems desirable, at least for the near future.

Generating Light

The Bell Labs computer S-SEEDs uses the semiconductor gallium arsenide because gallium arsenide can emit photons of light. The cheaper and much more familiar semiconductor silicon—as in "Silicon Valley"—simply bounces light off its surface while absorbing a little bit of its energy, just like most other materials. Although silicon can detect light, that is not enough. To be useful, it needs to produce light.

Finally in 1990 scientists found a way to make silicon emit light when either stimulated by an electric current or by light of a different wavelength. (Transparency, a familiar quality of the silicon compound we call glass, really involves emitting light of the same frequency with which the silicon compound is stimulated.)

Electromagnetic radiation at invisible wavelengths or as visible light is emitted from a solid when electrons from the conduction level of the solid fall into holes in the valence level, which releases electromagnetic radiation just the way billiard balls falling into pockets release sound. This normally cannot happen in silicon because electrons in it are in places where they cannot fall into holes by themselves. They need an energy boost. In the billiards analogy, they need enough energy to rise up over the edge of a barrier to reach the pocket. Although the resulting radiation may be infrared or ultraviolet, most physicists simply call the radiation light.

Leigh T. Canham of the UK Defense Research Agency announced in Sep 1990 that he has discovered that he can get silicon to produce light by etching away most of it to produce an array of tiny towers of silicon surrounded by four times as much empty space as there is silicon. The towers are on the order of 50 angstroms wide. Although no one knows exactly why, silicon treated this way emits light of one color when stimulated by light of a different wave length. Somehow, they propose, the towers give the silicon a fractal dimension that allows the electrons to reach the holes by falling directly into them.

A French group confirmed that Canham's methods work for silicon stimulated by a laser. Canham claims that he also can get light by stimulating the silicon sponge with an electric current, which might be important in applications.

A team from the University of San Diego (CA) found that the etched silicon could be broken into tiny bits, suspended in a liquid, and finally dried into a thin film that would still produce light when stimulated. This may turn out to be a more useful form in practical applications than the silicon sponge in its original bulk form.

Silicon is not the only material being used to generate light. One new kind is designed to make blue light. The "old-fashioned" semiconductor lasers made from gallium arsenide or other materials that maintain a direct path between conduction electrons and valence holes usually emit light near the red end of the visible spectrum, which means at long wavelengths for light. Shorter wavelengths—that is, bluer light—could store more information. This would be useful not only for optical computers but also for the more prosaic CDs and laser printers. In 1991, as reported in the Sep 9

Applied Physics Letters, Michael A. Haase and coworkers at 3M Co. in St Paul MN developed semiconductor lasers that emit blue-green light around 500 nanometers in wavelength. The lasers are mainly zinc selenide, layered with cadmium selenide.

Data Storage

One possibility for an all-optical system would be to use holograms to store information instead of patterns of magnetic domains, the common method used for memory storage on floppy or hard disks. Holography, although it has lasers in common with the CD storage that is just beginning to become commonplace, relies on an entirely different application of light. Magnetic storage and CDs store information one bit at a time and use mechanical means. A holographic system can store information in complete images, somewhat like the 3-dimensional images seen on credit cards; it can also store and retrieve those images without using moving parts. Each image is stored as an interference pattern in a particular location of a crystal or in a special transparent polymer (see p 35 for a description of interferometry). The interference is between a fixed reference beam produced by one laser and, in one holographic system at least, another beam that is one of many produced by a tiny laser on a chip. One such chip can contain as many as 10,000 lasers. Even tinier lasers, developed in 1991 by Sam McCall at Bell Labs, are expected to be used in the future to make optical computers a practical reality in this and a number of other applications (see "Tiny Lights and A Motor," p 600). Recovery of data from a hologram is almost instant since all that is needed to accomplish it is to turn on a particular laser. This can be much faster than the mechanical searching that must be done to recover data from even the fastest magnetic hard disk or CD.

Note that holographic storage does not need to wait for optical computers to become a reality. The method can be used with present electronic computers. In 1992, however, a principal barrier to the use of holographic data storage is that even though the actual storage location can be the size of a Scrabble tile, and the chips with the tiny laser on them are even smaller, the equipment used to operate the device is larger than most current laptop computers. This part of the operation needs to be miniaturized before it becomes practical. Costs are also high, but that is always the case for prototypes.

Other data storage systems that use laser beams, but not holograms, are also in the works. These are 3-dimensional versions of the CD, in which the power of lasers to direct energy to a tiny spot is the key. As with CDs, however, it is much easier to create the first version of data with such a system than it is to take data and revise or replace it in the same location. Such systems are better for permanent data storage than they are for replacements for current magnetic storage, which has the virtue (and sometimes the flaw) of being almost completely erasable.

Obtaining Data to Store

In addition to the data input devices already in use or under development for present electronic computers—keyboards, various species of mice, handwriting, voice, and so forth—optical computers can obtain data from optical means. The advantages of a computer that could see are obvious and immense.

In Oct 1991 computers that use a person's signature for a password first came to market. These use pressure and electronic information processing instead of optical processing. Banks, however, are working on development of devices that would use optical recognition of signatures to deal more efficiently than they now do with forged checks or other documents.

Less obvious is direct sensing of data by means of fiber-optic sensors. Increasingly industry is turning to fiber-optic sensors for measuring with exquisite sensitivity quantities such as pressure and temperature. Although more expensive than traditional sensors, the fiber-optic versions have advantages in size and the ability to transmit the measurement to a different location using a fiber-optic cable.

One example of how fiber-optic sensors can be used is in the manufacture of composite bodies for aircraft or other high-tech vehicles. Information about flaws and the state of curing in vehicle bodies can be obtained by directly embedding thousands of fiber-optic sensors in the material as the bodies are manufactured. After the job is complete the cables can be snapped off, leaving the sensors in place but out of touch; or the cables can be left in place themselves and connected to a monitor in the vehicle. Such a system might be especially helpful for supersonic vehicles or vehicles that operate high in or somewhat above Earth's atmosphere.

(Periodical References and Additional Reading: *Science* 2-9-90, p 679; *Discover* 1-91, p 63; *New York Times* 6-6-91, p F9; *Scientific American* 7-91, p 108; *Scientific American* 9-91, p 48; *New York Times* 9-8-91, p F9; *New York Times* 11-6-91, p D7; *New York Times* 11-8-91, p D4; *Science* 1-3-91, pp 7 & 66)

THE LIGHT TOUCH

Light is being manipulated in many ways, not just by computers. Humans as primates are uniquely tuned to using light. The ability to manipulate it through chemistry and by way of lasers and fiber optics has come recently, but our ancestors started making light with fire some million years ago.

Window Wonders

Some new ways to manipulate light may seem a bit frivolous. In 1991 the Taliq Corporation of Sunnyvale CA, a division of Raychem Corporation, was marketing windows based on liquid crystals that are transparent in the presence of an electric current but opaque when the switch is turned off. At a cost of about 20 to 30 percent more than plain plate glass—not to mention the cost of the electricity needed to keep the glass clear—such window panels are now used primarily as dividers and in a few office buildings. Taliq hopes to get costs down to where the panels will be common in homes and even automobiles.

The Advanced Environmental Research Groups of Davis CA developed a coating for windows that is based on principles of holography. As the Sun moves across half the sky, sunlight that passes through the window is directed toward the same place in a building. This can push light into dark recesses of a room and reduce electric costs.

Could the laser window be combined with the liquid crystal window to provide all the advantages of both without adding to the electric bill?

Superlight

Optical fibers are being used for more and more applications. Some of these applications are incredibly transparent. According to Ze Cheng of the Chinese Center for Advanced Study and Technology in Beijing, transparency can become even better. He predicts supertransparent materials that behave for light the way that superconducting materials behave for electric currents (see below). Although the materials would be different and not the light, physicists call this predicted phenomenon superlight.

The basic idea is that a crystal with very particular refracting properties could induce behavior in photons that is similar to the BCS behavior of paired electrons in conventional low-temperature superconductors (see "Superconductivity UPDATE," p 586). So far, however, the problem is that no one knows how to construct materials that would behave as described by Cheng's theory. Some physicists at least doubt that superlight will ever become reality.

Laser Tweezers and Optical Molasses

Since 1985 experimentalists have been using lasers to handle individual atoms or molecules and tiny clusters of atoms or molecules or even aggregations of some size. The basic idea is to use light to exert a force, which is possible because photons carry momentum. Light pressure works best when particles are already nearly motionless from extremely low temperatures. It also helps if the particles face few other interactions by being part of a rare gas or a colloidal suspension.

The cold temperature is the secret. Temperature lowering is accomplished by laser light also. The particle to be cooled is placed in the intersection of three laser beams that resist any motion an atom makes in any direction. The result, nicknamed "optical molasses," slows down the particles. Slower particles individually are cooler ones in bulk, according to the kinetic theory of heat. This method of cooling was first proposed as early as 1975 but it took ten years for the details to be worked out. When put into practice, optical molasses made atoms even cooler than predicted. Theory later caught up, finding several mechanisms that work together to promote laser cooling.

Once the particles are cool enough, the same kind of laser forces can be focused on individual atoms to manipulate them. Stephen Chu of Stanford University was among the first to use "optical tweezers" made from laser light to manipulate small particles. The idea soon became popular among biologists after it was shown that particles as large as single bacterium can be manipulated with optical tweezers.

Workers in the field hope soon to start building materials such as polymers one particle at a time, mainly to see exactly which part of the process causes specific properties to change.

HIGH-DEFINITION TELEVISION GOES DIGITAL

Although not strictly speaking a new way to use light nor a new use for light, the light that may have the most consumer impact in the 1990s is the one coming from the impact of electrons on the luminescent screens of television sets. In the 1980s television picture tubes that would produce pictures approximately as clear as cinema were touted as the key to unlocking a vast amount of new money, if not really very many new products. High-definition television (HDTV) was seen not so much as vital technological breakthrough but as an index of technological success.

Of course, Japan got there first. As early as Jan 3 1989 Japan was able to initiate regular broadcasts using its own HDTV standard, although sets for consumers were not available until Dec 1991, and they cost $34,000 each. By then the HDTV broadcasts were available for one hour each day. Eight hours a day of broadcasting are planned to start some time in 1992.

Despite Japan's lead, the United States may emerge with the better system. Although digital HDTV was thought impossible as recently as 1989, in Jun 1990 the Videocipher Division of General Instrument Corporation of Chicago IL announced that it had developed a working prototype of a digital system. Its success inspired three different varieties of digital systems. One of the digital types is likely to be licensed by the US Federal Communications System instead of the analog system used in Japan. The decision is not scheduled until Jun 1993, however, so anything can happen.

Digital television offers better possibilities for interaction than analog does, meaning that a home user might be able to order specific movies for home viewing at a convenient time from a catalog or use the television set as an interactive reference library.

(Periodical References and Additional Reading: *Science* 4-6-90, p 29; *Science News* 9-1-90, p 143; *New York Times* 12-6-90; *Science* 5-17-91, p 922; *Scientific American* 9-91, p 48; *New York Times* 8-18-91, p F7; *New York Times* 9-8-91, p F9; *Science News* 9-21-91, p 183; *New York Times* 10-16-91, p D9; *New York Times* 10-20-91, p F11; *Science* 11-29-91, p 1294; *New York Times* 12-3-91)

SUPERCONDUCTIVITY UPDATE

In the late 1980s the promise of warm-temperature superconductivity was a major story in newspapers and other mass media. When nothing superconducting and dramatic moved from the laboratory into the factory, home, or office, the story seemed to die down. That impression was deceptive, however, as progress in superconductivity continues. In the early 1990s, however, that progress was reported mostly in science journals. The rate of publications concerned with warm-temperature superconductivity, estimated by one observer at nearly 5000 papers on the topic each year in the late 1980s, continued unabated.

Better Superconductors

The first ceramic, or oxide, superconductors were discovered in 1964, at the tail end of an important round of activity in the field. They languished until 1986 when one of them was found to break the old temperature record for superconductivity by a substantial amount. After that discovery, there was a flurry of well-publicized advances in the production of new materials that could reach higher and higher superconducting temperatures. Today the problems are more in the development of materials that can carry higher electric currents or that behave better in magnetic fields or that can be shaped into useful devices. Ceramics are not metals, and there are physical problems in making wires, for example, as well as other obstacles.

The best ceramic superconductors are based on copper oxides that have some atoms of a metal mixture added to them. The three main families use yttrium and barium, lanthanum and barium, or bismuth and thallium. Kenneth Poeppelmeier of Northwestern University in Evanston IL and coworkers produced a fourth family of warm-temperature superconducting oxides by adding gallium to the yttrium-barium type and replacing about a third of the yttrium with calcium. The new family seems to be more stable than the other three, losing less oxygen when heated in air.

Background

In 1908 physicist Heinke Kammerleigh Onnes [Dutch: 1853-1926] liquefied helium, which as a by-product of the process resulted in temperatures only a few degrees above absolute zero. Experimenting with the very cold liquid, Kammerleigh Onnes discovered that an electrical current in frozen mercury or supercooled lead could proceed without facing any resistance at such cold temperatures, although this property disappears in the presence of a sufficiently large magnetic field. The newly discovered property was named superconductivity. In 1933 it was found that a magnetic field not large enough to destroy superconductivity is excluded from the interior of superconductors. The cause of superconductivity was finally explained after 46 years by John Bardeen, Leon N. Cooper, and John R. Schrieffer (the BCS theory) in terms of pair of electrons that are able to move through a material effortlessly because they do not interact with the atoms.

BCS theory explains why superconductivity can take place only in materials close to absolute zero, but it does not say how close. By 1967 the highest temperature of superconductivity had passed 20 K (-253° C or -423° F). This record was not to be substantially surpassed for almost 20 years, by which time most researchers had stopped looking for higher temperatures. When in 1986 Karl Alex Müller and Georg Bednorz discovered a ceramic oxide that is superconducting at 35 K (-238° C or -396° F), everyone was astonished. Soon other "warm-" or "high-" temperature superconducting oxides were found, although the record of about 125 K (-148° C or -234° F) from the later 1980s still stands as the warmest temperature, despite occasional claims to have surpassed it.

There also is hope for yet another class of superconductors, one based on doped buckyballs of carbon (see p 157). A thin buckyball film doped with rubidium and thallium prepared by scientists at Allied-Signal Inc. in Morristown NJ was superconducting at about 42 K (-231° C or -385° F), which seems to be the current leader among buckyball metal superconductors. Although higher than the previous record for old-fashioned metal superconductors, 42 K is still quite a bit less than the 125 K record for the new warm-temperature superconducting oxides. Some theories suggest that doping buckyballs can produce superconductivity at still higher temperature, perhaps even higher than those of the oxides.

Better Theories

A principal barrier to advances in superconductivity has been that theorists have had difficulty agreeing on an explanation of what causes the effect in the warm-temperature materials (superconductivity in materials near absolute zero was adequately explained by BCS in 1957 and validated by a Nobel Prize to Bardeen, Cooper, and Schrieffer in 1972).

One problem has been that warm-temperature superconductors have not in practice conducted much current, at least not in wires or other bulk materials; electric conductivity is much better for thin films. Theorists decreed that the problem in bulk materials is caused by wandering magnetic flux lines that slow down electron flow. Theory did not offer a clue as to why the flux lines in thin films do not similarly wander and impede electrons. The inability of bulk warm-temperature superconductors to conduct such currents has prevented this promising technology from being useful for real applications.

On Mar 18 1991 two groups independently offered their evidence that microscopic screwlike structures account for the ability of some kinds of superconducting films to conduct relatively high currents.The tiny screwlike structures were first reported to the journal *Science* by a research group from Los Alamos NM on Dec 27 1990 and separately reported to *Nature* by Georg Bednorz and coworkers from the IBM Research Division in Zurich, Switzerland, on Feb 15 1991. Internal publishing schedules resulted in the *Nature* report being published on Mar 28 and the *Science* one on Mar 29.

By Jul, however, problems with this theory had begun to surface. Experiments that put the defects into wires at high density did not make the wires conduct as well as thin films do. This experiment implies that something else must explain why thin films conduct more current that comparable wires

The Future

Since 1989 various laboratories have been developing warm-temperature superconducting microwave experiments to be flown in space, currently scheduled for 1992. A second round of experiments will fly around 1996. The main purpose will be simply to find out how the devices behave.

(Periodical References and Additional Reading: *Science* 1-19-90, pp 263 & 307; *Science* 2-9-90, p 656; *Science News* 2-10-90, p 95; *Science News* 3-31-90, p 207; *Science* 5-18-90, pp 791 & 840; *Science News* 6-2-90, p 341; *Science News* 6-9-90, p 367; *Science* 8-24-90, p 862; *Science* 9-28-90, pp 1479 & 1549; *Physics Today* 12-90, p 25; *New York Times* 3-28-91; *Science* 3-29-91, p 1565; *Physics Today* 6-91, pp 22, 24, 34, 44, 54, 64, & 74; *Scientific American* 6-91, p 20; *Science News* 6-8-91, p 358; *Science* 6-14-91, pp 1467, 1501, & 1509; *Science* 6-28-91, pp 1763 & 1829; *Science* 7-12-91, p 132; *Science* 7-26-91, p 373; *Science News* 8-3-91, p 77; *Science News* 8-10-91, p 84; *Science News* 8-24-91, p 127)

BIOTECHNOLOGY

TRANSFORMING GENES IN HIGHER PLANTS

The initial successes with genetic engineering involved inserting new genes into bacteria that could then be grown in tanks. The technique was similar to that used in familiar fermentation processes ranging from making vinegar to some forms of pickling. An even more familiar technique could be used for production when new genes were inserted in yeast, grown since ancient times as part of bread, beer, and wine production. Some desired genetic products, usually complex hormones, were too complicated for bacterial or yeast cells to manufacture, so they were grown in genetically re-engineered animal cells, a much more difficult and less familiar procedure.

A whole kingdom of organisms, however, is even better understood than bacteria and yeasts—plants (often called higher plants because older taxonomic conventions considered yeasts or even bacteria as simple, or lower, plants). Plants have been grown by humans since the agricultural revolution; indeed, almost every human being grows a plant of some kind at least once, while until recently the majority of humans devoted their lives to the production of useful materials and chemicals from plants, either as farmers or as workers in industries based directly on plants. Thus, an important goal of genetic engineering has been to modify plants both for improved ways of fulfilling traditional roles and for brand new ways of producing useful complex chemicals—for while a plant cell is very different from an animal cell, it is just as sophisticated. Plants have long specialized in making complex compounds for their own purposes, so they have cellular machinery that bacteria and yeasts lack.

But genetic engineering of plants is not easy. For one thing, the traditional laboratory organisms of research have been insects and other invertebrates and mammals, with zebra fish coming on strong in recent years. Many fewer basic researchers work with the traditional laboratory creatures of plant science, tobacco and maize—corn in the United States. Tobacco is easy to breed and grow; maize is extremely important agriculturally and can be bred easily because it has both male (tassels) and female (silk) reproductive organs. It was not until the early 1990s that genetic engineers began to succeed routinely with plants. The first technique used, starting in 1983, used the plant pathogen *Agrobacterium tumefaciens* to introduce new genes into plant cells, just as bacterial genes are often transformed by the viruses called phages. A pathogen used for this purpose is called a vector. One early problem encountered was that *A. tumefaciens* does not infect monocots, the source of virtually all grains consumed by humans.

MONOCOTS AND DICOTS

Nearly all plants that we see around us are angiosperms, or flowering plants. The most noticeable exceptions are conifers (narrow-leaved evergreens, such as pines and yews), ferns, and mosses. Angiosperms are called flowering plants because they are the only plants with true flowers, which when fertilized produce fruit and seeds.

The flowering plants, however, are divided into two large groups that are easily recognized most of the time as quite different from each other when one knows what to look for. The groups are named after a difference in the first leaves to come from the seed, leaves that actually develop in the seed before they emerge. Such seed leaves have the technical name cotyledons. Flowering plants either have one cotyledon, and are therefore monocotyledons or monocots for short, or two—dicotyledons or dicots. This difference is easily observed when the seeds sprout. Differences between the two kinds of flowering plants are greater than that, however. For example, the two seed leaves of the dicot usually store food for early plant growth (think of the fat seed leaves of a bean plant), while the single seed leaf of a monocot does not.

As they grow, the two kinds of flowering plants maintain many differences. The parallel veins of monocot leaves contrast with the netted veins of dicot leaves, for example. Flower parts of dicots are typically in groups of four or five, while the flower parts of monocots form in multiples of three (compare the monocot tulip with the dicot mountain laurel). There are various other subtle and not-so-subtle differences as well.

The classic monocots are the lilies, grasses, irises, and orchids. Almost all grains, including wheat, rice, and maize, are monocot grasses. There are many families of dicots. Of commercial interest are the fruits and berries, deciduous trees, and most vegetables (except members of the onion family).

Genes Shot from Guns

In 1988 scientists at Cornell University developed a method for inserting new genes into the DNA of cells of yeast and algae. It used a device called the gene gun because in its first form it fired tiny gold BBs coated with DNA into cells. The shot pokes holes in cell walls and some of the DNA coating sticks to the original DNA in the cell. Various researchers have since reported success with gene transformations based on the gene gun. By the early 1990s improved versions of the gene gun had become a common way to engineer plants genetically.

In the United States, the main goal of plant genetic engineering has been to improve maize, a cereal already so improved by human breeding from its primitive forebears of 10,000 years ago that it seems to be a totally different plant. The vast amount of maize seen growing in the US Midwest is not the variety we know as a vegetable. Instead, most maize is grown for animal fodder, for oil, for starch, for syrup, for alcoholic beverages, and for chemicals

In Jan 1990 Biotechnica, Inc., of Minnetonka MN reported that it was the first to succeed in transforming corn genes and achieving successful reproduction of the transformed plants, but they did not give any details and scientists were generally skeptical. It was not known what method Biotechnica used. In Apr 1990 Monsanto in St. Louis MO reported that *it* had succeeded in transforming corn genes for the first time, using a gene gun, but it later became apparent the this was based upon an earlier successful application of the same technique by a team led by Catherine J. Mackey at DeKalb Plant Genetics in Groton CT. The DeKalb results were published in detail in the Jul 1990 *The Plant Cell,* a peer-reviewed publication. Confirmation of the DeKalb success was published by US Department of Agriculture scientists in Sep 1990.

On Nov 1 1990, Pal Maliga, Zora Svab, and Peter Hajdukiewica of Rutgers University in Piscataway NJ reported that they had shot foreign genes into the chloroplasts of a plant, specifically the laboratory workhorse tobacco. The plants were seedlings that then grew with the new genes incorporated into chloroplasts in new cells. The genes used were marker genes that could be used to identify seedlings in which the transformation was successful. Serious application of the technique could be used to cause chloroplasts to produce wanted chemicals or to use nitrogen from the air as a fertilizer.

Chloroplasts are like mitochondria in that they have their own DNA. They probably were originally free-living single-celled organisms that were incorporated into plant cells early in the evolution of plants. Because the chloroplasts are not passed from generation to generation in pollen (just as mitochondria are passed only in egg cells; see "The African Eve and the Pygmy Adam," p 476), there is no chance that the foreign genes can pass to surrounding weeds or even to plants of the same species not in the direct line. On the other hand, the transformed chloroplasts are inherited though the seeds of the original plants.

(Periodical References and Additional Reading: *Science* 6-22-90, p 1493; *Science* 8-10-90, p 630; *Science News* 11-10-90, p 295)

LET BACTERIA EAT IT

In Mar 1991 Michigan State University and the Research Development Corporation of Japan announced that they were going to spend $15 million to develop microbes that can better clean up oil and toxic wastes. Michigan State's financing came from the US National Science Foundation.

Microbes and Oil

Many people learned for the first time that bacteria or other microbes can be used to dispose of petroleum wastes when microbes were used, with limited success, in Prince William Sound after the 1989 *Exxon Valdez* oil spill. In that case no new organisms were introduced, but naturally occurring bacteria were provided with fertilizer, a technique known as bioremediation. Both bacteria and yeasts that naturally consume hydrocarbons as a part of their diet have evolved whenever the chemicals are available. In the case of Prince William Sound, bacteria had evolved to eat hydrocarbons called terpenes that are released by the spruce trees that cover the shore. Estimates of the increase in oil consumption as a result of fertilization of waters and beaches ranges from zero by some scientists who have studied the problem to between five and ten times by Exxon scientists.

Another method involves direct release of microbes that eat oil, nearly always microbes that are foreign to the environment in which the cleanup is desired. This approach, called seeding, produced no results in Prince William Sound or in controlled tests in Galveston Bay.

Controlling Industrial and Municipal Pollution

Bioremediation and seeding are more successful in hazardous waste sites and landfills than in the open ocean or even in relatively enclosed regions of the sea. It is difficult to maintain optimum temperature, nutrients, or oxygen at sea.

A low-tech way of using bacteria on pollution is the most common method of cleaning up waste water. As the water sits in ponds, filters through sand or charcoal, or is aerated, different kinds of microorganisms colonize it naturally, consuming the waste and proliferating. Removal of the bacteria removes much of the original waste, while other pollution is turned into relatively harmless carbon dioxide or completely harmless water.

This same idea has been co-opted for other organic or hydrocarbon pollution. Biofilters or soil bioreactors, similar to the devices and processes used in municipal water treatment plants, are in use at twenty-some places in the United States. In essence, the biofilter is a pile of dirt through which a liquid or, rarely, a gas is passed. Microbes in the dirt eat pentane, butane, hydrogen sulfide, and other relatively exotic chemicals as well as plain old sewage. The main trick is to keep feeding the bacteria. As wastes are piped through the pile of soil, sand, or even volcanic ash, bacteria that enjoy those particular wastes proliferate, while those that prefer other kinds of food fail to thrive.

Hydrocarbons and other organics seem like a natural form of food for bacteria, but they will eat almost anything it seems, even such unpleasant chemicals as polychlorinated biphenyls, or PCBs. In 1991 the General Electric Company (GE) started trying the bacterial approach on the half-million pounds of PCBs that it dumped into the Hudson River between 1946 and 1977 and that are, for the most part, still there. PCBs, like their chemical cousin DDT, do not degrade easily. PCBs are also recognized by the US Environmental Protection Agency as carcinogenic. In Dec 1990 officials from GE announced that bacteria were stripping chlorine atoms from the PCBs, which makes the chemicals less toxic. The 1991 test was planned to see whether natural bacteria could be helped along. The basic method will be bioremediation by the addition of fertilizer to the test sites. If the test sites degrade faster than control sites with no bioremediation, GE will probably ask permission to use bioremediation in addition to or instead of dredging.

Other Uses of Consumption

Not all uses of bacteria to consume hydrocarbons are involved with cleaning up oil spills. Among the 3000 components of crude petroleum is a group of chemicals called steranes that is especially useful in determining the geologic history of an oil deposit; this in turn can be used to estimate its size. Often, however, the steranes are consumed by bacteria before the geologic history can be evaluated. A group of French scientists from Sanofi Elf-Biorecherches and Elf-Aquitaine corporations investigated this process by preparing a brew of seventy-three different species of bacteria and giving the bacteria sterane-rich crude from Utah to consume. The scientists were able to report in *Nature* in June 1991 that bacteria of the genus *Nocardia* were hungriest for steranes. It is presumed that today's oil rich in *Nocardia* was previously rich in steranes.

Other microbes have been found that turn uranium wastes from a form in which they are difficult to separate from other mine tailings or nuclear-power-plant residues into a form that is easily filtered out. The bacterium involved, GS-15, was discovered by scientists from the US Geological Survey and the find reported in Apr 1991 in *Nature*.

(Periodical References and Additional Reading: *Science News* 1-27-90, p 63; *Science* 3-30-90, p 1537; *Discover* 7-90, p 22; *Science* 7-13-90, p 120; *Science* 7-27-90, pp 339 & 380; *New York Times* 1-27-91, p E6; *New York Times* 3-13-91, p D8; *New York Times* 4-9-91; *Science* 4-12-91, p 205; *Science* 5-10-91, pp 755 & 830; *New York Times* 6-25-91, p C2; *Science* 6-28-91, p 1784; *Science* 7-26-91, p 385; *Scientific American* 10-91, p 102)

WHO OWNS PCR?

In Apr 1983, while driving up the California coast, Cetus Corporation biochemist Kary Mullis thought of a way to make multiple copies of any part

HOW PCR and LCR WORK

The codes for starting and stopping reproduction of a gene or other line of DNA code were discovered as part of the deciphering of the genetic code in the 1960s. In PCR a start code is attached to opposite ends of two formerly twisted together strands of the DNA double helix for the desired line of code. The modified line is put into a solution where it can react with a good supply of the four bases that make up the code and an enzyme, DNA polymerase, that causes the line of code to be copied. The start code is copied along with it. Thus, as soon as the first copy is made, there are two copies in the solution, which has the instructions and ability to copy each of them, resulting in four copies. After ten such passes there are over a thousand copies and after twenty passes there are over a million copies. PCR is also known as gene amplification.

The whole process requires warm temperatures to separate the strands of the helix and cooler ones for the actual copying process. In the first version of PCR the DNA polymerase was degraded by the heat, which meant that more had to be added after each copying cycle, but the original form of DNA polymerase was replaced in 1987 by a version of the enzyme used by bacteria that normally live in hot springs in Yellowstone National Park. With the enzyme surviving the process, it was possible to develop a machine that automated the entire chemistry.

Ligase chain reaction (LCR) is based on a different enzyme and a different technique for modifying the stretch of DNA to be copied. The beauty of LCR is that it not only amplifies the gene involved but it can be set up to amplify only a particular DNA sequence, eliminating the need to use other methods to determine exactly what DNA has been amplified.

The ligase enzyme, like DNA polymerase, is normally degraded by heat, so LCR also depends on a version from hot springs, one discovered by Francis Barany of Cornell University Medical School in New York NY in Aug 1991 according to one patent suit; or one discovered by Abbott Laboratories according to another.

of a DNA molecule. When the details had been worked out 4 years later, the new process, called PCR for polymerase chain reaction, quickly became applied to sensitive genetic tests, including the process known as DNA fingerprinting. PCR also has been used to study ancient forms of life from bits of DNA found in fossils and is essential for the Human Genome Project. Even faster than genetic engineering and monoclonal antibodies, PCR has become one of the pillars of the revolution in biotechnology. It is perhaps no surprise that such a pivotal tool is valuable. It was somewhat of a surprise to discover in the middle of 1991 that the rights to the process were virtually bought out by a giant pharmaceutical company. At that point it was revealed that Hoffmann-La Roche had paid $300 million for all rights to the process, including any future applications.

The rights were originally the property of Cetus, since Mullis was working there when the concept came to him. In 1989 Cetus sold the rights to use of PCR for in vitro tests for human genes. Another company, Perkin-Elmer, most familiar from its problems with the mirror on the Hubble Space Telescope, bought the rights to develop instruments involved with PCR, a role it expects to continue in collaboration with Hoffmann-La Roche.

The whole field was muddied when E.I. Du Pont de Nemours contended in Aug 1989 in a lawsuit that PCR had been described by Har Gobind Khorana as early as 1971, so Cetus could not claim it. The US Patent and Trade Office (PTO), however, disagreed with that argument. The case was still in dispute in the courts when Hoffmann-La Roche decided that what was good enough for the PTO was good enough for Hoffmann-La Roche.

Hoffmann-La Roche announced that it would cooperate with scientists using PCR in basic research and that it would soon profit from its investment by selling diagnostic kits for AIDS, Lyme disease, chlamydia, and other human infectious diseases, as well as for cancer and genetic disorders.

And Who Owns LCR?

Just as the PCR rights were becoming clear, similar legal problems began to surface in what many see as a successor to PCR, ligase chain reaction, or LCR. The basic ligase gene detection method goes back to 1988, when it was first described by Leroy Hood and coworkers from Caltech. With PCR already known, it was but a short step for scientists from the Beckman Research Institute to publish a suggestion that a heat-stable ligase might be an improvement over DNA polymerase in automated gene copying. But by then at least three other laboratories were at work on developing the process. By 1991 so many companies claimed rights to LCR that some dubbed the process litigase chain reaction. A decision by the PTO may be years away. In the meantime, Abbott Laboratories, at least, plans to market equipment to use the process.

(Periodical References and Additional Reading: *Discover* 7-90, p 77; *Science* 7-13-90, p 127; *Science* 2-15-91, p 739; *Science* 6-21-91, p 1643; *Discover* 7-91, p 28; *Science* 8-9-91, p 627; *Science* 11-29-91, p 1292)

S M A L L   A N D   D E E P

SCANNING TUNNELING MICROSCOPE FINDS NEW USES

In 1986, when Gerd Binning and Heinrich Rohrer of IBM won a Nobel Prize in physics for their 1981 invention of the scanning tunneling microscope (STM), the device was primarily a way to image surfaces closely, even in terms of individual molecules or atoms. By the early 1990s, however, the basic principles of the microscope were being applied in a great many ways that did not involve imaging anything at all. It might be more appropriate to call the new machines based on the main components of the STM—very sharp needles, piezoelectronic controls, and tunneling electrons—scanning tunneling devices (STDs) instead of microscopes. Even that modification in name does not go far enough, for some of the new applications of these devices do not even use scanning (also medical workers already use STD for "sexually transmitted disease"). Perhaps the best name is just "tunneling devices."

Also, in the late 1980s various researchers developed a dozen or so devices based on the STM that used scanning without tunneling. All of these devices scan surfaces with a tiny tip, but some press their tip against the surface, some use the tip to measure magnetic fields, and others use ions or sound. Like the STM itself, some of these scanning nontunneling microscopes are even available commercially. Probably the most versatile is the atomic force microscope (AFM), which rests the tip against the surface being studied and measures variations in the height of the surface directly as the tip scans it.

A Tele-Nanorobotic Manipulation System

The basic method used by the STM is to maintain a specific level as the tip moves over a surface. The amount of electron tunneling between the tip and the surface varies with the height of the surface, more over a hill and less over a valley. The electric current is then converted into an image on a computer screen.

Conversion of distances into an image of the surface is something like the process of converting a contour map—a map that shows the heights of each point as a numerical distance above sea level—into a relief map that shows approximate heights by using techniques to make a two-dimensional drawing look three-dimensional. A contour map can also be converted into actual relief by making a three-dimensional model of the surface described by the numbers on the map. With such a relief model, the surface can be sensed by touching it. You can even tell what the sur-

NANO-THINGS

Technically, the prefix nano- refers to a billionth of a measure, as in a nanosecond or nanometer. It has been appropriated by scientists who need a general prefix that means somrthing smaller than micro-, which had a long usage as "very small" even before it became the international prefix for a millionth of a measure. Thus, an oceanographer may refer to very small plankton as nanoplankton even if they are several orders of magnitude larger than a nanometer.

A scanning tunneling microscope often deals with structures that are in the nanometer range. For example, an individual atom is typically measured in tenths of a nanometer. Thus, the prefix nano- often turns up in terms connected with STM-related technology. A tenth of a nanometer is also known by the name angstrom, and some researchers prefer to express measurements at STM levels in terms of angstroms.

face is like in the dark. If the surface is a real mountain chain on Earth, the scale is too great for you to envision it by feeling the actual mountains. But you can get an idea of the shape of the system by feeling a relief model of the mountains. You need to get the mountains into a human scale.

Ralph L. Hollis, Septimiu E. Salcudean, and David W. Abraham of IBM's Thomas J. Watson Research Center in Yorktown Heights NY have developed a device they call the "magic wrist" that can be used in conjunction with an STM to feel a surface at atomic levels. The combination magnifies the atomic and molecular features of the surface to the human scale at least for the sense of touch. As the distances change between the STM tip and the surface, the magic wrist both multiplies them and converts them to movements of a platform on which a person can rest his or her fingers, thus enabling the person to feel the bumps and hollows of the surface.

The team hopes to use the combination as what they have named a "tele-nanorobotic manipulation system" to move the STM tip to specific places on the surface. As research outlined below indicates, the tip can then pick up or drag atoms or small clumps of atoms to designated portions of the surface. It could also be used with an AFM to manipulate nano-clusters of metals.

Sensing Molecules

Noel MacDonald, Susanne Arney, Jason Yao, and coworkers at Cornell are building the smallest possible STMs, but not for imaging. While the business end of an STM, the tungsten tip, consists of only a few atoms, the rest of the machinery used to control the tip's position over a sample is usually somewhere between the size of a VCR and a microwave oven. The Cornell team hopes eventually to get the entire tunneling device down to the size of the cut end of a human hair. Their goal is to use tiny tunneling devices to sense specific molecules or changes in light or motion in the environment. They plan to use two tips instead of one and to use piezoelectric crystals and tunneling to monitor the current between the tips. Changes in the current in a given environment could mean the presence of a desired molecule or some change of special interest to the user. Several such devices could be placed in an array to detect a range of changes of interest.

Other teams of researchers are also working on miniature tunneling devices with various purposes. For example, William J. Kaiser and coworkers at the Jet Propulsion Laboratory in Pasadena CA are developing small tunneling sensors that can be used as parts of navigational systems for spacecraft. They also envisage using tunneling devices as seismometers small enough to be carried on spacecraft to other planets. Small silicon-based seismometers could be utilized to monitor prospective earthquakes or volcanoes on Earth as well.

Rearranging Atoms

When the tip of a tunneling device is placed 3 to 5 angstroms from a silicon surface and an electric potential of about +3 volts is applied, some of the atoms of silicon form a small mound. The mound may contain a few silicon atoms, depending on the exact distance between the tip and the silicon, the size of the voltage, and the sharpness of the tip. Moving the tip slightly closer to the mound causes the top layer of atoms to part from the surface and adhere to the tip. The number involved can be as small as one or it can be of the order of ten or so.

If the tip is then moved to a different place above the surface and a voltage of the opposite polarity but same amount (that is, about -3 V) is applied, the silicon atom or cluster of atoms drops off the tip and sticks to the other silicon atoms of the surface. The inventors of this technique, In-Whan Lyo and Phaedon Avouris of IBM's Thomas J. Watson Research Center, envisage using it to create very tiny semiconductor devices, a process they have named nanoelectronics. Another possible use for their technique would be in the development or manufacture of the small tunneling sensors that Noel MacDonald's team is developing, since these sensors are built largely of silicon.

Smallest Switch Possible

Donald M. Eigler, Christopher P. Lutz, and William E. Rudge at IBM's Almaden Research Center in San Jose CA developed in 1991 a tunneling

Background

In short, tunneling is passing through a barrier that is like the gospel-song description—"too high, you can't get over it; too wide, you can't get around it; too deep, you can't get under it." But some particles get over, around, or under the barrier anyway. Tunneling was one of the first astonishing paradoxes of quantum theory, first explained by George Gamow in 1928. Any particle can tunnel, but as in most quantum effects, the smaller the particle, the easier the tunneling. Electrons are among the smaller particles, so they tunnel very easily.

Perhaps the easiest way of thinking of tunneling is to remember that everything has both a wave aspect and a particle aspect—and then to forget the particle aspect, since tunneling is a function of the wave. It may also help to remember that comparatively speaking, large particles are small waves and small particles are large waves. Thus a small particle like an electron is associated with a large wave.

Tunneling works with any kind of barrier, whether it is caused by the electromagnetic force (such as the barrier a positive charge effects on another positive charge), by a gap (such as a deep ditch around a castle), or by a solid wall (such as an insulator to an electric charge). A person cannot tunnel through a brick wall because a person is fairly large, making for a small wave, and a brick wall is rather thick. An electron can tunnel through a gap of several angstroms because the electron has a big wave and several angstroms is not far.

Because it is a wave, the electron is not located exactly at any point in space. Some of the wave penetrates into the barrier. If the barrier is small enough, some of the wave appears on the other side of the barrier. Here is where quantum weirdness comes in. The wave we are discussing is a probability wave, the probability of the electron being located (as a particle) at a particular point. So if some of the wave is on the other side of the barrier, there is a definite probability greater than zero that the electron will appear on the other side of that barrier. And sometimes it does. That's what tunneling is all about.

The frequency of tunneling can be predicted very accurately by quantum mechanics. For a given source of electrons, the amount of tunneling depends on the size of the barrier only. Thus, an STM can use tunneling to measure quite exactly the distance of a tip to the individual hills and hollows of a surface, even when the hills are individual atoms.

device that acts as a switch based on the location of a single atom of xenon. An ordinary macroswitch depends on the physical closing of a circuit so that electrons can flow through a conductor. An electric tube, or "valve" to use the more appropriate British name, uses a small change in electric current to affect the size of a larger current. The millions or even billions of switches in a computer chip similarly depend upon changing the state of a tiny region in a semiconductor from insulating to conducting, often in a single direction, like a one-way street. In the tunneling device switch, conduction is not caused by directing the motion of electrons through a conductor but by facilitating the electron's chance of disappearing on one side of an insulator and reappearing on the other—tunneling.

Like tubes or valves and like transistors, the tunneling device switch can be changed from one state to another by applying a small current. An electric current can have one of two polarities, which we usually think of as positive and negative. When one polarity is used with the tunneling device, its tungsten tip lifts a single atom of xenon from its residence on the surface of a nickel crystal. That turns the switch on. Changing the polarity of the tip causes the xenon to drop back onto the crystal, turning the switch off. In one configuration, electrons can tunnel from the crystal to the tip, while in the other they cannot. Thus, the two locations of the xenon atom are similar in effect to the open and closed locations of a switch based on conduction.

In a commentary in the Aug 15 1991 *Nature,* where a description of the new switch was first published, C.F. Quate of Stanford University claimed that in theory switches of this type could be used to put the entire collection of the US Library of Congress onto a single disk with a 30-cm (12-in.) diameter.

(Periodical References and Additional Reading: *Science* 2-9-90, p 634; *Science News* 3-17-90, p 165; *Science News* 4-7-90, pp 216 & 223; *Science* 4-27-90, pp 419 & 454; *Science News* 1-5-91, p 6; *Science* 3-8-91, pp 1155 & 1206; *Science* 5-31-91, p 1252; *Science News* 7-13-91, p 21; *Science* 7-12-91, pp 115 & 173; *Science News* 11-17-90, p 310; *New York Times* 8-15-91, p D23; *Science* 9-20-91, p 1405; *Science* 10-4-91, p 68; *Scientific American* 11-91, p 20; *Science* 11-29-91, p 1319)

TINY LIGHTS AND A MOTOR

The STM and its relatives are helping in the creation of extremely small devices of all kinds. Small lasers have many possible applications. Although many of the new devices employ special properties that are operative for only short distances and with a few molecules at a time, a new micromotor shows that it is still possible to produce very small versions of the kinds of devices we are familiar with at more human scales.

Tiny Lights

Among the new devices of microtechnology are tiny lasers. High-intensity lasers developed by Photonics Research Inc. of Boulder CO are only a fraction of the size of a human hair. These are soon going into production.

They are intended for use in such commercial devices as smaller laser printers, bar-code readers, and scanners.

Not so far along are even tinier low-power lasers designed by Samuel L. McCall at AT&T Bell Laboratories in Murray Hill NJ. The lasers are the smallest ever made, with some as tiny as 2 microns in diameter and only 400 atoms thick. Ten thousand of the lasers would fit on the head of a pin.

The lasers even look like the heads of pins, or at least like thumbtacks. In this configuration the light courses around the rim of the disk at the top of the tacklike shape until the light builds up laser power and escapes from both the top and the edges of the disk.

McCall's easily manufactured devices may be useful in optical computers and telecommunications switches, where the tradeoff of low power for speed in switching is well worth it. At present, however, the lasers operate continuously only at temperatures below 0° C (32° F).

The top of the tack is a disk of layers of indium gallium arsenide and layers of indium gallium arsenide phosphide, a fairly expensive combination. Since all the action takes place at the edges of the disk, the central part is not needed. McCall is now working on a way to produce similar lasers that omit the central part, resulting in a shape more like a washer than a thumbtack.

Another set of tiny lasers, developed by Lester Eastman and coworkers at Cornell University, are somewhat larger than McCall's lasers—about the diameter of a human hair—but much faster. Turning on and off at a rate of 28 billion times a second, or 28 gigahertz, the Cornell lasers hold the current record for switching speed. The team's goal is to produce lasers that reach a switching speed of 44 gigahertz, close to what is now believed to be the 60-gigahertz theoretical limit for this type of laser and an especially useful speed in devices.

Fast switching speeds are important in communications using fiber optics. A 28 gigahertz laser can be used to increase the carrying capacity of a typical state-of-the-art communications network by a factor of fifteen.

The fast speed of the Cornell lasers is directly related to their small size. They are versions of a type called strained quantum well lasers, which confine electrons in small regions called quantum wells. The wells are composed of thin layers of material that are less than 40 atoms thick.

A Record Small Motor

The Toshiba Corporation announced on Nov 5 1991 that it had built the world's smallest electromagnetic motor, with an outside diameter of 0.25 cm (0.1 in.). The motor is a standard type, complete with magnets, coils, and bearings. Although silicon devices that are electrostatic motors are much smaller, the Toshiba motor is more practical. For one thing, the

Toshiba motor needs only about a fiftieth of the amount of electricity to produce a reasonable amount of power as does an electrostatic motor .

(Periodical References and Additional Reading: *New York Times* 11-6-91, p D7; *New York Times* 11-8-91, p D4; *New York Times* 12-31-91, p C6; *Scientific American* 1-92, p 28)

SEARCHING FOR GAS AND OIL

A vast number of earth scientists are employed by petroleum companies to help find oil. Conventional theory holds that oil and natural gas are formed by remains of organisms that have accumulated in sediments. Eventually, these remains become transformed by heat and pressure into petroleum and natural gas.

It used to be thought that "eventually" means several million or hundred million years. But in 1990, researchers working in the Gulf of California discovered that under proper conditions—thick deposits of organic remains and a source of geothermal heat—"eventually" can be shortened to less than 5000 years. But such conditions exist only a few places on Earth. The job of the geologist is to identify the more common places where large amounts of oil and gas could have been produced by typically slow geologic processes. Then test wells are used to determine whether or not the oil or gas is actually present.

An Expert System

Among the principal clues as to what has occurred in the Earth's past are tiny fossils, called microfossils. Many of these are the shells of such microscopic protists as Foraminifera or hard parts of small aquatic animals such as those known as conodont animals. There are thousands of different kinds of microfossils. Even two microfossils from the same species may look quite different if they have been damaged in different ways during preservation. Highly specialized paleontologists currently spend years developing skills at identifying microfossils, and even they have to turn frequently to large reference books for aid.

Peter Swaby and coworkers at British Petroleum's Research International announced at the May 1990 conference on Innovative Applications of Artificial Intelligence in Washington DC that they had found a better way, one based on key features similar to those used by birders. A good birder does not have to see a whole bird clearly to identify it—a flash of color, a bit of beak, or a burst of song, combined with a characteristic habitat, is enough. The experts in identifying microfossils use the same methods; they begin by recognizing some specific characteristic feature, depending on what has been preserved in the fossil. This might be a sharp spike, if it made it through the ages, or a wavy surface. Swaby and his coworkers found out what these clues were and wrote a computer program incorporating them.

Unlike older expert systems, however, this one is not just a list of questions that lead to a branching program ("Is there a long spike attached to a wavy surface?" If yes, "Is the approximate length of the spike less than or greater than 1 mm?" and so forth). Instead, the power of computer graphics is used to display images of the typical microfossils. There is still a branching program that leads to specific identification, but the graphics allows the human eye and brain to compensate for the fuzziness of lengthy descriptions. Tests on novice postgraduate students suggest that the system works. It is expected to be introduced into the field in 1993 or 1994.

On the Other Hand

Not everyone believes that the conventional view of the origin of petroleum and natural gas is correct. Thomas Gold, emeritus professor of astronomy at Cornell University, thinks that natural gas has always abounded in the depths of Earth, left there when the planet formed. This is not completely unreasonable, since astronomers observe methane, the chief component of natural gas, on apparently lifeless bodies in the solar system. If Gold's theories are correct, the association of natural gas and petroleum with certain microfossils is purely a matter of chance. Gold sees no reason why one could not find large amounts of natural gas in certain formations of igneous rock, such as granite. Igneous rock contains no microfossils (or any other kinds of fossils) at all.

The Swedish State Power Board saw in Gold's theories a chance to solve a growing need for energy in Sweden. Sweden has few options for energy expansion. Virtually all the hydropower Sweden could produce is being utilized already. In 1980 Swedish voters approved a referendum to close down all Swedish nuclear plants by 2010. Its high latitude makes Sweden an unlikely place for solar sources of energy, so that leaves fossil fuels, such as coal, oil, or natural gas. But Sweden has little sedimentary rock of any kind and therefore, according to conventional theories, virtually no chance of finding domestic fossil fuels. Gold's idea, however, suggests that Sweden could find natural gas in the granite on which the nation rests.

Although a panel of experts told the Swedish State Power Board that Gold's theories are wrong, the Power Board proceeded to form a semi-independent organization, Dala Deep Gas, to go for the gas in the granite. Much of the money for Dala Deep Gas was raised from small and large investors, including money funneled into a holding company with the unlikely name of Anathema Oil. The location chosen was the site of an ancient meteor in the Siljan lake district, on the thought that the impact would have fractured the granite and allowed gas to collect in the cracks. Thus the well, officially known as Gravberg-1, came to be called "The Siljan Hole." Drilling started in Jul 1986.

Since then, the project has been marked by optimistic press releases and major problems. Although the hole reached 6800 m (22,310 ft) by Sep 1989, there was less methane in the rock than is frequently found in other drill sites. Evidence suggest that most methane in the hole came from the lubricants used on the drill.

Despite the lubricants, the drill bits were not traveling straight down. The bend in the hole finally got so great that drilling was stopped. Gold says they did not go far enough to get to the methane, which he believes is at about 7400 m (24,280 ft), just a few hundred meters from where they stopped. Experts in the business, of course, think he is wrong. At last account, plans were being made to get new financing and either go on with the present hole or drill another in the region. (See page 254 for another account of deep drilling projects.)

One surprise was the discovery of tiny grains of the mineral magnetite in deep sludge. Gold believes that these grains are produced by bacteria that live on the methane he theorizes is there—in fact, Gold expects that there is a "huge amount of bacterial life down there" hundreds of meters below the surface. In the case of the magnetite-bearing sludge, J. M. Suflita from the University of Oklahoma has been able to grow bacteria in cultures from the sludge, but has not yet succeeded in identifying them.

(Periodical References and Additional Reading: *Science* 3-9-90, p 1177; *Science* 6-1-90, p 1080; *Discover* 5-90, p 22; *Science* 6-28-91, p 1790)

THE NATIONAL INVENTORS HALL OF FAME

In 1973 the US National Council of Patent Law Associations (now the National Council of Intellectual Property Law) began the practice of naming inventors who hold US patents to The National Inventors Hall of Fame. Although many of those honored and listed below have several patents, a selection committee chooses one patent for each inventor as the occasion for the award, which they identify using the title of the original patent application, given in capital letters below. The date in parentheses at the end of each entry is the year of induction. Note that holding a US patent does not mean that the honoree is necessarily American. Some, such as Rudolf Diesel and Louis Pasteur, were lifelong citizens of other countries.

Alexanderson, Ernst F.W. [Swedish-US: 1878-1975]
HIGH FREQUENCY ALTERNATOR This basic device makes it possible for radio and television to transmit voices and music, not just dots and dashes; Alexanderson was also a pioneer in television and many other kinds of electrical equipment, receiving 322 patents. (1983)

Alford, Andrew [Russian-US: 1904-]
LOCALIZER ANTENNA SYSTEM With this and other inventions, Alford developed the radio system for airplane navigation and instrument landing systems.(1983)

Alvarez, Luis Walter [US: 1911-1988]
RADIO DISTANCE AND DIRECTION INDICATOR Despite his citation, Alvarez is far better known as the developer of specialized equipment for studying subatomic particles (which led to his 1968 Nobel Prize) and as a proponent—along with his son Walter—of the theory that dinosaurs became extinct as the result of the impact of a massive body on Earth 65 million years ago. (1978)

Armstrong, Edwin Howard [US: 1890-1954]
METHOD OF RECEIVING HIGH FREQUENCY OSCILLATIONS Armstrong's several inventions connected with radio broadcasting and reception made him rich and created the "radio days" of the 1920s through the 1940s; then, in 1939, he invented FM broadcasting and reception, which helped lead to another revolution in radio. (1980)

Baekeland, Leo Hendrik [Belgian-US: 1863-1944]
SYNTHETIC RESINS Baekeland's plastic (synthetic resin) that he named Bakelite was not the first plastic to be manufactured (that was celluloid), but it was the first to make people realize the potential of plastics in general; Baekeland also developed the first commercially successful photographic paper. (1978)

Bardeen, John [US: 1908-1991]
TRANSISTOR, Bardeen invented the transistor with William Shockley and Walter Brattain; the transistor is the essential semiconductor device that is used on microprocessors and other chips. (1974)

Beckman, Arnold O. [US: 1900-]
APPARATUS FOR TESTING ACIDITY Although there are many simple ways to determine acidity, a precise measuring instrument developed by Beckman became the foundation of the Beckman Instrument Co., a leading company in the scientific instrumentation field; Beckman's other inventions of precision instruments also contributed to the company's growth. (1987)

Bell, Alexander Graham [Scottish-US: 1874-1922]
TELEGRAPHY Despite the title, the invention was the telephone; another important Bell invention was the disk phonograph. (1974)

Bennett, Willard Harrison [US: 1903-1987]
RADIO-FREQUENCY MASS SPECTROMETER A mass spectrometer is a device to separate particles according to their masses. Prior to the Bennett spectrometer, patented in 1955, heavy magnets were used for this task; the Bennett device uses easy-to-generate radio waves. In addition to his invention, Bennett is noted for pioneer work in plasma physics. (1991)

Black, Harold Stephen [US: 1898-]
NEGATIVE FEEDBACK AMPLIFIER The basic principle of the negative feedback amplifier, the feedback of information to control a process, has become fundamental to many other devices since Black's first use of it to control distortion; Black also invented pulse-code modulation (1981)

Brattain, Walter H. [US: 1902-1987]
TRANSISTOR (See Bardeen, John)

Burbank, Luther [US: 1849-1926]
PEACH Burbank holds sixteen plant patents with numbers between 12 and 1041 (all issued posthumously); his work in developing more than 800 new varieties of plants was in part responsible for development of the plant patent program, which began in 1930. (1986)

Burroughs, William Seward [US: 1857-1898]
CALCULATING MACHINE Although a calculating machine had been built by Wilhelm Schickardt as early as 1623 (and other types by later inventors), Burroughs was the first to develop a practical device that could be mass produced and easily used. (1987)

Burton, William Meriam [US: 1865-1954]
MANUFACTURE OF GASOLINE The highlight of Burton's years in the oil business came when he developed the first commercially successful cracking process, a method that yields twice the amount of gasoline from crude oil as had previous methods. (1984)

Camras, Marvin [US: 1916-]
METHOD AND MEANS OF MAGNETIC RECORDING Before the tapes currently used to record sound and pictures, sound was recorded on the wire recorder that Camras invented in the 1930s; he went on to develop over 500 inventions, most connected with improvements in recording methods. (1985)

Carlson, Chester F. [US: 1906-1968]
ELECTROPHOTOGRAPHY Carlson invented the dry copying method used in most offices today, which he named xerography; although first patented in 1940, the dry

copier did not reach the market until 1958, by which time Carlson had many patents on improvements in the process. (1981)

Carothers, Wallace Hume [US: 1896-1937]
DIAMINE-DICARBOXYLIC ACID SALTS AND PROCESS OF PREPARING SAME AND SYNTHETIC FIBER Despite the formidable title of his patent, Carothers's invention of nylon has contributed an important fiber to the world; he also developed the first commercially synthetic rubber. (1984)

Carrier, Willis Haviland [US: 1876-1950]
APPARATUS FOR TREATING AIR Not only did Carrier invent the first really workable air-conditioning system, he also invented many of the techniques used in modern refrigerators. (1985)

Carver, George Washington [US: 1864-1943]
PRODUCTS USING PEANUTS AND SWEET POTATOES Carver was a successful black scientist in Iowa when he accepted an invitation to return south to Tuskegee AL to help other blacks advance. He recognized that traditional agricultural practices were destroying the soil. To encourage farmers to plant regenerative plants instead of cotton and tobacco, Carver developed over 300 new uses for the peanut and 118 sweet potato by-products, taking no personal profits from any of these inventions. (1990)

Colton, Frank B. [Polish-US: 1923-]
ORAL CONTRACEPTIVES Colton not only pioneered the first birth control pill, in 1960, he also was an early developer of anabolic steroids. (1988)

Conover, Lloyd H. [US: 1923-]
TETRACYCLINE Before Conover created tetracycline (patented Jan 11 1955), no one thought that a natural drug could be chemically modified to improve its action; tetracycline remains the drug of choice for tick-spread diseases such as Rocky Mountain spotted fever and Lyme disease. (1992)

Coolidge, William D. [US: 1873-1974]
VACUUM TUBE The "Coolidge tube" is actually an X-ray generator; among Coolidge's many other inventions is the modern tungsten-filament electric light. (1975)

Cottrell, Frederick G. [US: 1877-1948]
ELECTROSTATIC PRECIPITATOR Cottrell's invention, patented Aug 11 1908, uses high-voltage electricity to capture the particulates, including fly ash, dust, and droplets of acids or other chemicals found in smoke from the burning of fossil fuels and many industrial processes; the tons of waste can then be removed from smoke instead of being spread around the countryside. (1992)

Damadian, Raymond V. [US: 1936-]
APPARATUS AND METHOD FOR DETECTING CANCER IN TISSUE Damadian was the first to realize that the nuclear magnetic resonance technique could be used on living creatures (it was already a success as a laboratory tool used by chemists) and that it could detect cancer cells. (1989)

Deere, John [1804-1886]
PLOW Anyone who grew up near a farm knows the name John Deere from the company he founded, which still makes farm tools; few remember that a vastly improved plow was the start of his whole enterprise. (1989)

DeForest, Lee [US: 1873-1961]
AUDION AMPLIFIER Although he eventually acquired more than 300 patents related to radio, De Forest's invention of the triode was the key to modern radio and later developments in amplification of signals. (1977)

Diesel, Rudolf [German: 1858-1913]
INTERNAL COMBUSTION ENGINE The pressure-ignited heat engine is still called the diesel engine. (1976)

Djerassi, Carl [Austrian-US: 1923-]
ORAL CONTRACEPTIVES Djerassi has been a major influence on modern organic chemistry, although his interests within that field have been so broad, it is difficult to single out any one of them. (1978)

Dow, Herbert Henry [Canadian-US: 1866-1930]
BROMINE Besides new methods of extracting bromine and chlorine from naturally occurring salt deposits, Dow patented over ninety inventions and founded the Dow Chemical Company. (1983)

Draper, Charles Stark [US: 1901-1987]
GYROSCOPIC EQUIPMENT Because a spinning gyroscope maintains the same relative position if it is free to do so, a gyroscope can be used to maintain a missile, a plane, or a ship on its course. In some ways a gyroscope is better than a compass; for example, it can be set to point true north and it is not affected by variations in magnetism. Draper's gyroscopic stabilizer helped both antiaircraft guns and falling bombs hit their targets in World War II. Later he developed systems for navigation and for guiding missiles. (1983)

Durant, Graham J. [English-US: 1934-]
CIMETIDINE (TAGAMET^R) With John C. Emmet and C. Robin Ganellin, Durant developed the major drug used as an acid suppresser for stomach and intestinal ulcers. Introduced in the United Kingdom in 1976 and in the United States in 1977, by 1980 Tagamet^R was the best-selling drug in America. (1990)

Eastman, George [US: 1854-1932]
METHOD AND APPARATUS FOR COATING PLATES FOR USE IN PHOTOGRAPHY Before Eastman's work, photographers had to use cumbersome wet plates. Eastman learned that it is possible to make a dry plate, and developed the necessary equipment to prepare the plates commercially. He also invented the transparent roll film that was the basis of the first Kodak box camera and a stronger motion picture film for use in the newly invented cinema. (1977)

Edgerton, Harold E. [US: 1903-]
STROBOSCOPE Although the idea of using a flash of bright light to stop action in a photograph goes back to at least 1908, in the 1930s Edgerton created the special device called the stroboscope to produce such flashes on a regular basis. His classic photograph of the crown produced by a drop of milk falling into a bowl of milk dates from the 1930s also, but Edgerton went on to make many other analyses of events using the stroboscope to stop motion. He also contributed inventions to underwater photography. (1986)

Edison, Thomas Alva [US: 1847-1931]
ELECTRIC LAMP In addition to the carbon-filament electric lamp, Edison patented

a phonograph, the mimeograph, the fluoroscope, and motion picture cameras and projectors. (1973)

Elion, Gertrude Belle [US: 1918-]
DNA-BLOCKING DRUGS Elion, with her partner George H. Hitchings, both working at the Burroughs Wellcome Company in the 1940s, created artificial variations on the genetic material DNA that could block replication of cells; the chemicals became the basis for drugs to treat leukemia, septic shock, transplant rejection, herpes, and other diseases. She is the first woman to enter the National Inventors Hall of Fame. (1991)

Emmet, John C. [English: 1938-]
CIMETIDINE (TAGAMET^R) (See Durant, Graham J.)

Farnsworth, Philo Taylor [US: 1906-1971]
TELEVISION SYSTEM While still a high-school student, Farnsworth conceived of an all-electronic television system. Crude pictures had been transmitted previously, but the most common system relied on spinning mirrors. Farnsworth patented many of the components of all-electronic television and was a pioneer in electronic microscopes, radar, ultraviolet light for seeing in the dark, and nuclear fusion. (1984)

Fermi, Enrico [Italian-US: 1901-1954]
NEUTRONIC REACTOR Fermi's nuclear reactor is the basis of nuclear power today. His many contributions to modern physics include basic theoretical work as well as experimental physics. (1976)

Ford, Henry [US: 1863-1947]
TRANSMISSION MECHANISM Many of Ford's "inventions" that revolutionized society, such as the automobile assembly line, vertical integration of manufacturing, inexpensive automobiles, and the $5-a-day wage (in 1914) were not patentable. Ford did invent and patent numerous mechanisms used in automobiles. (1982)

Forrester, Jay W. [US: 1918-]
MULTICOORDINATED DIGITAL INFORMATION STORAGE DEVICE A pioneer in the development of electronic computers after World War II, Forrester's main invention was magnetic storage of information. Most computers today, from giant mainframes to tiny laptops, still use magnetic storage to retain data even when the computer has been shut off. (1979)

Ganellin, C. Robin [English: 1934-]
CIMETIDINE (TAGAMET^R) (See Durant, Graham J.)

Ginsburg, Charles P. [US: 1920-]
VIDEOTAPE RECORDER The ubiquitous VCR, now known mostly from Japanese imports, was actually developed in the United States by an engineering team led by Ginsburg. (1990)

Goddard, Robert Hutchings [US: 1882-1945]
CONTROL MECHANISM FOR ROCKET APPARATUS The father of American rocketry, Goddard's experiments with liquid-fueled rockets between the two world wars were often derided, notably in a famous *New York Times* editorial. During both of the wars, however, the military accepted his offer of help, and he devised successful rocket weapons and rocket-assisted mechanisms for carrier-based airplanes. He obtained 214 patents on various aspects of rocketry. (1979)

Goodyear, Charles [US: 1800-1860]
IMPROVEMENT IN INDIA-RUBBER FABRICS Goodyear's discovery of how to make rubber that could withstand heat and cold is an illustration of the principle that people who find new ideas by accident are usually looking for that idea in the first place. While Goodyear was working on ways of making rubber more practical in 1844, he accidentally dropped rubber mixed with sulfur on a hot stove. The result, which Goodyear named vulcanized rubber, was what he had been seeking. Although Goodyear patented vulcanization and other ways to improve rubber, his patents were constantly infringed upon, and he died poor. (1976)

Gould, Gordon [US: 1920-]
OPTICALLY PUMPED LASER Although Gould developed the idea for the laser slightly before anyone else, he was not the first to get a patent filed. Despite this, after a 20-year battle, he finally obtained patent rights in 1977. By then lasers were already in use in a host of applications. He also holds patents in other fields, notably fiber optics. (1991)

Greatbatch, Wilson [US: 1919-]
MEDICAL CARDIAC PACEMAKER Greatbatch's pacemaker has helped millions of people with heart disease; he is also the inventor and manufacturer of the batteries that can be implanted along with the pacemaker to keep the machinery running without the adverse effects sometimes caused by the chemicals. (1986)

Greene, Leonard M. [US: 1918-]
AIRPLANE STALL WARNING DEVICE In the 1950s, when Greene's stall warning device was just coming into use, more than half of all aviation accidents occurred because pilots failed to recognize that the plane was about to stall. Subsequently, Greene patented more than sixty air-safety features, including a device that warns of low-altitude wind shears. (1991)

Hall, Charles Martin [US: 1863-1914]
MANUFACTURE OF ALUMINUM After his college chemistry teacher remarked that discovering a way to make cheap aluminum (then selling at five dollars—1886 dollars!—a pound) would make a person rich and famous, Hall set himself to the task. In only 8 months, he found the method and indeed became rich and famous. That same year, the French metallurgist Paul-Louis-Toussaint Héroult discovered the same process. Patent litigation between the two independent discoverers was eventually resolved amicably. (1976)

Hanford, W.E. "Butch" [US: 1908-]
POLYURETHANE Hanford and his partner Donald F. Holmes developed their first form of polyurethane in 1937 while both were at E.I. du Pont Nemours & Company, but they are recognized for an improved form they patented in 1942. (1991)

Hewlett, William R. [US: 1913-]
VARIABLE FREQUENCY OSCILLATION GENERATOR The first invention of one of the founders of Hewlett-Packard was the audio oscillator in 1939 (not patented until Jan 6 1942), a device for generating high-quality audio frequencies that could be used for many different purposes; among the first uses was production of special sounds for the movie *Fantasia.* (1992)

Higonnet, René Alphonse [French: 1902-1983]
PHOTO COMPOSING MACHINE Along with Louis Marius Moyroud, Higonnet developed the first machine to set type by recording the images of letters on film. Since this invention in 1946, film composition has become the standard way of setting type, replacing type set from hot metal. (1985)

Hillier, James [Canadian-US: 1915-]
ELECTRON LENS CORRECTION DEVICE Although Hillier was not the first to make a microscope using electrons, his microscopes became the standard in the field. Electron microscopes can enlarge much smaller details than light microscopes because the wavelength of an electron is much smaller than the wavelength of a photon of visible light. (1980)

Hollerith, Herman [US: 1869-1929]
STORAGE AND PROCESSING OF NUMERICAL DATA The punched cards and readers Hollerith developed for the 1890 US census became the basis of data processing. The company he founded in 1896, along with three other companies, in 1924 became IBM. (1990)

Holmes, Donald Fletcher [US: 1910-1980]
POLYURETHANE (See Hanford, W.E. "Butch")

Houdry, Eugene J. [French: 1892-1962]
CATALYTIC CRACKING OF PETROLEUM Houdry's patent recognition is for a process by which high-grade gasoline and airplane fuel is made from crude oil, but he also developed many other important catalytic processes and devices, ranging from a method of making synthetic rubber to the basic catalytic converter used in modern automobile mufflers. (1990)

Julian, Percy F. [US: 1899-1975]
SYNTHESIS OF CORTISONE AND OTHER HORMONES Despite a Ph.D. from the University of Vienna and an already brilliant record in chemistry, Julian was rejected for a professorship at DePauw University, presumably because he was black. In private industry, he developed many important industrial products based on soybeans before discovering that soybeans can be used as the basis for synthesizing cortisone, an important hormone with many medical uses. (1990)

Kettering, Charles Franklin [US: 1875-1958]
ENGINE STARTING DEVICES AND IGNITION SYSTEM Although one tends to think of Kettering as the genius behind many developments in automotive engineering at General Motors, his inventing career actually began with the electric cash register at National Cash Register. Kettering's Dayton Engineering Laboratories Co. (Delco) produced the self-starter for automobiles and the first small electric generator that could be used on isolated farms before rural electrification. After he sold Delco to General Motors, Kettering continued to run a research laboratory. In addition to automobile-related inventions, Kettering's laboratory developed diesel locomotive engines. (1980)

Kilby, Jack S. [US: 1923-]
MINIATURIZED ELECTRONIC CIRCUITS A number of people worked on putting several transistors and other solid-state electronic devices on a single chip, but the

monolithic integrated circuit that Kilby developed for Texas Instruments in 1959 was the beginning of the modern integrated circuit. (1982)

Kolff, Willem J. [Dutch-US: 1911-]
SOFT-SHELL, MUSHROOM-SHAPED HEART Although Kolff's American patent that is cited is for an early version of an artificial heart, his most important work was the development of the artificial kidney dialysis machine while he was still in the Netherlands. (1985)

Land, Edwin Herbert [US: 1909-1991]
PHOTOGRAPHIC PRODUCT COMPRISING A RUPTURABLE CONTAINER CARRYING A PHOTOGRAPHIC PROCESSING LIQUID Land's first success was not the instant camera for which he became world famous, but the development of substances that polarize light and applications of those substances. He also made important contributions to the theory of color vision. (1977)

Langmuir, Irving [US: 1881-1957]
INCANDESCENT ELECTRIC LAMP The original Edison-Swan light bulbs relied on a vacuum to keep the filament from burning too fast, but in 1913 Langmuir realized that filling the bulb with a nonburning gas would result in a longer-lasting light. He also made many basic scientific discoveries, including those stemming from work with the chemistry of surfaces that won him the 1932 Nobel Prize in chemistry. (1989)

Lawrence, Ernest Orlando [US: 1901-1958]
METHOD AND APPARATUS FOR THE ACCELERATION OF IONS Although Lawrence did not develop the very first particle accelerator (popularly known as an atom smasher), his 1930 cyclotron, with many refinements and contributions by other inventors, has set the basic pattern for the most successful and most powerful machines of its type since. (1982)

Ledley, Robert S. [US: 1926-]
WHOLE-BODY CT SCANNER When Ledley proposed his revolutionary device that would make three-dimensional images of living tissue, manufacturers were not interested—but radiologists were. Ledley had to form his own company to make the device, which soon became a successful part of the medical marketplace. (1990)

Maiman, Theodore Harold [US: 1927-]
RUBY LASER SYSTEMS There has been a lot of dispute about the invention of the laser, but it is clear that Maiman's ruby laser was the first to be recognized worldwide and to be commercially successful. (1984)

Marconi, Guglielmo [Italian: 1874-1937]
TRANSMITTING ELECTRICAL SIGNALS Marconi's patent was for using radio waves to carry coded messages—also known as wireless telegraphy. (1975)

McCormick, Cyrus [US: 1809-1884]
REAPER McCormick developed his machine for harvesting grain in 1831 and patented it in 1834, when he learned that there might be competition in the field. By 1847 his factory was turning out the machines that would be among the first part of the revolution in agriculture in the United States. (1976)

Mergenthaler, Ottmar [German-US: 1854-1899]
MACHINE FOR PRODUCING PRINTING BARS Mergenthaler's invention, known as the Linotype, constituted the first major improvement in setting type since Guten-

berg began using movable type about 1440. The Linotype machine casts each line of type from melted metal, based on instructions from a keyboard. Modern versions of the Linotype are still in use, but most typesetting today is done by photographic processes. (1982)

Morse, Samuel F.B. [US: 1791-1872]
TELEGRAPH SIGNALS The telegraph that Morse developed was the first that was commercially successful. Scientist Joseph Henry [US: 1797-1878] was the genius behind the electricity-based devices that made the telegraph work, but promoter Morse and his dot-dash code made long-distance, instantaneous communications possible. (1975)

Moyer, Andrew J. [US: 1899-1959]
METHOD FOR PRODUCTION OF PENICILLIN World War II brought a great need for medicines that would halt infections. Although penicillin was known to achieve the desired results, no one knew how to make it in quantity. Moyer, a microbiologist at the US Department of Agriculture's Northern Regional Research Laboratory in Peoria IL, solved the problem. Combining a strain of mold found on a rotting muskmelon with a new growing medium (a by-product of the manufacture of corn starch) solved the problem. The basic method continues to be used today in the manufacture of many other antibiotics and substances produced by microorganisms. (1987)

Moyroud, Louis Marius [French: 1914-]
PHOTO COMPOSING MACHINE (See Higonnet, René Alphonse)

Noyce, Robert N. [US: 1927-1990]
SEMICONDUCTOR DEVICE-AND-LEAD STRUCTURE Noyce was at the center of development at two seminal semiconductor producers, Fairchild and Intel, both of which he helped found. Intel today makes the most widely used microprocessor chips for personal computers, those at the heart of various IBM models and their clones. (1983)

Olsen, Kenneth H. [US: 1926-]
IMPROVED MAGNETIC CORE MEMORY After leaving MIT, Olsen founded Digital Equipment Corporation (DEC) to manufacture computers based on his new memory devices. DEC, with Olsen still president, soon became one of the leaders in electronic computing and remains so today. Olsen also contributed extensively to the development of the minicomputer, a practical computer smaller than the room-sized main frames, but still far too large for a desk top. (1990)

Otis, Elisha Graves [US: 1811-1861]
IMPROVEMENT IN HOISTING APPARATUS Modern skyscrapers would have been impossible without the safety elevator that Otis devised in 1853 when his employer asked him to build a hoist to lift heavy equipment. Eight years later the first Otis passenger elevators were being installed. (1988)

Otto, Nikolaus August [German: 1832-1891]
GAS MOTOR ENGINE While Otto's four-stroke engine of 1876 is the basis of the modern internal combustion engine, it ran on compressed natural gas instead of gasoline. The gasoline-powered engine was a further development of the Otto engine pioneered by Gottlieb Daimler and Wilhelm Maybach for use in automobiles in 1889. (1981)

Parker, Louis W. [Hungarian-US: 1906-]
TELEVISION RECEIVER Parker not only invented the basic type of television receiver that is in common use today, but also the basic type of color television transmission and reception that is most commonly used. (1988)

Pasteur, Louis [French: 1822-1895]
BREWING OF BEER AND ALE Louis Pasteur is not usually thought of as an inventor (and his work on beer and ale is generally thought to be unsuccessful), but he did invent vaccines for several diseases, although these very important inventions were not patented, and pasteurization, the process of protecting beverages and food from contamination by microbes through heating. (1978)

Plank, Charles J. [US: 1915-]
CATALYTIC CRACKING OF HYDROCARBONS WITH A CRYSTALLINE ZEOLITE CATALYST COMPOSITE Along with Edward J. Rosinski, Plank discovered in the early 1960s that zeolites (various aluminum silicates, a fairly common type of mineral, see p 171) could be used to improve the production of gasoline and other petroleum products. (1979)

Plunkett, Roy J. [US: 1910-]
TETRAFLUOROETHYLENE POLYMERS In 1938, Plunkett discovered the tetrafluoroethylene polymer we know as Teflon. He later developed many of the chlorofluorocarbons (Freons) that are causing considerable concern today, since they are eroding the atmosphere's protective layer of ozone. (1985)

Rosinski, Edward J. [US: 1921-]
CATALYTIC CRACKING OF HYDROCARBONS WITH A CRYSTALLINE ZEOLITE CATALYST COMPOSITE (See Plank, Charles J.)

Rubin, Benjamin A. [US: 1917-]
BIFURCATED VACCINATION NEEDLE Rubin's needle, patented Jul 13 1965, enabled easy use of small amounts of smallpox vaccine, making it possible for vaccine supplies to be stretched and helping the World Health Organization to eliminate smallpox as a disease by May 8 1980. (1992)

Sarett, Lewis Hastings [US: 1917-]
THE PROCESS OF TREATING PREGNENE COMPOUNDS Pregnene compounds are the predecessor chemicals of certain steroids. In 1944 Sarett found a way to produce cortisone as artificial steroid from its predecessor chemicals. By 1949 Sarett and his collaborators had learned to make cortisone from simple inorganic chemicals. Cortisone and closely related steroids are widely used in medicine today for treatment of conditions from arthritis to psoriasis. (1980)

Shockley, William Bradford [English-US: 1910-1989]
TRANSISTOR (See Bardeen, John)

Sikorsky, Igor I. [Russian-US: 1889-1972]
HELICOPTER CONTROLS Although known primarily today for his invention of the first commercially successful helicopter, Sikorsky designed and built many early airplanes that were widely used. In 1931 he made a critical breakthrough in helicopter design, which he had worked on for years; his continued developments led to the helicopter of today. (1987)

Sperry, Elmer A. [US: 1860-1930]
SHIP'S GYROSCOPIC COMPASS Sperry used the idea behind what was until then a child's toy to develop a compass that remains pointed in the same direction without constant correction for the difference between magnetic north and true north. In addition to the gyroscopic compass, he also invented mining and railroad equipment. (1991)

Steinmetz, Charles Proteus [German-US: 1865-1923]
SYSTEM OF ELECTRICAL DISTRIBUTION Steinmetz was an important theoretician as well as an inventor, and his most significant work was the development of the theory of alternating current (AC) that makes our present power grids possible. Among his inventions was a machine that produced "lightning in the laboratory," a device that made Steinmetz a household word near the end of his life. (1977)

Stibitz, George R. [US: 1904-]
COMPLEX COMPUTER Stibitz was one of several scientists who developed electromechanical computers in the late 1930s and during World War II. His innovations at Bell Telephone Laboratories and in the US Office of Scientific Research and Development include floating decimal-point arithmetic and taped computer programs. (1983)

Tabern, Donalee L. [US: 1900-1974]
THIO-BARBITURIC ACID DERIVATIVES Along with Ernest H. Volwiler, Tabern in 1936 developed Pentothal, the anesthetic of choice for short surgical procedures and for use prior to the administration of a general anesthetic. Tabern later introduced the therapeutic use of radioactive chemicals. (1986)

Tesla, Nikola [Croatian-US: 1857-1943]
ELECTRO-MAGNETIC MOTOR Tesla's induction motor was simpler than previous electric motors and was powered by AC current, which can more easily be distributed over long distances than direct current. (1975)

Tishler, Max [US: 1906-]
RIBOFLAVIN AND SULFAQUINOXALINE In the late 1930s, Tishler developed an economical method for synthesizing riboflavin, also known as vitamin B_2. Later he and his coworkers developed a way to produce commercially sulfaquinoxaline (an antibiotic that prevents and cures a common disease of poultry). (1982)

Townes, Charles Hard [US: 1915-]
MASERS The maser, which preceded the better-known laser, is essentially a laser that works at microwave wavelengths instead of at the shorter wavelength of visible light. Masers are used in many applications. Townes, who was solely responsible for the maser, also contributed to the development of the laser. (1976)

Volwiler, Ernest H. [US: 1893-]
THIO-BARBITURIC ACID DERIVATIVES (See Tabern, Donalee L.)

Wang, An [Chinese-US: 1920-1990]
MAGNETIC PULSE CONTROLLING DEVICE Although best known for his state-of-the-art word processor of the 1960s and 1970s, Wang contributed many fundamental ideas to the development of electronic computers, including the principle upon which magnetic core memory is built. (1988)

Westinghouse, George [US: 1846-1914]
STEAM-POWERED BRAKE DEVICES Westinghouse specialized in improving rail transportation at the time of its greatest expansion, just after the US Civil War. In 1869 he patented his first air brake for locomotives, his most important contribution to railroad safety. Later he worked on improving signals and switches, work that led him to form Westinghouse Electric Company in 1884, chiefly to implement the possibilities of alternating current. (1989)

Whitney, Eli [US: 1765-1825]
COTTON GIN By making it possible to remove seeds from cotton mechanically, the gin made large-scale cotton farming possible; Whitney also introduced interchangeable parts, the beginning of mass production. (1974)

Williams, Robert R., Jr. [US: 1886-1965]
ISOLATION OF VITAMIN B_1 (THIAMINE) Son of Baptist missionaries to India, Williams gained early familiarity with beriberi, the disease caused by thiamine deficiency. After World War I, Williams began vitamin research in his spare time, but it was not until 1933 that he isolated pure thiamine. Two years later he synthesized the vitamin, which almost immediately went into commercial production. Money from his patents has been used to encourage vitamin enrichment of foods as a way to stop deficiency diseases. (1991)

Wright, Orville [US: 1871-1948]
and Wilbur [US: 1867-1912]
FLYING MACHINE Not only did the Wright brothers invent the first airplane, they also popularized, manufactured, and sold the new machines. For the first few years after their first 1903 flight, people took little notice of their work. In 1908, however, Orville demonstrated a flight of an hour's duration that was widely reported. By World War I, airplanes were regularly used by the armed services of the combatants. (1975]

Zworykin, Vladimir Kosma [Russian-US: 1889-1982]
CATHODE RAY TUBE The cathode tube that Zworykin invented in 1928 is the kinescope, the basic picture tube used in modern television. Ten years later he developed the iconoscope, the first practical television camera. His later work on electron microscopes created the type of electron microscope that is commonly used today, although not the first electron microscope to be developed. (1977)

Appendices

APPENDIX A

USEFUL BOOKS FOR FURTHER READING

A list of all the books consulted in preparing *Current Science* would be long and tedious. Here are some that are especially new, interesting, and useful. Although most are primarily reference books, each can be read for enjoyment as well. The broad subject area that each falls into, as defined in this volume, is indicated along with the difficulty level within a broad range: books for <u>anyone</u> who might be a reader of *Current Science,* books for the <u>scientifically literate,</u> and books meant to be read by <u>scientists</u> in a particular field. There are many wonderful science books not listed here, mainly because they are not especially useful as reference tools.

The Astronomy and Astrophysics Encyclopedia. Maran, Stephen P. (Ed.). Van Nostrand Reinhold: New York, 1992. Astronomy and Space. <u>Scientifically literate.</u>

The Atomic Scientists. Boorse, Henry A., Lloyd Motz, and Jefferson Hane Weaver. Wiley: New York, 1989. Physics. <u>Anyone.</u>

Encyclopedia of Astronomy and Astrophysics. Meyers, Robert A. (Ed.). Academic Press: San Diego, 1989. Astronomy and Space. <u>Scientists.</u>

Encyclopedia of Human Evolution and Prehistory. Tattersall, Ian, Eric Delson, and John van Couvering (Eds.). Garland: New York, 1988. Life Science. <u>Scientifically literate.</u>

A History of Mathematics, 2nd Edition. Boyer, Carl B., revised by Uta C. Merzbach. Wiley: New York, 1991. Mathematics. <u>Scientifically literate.</u>

The Lever of Riches. Mokyr, Joel. Oxford University Press: New York, 1990. Technology. <u>Scientifically literate.</u>

Mapping the Next Millenium. Hall, Stephen S. Random House: New York, 1992. The State of Science. <u>Anyone.</u>

The Merck Index: An Encyclopedia of Chemicals, Drugs, and Biologicals. Budavari, Maryadel, N. O'Neil, Ann Smith, and Patricia E. Heckleman (Eds.). Eleventh Edition. Merck & Co., Inc, 1989. Chemistry. <u>Scientifically literate.</u>

Natural Hazards. Bryant, Edward. Cambridge University Press: Cambridge, UK, 1991. Environment and Issues. <u>Anyone.</u>

A Neotropical Companion: An Introduction to the Animals, Plants, and Ecosystems of the New World Tropics. Kricher, John C. Princeton University Press: Princeton NJ, 1989. Life Science. <u>Scientifically literate.</u>

Origins: The Lives and Worlds of Modern Cosmologists. Lightman, Alan and Roberta Brawer. Harvard University Press: Cambridge MA, 1990. Astronomy and Space. <u>Anyone.</u>

Paleoclimatology. Crowley, Thomas J. and Gerald R. North. Oxford University Press: New York, 1991. Earth Science. <u>Scientifically literate.</u>

Stars. Kaler, James B. Scientific American Library: New York, 1992. Astronomy and Space. <u>Scientifically literate.</u>

The Timetables of Science, New, Updated Edition. Hellemans, Alexander and Bryan Bunch. Simon & Schuster: New York, 1991. The State of Science. <u>Anyone.</u>

APPENDIX B

OBITUARIES: 1990–1991

Adler, Benjamin [Nov 10 1903–Apr 16 1990] US electronic engineer who in 1947 founded Adler Electronics, Inc., which built the first UHF station in 1953. UHF, or ultra-high-frequency television, provides additional channels and is also used to service mountainous areas not penetrated by conventional high-frequency channels. His company also developed the first VHF-UHF translator in 1956.

Adler, Richard Brooks [May 9 1922–Feb 6 1990] US engineering professor who established the Semiconductor Electronics Education Committee, a group of industrial and educational leaders who produced material that revamped undergraduate engineering education.

Anderson, Carl D. [Sep 3 1905–Jan 11 1991] Nobel Prize winning US physicist who discovered the positron (the antiparticle of the electron, sometimes called the antielectron) in 1932.

Anderson, Charles A. [Jun 6 1902–Jan 16 1990] US geologist whose contributions to the knowledge of mineral resources led to the naming of the mineral andersonite in his honor.

Anderson, Thomas F. [Feb 7 1911–Aug 11 1991] US biophysical chemist and geneticist who was one of the first scientists to use the electron microscope to study viruses. He developed the "critical point" method, in which specimens are dried to eliminate the boundary between liquid and gaseous phases, making it easier to study how certain bacteria enter their host cells.

Aston, John G. [Dec 30 1902–Aug 6 1990] US chemist who led the Low Temperature (now Cryogenics) Laboratory at Pennsylvania State University, which was known for producing liquid hydrogen and helium, conducting experiments in nuclear cooling, and being the first laboratory to achieve temperatures close to absolute zero.

Aurbach, Gerald D. [Mar 24 1927–Nov 4 1991] US health researcher who was elected to the National Academy of Sciences in 1986 for his work in parathyroid diseases. He was the first to isolate a hormone produced by parathyroid glands that regulates blood calcium in adults.

Bagnold, Ralph [Apr 3 1896–May 28 1990] US physicist and brigadier who was known for his study of the physics of desert sands and wind, which he began in

World War II. In 1970 he was awarded the Penrose Medal, the Geological Society of America's highest honor, for his research on sand dune formations and movement.

Bardeen, John [May 23 1908–Jan 30 1991] Two-time winner of the Nobel Prize in physics who with Walter Brattain and William Shockley invented the first semiconductor transistor in 1947. He also developed a low-temperature superconductivity theory that led to technologies to make medical diagnoses and made it possible to devise alloys that become superconducting at less extreme temperatures.

Barrett, Alan H. [Jun 7 1927–Jul 3 1991] US physicist who detected the presence of hydroxyl in interstellar space, the first time a molecule had been found in the Milky Way. He also designed the microwave detection equipment used on the 1962 Mariner I and II space missions that found surface temperatures on Venus to be too high to support life as it is known on Earth.

Barry, Arthur J. [Mar 11 1909–Feb 3 1990] US chemical researcher who was credited for developing silicon products used for waterproofing.

Bég, Mirza Abdul Baqi [Sep 20 1934–Jan 30 1990] US physicist who helped verify that elementary particles such as protons and neutrons are made up of quarks.

Bell, John Stewart [Jul 28 1928–Oct 1 1990] Physicist at CERN (European Nuclear Research Center), born in Belfast, Northern Ireland, who was known for his "Bell's theorem" of quantum physics, which has shown that particles communicate with each other even when they have flown far apart because the apparent coordination arises from the action of measurement.

Bennett, Dorothea [Dec 27 1929–Aug 16 1990] US geneticist who was known for her research in the genetics of early mammalian development and the first definition of mammalian sperm surface structures thought to control sperm development and fertilization.

Berko, Stephan [Dec 16 1924–May 15 1991] Romanian-US experimental physicist who introduced the use of positrons to study condensed matter.

Bernstein, Richard B. [Oct 31 1923–Jul 8 1990] US chemist who was a pioneer in "femtochemistry," the study of chemical events that occur only in approximately one-quadrillionth of a second, or femtosecond; this branch has been helpful in studying engine fuel combustion, explosions, and other chemical processes. He was awarded the US National Medal of Science in 1989 for developing the science of molecular beam scanning.

Bloch, Herman S. [Jun 15 1912–Jun 16 1990] US chemist who in the 1950s developed the catalytic converter, a device that removes pollutants from automobile exhaust, and held over 200 patents for other inventions.

Blumenthal, Sidney [Jun 24 1909–Jun 19 1990] US cardiologist who was a former director of the heart and vascular disease division of the National Heart, Lung and Blood Institute, one of the US National Institutes of Health.

Bock, Robert M. [Jul 26 1923–Jul 2 1991] US molecular biologist and former graduate school dean at the University of Wisconsin in Madison who conducted basic studies that helped decipher the genetic code and was the first to crystallize tRNA, a keystone molecule in the process of protein synthesis.

Borgstrom, Georg [Apr 5 1912–Feb 7 1990] Swedish scientist, geographer and ecologist who was an expert on world hunger and edited more than 30 books on the subject, including *The Hungry Planet*, published in 1965.

Bostick, Winston H. [Mar 5 1916–Jan 19 1991] US atomic physicist who in the mid-1950s developed a "plasma gun" that shot bursts of atomic particles at high speeds through a magnetic field for research purposes.

Bowles, Edward L. [Dec 9 1897–Sep 5 1990] US electrical engineer who during World War II developed radar for the armed forces and fitted Army aircraft with radar equipment, and also pioneered in electronics with his development of the multivibrator to maintain radio frequency.

Brody, Michael J. [Aug 16 1934–Dec 3 1990] US pharmacologist who was known for his work in showing the role played by the central nervous system in the development of hypertension.

Brophy, James John [Jun 6 1926–Dec 14 1991] US physicist and electrical engineer who held several patents and in 1989 led the University of Utah's unsuccessful research into duplicating nuclear cold fusion previously claimed by B. Stanley Pons of the same university and Martin Fleishmann of England, in the hope of finding a cheap energy source.

Brown, William Lacy [Jul 16 1913–Mar 8 1991] US geneticist who directed crop breeding programs in maize, sorghum, soybeans, and wheat, and was a former president of Pioneer Hi-Bred International, Inc., one of the world's largest hybrid seed corn producers.

Browning, Iben [Jan 9 1918–Jul 25 1991] US climatologist who warned of the 1989 San Francisco earthquake a week before it occurred as well as the eruption of the Mt St Helens volcano in 1980 but wrongly predicted that New Madrid, MO would be the epicenter of a disastrous earthquake in 1990.

Brownlow, James Merritt [May 20 1920–Apr 11 1991] US research scientist who helped develop memory cores of ceramic materials for computers and integrated semiconductor circuits for I.B.M.

Bull, Gerald V. [Mar 9 1928–Mar 22 1990] Canadian-born rocket scientist who violated the arms embargo by providing weapons to South Africa. His defense research included the development of a project designed to use two 16-inch US naval guns to launch small missiles into space.

Bunzel, Ruth L. [Apr 18 1898–Jan 14 1990] US anthropologist who was known for her study of the Zuni Indians. In the 1920s, she lived in a Zuni household and learned their language.

Castle, William Bosworth [Oct 21 1897–Aug 9 1990] US hematologist who discovered that pernicious anemia, a disease caused by a decrease in the blood's red corpuscles, is caused by a deficiency in the body's digestive system and can be treated by a proper diet.

Chang, M.C. [Feb 1 1908–Jun 5 1991] US scientist, born in Taiyuan, China, who was a codeveloper of the birth control pill and a pioneer in vitro fertilization, or "test tube babies."

Child, Charles G. [Feb 1 1908–Jun 23 1991] US surgeon who was the lead investigator for federally funded research on portal hypertension, or high blood pressure in the abdominal veins.

Chin, Gilbert Y. [Sep 21 1934–May 5 1991] US researcher, born in China who headed a Bell Labs team that developed the Chromindur family of alloys which, when formed into magnets, were an improvement over the hot stamp magnets formerly used in the conversion of electrical signals to speech in telephone handsets.

Clapp, Roger Edge [Oct 9 1919–Feb 19 1991] US physicist who worked on the development of radar at MIT from 1942 to 1946 and held patents for inventions in radioastronomy, radiometric receivers, and nonimpact printing.

Clarke, J.F. Gates [Feb 22 1905–Sep 17 1990] US entomologist and research associate at the Smithsonian Institution's Museum of Natural History who was an authority on moths and helped develop a collection of more than 30 million insects for the museum.

Compton, Walter Ames [Apr 22 1911–Oct 11 1990] US pharmaceutical researcher who was the head of Miles Laboratories, where he helped develop One-A-Day multivitamin supplements and Bactine, a nonstinging antiseptic, and created Clinitest and Clinistix, both used to detect sugar in urine.

Cook, Robert Carter [Apr 9 1898–Jan 7 1991] US geneticist and demographer who was director of the Population Reference Bureau and was a proponent of birth control as a means of maintaining population balance.

Croxton, Frederick E. [May 23 1899–Jan 10 1991] US statistician and educator who was the coauthor of several widely used textbooks, including *Practical Business Statistics* and *Applied General Statistics*.

Curtis, Doris Malkin [Jan 12 1914–May 26 1991] US geologist who had been president of several organizations, including the Geological Society of America and headed the national committee on geology at the US National Academy of Sciences.

Dammin, Gustave J. [Sep 17 1911–Oct 11 1991] US pathologist who in 1954 helped participate in the first kidney transplant and was an authority on organ transplant pathology. In response to his later research on Lyme disease and babesiosis, which is similar to malaria, the deer tick that causes these diseases was named *Ixodes dammini* in his honor.

Dana, Leo I. [Nov 1 1895–Aug 16 1990] US physicist who, while at Union Carbide, developed thermal superinsulation to ease the shipment of liquefied gases used in the metallurgical and chemical industries, synthetic sapphires, and automotive antifreeze products.

Decker, David [Sep 14 1917–Jul 17 1990] US researcher who developed cisplatin and cyclophoshamide, two drugs widely used for the treatment of cancer.

DeJong, Russell Nelson [Mar 12 1907–Aug 20 1990] US neurologist who was founding editor of *Neurology*, the journal of the American Academy of Neurology, and wrote over 200 books and articles on the subject.

de Mestral, Georges [1907–Feb 11 1990] Swiss engineer who in 1948 invented Velcro, a clothing fastener that is made of nylon loops and hooks that can be pressed together for holding and subsequently peeled apart.

DeWire, John W. [Jun 12 1916–Sep 17 1990] US physicist who worked on the physics of uranium fission during World War II while at Los Alamos Laboratory in New Mexico and participated in the first atomic test explosion there. The Cornell professor also helped develop electron accelerators and was head of the Fermi National Laboratory in 1967.

Dock, William [1898–Oct 17 1990] US cardiologist who often took opposing positions on medical practice and was known as the author of Sutton's Law, which claims that a single test is more likely to provide a true diagnosis than a series of expensive routine examinations.

Doll, Henri-Georges [Aug 13 1902–Jul 25 1991] US engineer, born in Paris, who held more than 70 patents in oil research, and received promotion to officer of the Legion d'Honneur, France's highest decoration, in 1987.

Downs, Wilbur George [Aug 7 1913–Feb 17 1991] US expert on tropical medicine who directed a malaria-control program in Mexico for the Rockefeller Foundation, where he was among the first to question the use of insecticides such as DDT in trying to control the disease. He was also director of the Yale Arbovirus Research Unit, where he isolated and characterized viruses, including the deadly Lassa fever virus.

DuBois, Cora [Oct 26 1903–Apr 7 1991] US anthropologist whose accomplishments include a social and psychological study of the people of Alor, an island in the Dutch East Indies. She was chief of research and analysis for the US Army's Southeast Asia Command during World War II and was awarded the Exceptional Civilian Award.

Edgerton, Harold E. "Doc" [Apr 6 1903–Jan 4 1990] US inventor who in 1931 developed a repeatable short-duration electronic flash for photography, which is now built into most cameras.

Egbert, Robert Baldwin [Dec 13 1916–Jul 10 1991] US chemical engineer who won the 1942 William H. Walker Award of the American Institute for Chemical Engineers for his study of global warming, and whose research in polyesters helped greatly reduce the expense of manufacturing Dacron and Mylar.

Elsasser, Walter M. [Mar 20 1904–Oct 14 1991] US geophysicist and refugee from Nazi Germany whose research earned him the US National Medal of Science in 1987. His work on the cooling of earth's core as a cause of its magnetic field helped establish the theory of plate tectonics and continental drift. He also led the earth-science field in quantum mechanics research.

Erim, Kenan T. [Feb 13 1929–Nov 2 1990] US archaeologist, born in Istanbul, Turkey, who for 30 years led the excavations of the ancient all-marble Roman city of Aphrodisias in southwestern Turkey.

Fairchild, Ralph G. [Sep 24 1935–Dec 17 1990] physicist and medical researcher at the Brookhaven National Laboratory who was a leading American investigator in boron-neutron-capture therapy, in which radiation is focused on to cancer cells, destroying them while leaving neighboring healthy cells undamaged.

Finch, William G.H. [Jun 28 1895–Nov 12 1990] US radio engineer who held hundreds of patents for inventions in radio technology, including the design of the first fascimile machine to transmit printed matter and photographs over telephone lines, which he patented in 1938. The invention but received little attention because of the preoccupation with World War II.

Fireman, Edward L. [Mar 23 1922–Mar 29 1990] US astrophysicist who discovered the length of time meteorites have been traveling in space by analyzing their radioactive isotopes, and who also devised a method for measuring the ages of Antarctic ice samples, which helped refine Earth's climatic record over the last 200,000 years.

Fletcher, James Chipman [June 5 1919–Dec 22 1991] US physicist who was head of the National Aeronautics and Space Administration during the Skylab missions and Viking probes on Mars in the 1970s and from 1986 to 1989, after the explosion of the space shuttle *Challenger.*

Forsyth, George H. [1901–Jan 26 1991] US architectural historian whose archaelogical expeditions included the discovery of religious relics in the ruins of a church in France built in the ninth century by a son of Charlemagne and the excavation of St Catherine's, the oldest existing Greek Orthodox monastery, located at the foot of Mt Sinai.

Frank, Ilya M. [Oct 23 1908–Jun 24 1990] Soviet physicist and professor at Moscow University who with Pavel A. Cherenkov and Igor Y. Tamm won the Nobel Prize in physics in 1958 for his work in neutron physics, and was a specialist in physical optics.

Freese, Ernst [Sep 27 1925–Mar 30 1990] US biologist, born in Germany, who was chief of the molecular biology laboratory and director of the basic neurosciences program at the National Institute of Neurological Disorders and Stroke, one of the National Institutes of Health. There he conducted research on cancer-causing properties of chemicals and on the causes of Parkinson's and Alzheimer's diseases.

Friedman, Maurice Harold [Oct 27 1903–Mar 8 1991] US medical doctor and reproductive-physiology researcher who developed the "rabbit test," in which pregnancy is detected by injecting a woman's urine sample into a female rabbit to determine whether the woman's hormones cause formations in the rabbit's ovaries.

Frost, John K. [Mar 12 1922–Aug 29 1990] US physician who had been described by the International Academy of Cytology as "the most prominent educator and teacher of cytopathology [the study of the life history of cells and how they cause disease] in the United States." He successfully argued against the American Cancer Society's opposition to X-ray screening for lung cancer because his findings showed that such screening detects the disease earlier.

Germeshausen, Kenneth J. [May 12 1907–Aug 16 1990] US engineer and inventor whose work in high-speed photography and stroboscobic devices provided the basis for developing radar modulators and firing mechanisms for atomic weapons.

Gollnick, Philip [Nov 22 1934–Jun 26 1991] US physiologist whose research into muscular adaptation included taking samples from astronauts who flew in the space shuttle *Columbia* in order to study the effect of weightlessness.

Goodwin, Harold Leland [1914–Feb 18 1990] US author who wrote science-adventure books for children, including *The Magic Talisman* in 1990 and the Rick Brant Science Adventure Series. He also worked at the US National Aeronautics and Space Administration and US National Oceanic and Atmospheric Administration.

Goulding, Robert, Jr. [Jun 29 1920–Mar 15 1991] US entomologist who placed time-release pesticides on resin to develop the flea collar for dogs and cats in 1964.

Grob, Howard S. [Jun 18 1932–Jul 20 1990] US biologist whose 1966 research on ovarian cells led to a method of removing follicles—the eggs and surrounding cells—from ovaries so they could survive in vitro fertilization.

Gronquist, Carl Harry [1903–Jun 16 1991] US civil engineer who helped design some major bridges, including the Mackinac Straits Bridge in Michigan, one of the world's longest single-unit suspension bridges, and the Kingston-Rhinecliff Bridge over the Hudson River. He was also involved in work on the Golden Gate Bridge, the Henry Hudson Bridge and the Sault-Ste Marie International Bridge.

Gueron, Jules [Jun 2 1907–Oct 11 1990] French nuclear scientist who was director of general reserach for the European Atomic Energy Community, or Euratom, wrote several books and articles on atomic energy, and was a chevalier of the French Legion of Honor.

Guth, Eugene [1905–Jul 5 1990] Hungarian specialist in physics and polymer science who helped develop the kinetic theory of rubber elasticity.

Haagensen, Cushman D. [Jul 6 1900–Sep 16 1990] US cancer specialist who was known for advocating radical mastectomies for breast cancer patients and opposing breast reconstruction, arguing that cancer could be spread by another operation and that a successful operation could not leave enough skin for an implant.

Hafner, Theodore [Oct 4 1901–Mar 26 1990] US electronics physicist born in Vienna whose sound and video research included the development of the G-line, a transmission device that carries television and other microwaves through a tube of air and allows communication with vehicles without the use of overhead wires.

Hagen, John P. [Jul 31 1908–Aug 26 1990] US astronomer who in the 1950s was director of the first US space program, Project Vanguard. Although the first space probe exploded during a 1957 launching, the Vanguard satellite was successfully launched later and has become the oldest artificial object in space. He was also associate director of the Office of Plans and Program Evaluation at NASA.

Hale, Mason E. [Sep 23 1928–Apr 30 1990] US botanist who worked at the Smithsonian Institution's National Museum for 33 years, where he was known for his research on lichens and built the third largest lichen collection in the world.

Hall, John Scoville [Jun 20 1908–Oct 15 1991] US astronomer whose research at MIT during World War II developed radar equipment used on aircraft carriers. He was also a pioneer in photoelectric photometry of stars in the infrared portion of the spectrum and a codiscoverer of interstellar polarization in the Milky Way.

Harker, David [Oct 19 1906–Feb 27 1991] US scientist who was a pioneer of X-ray crystallography, a method of cell research that has led to the deciphering of the molecular structure of hormones, proteins, drugs, and antibiotics, which has helped alter these substances to make them more effective.

Harris, Milton [Mar 21 1906–Sep 12 1991] US chemist who helped develop a method to prevent wool from shrinking and promoted the use of oil from the jojoba plant as a substitute for sperm whale oil, thus helping to prevent the whale's extinction. He was also a former president of the American Institute of Chemists and the American Chemical Society.

Heidelberger, Michael [Apr 29 1888–Jun 25 1991] US pathologist, known as the father of immunology, whose used quantitative research methods to discover that antibodies are proteins, thus developing the field of immunochemistry. His honors included the US National Medal of Science and the Louis Pasteur Gold Medal of the Swedish Medical Society.

Hertig, Arthur Tremain [May 12 1904–Jul 20 1990] US pathologist whose pioneering research in embryology shed light on the nature of infertility and led to the development by others of the contraceptive pill and in vitro fertilization.

Hirschfelder, Joseph Oakland [May 27 1911–Mar 30 1990] US physicist who helped develop the atomic bomb at Los Alamos NM and who was also chief phenomenologist at the atomic bomb tests at Bikini.

Hoeffding, Wassily [Jun 12 1914–Feb 28 1991] Internationally known statistician who developed U-statistics, or unbiased estimator Statistics.

Hofstadter, Robert [Feb 5 1915–Nov 17 1990] US physicist and former Stanford University professor who was awarded the Nobel Prize in physics in 1961 for research that led to the precise determination of the size and shape of the proton and neutron, the minute fundamental particles that constitute the nuclei of atoms.

Hsu, Hsi Fan [Mar 9 1906–Feb 2 1990] US parasitologist, born in Huangyan, China, who with his wife, Shu Ying Li, developed a vaccine to control the parasitic disease schistosomiasis.

Huggins, Charles Edward [May 7 1929–Apr 18 1990] US physician and cryobiologist who developed a process for freezing and reusing red blood cells by using a solution of glycerol, which prevents the cells from being damaged when thawed.

Johnson, Clarence L. "Kelly" [Feb 27 1910–Dec 21 1990] US aircraft designer who headed Lockheed's "Skunk Works," the advanced products unit that produced the F-80 Shooting Star jet fighter, the U-2 high-altitude reconnaissance plane, and the SR-71 Blackbird reconnaissance plane, as well as many other military aircraft.

Johnson, Frank H. [Jul 31 1908–Sep 22 1990] US biologist who helped separate luciferin, a light-emitting compound, from a South Pacific luminescent fish, isolated the luciferin crystals, and conducted research that led to the development of the most sensitive test for calcium in muscle tissue.

Kaplan, Joseph [Sep 8 1902–Oct 3 1991] US geophysicist, born in Austria-Hungary, who studied auroras, airglow, and other atmospheric phenomena, and served as a scientific adviser to the military in the 1940s and 1950s. He was chairman of the US National Committee for the International Geophysical Year in 1953.

Karplus, Robert [Feb 23 1927–Mar 22 1990] US physicist who played a major role in the Science Curriculum Improvement Project, which produced teaching kits for thousands of classrooms and has been applied to teaching in Canada, Sweden, Japan, and Korea.

Kass, Edward H. [Dec 20 1917–Jan 17 1990] US epidemiologist who in 1985 led researchers who discovered certain fibers in some tampons that help produce the bacterial toxin that causes toxic shock syndrome. He was also a pioneer in the research of hypertension and urinary infections.

Keefe, Denis [1930–Mar 11 1990] US experimental physicist who worked with particle accelerators. Earlier in his career he was among the last to study cosmic rays as a source of particle physics, taking part in the experiments that found the kaon.

Kelly, Dennis D. [Jun 18 1938–Jun 13 1991] US physiological psychologist who at the time of his death was serving a one-year term as president of the New York Academy of Sciences.

Kerr, John Austin [Jul 15 1900–Dec 12 1990] US research physician with the Rockefeller Foundation who directed research to help eliminate the spread of yellow fever.

Kidd, John Graydon [Jul 20 1908–Jan 28 1991] US pathologist who conducted animal research into the relationship between viruses and cancer that led to the discovery of the enzyme asparaginase, an agent used against cancer.

Klopsteg, Paul E. [May 30 1889–Apr 28 1991] US scientist who held several posts during his lifetime, including president of the American Association for the Advancement of Science and associate director of the US National Science Foundation, the latter of which he helped organize. As an inventor he held more than 50 patents for instruments used in scientific research and teaching and for prosthetic devices.

Kolchin, Ellis R. [Apr 18 1916–Oct 30 1991] US mathematician and former Columbia University professor who was known for making the field of differential equations more accessible to mathematics learners.

Konopinski, Emil [Dec 25 1911–May 26 1990] US physicist who conducted research with J. Robert Oppenheimer and Edward Teller on the first atomic bomb at Los Alamos NM during World War II. He calculated that a thermonuclear explosion would not spread to destroy the entire earth.

Land, Edwin H. [May 7 1909–Mar 1 1991] US inventor of instant photography and founder of the Polaroid Corporation. He held 533 patents, which included instant X rays and Polaroid sunglasses, and was awarded honorary doctorates from Harvard and other institutions, although he never earned a university degree.

Landes, Ruth [1908–Feb 11 1991] US anthropologist whose best-known book was *The City of Women*, which focuses on life in Bahia, Brazil, where women play the dominant roles in a society of descendants of African slaves. She was also praised for her studies of Indians in North Dakota and Hispanic Americans in the American Southwest.

Lanier, Gerald N. [Dec 9 1937–Jan 28 1990] US epidemiologist who devised ways to trap and eradicate bark beetles that cause Dutch elm disease.

Laskar, Amylya Lal [Jun 11 1931–Jul 19 1991] US physicist, born in Dhaka, India, whose research advanced the development of solid-state batteries such as the miniature ones used in pacemakers.

Lautman, Don A. [Apr 3 1930–Aug 9 1991] US astronomer who helped plot the orbit of the Soviet satellite Sputnik in 1957.

Lawrence, John H. [Jan 7 1904–Sep 7 1991] US medical researcher who in the 1930s discovered the harmful effects of neutrons on tissue and developed treatments using radioactive isotopes and high-energy particle beams; for example, he and his colleagues at the University of California at Berkeley used neutron beams on cancer and particle beams on acromegaly and several other illnesses.

Lesher, George Y. [Feb 22 1926–Mar 17 1990] US research scientist who identified a new class of antibacterial agents and later discovered another class of therapeutic agents that are used in the treatment of heart disease.

Lester, David [Jan 22 1916–Sep 15 1990] US medical researcher who, as scientific director of the US National Alcohol Research Center at Rutgers University, conducted extensive research on alcoholism, including a 30-year study that included research on 13-year-olds and their development.

Levi, Doro [1898–Jul 3 1991] Italian archaeologist who worked with the US government in World War II to save Italian monuments from Allied bombardment.

Levinthal, Cyrus [May 2 1922–Nov 4 1990] US scientist who conducted research in genetics that directly linked genes to the proteins they encode, and who was the first to measure the size of a pair of chromosomes.

Licklider, Joseph C.R. [Mar 11 1915–Jun 26 1990] US computer scientist whose research in "man-computer symbiosis" helped improve the user's dialogue with the computer, thus making computers more applicable at libraries, universities, and other institutions.

Lorch, Edgar R. [Jul 22 1907–Mar 5 1990] US mathemetician, born in Nyon, Switzerland, who was considered a leader in the development of mathematics theory and who wrote several textbooks.

Lorente de No, Rafael [Apr 8 1902–Apr 2 1990] US neural researcher, born in Zaragoza, Spain, who is credited with determining the structure and function of the cerebral cortex, the section of the brain that is responsible for sensations, action, memory, and intelligence.

Loveless, Mary Hewitt [Apr 28 1899–Jun 2 1991] US physician who developed a venom injection to immunize patients against allergic reactions to the stings of bees and wasps, and who discovered blocking antibodies for sufferers of pollen allergies and hay fever.

Luria, Salvador E. [Aug 13 1912–Feb 6 1991] US biologist and physician, born in Turin, Italy, who in 1969 shared the Nobel Prize in physiology or medicine for discoveries that led to a better understanding of the replication mechanism and genetic structure of viruses. He was also known for his involvement in the peace movement and other political activities, some of which led to his being blacklisted by the US National Institutes of Health.

MacWilliams, Florence Jessie [1916–May 27 1990] US mathematician who, while working at AT&T Bell Laboratories, developed digital coding to correct telephone and satellite communication errors. She also discovered a set of fundamental equations known as the MacWilliams Identities.

Maguire, Bassett [Aug 4 1904–Feb 6 1991] US botanist who was former head curator of the New York Botanical Garden in the Bronx. He explored remote areas of northern South America and brought back hundreds of thousands of exotic plant specimens. He also discovered a botanically rich mountain on the Brazil-Venezuela border, which he named Cerro de la Neblina, or "Mountain of the Clouds."

Manley, John H. [Jul 21 1907–Jun 11 1990] US physicist who assisted in the Manhattan Project to build the atomic bomb at Los Alamos National Laboratory, where he was also the associate director from 1946 to 1951 and research adviser from 1957 to 1972.

Massie, Edward [Nov 21 1910–Feb 5 1990] US cardiologist who with Bernard Lipman wrote *Clinical Electrocardiography*, a standard text in the field.

Masursky, Harold [Dec 23 1923–Aug 24 1990] US astrogeologist who in the 1970s led the first team to map the planet Mars, a task useful in selecting a site to land the Viking spacecraft in 1976. He also helped lead the lunar exploration program in the 1960s and in the 1980s played a major role in the Magellan project for mapping the planet Venus.

McCredie, Kenneth B. [Jul 2 1935–Mar 30 1991] US researcher who pioneered the use of combinations of anticancer drugs and biological agents such as interferon in the treatment of leukemia.

McMahon, Howard D. [Sep 16 1914–Aug 5 1990] US scientist and inventor who was president of Arthur D. Little, Inc. and in 1951 codeveloped the Collins Helium Cryostat, which liquefies helium gas at a temperature of -269° C (-452° F).

McMillan, Edwin [Sep 18 1907–Sep 7 1991] US physicist who won the Nobel Prize in chemistry in 1951 for codiscovering the transuranic elements plutonium and neptunium. He also invented the synchrotron, a form of particle accelerator used in high-energy physics research. He was awarded the US National Medal of Science in 1990.

Migdal, Arkady [Mar 11 1911-Feb 9 1991] Russian theoretical physicist who was among the first to apply quantum field theory to condensed matter and to provide a rigorous treatment of strong interactions inside atomic nuclei.

Meyer, Karl [Sep 4 1899–May 18 1990] US biochemist, born in Kerpen, Germany, who was one of the first scientists to recognize the significance of bacterial enzymes as a tool in analyzing the structure of animal tissue components.

Moss, N. Henry [Apr 9 1925–Dec 26 1990] US surgeon who was president of both the Academy of Sciences in Philadelphia and the New York Academy of Sciences.

Murphy, Sheldon D. [Jul 16 1933–Apr 30 1990] US toxicologist who helped develop this discipline, which is the study of poisons. He was a charter member of the Society of Toxicology and founding organizer of the International Union of Toxicology.

Nathan, Henry C. [Aug 16 1924–Jul 6 1991] US biologist who helped develop Ornodyl, a drug introduced in 1991 to treat African sleeping sickness.

Nelson, Norton [Feb 6 1910–Feb 4 1990] US biochemist who was an early opponent of smoking and helped create the US Environmental Protection Agency. He also formed the National Institute of Environmental Services, part of the US National Institutes of Health.

Noyce, Robert N. [Dec 12 1927–Jun 3 1990] US physicist who revolutionized the electronics industry with his invention of the microchip, or integrated circuitry, a system of interconnecting transistors that is used to build computers, pocket calculators, and other appliances. He was awarded the US National Medal of Science in 1979 and the National Medal of Technology in 1987.

Oxtoby, John C. [Sep 14 1910–Jan 2 1991] US mathematician known by scholars and graduate students for his book *Measure and Category*. At the time of his death he was professor emeritus at Bryn Mawr College in Pennsylvania.

Page, Irvine H. [Jan 7 1901–Jun 10 1991] US physician whose research led to the discoveries of compounds in the body that affect blood pressure. In 1937 he developed a treatment for malignant hypertension, the most severe form of high blood pressure.

Parkinson, David B. [May 16 1911–Mar 17 1991] US engineer who won a Presidential Award and Franklin Institute Medal for developing an electromechanical device to guide a marking stylus to control antiaircraft guns, thus improving the accuracy of weapons.

Parr, Albert E. [Aug 15 1900–Jul 24 1991] US oceanographer born in Bergen, Norway, who was director of the American Museum of Natural History from 1942 to 1959. In this post he was dedicated to seeking greater understanding of how human beings interact and affect the rest of nature.

Penney, William George [Jun 24 1909–Mar 3 1991] British scientist, considered the father of the British atomic bomb. He assisted in the production of the first bomb at Los Alamos NM in addition to the one that was dropped on Hiroshima, Japan, on Aug 6 1945.

Perlis, Alan J. [Apr 1 1922–Feb 7 1990] US computer scientist who helped found the computer science departments at Yale and Carnegie-Mellon universities and helped develop algebraic language compilers and assemblers that convert programming instructions into computer operations.

Perlman, Isador [Apr 12 1915–Aug 3 1991] US nuclear chemist who discovered several artificial isotopes, researched alpha particle decay, one of three forms of radioactive disintegration of atoms, and advanced understanding of the energy-level structure of radioactive elements.

Phillips, Samuel Cochran [Feb 19 1921–Jan 31 1990] US Air Force lieutenant general who directed the Apollo space program through the lunar landing of Apollo 11 in 1969.

Piotrovsky, Boris P. [1908–Oct 15 1990] Russian archaeologist and director of the Hermitage in St Petersburg who discovered the ancient civilization of Urartu at Karmir Blur in present-day Armenia, and helped analyze the remains of King Tutankhamen's tomb in Egypt.

Pollack, Herbert [Jun 29 1905–Jan 2 1990] US physician and government consultant who testified before the Senate in 1977, stating that US Embassy employees in Moscow did not suffer long-term effects from the radiation directed at them by the Soviets.

Racker, Efraim [Jun 28 1913–Sep 9 1991] US physiologist, born in Poland, whose discoveries included how sugar-fermenting enzymes cause dangerous lactic acid imbalance in cancer cells, and how an unusual bacterium that thrives in salt converts sunlight into chemical energy by passing protons through its membranes. He was awarded the US National Medal of Science in 1977.

Revelle, Roger [Mar 7 1909–Jul 15 1991] US researcher and administrator who headed the Scripps Institution of Oceanography in La Jolla CA from 1951 to 1964, where his study of the increase of carbon dioxide from the use of fossil fuels helped predict global warming, and his discovery of hot material flowing under the oceans led to the theory of plate tectonics. He was awarded the US National Medal of Science in 1990.

Rich, Arthur [Aug 30 1937–Aug 25 1990] US physicist who as a specialist in sub-atomic particles coinvented the positron transmission microscope, which uses the beams of positrons to produce images of the structure of materials, and who led research to develop a positron beam to explore the chemical origins of life.

Roberts, Walter Orr [Aug 20 1915–Mar 12 1990] US scientist who founded the National Center for Atmospheric Research and was one of the first to warn that technology is a factor in global warming. He was also a former president of the American Association for the Advancement of Science.

Robinson, Robert A. [Jan 9 1914–Aug 16 1990] US surgeon whose electron microscope research of bones led him to be a founder of the specialty of spinal surgery, and who codeveloped the anterior cervical fusion operation.

Roe, Kenneth Andrew [Jan 31 1916–Jun 3 1991] US mechanical engineer who was chairman of Burns & Roe Enterprises, a consulting firm that pioneered advanced technological projects like Project Mercury and the Gemini space programs as well as the construction of two nuclear generating plants.

Rogers, Eric M. [Aug 15 1902–Jul 1 1990] US physics professor, born in Bickley, England, who was known for his ability to make the complex principles of physics more understandable. He received several awards, and his 1960 textbook *Physics for the Inquiring Mind* was used in many colleges.

Rose, Albert [Mar 30 1910–Jul 26 1990] US research scientist who invented the image orthicon television camera tube for military purposes during World War II. The orthicon tube was used as the electric eye in all television cameras during the early years of broadcasting. He then led a team of RCA scientists to develop the first photoconductive sensor, which is now used in television studio broadcasting.

Rosen, Ora M. [Oct 26 1935–May 31 1990] US scientist who studied the development of cells and is known for research on the human insulin receptor; she and her colleagues cloned the gene, which led to further investigation on how insulin affects normal cell function as well as diabetes and other disorders.

Rubinstein, Lucien J. [Oct 15 1924–Jan 22 1990] US neuropathologist, born in Brussels, Belgium, who was known for the reference book *Tumors of the Central Nervous System*, written with Dorothy Russell in 1972. He was head of the American Association of Neuropathologists and received its Merit Award in 1989.

Samuel, Arthur L. [Dec 5 1901–Jul 29 1990] US computer scientist who supported his hypothesis that computers can be designed to learn as humans do through development of a program enabling computers to successfully compete with checker champions; seminal techniques in rote and generalization learning; and work with time-sharing systems at Stanford University and with early word processing programs.

Sarnoff, Stanley J. [Apr 5 1917–May 23 1990] US cardiologist and inventor who held more than 40 patents for medical devices, including an automatic injector used by patients to self-administer drugs and a small electronic transmitter that sends an electrocardiogram by telephone to an emergency medical center where heart attacks can be diagnosed.

Scheie, Harold G. [Mar 24 1909–Mar 5 1990] US ophthalmologist who developed several surgical techniques, including the treatment of congenital cataracts and glaucoma, as well as various ophthalmological tests and procedures.

Schmidt, Fred [Sep 12 1915–Jan 17 1991] US physicist who was on the Manhattan Project, which developed the atomic bomb, and was known for his book *The Fight Over Nuclear Power.*

Schneiderman, Howard A. [Feb 9 1927–Dec 5 1990] US scientist who led the Monsanto Company, a chemical producer, to increase its profits through the genetic engineering of plants and animals.

Seibert, Florence B. [1897–Aug 23 1991] US scientist who invented a tuberculosis test that uses isolated tuberculosis protein molecules in a skin reaction test; this method has been the standard one in the United States since 1941. She also developed a method for IV transfusion that does not cause sudden high fevers.

Shapiro, Harry Lionel [Mar 19 1902–Jan 7 1990] US anthropologist and former curator at the American Museum of Natural History who was known for his genetic study of Pitcairn Islanders in the South Pacific who practiced racial inter-marriage, as well as for his advanced studies in forensic anthropology, in which human remains are studied for military and criminal justice purposes.

Shelley, Edwin F. [Feb 19 1921–Jan 4 1990] US inventor and engineer who developed commercial production-line robots, automatic mail-sorting systems and automated training equipment for teaching clerical skills, all while he was an executive for US Industries, Inc., between 1957 and 1964.

Shemin, David [Mar 18 1911–Nov 26 1991] US biochemist whose research in the 1940s shed light on how the body produces hemoglobin, the constituent in the blood that carries oxygen from the lungs to the other parts of the body.

Sitterly, Charlotte M. [Sep 24 1898–Mar 3 1990] US astrophysicist who found the element technitium in the spectrum of sunlight, thus showing that technitium exists in nature.

Slichter, William P. [Mar 31 1922–Oct 25 1990] US scientist who was executive director at AT&T Bell Laboratories, where he performed research on polymers and on semiconductors used in computers and telephones. He was awarded the American Physical Society Prize in 1970.

Smith, Harlan J. [Aug 25 1924–Oct 17 1991] US astronomer who was director of the University of Texas's McDonald Observatory, where he helped build a reflector telescope that was 272 cm (107 in.) in diameter, the world's third largest in 1965 when it was first used. He was also known for discovering the variability of quasars, the influence of solar wind on radio emissions from Jupiter, and the existence of variable stars called dwarf cepheids.

Smyth, Charles P. [Feb 10 1895–Mar 18 1990] US chemist who worked on the Manhattan Project to develop the atomic bomb during World War II and spent his career investigating the electrical properties of matter and their relation to the structure of molecules.

Sperry, Warren M. [Jun 30 1900–Jul 11 1990] US neurochemist who in the late 1960s developed the "Sperry Method," a standard technique for measuring cholesterol in blood plasma.

Sperti, George Speri [Jan 17 1900–Apr 20 1991] US research scientist who held over 120 patents for medicines including Preparation H for hemorrhoid treatment, Aspercreme for arthritis relief, a sunlamp, and frozen orange juice concentrate.

Spurr, Stephen H. [Feb 14 1918–Jun 20 1990] US ecologist who was known as an authority on forest ecology and a controversial president of the University of Texas in the 1970s. He served on the US President's Advisory Panel on Timber and the Environment, and in the 1940s invented the parallax wedge and the multiscope, tools used in forestry.

Stallings, James Henry [Sep 20 1892–Dec 21 1990] US soil scientist who helped develop no-tillage farming, a system designed to protect soil against erosion by controlling weeds with herbicides.

Steer, Charles M. [Feb 13 1913–Feb 22 1990] US obstetrician who developed the electrohysterograph, which measures electrical activity in a pregnant woman's uterus to determine whether her labor will be normal.

Steinberg, Ellis P. [Mar 26 1920–Dec 22 1991] US physicist whose research included his study of nuclear fission reactions at the Chicago Metallurgical Laboratory, the secret World War II project in which Enrico Fermi, Arthur Compton, and other physicists produced the first controlled nuclear chain reaction in 1942.

Stetten, DeWitt, Jr. [May 31 1909–Aug 28 1990] US biochemist who held several administrative posts, including deputy director of the US National Institutes of Health, and in the 1970s led a committee that drafted guidelines on genetic engineering.

Sugarman, Nathan [Mar 3 1917–Sep 6 1990] US nuclear chemist who while with the Manhattan Project led a group of chemists who determined the efficiency of the first nuclear explosion. He also won the University of Chicago's Quantrell Award for excellence in teaching.

Suits, Chauncey Guy [Mar 12 1905–Aug 14 1991] US researcher who in 1962 helped develop a process for making synthetic diamonds by compressing carbon while heating it to 5000° C (9000° F), eliminating the use of a metal catalyst.

Talbott, John Harold [Jul 10 1902–Oct 10 1990] US medical researcher and author who was editor of the *Journal of the American Medical Association* from 1959 to 1971.

Throckmorton, Peter [Jan 20 1927–Jun 5 1990] US underwater archaeologist who was known for excavating the two oldest sunken ships on record: a cargo ship in the Aegean Sea believed to date back to 2500 BC during the Bronze Age, and a wreck in the Mediterranean Sea off Turkey dating from 1300 BC.

Trapido, Harold [Dec 10 1916–Jul 25 1991] US biologist who took part in the first field testing in the western hemisphere of DDT to kill malaria-carrying mosquitoes.

Trout, G. Malcolm [Mar 7 1896–Nov 1 1990] US food scientist who in the 1930s linked the processes of pasteurization and homogenization to make homogenized milk easier to produce.

Ulfelder, Howard [Aug 15 1911 1908–Apr 29 1990] US gynecologist and cancer specialist who in 1971 helped discover the link between adenocarcinoma, a rare form of vaginal cancer, and DES, or diethylstilbestrol, a drug prescribed to pregnant women to prevent miscarriages.

van der Ziel, Aldert [Dec 12 1910–Jan 29 1991], Dutch-US physicist who was a specialist in electronic noise and who discovered the principle of parametric amplification in 1948.

Vogel, Orville Alvin [May 19 1907–Apr 12 1991] US agronomist and National Medal of Science winner for his team's development of semidwarf wheat, released to farmers in 1961; the wheat produces heavy yields without "lodging," falling over from the weight of the grain. He also contributed to the green revolution by developing equipment used to automate planting and harvesting of research plots.

Wagley, Charles [Nov 9 1913–Nov 25 1991] US anthropologist who was president of the American Anthropological Association and the American Ethological Society as well as an expert on the people of the Amazon basin of Brazil.

Wakelin, James Henry, Jr. [May 6 1911–Dec 21 1990] US oceanographer and physicist who helped establish the Office of Naval Research and who held numerous US federal government posts related to oceanography, including assistant navy secretary and assistant secretary of commerce for science and technology.

Wang, An [Feb 7 1920–Mar 24 1990] US inventor, born in Shanghai, China, who held 40 patents, including one for a ring of iron that served as the core of computer memory until the microchip came along. He was the founder of Wang Laboratories and received 23 honorary degrees.

Warren, James [Jul 1 1915–Feb 15 1990] US physician whose contributions to cardiovascular research led to methods of controlling blood pressure in space flight and the use of cardiac catherization for diagnosing heart problems. His many awards included four from the American Heart Association.

Waters, Aaron C. [May 6 1905–May 18 1991] US earth scientist who in the mid-sixties was chosen by the National Aeronautics and Space Administration to conduct geological training for a group of astronauts who later took part in lunar landings.

Williams, Carroll M. [Dec 2 1916–Oct 11 1991] US biologist whose pioneering work on insect physiology included the discovery of the features of juvenile hormone, chemical communication between insects, and chemical defenses used by plants to ward off attacking insects.

Wilson, Allan C. [Oct 18 1934–Jul 21 1991] US biochemist, born in Ngaruawahia, New Zealand, who developed the idea of a "molecular clock" determining human evolution. He and his fellow researchers hypothesized that all modern humans can be traced back to a female ancestor who lived in Africa 200,000 years ago.

Wood, Francis C. [Oct 1 1901–Dec 16 1990] US medical educator born in Wellington, South Africa, who contributed to the development of the electrocardiograph in the 1930s and was at his death chairman emeritus of the department of medicine at the University of Pennsylvania Medical Center, where he had taught from 1947 until 1984.

Wood, Harrison F. [May 14 1919–Jul 26 1991] US physician who discovered that continuous drug therapy reduces recurrences of rheumatic fever in children. He also participated in a study that found in 1961 that continuous doses of antibiotics can ward off infections and thus prevent the disease.

Wood, Lysle A. [Feb 23 1904–May 18 1991] US vice president of the Boeing Company who helped design bombers and aircraft, including the Model 314 Clipper flying boat and the 307 Stratoliner, the first pressurized commercial aircraft.

Woodhouse, John C. [Apr 29 1898–Feb 17 1991] US inventor whose 70 patents included ones for chemicals that form "sudsless" detergent soaps, hydrolic brake fluid, and soft plastic lenses for the eyes.

Zubin, Joseph [Oct 9 1900–Dec 18 1990] US research psychologist born in Raseiniai, Lithuania, who was president of the American Psychopathological Association as well as an authority on the diagnosis of schizophrenia and other diseases.

APPENDIX C

UNITS OF MEASURE

SYSTEMS OF MEASURE

There are two measurement systems that are widely used. Most of the world uses a system often still called the metric system, a simplification of the International System of Units (abbreviated SI from *Systeme Internationale,* its name in French). The true metric system was the French ancestor of SI. The United States continues to use a system that is called US customary measure.

THE INTERNATIONAL SYSTEM

The International System of Units represents a serious effort to make people measure in a rational and consistent fashion. Its origins are in the rational idealism of the French Revolution. But people refuse to be rational or consistent, so there are various accommodations that have to be made.

The idea of SI is to have a limited number of basic units that can then be combined with a fixed set of prefixes to indicate multiplication or division by various powers of ten, giving a reasonably sized unit for every measurable quantity. Problems that arise in working this out include the choice of a base unit that already has a prefix (such as the kilogram, kept as a base unit in SI because it was a base unit for the French metric system), the use of units that derive from nature (such as the speed of light or the mass of the electron), and the use of units that derive from SI units in such a complicated fashion that they must be used with special names.

Furthermore, the official SI units are often awkward in size for the object being measured, whereas a former metric or customary unit may be more convenient.

The official seven base units are as follows:

Quantity to be measured	Base unit	Definition
Time	second (s)	Time it takes hot cesium atoms to vibrate 9,192,631,770 times
Length	meter (m)	Distance light travels in a vacuum in 1/299,792,458 of a second
Mass	kilogram (kg)	Mass of a platinum-iridium cylinder kept in Sèvres, France, called the International Prototype Kilogram
Electric current	ampere (A)	A current that if maintained in two infinite, parallel conductors of negligible cross-section that are 1 meter apart would produce a force between the conductors equal to 10^{-7} newton
Temperature	kelvin (K)	1/273.16 of the thermodynamic temperature of the triple point of water
Amount of substance	mole (mol)	The amount of a substance that contains 6.0225×10^{23} (Avogadro's number) of basic units, such as atoms or molecules
Luminous intensity	candela (cd)	1/600,000 of the light produced by a 1-square-meter cavity at the temperature of freezing platinum (2042 K)

For reasons known only to the General Conference of Weights and Measures, measures for plane and three-dimensional angles are termed supplementary, although they are at least as basic as the seven base measures. The radian (rad) for plane angles is most easily thought of as the measure of the central angle of a circle that intercepts the radius of the circle. A solid angle is measured in steradians (sr), a three-dimensional central angle that intercepts a region measuring $\pi/4$ on a sphere whose radius is 1.

These units are modified by a set of standard prefixes to produce smaller or larger measures. For example, centi- is the prefix for $\frac{1}{100}$ (10^{-2}) and is abbreviated c, so $\frac{1}{100}$ of a second is a centisecond (cs) and $\frac{1}{100}$ of a candela is a centicandela (ccd). Since the base unit for mass, the kilogram, already uses the prefix that means 1000 (10^3), prefixes for measures of mass are attached to the gram, so a mass of 1000 kilograms is not labeled a kilokilogram, but is instead a megagram (Mg).

Factor	Scientific notation	Prefix	Symbol
1/1,000,000,000,000,000,000	10^{-18}	atto-	a
1/1,000,000,000,000,000	10^{-15}	femto-	f
1/1,000,000,000,000	10^{-12}	pico-	p
1/1,000,000,000	10^{-9}	nano-	n

(Continued)

Factor	Scientific notation	Prefix	Symbol
1/1,000,000	10^{-6}	micro-	μ
1/1000	10^{-3}	milli-	m
1/100	10^{-2}	centi-	c
1/10	10^{-1}	deci-	d
10	10^1	deka-	da
100	10^2	hecto-	h
1000	10^3	kilo-	k
1,000,000	10^6	mega-	M
1,000,000,000	10^9	giga-	G
1,000,000,000,000	10^{12}	tera-	T
1,000,000,000,000,000	10^{15}	peta-	P
1,000,000,000,000,000,000	10^{18}	exa-	E

In the table below, the most commonly used SI units are defined in terms of each other instead of strictly in terms of the base units. Also included are derived SI units and units commonly used by scientists that are not SI units. Following this listing is a brief table of US customary units. At the end of this appendix there are tables converting SI or related units into US customary units and vice versa.

Length or Distance

decimeter (dm) = 10 centimeters = 0.1 m
centimeter (cm) = 0.01 m
millimeter (mm) = 0.1 cm = 0.001 m
micron (μ) = 0.001 mm = 0.0001 cm = 0.000001 m
angstrom (Å) = 0.0001 micron = 0.0000001 mm
dekameter (dam) = 10 m
hectometer (hm) = 10 dekameters = 100 m
kilometer (km) = 10 hectometers = 100 dekameters = 1000 m

Area

1 square millimeter (mm^2) = 1,000,000 square microns
1 square centimeter (cm^2) = 100 mm^2
1 square meter (m^2) = 10,000 cm^2
1 are (a) = 100 m^2

1 hectare (ha) = 100 a = 10,000 m^2

1 square kilometer (km^2) = 100 ha = 1,000,000 m^2

Volume

1 cubic centimeter (cm^3) = 1000 cubic millimeters (mm^3)

1 cubic decimeter (dm^3) = 1000 cm^3

1 cubic meter (m^3) = 1000 dm^3 = 1,000,000 cm^3

Cubic centimeter is sometimes abbreviated cc and is used in fluid measure interchangeably with milliliter (mL).

Fluid volume measurements are directly tied to cubic measure. One *milliliter* of fluid occupies a volume of 1 cubic centimeter (cm^3 or cc). A *liter* of fluid (slightly more than the customary quart) occupies a volume of 1 cubic decimeter or 1000 cubic centimeters. Textbook writers have sensibly decided to use L as the abbreviation for liter, although most dictionaries still use l, which is easily confused with 1.

1 centiliter (cL) = 10 milliliters (mL)

1 deciliter (dL) = 10 cL = 100 mL

1 liter (L) = 10 dL = 1000 mL

1 dekaliter (daL) = 10 L

1 hectoliter (hL) = 10 daL = 100 L

1 kiloliter (kL) = 10 hL = 1000 L

Mass and Weight

Mass and weight are often confused. Mass is a measure of the quantity of matter in an object and does not vary with changes in altitude or in gravitational force (as on the moon or another planet). Weight, on the other hand, is a measure of the force of gravity on an object and so does change with altitude or gravitational force.

The International System generally is used to measure mass instead of weight. The early metric system's basic unit for measurement of mass was the gram, which was originally defined as the mass of 1 milliliter (= one cubic centimeter) of water at 4 degrees (4°) Celsius (about 39° Fahrenheit). Today the official standard of measure is the kilogram (1000 grams), a metal mass that reflects the older water standard.

1 centigram (cg) = 10 milligrams (mg)

1 decigram (dg) = 10 centigrams = 100 milligrams

1 gram (g) = 10 decigrams = 100 centigrams = 1000 milligrams

(Continued)

1 kilogram (kg) = 10 hectograms (hg) = 100 dekagrams (dag) = 1000 grams

1 metric ton (t) = 1000 kilograms

Time

The International System adopted in 1967 a second that is based on the microwaves emitted by the vibrations of hot cesium atoms. A second is the time it takes the atoms to vibrate exactly 9,192,631,770 times. In the customary measure of time, a second is $\frac{1}{86,400}$ of 1 rotation of Earth, or 1 day. For most nonscientific purposes, an International System second (s) and a customary second can be treated as the same. Since Earth's rotation is gradually slowing, scientists since 1972 have also periodically added a second to a day to keep astronomical time in step with SI clocks—in 1972 it was actually two seconds. Through 1992 seventeen such seconds have been added, with the second in 1992 added at 7:59:59 p.m. Eastern daylight time on Jun 30, making the last second before 8:00 p.m. 2 seconds long.

In SI, decimal fractions of time are used to measure smaller time intervals, typically the following:

millisecond (ms) = 0.001 second (10^{-3})

microsecond (μs) = 0.000001 second (10^{-6})

nanosecond (ns) = 0.000000001 second (10^{-9})

picosecond (ps) = 0.000000000001 second (10^{-12})

Although not part of SI, longer intervals are measured in minutes (60 seconds) or hours (3600 seconds) or days of 24 hours (86,400 seconds).

A calendar year, technically called the *tropical year,* is 365.2422 days (31,556,926 seconds). Astronomers also use the *sidereal year,* based on the position of the Sun with respect to the background of stars. It is 365.2564 days (31,558,153 seconds).

Temperature

The International System measures in the Kelvin (sometimes called the absolute) scale denoted by the symbol K. However, for temperatures in the range at which humans live, many scientists still use the Celsius (formerly centigrade) system, whose symbol is C. In the Celsius scale, water freezes at 0° C and boils at 100° C. The temperature change denoted by 1 degree is the same size in the Kelvin and Celsius scales, but the zero point for Kelvin is set at *absolute zero.* A third system, called Fahrenheit, is commonly used in the United States. Its symbol is F, and it has a degree $\frac{5}{9}$ the size of a Kelvin degree. The equivalences between Kelvin, Celsius, and Fahrenheit are as follows:

	K	C	F
Absolute zero	0	-273.15	-459.7
Freezing point, water	273.15	0	32
Normal human body temperature	310.15	37	98.6
Boiling point, water	373.15	100	212

Force, Work/Energy, Power

In physics, compound measurements of force, work or energy, and power are essential. There are two parallel systems using International System units:

The *meter/kilogram/second* (mks) system is the official SI system. Many scientists, however, are used to working in the *centimeter/gram/second* (cgs) system, especially if they regularly measure small quantities. The systems are described below.

Measurement of Force

mks unit	newton (N)	The force required to accelerate a mass of 1 kg 1 m/s^2
cgs unit	dyne (dy)	The force required to accelerate a mass of 1 g 1 cm/s^2

Measurement of Work or Energy

mks unit	joule (j)	The newton-meter, i.e., the work done when a force of 1 newton produces a movement of 1 m (=10,000,000 ergs)
cgs unit	erg	The dyne-centimeter, i.e., the work done when a force of 1 dy produces a movement of 1 cm

Measurement of Power

mks unit	watt (w)	The joule/second, i.e., a rate of 1 joule per second (= 10,000,000 erg-seconds)
cgs unit	erg/second	A rate of 1 erg per second

Heat energy is also measured using the *calorie* (cal) or gram calorie, which is defined as the energy required to increase the temperature of 1 cubic centimeter (1 mL) of water by 1 degree K. One calorie is equal to about 4.184 joules. The *kilocalorie* (Kcal or Cal) is equal to 1000 calories and is the unit in which the energy values of food are measured. This more familiar unit, also commonly referred to as a Calorie, is equal to 4184 joules.

Electrical Measure

The basic unit of quantity in electricity is the ampere (defined above), but in the older metric system it was the coulomb. A coulomb is equal to the passage of 6.25×10^{18} electrons past a given point in an electrical system.

When the coulomb is used as a base unit, the ampere is equal to a coulomb-second, i.e., the flow of 1 coulomb per second. The ampere is analogous in electrical measure to a unit of flow such as gallons-per-minute in physical measure. In SI, where the ampere is the base unit, the coulomb is defined as an ampere-second.

The unit for measuring electrical power in SI is the *watt* (W), defined as one joule per second.

Since the watt is such a small unit for many practical applications, the *kilowatt* (= 1000 watts) is often used. A kilowatt-hour is the power of 1000 watts over an hour's time.

The unit for measuring electrical potential energy is the *volt*, defined as one watt per ampere. This can also be translated into 1 joule/coulomb, i.e., 1 joule of energy per coulomb of electricity. The volt is analogous to a measure of pressure in a water system.

The unit for measuring electrical resistance is the *ohm*, whose symbol is Ω. The ohm is the resistance offered by a circuit to the flow of 1 ampere being driven by the force of 1 volt, which can also be interpreted as 1 volt per ampere. It is derived from Ohm's law, which defines the relationship between flow or current (amperes), potential energy (volts), and resistance (ohms).

Measures of Angles and Arcs

Angles are measured by systems that are not exactly part of either the customary or International systems, although degree measure might be described as the customary system for most people most of the time. Arcs of a circle can be measured by length, but they are also often measured by angles. In that case, the measure of the arc is the same as the measure of an angle whose vertex is at the center of the circle and whose sides pass through the ends of the arc. Such an angle is said to be subtended by the arc.

The most commonly used angle measure is *degree measure.* One degree is the angle subtended by an arc that is 1/360th of a circle. This is an ancient system of measurement that was probably originally developed by Sumerian astronomers. These astronomers used a numeration system based on sixty ($60 \times 6 = 360$), as well as a 360-day year. They divided the day into twelve equal periods of thirty smaller periods each ($12 \times 30 = 360$), and used roughly the same system for dividing the circle. Even when different years and numeration systems were adopted by later societies, astronomers continued to use a variation of the Sumerian system.

1 degree (1°) = 60 minutes (60') = 3600 seconds (3600")

1 minute = 60 seconds

When two lines are perpendicular to each other, they form four angles of the same size, called *right angles*. Two right angles make up a line, which in this context is considered a *straight angle*.

1 right angle = 90°
1 straight angle = 180°

While this system is workable for most purposes, it is artificial. Mathematicians discovered that using a natural system of angle measurement produces results that make better sense in mathematical and many scientific applications. This system is called *radian measure*. The SI system accepts this, but calls radian measure supplementary instead of basic. One radian is the measure of the angle subtended by an arc of a circle that is exactly as long as the radius of the circle.

1 radian is about 57° 17' 45"

The circumference, *C*, of a circle is given by the formula $C = 2\pi r$, where π is a number (approximately 3.14159) and *r* is the radius. Therefore, a semicircle whose radius is 1 is π units long, which implies that there are π radians in a straight angle. Many of the angles commonly encountered are measured in multiples of π radians.

0° = 0 radians
30° = $\pi/6$ radians
45° = $\pi/4$ radians
60° = $\pi/3$ radians
90° = $\pi/2$ radians
180° = π radians
270° = $3\pi/2$ radians
360° = 2π radians

To convert from radians to degrees, use the formula *t* radians = $(180/\pi)t°$. To convert from degrees to radians, use the formula $d° = (\pi/180)d$ radians.

The US artillery uses the *mil* to measure angles. A mil is the angle that is subtended by an arc that is ⅟₆₄₀₀ of a circle.

1 mil = 0.05625° = 3' 22.5"
1 mil is almost 0.001 radian

Astronomical Distances

One very large measure of distance useful in astronomy is the *light-year*. It is defined as the distance light travels through a vacuum in a year (approximately 365¼ days). Light travels though a vacuum at the rate of about 186,250 mi an hour (exactly 299,792,458 m per s, exact because the meter is defined in terms of the speed of light in a vacuum). It is approximately equivalent to 5,880 billion miles (9,460 billion km).

Astronomers also use a measure even larger than the light-year, the *parsec,* equal to 3.258 light-years, or about 19,180 billion miles (30,820 billion km).

A smaller unit, for measurements within the solar system, is the *astronomical unit,* which is the average distance between Earth and the Sun, or about 93 million mi (150 million km).

US CUSTOMARY MEASURES

1 foot (ft) = 12 inches (in.)
1 yard (yd) = 3 ft = 36 in.
1 rod (rd) = 5 ½ yd = 16 ½ ft
l furlong (fur) = 40 rd = 220 yd = 660 ft
1 mile (mi) = 8 fur = 1760 yd = 5280 ft

An International Nautical Mile has been defined as 6,076.1155 ft.

1 square foot (sq ft) = 144 square inches (sq in.)
1 square yard (sq yd) = 9 sq ft
1 square rod (sq rd) = 30 ¼ sq yd = 272 ½ sq ft
1 acre (A.) = 160 sq rd = 4840 sq yd. = 43,560 sq ft
1 square mile (sq mi) = 640 A.
1 section = 1 mile square (that is, a square region 1 mi on a side)
1 township = 6 miles square, or 36 sq mi

Note that the US customary acre is abbreviated with a period (A.), while the SI ampere has no period (A). In practice it is easy to tell the two apart from context.

1 cubic foot (cu ft) = 1728 cubic inches (cu in.)
1 cubic yard (cu yd) = 27 cu ft

A gallon is equal to 231 cu in. of liquid or capacity.

1 tablespoon (tbs) = 3 teaspoons (tsp) = 0.5 fluid ounce (fl oz)
1 cup = 8 fl oz
1 pint (pt) = 2 cups = 16 fl oz
1 quart (qt) = 2 pt = 4 cups = 32 fl oz
1 gallon (gal) = 4 qt = 8 pt = 16 cups
1 bushel = 8 gal = 32 qt

In customary measure, it is more common to measure weight than mass. The most common customary system of weight is the *avoirdupois.*

1 pound (lb) = 16 ounces (oz)

1 (short) hundredweight = 100 lb

1 (short) ton = 20 hundredweights = 2000 lb

1 long hundredweight = 112 lb

1 long ton = 20 long hundredweights = 2240 lb

A different system called *troy weight* is used to weigh precious metals. In troy weight, the ounce is slightly larger than in avoirdupois, but there are only 12 ounces to the troy pound.

Customary measure is often said to include the Fahrenheit scale of temperature measurement (abbreviated F). In this system, water freezes at 32° F. and boils at 212° F. Absolute zero is -459.7° F.

The *foot/pound/second* system of reckoning includes the following units:

poundal = fundamental unit of force

slug = the mass to which a force of 1 poundal will
give an acceleration of 1 ft per second (= approximately 32.17 lb)

foot-pound = the work done when a force of 1 poundal
produces a movement of 1 ft

foot/pound/second = the unit of power equal to 1 foot-pound per
second

Another common unit of power is the *horsepower,* which is equal to 550 foot-pounds per second.

Thermal work or energy is often measured in *British thermal units* (Btu). One BTU is defined as the energy required to increase the temperature of a pound of water 1 degree F.

The Btu is equal to 0.778^+ foot-pound.

CONVERSIONS

From time to time, the US government has taken steps to change from the customary system to the International System, but these efforts have failed. Metric measure is legal in the United States, but nearly everyone continues to use the customary system. One result is that there is a need in the United States to be able to convert back and forth from one system to another. The following tables include methods of making conversions for each of the major forms of measure.

Length

In 1959 the relationship between customary and SI measures of length was officially defined as follows:

0.0254 meter (exactly) = 1 inch

0.0254 meter × 12 = 0.3048 meter = 1 *international foot*

This definition, which makes many conversions simple, defines a foot that is shorter (by about 6 parts in 10,000,000) than the *survey foot*, which had earlier been defined as exactly $^{1200}/_{3937}$ or 0.3048006 meters.

Using the *international foot* standard, the following are the major equivalents:

1 inch = 2.54 cm = 0.0254 m
1 foot = 30.48 cm = 0.3048 m
1 yard = 91.44 cm = 0.9144 m
1 mile = 1609.344 m = 1.609344 km

1 centimeter = 0.3937 in.
1 meter = 1.093613 yd = 3.28084 ft = 0.00062137 mi

Area

1 square inch = 6.4516 cm^2
1 square foot = 929.0304 cm^2 = 0.09290304 m^2
1 square yard = 8361.2736 cm^2 = 0.83612736 m^2
1 acre = 4046.8564 m^2 = 0.40468564 hectares
1 square mile = 2,589,988.11 m^2 = 258.998811 ha = 2.58998811 km^2

1 square centimeter = 0.1550003 sq in.
1 square meter = 1550.003 sq in. = 10.76391 sq ft = 1.195990 sq yd
1 hectare = 107,639.1 sq ft = 11,959.90 sq yd = 2.4710538 A.
   = 0.003861006 sq mi
1 square kilometer = 247.10538 A. = 0.3861006 sq mi

Volume

1 cubic inch = 16.387064 cm^3
1 cubic foot = 28,316.846592 cm^3 = 0.0028316847- m^3
1 cubic yard = 764,554.857984 cm^3 = 0.764554858- m^3

1 cubic centimeter = 0.06102374 cu in.
1 cubic meter = 61,023.74 cu in. = 35.31467 cu ft = 1.307951 cu yd

1 fluid ounce = 29.573528 mL = 0.02957 L
1 cup = 236.588 mL = 0.236588 L
1 pint = 473.176 mL = 0.473176 L
1 quart = 946.3529 mL = 0.9463529 L
1 gallon = 3785.41 mL = 3.78541 L

1 milliliter = 0.0338 fl oz
1 liter = 33.814 fl oz = 4.2268 cup = 2.113 pt = 1.0567 qt = 0.264 gal

1 pint, dry = 33.600 cu in. = 0.551 L

1 quart, dry = 67.201 cu in. = 1.101 L

Mass and Weight

Since mass and weight are identical at standard conditions (sea level on Earth), grams and other International System units of mass are often used as measures of weight or converted into customary units of weight. Under standard conditions

1 ounce = 28.3495 g

1 pound = 453.59 g = 0.45359 kg

1 short ton = 907.18 kg = 0.907 metric ton

1 milligram = 0.000035 oz

1 gram = 0.03527 oz

1 kilogram = 35.27 oz = 2.2046 lb

1 metric ton = 2204.6 lb = 1.1023 short ton

Temperature

The simplest means of converting is by formula.

To convert a Fahrenheit temperature to Celsius on a calculator, subtract 32 from the temperature and multiply the difference by 5; then divide the product by 9: the formula is

$$C = 5/9 \times (F - 32)$$

To convert a Celsius temperature to Fahrenheit on a calculator, multiply the temperature by 1.8, then add 32; the formula is often given with the fraction 9/5 instead of the equivalent decimal, 1.8.

$$F = 9/5 \, C + 32 = 1.8C + 32$$

To convert to Kelvin, find the temperature in Celsius and add 273.15.
$K = C + 273.15$

Measurement of Force

1 poundal = 13,889 dynes = 0.13889 newton

1 dyne = 0.000072 poundal

1 newton = 7.2 poundals

Measurement of Work/Energy

1 foot-pound = 1356 joules

British thermal unit = 1055 joules = 252 (gram) calories

1 joule = 0.0007374 foot-pounds

1 (gram) calorie = 0.003968 Btu

1 (kilo) Calorie = 3.968 Btu

Measurement of Power

1 foot-pound/second = 1.3564 watts

1 horsepower = 746 watts = 0.746 kilowatts

1 watt = 0.73725 ft-lb/sec = 0.00134 horsepower

1 kilowatt = 737.25 ft-lb/sec = 1.34 horsepower

SIMPLIFIED CONVERSION TABLE
(ALPHABETICAL ORDER)

To convert	into	multiply by
centimeters	feet	0.03281
centimeters	inches	0.3937
cubic centimeters	cubic inches	0.06102
cubic feet	cubic meters	0.02832
degrees	radians	0.01745
feet	centimeters	30.48
feet	meters	0.3048
gallons	liters	3.785
gallons of water	pounds of water	8.3453
grams	ounces	0.03527
grams	pounds	0.002205
inches	centimeters	2.54
kilograms	pounds	2.205
kilometers	feet	3280.8
kilometers	miles	0.6214
knots	miles per hour	1.151
liters	gallons	0.2642
liters	pints	2.113
meters	feet	3.281
miles	kilometers	.609
ounces	grams	28.3495
ounces	pounds	0.0625
pounds	kilograms	0.4536

Index